ENCYCLOPEDIA OF INORGANIC CHEMISTRY

ENCYCLOPEDIA OF INORGANIC CHEMISTRY

Vol. 2

By

Sananda Chatterjee

DAE, DCA, CMCNet

Scientific Consultant

Institute for Natural Sciences

Kolkata

(West Bengal)

D P H

DISCOVERY PUBLISHING HOUSE PVT. LTD.

NEW DELHI-110 002

Published by:
Tilak Wasan
DISCOVERY PUBLISHING HOUSE PVT. LTD.
4383/4A, Ansari Road, Darya Ganj
New Delhi-110 002 (India)
Phone : +91-11-23279245, 43596064-65
Fax : +91-11-23253475
E-mail : parul.wasan@gmail.com
discoverypublishinghouse@gmail.com
web : www.discoverypublishinggroup.com

***First Edition:* 2012**

ISBN: 978-81-8356-874-6

Encyclopedia of Inorganic Chemistry

Printed at:
Shree Balaji Art Press
Delhi

Preface

Chemistry is limitless subject but full of experiments and surprises, which is interesting indeed, especially for a new student, though some mathematical and theoretical calculations are there, in spite of all these, chemical world is world of lure.

The origin of chemistry is very deep-rooted and comes from a very early time, the time of the mediaeval history rather from the time of the Alchemists.

Egyptians, whom are thought to be the father of the alchemistry, used the word "kimiya". Kimiya moved from Egypt to Arabia and it flourished in whole world, in England it was named as "Alchemistry" (in India it was named as "*Ap-Rasayana*"). From this alchemistry our modern chemistry has been developed.

Medicines, drugs, paints etc. are all chemical products. Some processes such as distillation, fractionation, and purification, which we use widely now in modern chemistry are all have been given by the ancestors of chemistry, the Alchemistry.

The Modern Periodic Table developed by Dimitri Ivanovich Mendeleev is the modernized systematic formed of the elements (among them some were discovered by the Alchemists). The original table consists of only 107 elements but in this book all 119 elements have been discussed.

In this modern world of chemistry when I were engaged in my chemical lab, I were thinking that there should be something from which a student of chemistry will easily get all the information of chemistry.

Also a man who is anxious of chemistry, will start loving this subject, so to make the entrance of this chemical world easy and nice, I thought to write a book on chemistry, providing all the necessary information about chemistry [though the inorganic part].

While writing this book, I must first thank to the Supra-cosmic force for giving me the power to write this book, after this I must write the name of that person (My Lord) BABA

(Swami Arupanandaji), without whom the book would have been remained as an ordinary work.

And my respected Fairy Godmother my beloved Mummy (Mrs. Dianna Robinson) who gave me different ideas and thoughts.

While writing this book, some informations was needed and that is obvious so here comes the name of Dr. Paul Anderson, (my friend) Director Research Coordinator IUPAC, without whom the book would not have been finished.

I got an immense help from Dr. Palash Gangopadhyay School of Chemistry Central University, Hyderabad for the drawings and different applications such as CHEMSketch, Origin 6.0 and MathLab.

Also must thank to Mr. Dipak Dey and Mr Dipankar Dey of D.K. Scientific, (C-51 College Street Kol-7) for supplying the instruments and chemicals to me.

I am greatful to my three students Miss Ria Banerjee, Miss Moumita Banerjee and Miss Subhomita Banerjee for taking this colossal effort to publish this book.

And last but not the least my father, mother (for the computer) grand-mother and all other members of my family to finish this entire project.

At the end I must only appeal to my audience that, if you are interested in this book that will be my actual reward, I think and also do believe that, if any person is interested in the experimental chemistry, then from the guidance of A chemical Analyser's guide he or she will be able to set up his or her own chemistry lab). If you find any thing interesting at all then I will find that golden opportunity of sharing the secret of divine happiness and joy.

SANANDA CHATTERJEE

Contents

Induction Heating? —What is Induction Heating ? —Major Uses of Induction Heat Treatment—Properties—Physical Properties—Chemical Properties—Compounds of Scandium—Titanium—Occurrence—Extraction—Uses of Titanium—Metallic Titanium—Physical Properties—Compounds of Titanium—Vanadium—History—Occurrence and Extraction—Elementary Vanadium—Physical Properties—Compounds of Vanadium—Chlorides—Chromium—History—Occurrence and Extraction—Properties—Chemical Properties—Chemical Properties—Industrial Application of Chromium—Compounds of Chromium—Oxides of Chromium—Chromous Compounds—Chromic Compounds—Chromates and Dichromates—Detection and Estimation—Manganese—History and Occurrence—Extraction—Properties—Physical Properties—Compounds of Manganese—Properties of Potassium Permanganate—Complex Cyanides of Manganese—Detection and Estimation of Manganese—Iron—History and Occurrence—Varieties of Iron—Manufacture of Pig Iron and the Description of Blast Furnace—The Blast Furnace Temperatures and Reactions—Cast Iron—Composition of Cast Iron—Malleable Cast Iron—Wrought Iron—Steel—Lintz-Donawitz Process (L-D Process)—Open Hearth Furnace Process—Acid Hearth Process—Electric Arc Furnace—Description of Electrical Arc Furnace—Physical Properties of Iron—Chemical Properties—Rustion of Iron—The Corrosion Process—The Reactions Related with Rusting—Galvanic Series in Sea Water—Corroded End (Anodic)—Protected End (Cathodic)—Pitting and Crevice Corrosion—Protecting Metals from Corrosion—Thermal Spray Coatings for Corrosion Protection—Anodic Coatings—Cathodic Coatings—Neutral Coatings—Compounds of Iron—Uses—Uses—Precaution—Cobalt—History and Occurrence—Extraction—Reaction with Acids—Reaction with Non-Metals—Alloys—Grouping of Cobalt Alloys—Identification of Cobalt High-temperature Alloys—Compounds of Cobalt—Cobalt Complexes—Cobalt Carbonyls—Hydrolysis and Reactions with Acids and Bases—Reaction with Halogens—Reaction with Hydrogen—Reaction with Ammonia and the Amines—Reaction with Nitric Oxide—Detection and Estimation of Cobalt—Nickel—Isolation and Occurrence—Ores—Smelting and Refining—Alloys of Nickel—Physical Properties—Chemical Properties—Electronic Configuration and Oxidation State—Compounds of Nickel—Detection and Estimation—Copper—History—Cupric Compounds—Zinc—Recycling of Zinc—Yttrium—Palladium—Silver—History and Occurrence—Extraction—Physical Properties—Chemical Properties—Reduction of Silver Nitrate to Metallic Silver—Uses of Silver—Compounds of Silver—Action of Light upon Silver Salts—Detection and Estimation—Cadmium—History and Occurrence—Extraction—Physical Properties—Chemical Properties—Compounds of Cadmium—Detection and Estimation—Tin—History and Occurrence—Extraction—Physical Properties—Chemical Properties of Tin—Compounds of Tin—Detection and Estimation—Tungsten—History and Occurrence—Extraction—Physical Properties—Chemical Properties—Compounds of Tungsten—Estimation and

CHAPTER 14

Carbon and Compounds

Introduction

Carbon is one of the most important elements, which is freely available in nature. It is available in many forms. Among them coal or charcoal is the chief source of carbon.

Occurrence

Carbon occurs in nature both as free element and in the form of compounds. Carbon as an element is found as diamond form also as graphite form. It is also found as gas called carbon dioxide present in the atmosphere (0.03%). Carbon is found combined with hydrogen forming the well-known group of compounds; 'Hydrocarbons' used as fuel. It is also appears in most of the important minerals, *e.g.,* Limestone, marble Calcium Carbonate and Magnesium Carbonate. Carbon is also present in all living matter as one of the constituent of glucose. The product of the decay of plants and animals also consists of carbon as one of the chief ingredient.

About one million or more than nineteenth of all known compounds contain carbon. Because of vast number and diversity of carbon containing compounds another of chemistry known as organic chemistry is developed solely for their study.

Allotropes of Carbon

Carbon consists in various allotropic forms, of which diamond and graphite are in *crystal state,* while coal, wood charcoal, animal charcoal lamp black coke and gas carbon are in *amorphous* form. The X-Ray analysis has revealed, however, that so-called amorphous carbon in reality in crystalline in structure. However they consist of entirely fine crystals. *These carbons are therefore termed as micro-crystalline form.*

The chart below shows various crystalline and amorphous forms of carbon.

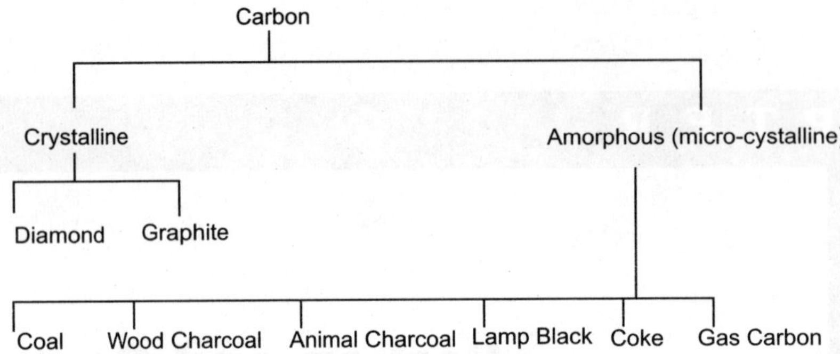

Diamond: Diamond was discovered for the first time in India was known as diamond producing country. The famous *Kohinoor* and the reagent or Pitt (studded in the Napoleon's State Sword) was found in India near Krishna river (State of Andhra Pradesh). But at present India occupies very insignificant place in the diamond map. Diamonds are chiefly found in the South Africa, the Belgian Congo, and Brazil.

Diamond is a big three-dimensional polymer comprising a large number of carbons arranged tetrahedrally, by utilizing their Sp3 hybrid orbitals. The structure of diamond is shown in the Fig. 14.1.

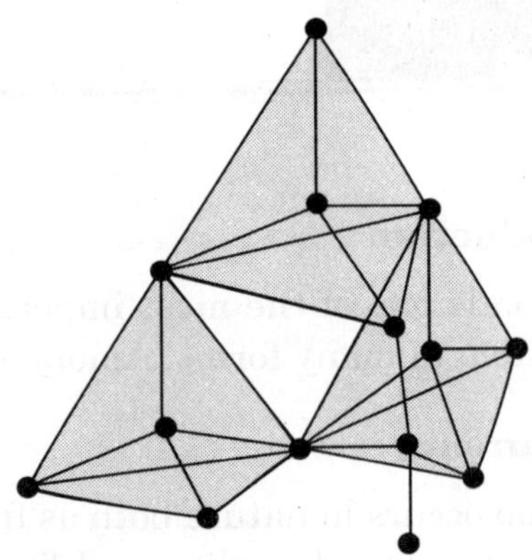

Fig 14.1. The Structure of Diamond

In the structure (Fig. 14.1) each carbon atom is tetrahedrally surround by four other carbon atoms and the C - C bond length is 1.54Å. The structure extends throughout the entire crystal in a great molecule fashion. The physical characteristics of diamond would involve the breaking of numerous covalent bonds, hence *extreme hardness* that the diamond possesses. *Lack of electrical conductivity* in diamond is due complete pairing of all the available electrons, so that none are left for the purpose of conduction.

Artificial Diamonds: These were prepared for the first time by Mossian in 1883. He heated pure sugar charcoal and iron in a graphite crucible to a temperature of 3000°C in an electric arc furnace. At this temperature, the iron dissolved an appreciable amount of carbon. The graphite crucible containing the molten mass was cooled rapidly by placing in a molten mass of lead (at 327°C). As a result of the solidification of the external surface of iron an enormous expansion occupied in the internal surface (cast iron expands on solidification). Under there condition carbon was thrown out in the form of minute diamonds. The mass cooled and treated with hydrochloric acid in order to dissolve iron. The residual carbon was found to contain a number of small diamonds (not more than 0.8 mm in diameter) some transparent some black.

Properties: Diamond is the purest form of carbon and is the extreme hardest substance known. It has as very high refractive index (2.417) and a very high density (3.51 and 15). It is transparent to X-Rays and is a bad conductor of electricity.

Uses of Diamond: Diamond is used (*i*) as gem and (*ii*) Black Diamond called Carbonate (valueless as gems) are used for cutting glass and drilling rocks.

Diamond is also used for making diamond anvils, which a base or platform on which hydrogen gas is placed and pressure is applied for solidification.

Graphite: This is the second crystalline form of carbon found in Travancore, Ajmer in India, graphite also occurs in Ceyleone U.S.A. Graphite is not freely available in nature, so most of the graphite is manufactured from coke and anthracite coal. The structure of graphite is shown in Fig. 14.2.

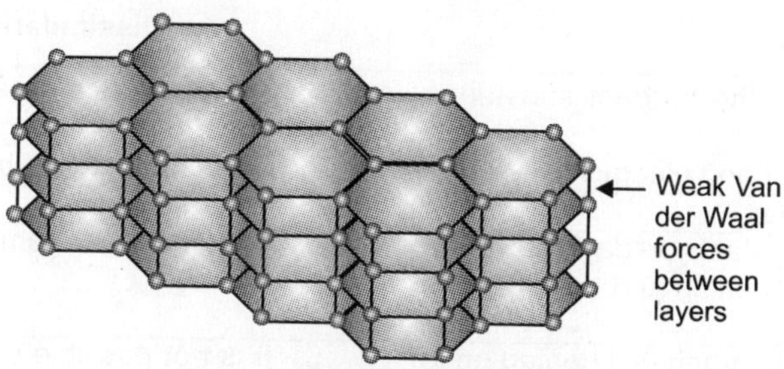

Fig. 14.2. Structure of Graphite

Industrial Preparation of Graphite

Powdered anthracite or coke is heated up to 3600°C in an electric arc furnace along with a little sand (used as a catalyst). Through an intermediate compound silicon carbide the amorphous form of carbon is changed to crystalline form. Heating is to be done for 24 hours Fig. 14.3.

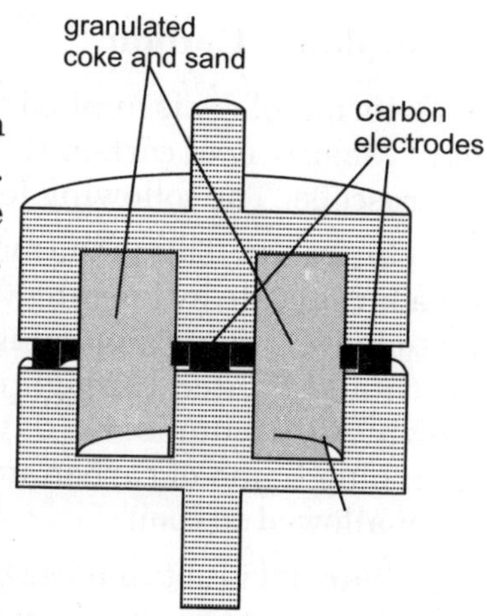

Fig. 14.3

Reaction Equations

Sand + Coke ⟶ Silicon Carbide + Carbon Monoxide

$$SiO_2 + 3C \longrightarrow SiC + 2CO$$

Silicon Carbide ⟶ Graphite + Silicon (volatilizes away)

$$SiC \longrightarrow C + Si$$

Properties: Soft dark solid with a metallic lusture. It is a good conductor of heat and electricity and has specific gravity 2.2. Marks the paper black. On strong heating with potassium dichromate acidulated with sulphuric acid it produces carbon dioxide.

Uses: Graphite is used:

1. in making pencil under the name black lead or 'plumage' lead of the lead pencil is mainly graphite and clay. Soft pencils contain more graphite and hard clay.
2. in making lubricates for machines running at high temperature *i.e.*, graphite and petroleum jelly or aquadag and oildag (suspension of graphite in water and oil).
3. in making refractory bricks and crucibles.
4. in making black paints and polishes.

Similarities and Dissimilarities Between Graphite and Diamond

Diamond	Graphite
Similacities	**Similacities**
1. Crystalline structure	1. It is also crystalline in mature.
2. Mayor may not have brightness.	2. It is has a metallic lusture.
3. Inert, burns in air at 900°C to to carbon dioxide.	3. Less inert, burns in air at a relatively low temperature (690°C) to carbon dioxide.
Dissimilarity	**Dissimilarity**
1. Hard and brittle. It is the hardest substance known.	1. Soft and has a greezy feel
2. Bad conductor of heat and electricity.	2. God conductor of heat and electricity
3. It does not mark paper	3. It marks paper.
4. Used as gems. But black or dark coloured varieties are used for cutting and drilling rocks and glass.	4. Unsuitable for gems and also for cutting and drilling.
5. It can be transferred to graphite if heated up to 1880°C	5. It is not possible to convert it into diamond.

Amorphous Carbon

This terminology is applied to a large number of materials (*e.g.*, charcoal, coke lamp black etc.) consisting of carbon that does not appear to be crystalline to the naked eye or under microscope. The following few are the more important examples of amorphous carbon.

Charcoal: This is the most common variety of amorphous carbon, which is the obtained on a large scale by burning wood in a limited supply of air. The wood is piled in such a way as to leave a central opening and a few air holes around to base. The pile is covered with earth and lighted through one of the air hole at the base. The heap is allowed to burn slowly. After a number of days the luminous flame at the top of the central shaft is replaced by a blue flame of carbon monoxide. All the air holes are then closed and the charcoal is then allowed to cool.

Charcoal is obtained as the product of this destructive distillation of wood in closed iron retorts (known as wood charcoal). The other products are methyl alcohol, acetic acid and acetone.

Charcoal is a black, porous substance. It possesses a remarkable power of adsorbing gases. For use in gas masks, employed for the adsorption of poisonous vapours the charcoal is activated by heating for a considerable time in a current of steam, which removes absorbed hydrocarbons.

Coal: Coal occurs in enormous quantities in certain localities. It has been shown to be produced by the decomposition of vegetable matter of the remote part in a limited supply of air. The different varieties of coal represent the various stages of transformation that the vegetable matter has reached. Peat is a loose and light variety representing the first stage of decomposing. Lignite or 'brown coal', which is the common variety, represents the third stage of transformation. It burns with a bright smoky flame. Its main use is for making coal gas. The forth and the last stage is represented by Anthracite, which is a hard, black brittle material. It burns without smoke giving out intense heat. It is therefore, used for steam boilers.

Name	Percentage of Carbon	Characteristics
1. Wood	40	———
2. Peat	60	Loose and light used for fuel.
3. Lignite	70	More compact, brown lustrous als used as fuel.
4. Bituminous (Common Coal)	80	Black stony mass burns with bright yellow flame and smoke is also produced.
5. Anthracite (Steam Coal)	90	Hard Brittle, non-smoky (steam raising very hot and short flame)

(iii) Coke: Coke is left behind as residue in the retorts during the dry distillation of coal for the manufacture of coal gas industry. It is used as a fuel and as a reducing agent in the metallurgy of iron and certain other metals.

(iv) Gas Carbon: This too, is by product of the coal gas industry it is formed as a hard deposit on the roofs and the sides of the retort. Gas carbon is nearly pure carbon and is very hard it is, sed for making the "carbon of arc lamps".

(iv) Lamp Black: It is also known as the carbon black, is the soot formed during the burning of oily materials rich in carbon, e.g. petrol turpentine oil etc. The burning is carried out in the limited supply of air, so as to get the maximum quantity of smoke. The latter is conveyed into chambers in which coarse blankets hung up. Lamp black gets deposited on the blankets and are subsequently removed. Lamp black contains appreciable proportion (up to 25%) of oily impurities and is used for making printer's ink, Indian ink, Black and Grey Paints.

Activated Carbon

This is an amorphous carbon that has specially treated to produce a highly developed pure substance with a large surface area. The large area enables the activated carbon to absorb large quantities of gases and vapours from gases or dissolved types of activated carbon, (*i*) Gas absorbance carbon (*ii*) Decolorizing carbon. The first one is used for gas absorption and the second one is used for absorption of liquids.

Methods of Preparation

The materials consisting of carbon are used for the preparation of activated carbon include wood, sawdust, peat lignite, lignin (paper pulp waste) beans, coconut shells, rice halls etc. the important processes of activation are as follows:

1. *Carbonization followed by Ignition:* The first step here is carbonization that involves heating the material at temperature below adsorptive level power. To bring about an improvement in this property, the char ignited at higher temperature to remove volatile impurities. The process is particularly useful in the production of bone black.
2. *Incorporation of Mineral Salts:* A typical example of this method is the production of active carbon from sawdust, paddy dust, etc. by first mining the material with a concentrated solution of zinc chloride. The heating of the mixture is carried out in rotary kilns at a temperature of 600°C-700°C in the absence of air for several hours. The product is washed with water to dissolve out zinc chloride. For re-use.
3. *Controlled Oxidation:* This method involves controlled, partial oxidation of carbon with steam or carbon dioxide at 800°C to 900°C on the degree of activation desired. Since oxidation of carbon with steam or carbon dioxide is an *endothermic reaction,* it is necessary to apply external heat. Alternatively, steam or carbon dioxide is mixed with sufficient air to provide an exothermic reaction, which serves to maintain the temperature at the desired level.
4. *Use of Chemical that Release Oxidizing Gases:* A typical example of such a procedure consists in mining the crushed carbonaceous source with dolomite $(Ca, Mg)CO_3$. After carbonization, the moisture is heated at 800°C - 900°C when carbon dioxide is evolved that cause selective oxidation of the carbon surface. After example is the use of phosphoric acid. The source, e.g. saw dust is saturated with syrupy phosphoric acid and then heated for several hours at 500°C. The resulting char is wasted with water to recover phosphoric acid to re-used. Alternatively, the char is further heated at 800°C- 900°C, at which temperature carbon is partially oxidized by phosphoric acid, which gets reduced to phosphene and then heated for several hours at 500°C. the resulting char is washed with water to recover phosphoric acid to be re-used. Alternatively, the char is further heated at 800°C - 900°C, at which temperature carbon is partially oxidized by phosphoric acid, which gets reduced to phosphene and elemental phosphorous.

Applications

1. *Sugar Refining:* The decolourization of sugar solution is generally carried out by passing it through a permeable granular animal charcoal or bone black in char filter. When the filter gets exhausted of its adsorptive power, the char is washed with water, dried and heated to redness is closed retorts to destroy the organic impurities and thus to restore its adsorptive power. The boneblack thus can be used again and again for as long as three to five years.
2. *Water Purification:* The municipal water that has been sterilized by treatment with chlorine processes an undesirable taste and disagreeable odour due to the products of reaction of chlorine on microorganism and treatment of chlorinated water did not succeed much in eliminating bad taste and odour. A break through was achieved in 1930 used activated carbon for this purpose with such that method

is new accepted as a standard final step in treatment of water for drinking and industrial purpose.

3. *Solvent Reclamation:* In the process of dry cleaning of dirty cloths, the solvent used (phenol etc.) becomes dark coloured due to the solid cloths. It thus becomes unfit for further use. Recovery of the solvent by distillation, through the preferred method is very expensive and moreover it does not completely remove the unpleasant odour. Treatment with powdered activated carbon has been proved to be much effective in removing both colours and odours, and if used often enough, prolongs solvent life and greatly reduces the frequency of distillation.
4. *Recovery of Volatile Solvents:* Vapours of solvents, such as either, alcohol, acetone etc., from manufacturing process, such as nitrocellulose, silk, rayon, plastics etc., are recovered by the use of a solvent is recovered by distillation.
5. *Gas Purification and Separation:* Activated carbon is also employed for the treatment of industrial gases to remove impurities or recover valuable constituents, thus granulated activated carbon is employed to recover petrol and gasoline from natural gas and light oils from coal gas. It also finds use for removing industrial odours and for removing cigarette smoke and cooling odours etc., from air circulated within an air conditional space. An important new application of activated carbon is the removal of small quantities of radioactive contaminants from nuclear reactors.
6. *Catalyst Application:* Considerable quantities of activated carbon find its uses as catalyst in chemical industries. The following are few examples of this category:
   (*i*) *Oxidation of certain salts,* such as ferrous sulphate, sodium arsenite, etc. in aqueous solution by means of air in the presence of activated carbon. This is believed to be due to irreversible adsorption of oxygen by activated carbon surface. The adsorbed oxygen acts powerfully acts as the oxidizer.

   (*ii*) *Hydrogen Peroxide is catalytically decomposed,* by activated carbon with liberation of oxygen.

   (*iii*) *In the manufacture of carbonyl chloride:* From carbon monoxide and chlorine and of sulphuryl chloride from sulphur dioxide and chlorine.

   (*iv*) In the manufacture of Vinyl chloride and vinyl acetate.

   (*v*) *As a first aid in poisoning and overdoes of drugs.* It can save life by adsorbing poisons like arsenic, strychnine, and kerosene and overdoes of aspirin, barbiturates and anti-depressant drugs. It is however, ineffective against cyanide, alcohols and mineral acids and caustic hydroxides.

Compounds of Carbon

Carbon forms a number of compounds; they are oxides, chlorides, sulphides etc. We shall first discuss the oxides of carbon.

CARBON MONOXIDE

Preparation

(I) By the Reduction of Carbon Dioxide with Red Hot Carbon: Carbon dioxide is passed through a silica tube packed with charcoal and made red-hot.

Carbon Dioxide + Carbon → Carbon Monoxide

$$CO_2 + C \rightarrow 2CO$$

The resulting gases are scrubbed with a strong solution of caustic soda to remove carbon dioxide.

(II) By the action of Sulphuric Acid on Formic Acid: When formic acid crystals are heated with sulphuric acid, carbon monoxide is set free, sulphuric acid is acting as a dehydrating agent.

Formic Acid + Sulphuric Acid → Carbon Monoxide + (Water + acid)

$$HCOOH + H_2SO_4 \rightarrow CO + (H_2O + H_2SO_4)$$

Exp: In a test tube some sulphuric acid is taken and few crystals of formic acids are also taken the tube is heated, then the mouth of the tube is brought near the burner flame, the mouth of the tube catches fire and burns in a pale blue flame.

Laboratory Preparation of Carbon Monoxide

Theory: In laboratory carbon monoxide is prepared by the reaction of sulphuric acid and oxalic acid. The gas is filtered from carbon dioxide by bubbling it through a solution of sodium hydroxide (see Fig 14.4).

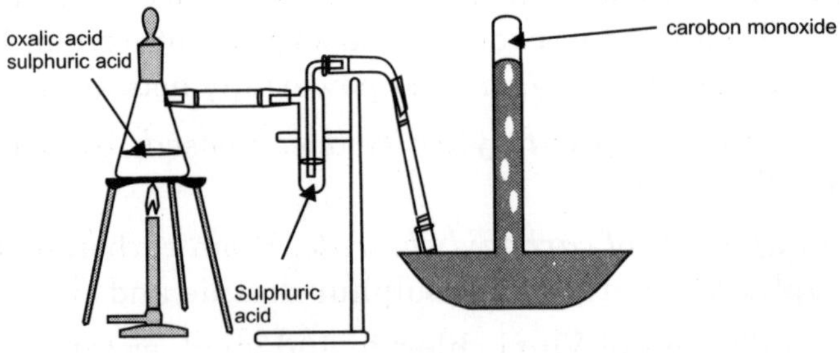

Fig. 14.4. The Lab Preparation of Carbon Monoxide

Reaction Equation

Oxalic Acid + Sulphuric Acid → Carbon Monoxide + Carbon Dioxide + (Water + acid)

$$(COOH)_2 + H_2SO_4 \rightarrow CO + CO_2 + (H_2O + H_2SO_4)$$

Precaution: *Carbon monoxide is a poisonous gas therefore should not be inhaled directly.*

Carbon Monoxide is also prepared from the potassium ferrocyanide crystals when heated with concentrated sulphuric acid.

Potassium Ferrocyanide	+	Sulphuric Acid	+	Water	$\rightarrow$	Potassium Sulphate	+	Ferrous Sulphate	+	Carbon Dioxide	+	Ammonium Bisulphate
$K_4[Fe(CN)_6]$	+	$6H_2SO_4$	+	H_2O	$\rightarrow$	$2K_2SO_4$	+	$FeSO_4$	+	$6CO$	+	$3(NH_4)_2$

Precaution: *Carbon monoxide is a poisonous gas therefore should not be inhaled directly.*

Properties of Carbon Monoxide

Carbon monoxide is a colourless gas having no taste, but with a faint acidic smell. It is slightly soluble in water and highly poisonous

Physical Properties

Melting Point	–200°C
Boiling Point	–191°C
Critical Temperature	–139°C
Critical Pressure	34.6 atmosphere
Standard Enthalpy of Formation	–110 kJ/mol
Standard Gibb's Energy of Formation	–137.02 kJ/mol
Standard Entroy	197.7 J/mol K
Heat of Capacity	29.1 J/mol K
Heat of Fusion (at melting point)	.83 kJ/mol

Chemical Properties

(I) Combustibility: Carbon monoxide is a combustibility gas and readily in atmosphere or oxygen, with a blue flame, forming carbon dioxide.

$$\text{Carbon Monoxide} + \text{Oxygen} \longrightarrow \text{Carbon Dioxide}$$

$$2CO + O_2 \longrightarrow 2CO_2$$

(II) Reducing Agent: Because of so much oxygen affinity it acts as a reducing agent, when passed over hot metallic oxides it reduces then to the metallic state.

$$\text{Ferrous Oxide} + \text{Carbon Monoxide} \longrightarrow \text{Iron} + \text{Carbon Dioxide}$$

$$Fe_2O_3 + 3CO \longrightarrow 2Fe + 3CO_2$$

Under the influence of ferric oxide

$$\text{Water} + \text{Carbon Monoxide} \longrightarrow \text{Carbon Doxide} + \text{Hydroen}$$

$$H_2O + CO \longrightarrow CO_2 \quad H_2$$

(III) Reaction with other Elements and compounds:

(a) With chlorine: Carbon monoxide and chlorine on exposure to sunlight produces an *Extremely poisonous gas* called Carbonyl Chloride. John Davy discovered this compound in the year 1811. It can be readily liquefied by cooling when a colourless liquid is produced with a boiling point of 8.2°C. The gas does not fume in moist air but is readily hydrolyzed by water.

Carbonyl Chloride + Water $\longrightarrow$ Carbon Dioxide + Water + Hydrogen Chloride

$$COCl_2 + 2H_2O \longrightarrow CO_2 + H_2O + HCl$$

This is the diamide of carbonic acid. Both urea and ammonium chloride are precipitated, but may be separated by warming with alcohol, in which urea is soluble. The alcoholic solution deposits crystals of urea on evaporation and cooling.

(b) With Alkalies: Carbon Monoxide is a neutral compound, but when it is heated with sodium hydroxide under pressure sodium formate is produced.

Sodium Hydroxide + Carbon Monoxide $\xrightarrow{140°C}$ Sodium Formate

$$NaOH + CO \longrightarrow HCOONa$$

Carbon Monoxide Compounds

(c) With Cuprous Chloride: Solution of cuprous chloride in dilute hydrochloride acid or ammonium hydroxide readily absorbs carbon monoxide forming and addition compound of $CuCl_2$ $CO.2H_2O$. The reaction is generally used in the estimation of carbon monoxide gas in a gas mixture.

(d) With Metals: Carbon monoxide reacts with hot nickel to form nickel carbonyls.

Nickel + Carbon Monoxide $\xrightarrow{100°C}$ Nickel Carbonyl

$$Ni + 4CO \longrightarrow Ni(CO_4)$$

Which decomposes at higher temperature

Nickel Carbonyl $\xrightarrow{180°C}$ Nickel + Carbon Monoxide

$$Ni(CO_4) \longrightarrow Ni + 4CO$$

(e) With Haemoglobin: Haemoglobin the red colouring matter of the blood, combines with carbon monoxide forming carbonyl hæmoglobin. The blood looses its poser of absorbing oxygen, small portion of inhalation of this gas causes headache, but a large amount can cause death.

POISON

Uses of Carbon Monoxide

Carbon monoxide is used as a fuel (producer gas, water gas). In the manufacture of methyl alcohol, formates and synthetic petrol. In the use of metallurgy as reducing agent and manufacturing of carbonyl chloride.

Testing of Carbon Monoxide

It burns with a blue flame. Paper soaked in platinic chloride (PtC_4) turns pink, green or black.

Formula of Carbon Monoxide

Formula of the carbon monoxide is determined by passing electrical sparks through a mixture of known volume carbon monoxide and oxygen, in a eudiometer tube, carbon monoxide is oxidized to carbon dioxide. On introducing to solution of potassium hydroxide reduction in volume occurs, representing the volume of carbon dioxide remaining gas is tasted to be oxygen. It has been determined *that two volumes of carbon monoxide gas combines with one volume of oxygen to form two volumes of oxygen to form two volumes of carbon dioxide.*

Now Applying Avogadro's Hypothesis:

2 mols of Carbon Monoxode + 1 mol of Oxygen → 2moles of Carbon dioxide

This is only possible if carbon monoxide is CO($2CO + O_2 \rightarrow 2CO_2$)

The vapour density of the gas 14 ∴ Molecular weight is 28(12 + 16).

CARBON DIOXIDE

Introduction

Van Helmont first prepared carbon dioxide in the year 1630; he called it as 'Gas Silvestre'. It was examined by Joseph Black in the year 1755 and was named as 'Fixed Air'. Bergman in the year 1774 did examine the gas fully. Until Lavoisier was the first to determine the composition by burning the charcoal in oxygen also showed that it combines with different oxides, *e.g.* calcium oxide, magnesium oxide etc, forming carbonates and thus called it 'Acidus Carbonique'. It was also called 'Carbonic acid gas.

Occurrence

Carbon dioxide largely occurs in the 'Poison Valley' (Java) and the Grotho del Cane (Naples) and the gas from such sources, as well as that collecting in a cellar (which extinguishes a candle) maintained by Pliny. It occurs in many mineral waters, also in soft drinks.

Atmospheric Carbon Dioxide

Normal outdoor air contains 3 volume of carbon dioxide per 10,000. The average percentage of the carbon dioxide in a crowded town or an unventilated room is about 0.04 – 0.3 per cent by volume. The continued breathing of air containing 2% of carbon dioxide is injurious. The 'stuffiness' of badly ventilated room is mainly due to the extraction of water vapour from the lungs, which tends to saturate the stagnant air and impedes the evaporation of perspiration.

The total amount of carbon dioxide in the atmospheric corresponds with about 600,000 million tons of carbon. The source of atmospheric carbon dioxide is the respiration of plants and animals, combustion, fermentation, decay of organic matter, mineral springs, volcanic activity and lime burning. Atmospheric carbon dioxide is diminished by the absorption by the sea and may due to the photosynthesis of plants and the weathering of silicate rocks (1.62×10^9) tons of carbon dioxide per annum. On the whole, the carbon dioxide in the atmospheric seems to be slowly increasing and the slight change of the climate may be due to this cause.

Carbon Dioxide Cycle in Nature

Carbon dioxide in the atmosphere is taken *p* by atmosphere organism (plants) and converted into carbohydrate (mainly glucose). This carbohydrate becomes the main food source of the hetrotrops (mainly the animals). All the animals return carbon dioxide to the air as a product of respiration and of decay. The burning of fossil fuel also releases carbon dioxide. In water, photosynthesis (see Fig 14.5, the carbon dioxide cycle).

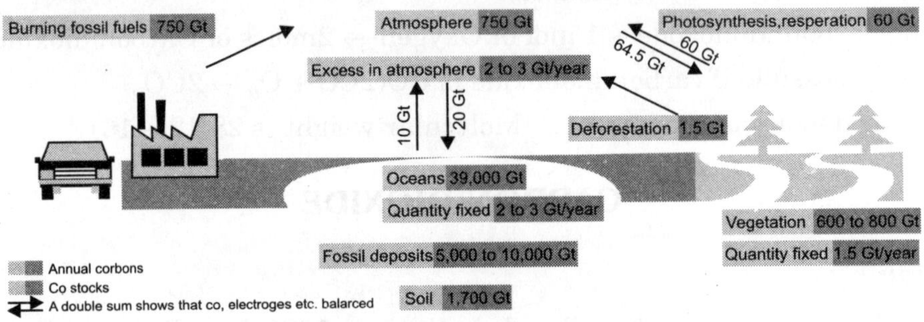

Fig 14.5. The Carbon dioxide Cycle in nature.

Preparation

1. Carbon dioxide is prepared by the complete combustion of carbon by heating metallic carbonates.

$$\text{Carbon} + \text{Oxygen} \longrightarrow \text{Carbon Dioxide}$$
$$C + O_2 \longrightarrow CO_2$$

2. When bicarbonate of any element like sodium is heated carbon dioxide is produced.

$$\text{Sodium Bicarbonate} \longrightarrow \text{Sodium Carbonate} + \text{Carbon Dioxide} + \text{Water}$$
$$NaHCO_3 \longrightarrow NaCO_3 + CO_2 + H_2O$$

3. When hydrochloric acid reacts with sodium bicarbonate carbon dioxide is produced.

$$\text{Sodium Bicarbonate} + \text{Hydrochloric Acid} \longrightarrow \text{Sodium Chloride} + \text{Carbon Dioxide} + \text{Water}$$
$$NaHCO_3 + HCl \longrightarrow NaCl + CO_2 + H_2O$$

Laboratory Preparation of Carbon Dioxide

Theory: In the laboratory carbon dioxide is prepared by the reaction of hydrochloric acid on marble chips. The reaction is carried out in a Woolf's bottle; the gas is collected by the upward displacement of air (see Fig 14.6).

Reaction Equation

$$\text{Marble Chips} + \text{Hydrochloric Acid} \longrightarrow \text{Carbon Chloride} + \text{Carbon Dioxide} + \text{Water}$$
$$CaCO_3 + 2HCl \longrightarrow CaCl_2 + CO_2 + H_2O$$

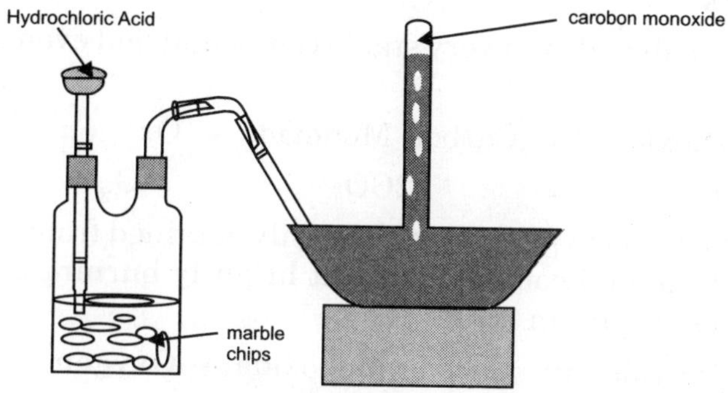

Fig. 14.6. The Lab Preparation of Carbon Dioxide

Precaution: *Sulphuric acid is not used instead of hydrochloric acid because it produces calcium sulphate which gets deposited on marble chips ultimately stops the reaction.*

Properties of Carbon Dioxide

Physical Propeties

1. Carbon dioxide is a gas at normal temperature and pressure, with a faint pungent smell and slightly acidic taste. The detail physical properties are being listed below:

Melting Point	–78.5°C
Boiling Point	–78.2°C
Density (at –37°C)	–1.101 g/cc
Standard Enthalpy of Formation	393.5 kJ/mol
Heat of Capacity	41.5 J/mol K
Vapour Pressure at –25°C	1684 Kpa
Heat of Vapourization at Boiling Point	6.04 kJ/mol
Heat of Fusion at Melting Point	9.2 kJ/mol
Critical Temperature	31.1 °C
Specific Volume (at 21°C)	140.32206 kg/m^3

2. It is 1 times heavier than air, and hence it sinks in the pits of out of use wells and other low lying areas, particularly in limekilns and manure heaps. It is non-poisonous but animals die in excess carbon dioxide. It is heavier than air that can proved by an experiment.

   Exp: Two balloons having same of huge quantities of gas under pressure and the same released with effervescence when pressure is released as in the case of soft drinks bottle.

3. At any temperature below 31°C carbon dioxide can be liquefied by pressure alone. 31°C is the critical temperature of the gas above this temperature; the gas cannot be converted to liquid by applying any amount of pressure.

Chemical Properties

(a) Stability: Carbon dioxide is a very stable compound and even at 2000°C dissociation is less than 2%,

$$\text{Carbon Dioxide} \rightleftharpoons \text{Carbon Monoxide} + \text{Oxygen}$$
$$CO_2 \rightleftharpoons 2CO \quad O_2$$

Because of this stability oxygen cannot be easily obtained from carbon dioxide and it is also non-supporter of life and combustion. But brightly burning sodium, magnesium or phosphorous continues to bum in it.

$$\text{Carbon Dioxide} \rightleftharpoons \text{Carbon Monoxide} + \text{Oxygen}$$
$$CO_2 \rightleftharpoons 2CO \quad O_2$$

(b) Acidic Nature: Moist carbon dioxide turns blue litmus red and it reacts with water to form carbon acid.

For this reason CO_2 is called an anhydrous carbon dioxde.

$$\text{Carbon Dioxide} \rightleftharpoons \text{Water} + \text{Carbonic Acid}$$
$$CO_2 \rightleftharpoons H_2O \quad H_2CO_3$$

Carbonic acid is very much unstable still it forms two types of salts:

1. Carbonates and

2. Bicarbonates

(*a*) When sodium hydroxide reacts with carbon dioxide sodium carbonate is produced.

$$\text{Sodium Hydroxide} + \text{Carbon Dioxide} \longrightarrow \text{Sodium Carbonate} + \text{Water}$$
$$NaOH + CO_2 \longrightarrow 2NaHCO_3 + H_2O$$

(*b*) Secondary when the sodium carbonate and water reacts with carbon dioxide sodium bi-carbonate is produced.

$$\text{Sodium Carbonate} + \text{Water} \longrightarrow \text{Carbon Dioxide} + \text{Sodium Bicarbonate}$$
$$NaHCO_3 + H_2O \longrightarrow CO_2 + 2NaHCO_3$$

3. Action of Lime Water: Calcium hydroxide in the same way reacts with carbon dioxide to form an insoluble precipitate (white) called calcium carbonate.

It is said that lime water turns milky

$$\text{Calcium Hydroxide} + \text{Carbon Dioxide} \longrightarrow \text{Calcium Carbonate} + \text{Water}$$
$$Ca(OH)_2 + CO_2 \longrightarrow CaCO_3 + H_2O$$

4. Reduction of Carbon Dioxide: When carbon dioxide is passed over hot zinc it is transferred from carbon dioxide to carbon monoxide.

$$\text{(Hot) Zinc} + \text{Carbon Dioxide} \longrightarrow \text{Zinc Oxid} + \text{Carbon Monoxide}$$
$$Zn + CO_2 \longrightarrow ZnO + CO$$

Or when it is passed over red-hot coke carbon dioxide is also transferred to carbon monoxide.

Industrial Process of Carbon Dioxide Gas

Carbon dioxide is prepared from the coke in the industry.

Reaction Equation

(Combustion): Carbon (atmorphous) + Oxygen (air) $\longrightarrow$ Carbon Dioxide

$$C + O_2 \longrightarrow CO_2$$

(Thermal Decompostion):

Calcium Carbonate + Heat $\longrightarrow$ Calcium Oxide + Carbon Dioxide

$$CaCO_3 + \text{Heat} \longrightarrow CaO + CO_2$$

(Recovery): Carbon Dioxide + Sodium Carbonate + Water $\longrightarrow$ Sodium Bicarbonate

$$CO_2 + NaCO_3 + H_2O \longrightarrow 2NaHCO_3$$

Sodium Bicarbonate + Heat $\longrightarrow$ Sodium Carbonate + Water + Carbonate Dioxide

$$2NaHCO_3 + \text{Heat} \longrightarrow Na_2CO_3 \quad H_2O + CO_2$$

Yield: The yield is 100%.

Process: Coke is burned in the furnace at 232°C temperature where fuel gas is produced. The gas is produced in such a way that it consists about 17 to 18% of carbon dioxide. At the same time steam is also produced. This gas is sent through a heat exchanger, where the temperature of the gas is reduced to 121°C the excess heat is taken up by continuously flowing strong lye. Then the gas is sent to scrubber where sulphur dioxide and dust particles are removed the temperature is again reduced to 37°C. The dilute carbon dioxide is then fed into the bottom of coke filled absorption tower, where it passes countercurrent to an aqueous solution of sodium carbonate called 'weak lye'. After absorbing the carbon, the solution of bicarbonate, known as 'strong lye' is pumped through heat exchangers operating at 104.4°C to 121°C where the solution is heated before entry in the lye boiler.

The absorber, operating at about 29.0°C, remove all but 9% of the carbon dioxide in the fuel gas, which is then released to the atmosphere.

The preheated solution of sodium bi-carbonate is heated with exhaust steam in the lye boiler. Here, at a temperature of 118.3°C, the bicarbonate is decomposed into sodium carbonate (weak lye) which is returned through the heat exchanger to the absorber for further re-use. The liberated gas consisting of 99% of carbon dioxide escapes trough the lye boiler. The gas is allowed to pass through a water cooler, where the temperature is further lowered and the moisture in the saturated gas is condensed and returned to the weak lye. The cooled gas is collected in the gasholder.

Solid Carbon Dioxide Dry Ice

Dry ice the name given to solid carbon dioxide, because at normal atmospheric pressure at changes directly from solid state to gaseous state. This behaviour is shown in the Pressure-Temperature relationship shown in the Fig. 14.7.

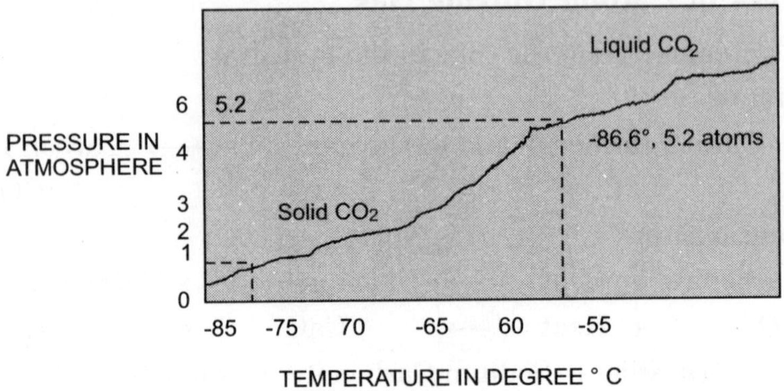

Fig. 14.7. The Pressure-temperature Relationship of Carbon Dioxide Solid and Liquid

At –56.6°C and 5.2 atmospheric, solid carbon dioxide is in a state of equilibrium with liquid carbon dioxide and gaseous carbon dioxide. Below this temperature and pressure solid carbon dioxide is in equilibrium with gaseous carbon dioxide (dry ice) can easily be obtained by cooling the gas even at ordinary pressure.

At normal atmospheric pressure solid carbon dioxide vapourizes without melting and the temperature remains -78°C during this evaporation, as shown in the Fig. 14.6. It can be easily seen that liquid carbon dioxide can exist above 5.2 atmospheric pressure.

Manufacture of Dry Ice

Dry ice is manufactured by evaporation of the liquid carbon dioxide at 20°C and 70 atmospheric pressure to a pressure less than 5.2 atmospheric. The diagrammatic representation of the dry ice plant is shown in the Fig. 14.8.

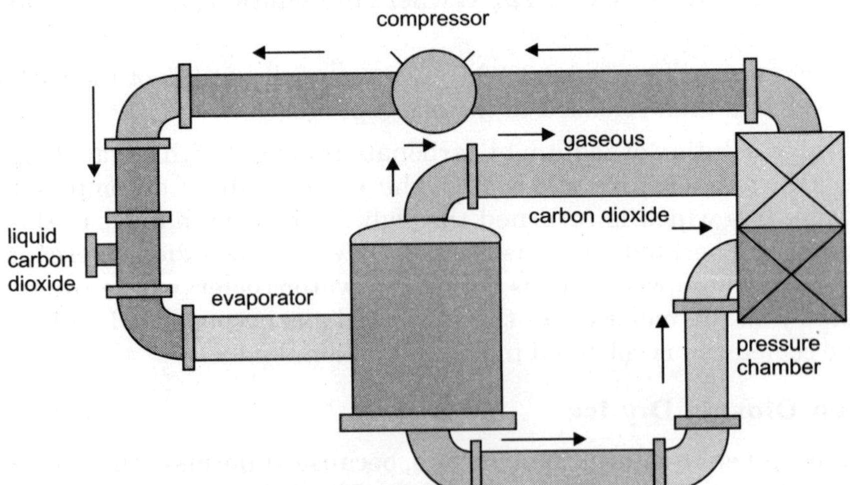

Fig. 14.8. The Manufacture of Dry Ice

Carbon dioxide gas obtained by any of the methods is at first set free from any impurities and moisture it is compressed to a pressure of about 70 atmosphere when it condenses to a liquid.

The liquid carbon dioxide obtained above is released through a jet into a chamber *E* (see Fig 14.8) called evaporator. The pressure here falls from about 70 atmosphere to 35 atmosphere and the gas cools down to about O°C due to Joule-Thomson effect. *Since the pressure is about* 5.2 *atmospheric the gas will remain liquid.*

The liquid carbon dioxide from the evaporator *E* is introduced to another chamber *P*, called *pressure chamber* through a jet. The gas further cooled. When the pressure falls below 5.2 atmospheric pressure the gas solidifies to give dry ice.

Properties of Dry Ice

Dry ice has its property in varieties of uses. It is used as a refrigerent under the popular commercial name *drink hold.* It is also used for the manufacture. It is also used for the manufacture of the ice cream. It is also used for transportation of perishable goods etc. It was also used in artificial rain.

Composition of Carbon Dioxide

Composition of carbon dioxide by weight is determined by passing a stream of pure dry oxygen over weighted quantity of purified sugar charcoal or diamond on a platinum boat, heated strongly in a hard glass tube in a combustion furnace. The tube is placed with a long column of granular copper oxide to oxidize any carbon monoxide to dioxide, which is then absorbed and weighted in potash bulbs (see Fig. 14.9). Any ash left behind in the boat is weighted and the weight deducted from the weight of charcoal taken. It is found that 3 g of carbon unite with 8 g of oxygen to yield 11gms of carbon dioxide. The density of the carbon dioxide is 22. Hence, the molecular weight of the gas is 44. Carbon dioxide therefore contains 12 parts of carbon and 32 parts of oxygen and the formula is CO_2.

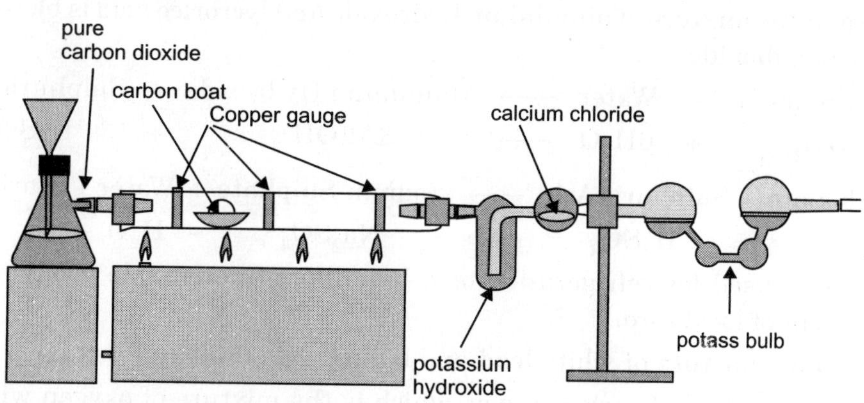

Fig. 14.9. The Experiment for the Determination of the Composition of Carbon Dioxide

Uses of Carbon Dioxide

Carbon dioxide has many uses they are as follows:

1. *Household fire extinguisher:* Air containing about 155 of carbon dioxide does not at all support combustion, carbon dioxide is therefore used as fire extinguisher.

(*a*) Dry fire extinguishers contain sand and sodium bicarbonate mixture, which is thrown over the fire where the bicarbonate of soda gets decomposed by heat and carbon dioxide is liberated.

(*b*) Washing soda, sulphuric acid type fire extinguisher contains a bottle of sulphuric acid supported in a metallic cylinder fitted with sodium carbonate (washing soda) solution, see Fig. 14.10. On striking the knob, the acid bottle breaks and the carbon dioxide is liberated as the result of the reaction.

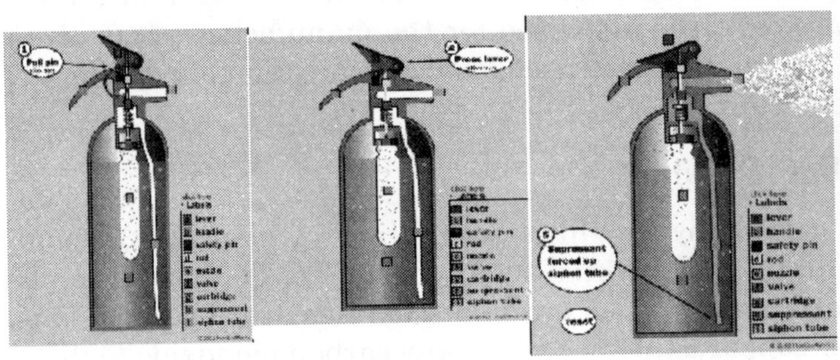

How Fire Extinguisher Works

Photograph of fire extinguisher

Fig 14.10. The Fire Extinguisher Working Principle and the Actual Photograph

(*c*) In foam type fire extinguisher, sulphuric acid is replaced by aluminium sulphate and lycrorice extract. Aluminium sulphate liberates carbon dioxide from washing soda and the mixture of aluminium hydroxide and lycrorice acid is blown into foam by carbon dioxide.

Aluminium Sulphate + Water ⟶ Aluminium Hydroxide + Sulphuric Acid

$$Al_2(SO_4)_2 + 6H_2O \longrightarrow 3Al(OH)_3 + H_2SO_4$$

Sodium Carbonate + Sulphuric Acid ⟶ Sodium Sulphate + Water + Carbon Dioxide

$$Na_2CO_3 + H_2SO_4 \longrightarrow Na_2SO_4 + H_2O + CO_2$$

2. Dry ice is used for refrigeration (*e.g.* handling of perishable goods) and for the treatment of local sore.

3. In the manufacture of white lead and baling soda (Solvay Process)

4. In the generation of carbogen gas, which is the mixture of oxygen with 5 to 10% carbon dioxide, given to the patient of pneumonia and the victim of carbon monoxide poisoning for artificial respiration, carbon dioxide stimulates breathing.

Hydrides of Carbon

We shall now study the compounds of carbon and hydrogen (only inorganic) often called as *hydrocarbons.*

METHANE 'MARSH' GAS

Occurrence

Methane occurs in natural gas coming out from petroleum wells. The percentage of methane is about 90. The firedamp of the coal mine also consists of this gas. Bacterial decay of the food and organic matter especially cellulose at the bottom of the marshy ponds produces methane which often bubbles up when the mud is distributed and comes on the surface catches fire. Coal gas consists of about 35% methane by volume.

Preparation

1. By sending an electrical arc between carbon electrodes in the presence of hydrogen gas.

$$\text{Carbon} + \text{Hydrogen} \longrightarrow \text{Methene}$$
$$C + 2H_2 \longrightarrow CH_4$$

2. By passing a mixture of carbon disulphide vapour and hydrogen sulphide vapour in the presence of hot copper.

$$\text{Carbon Disulphide} + \text{Hydrogen Sulphide} + \text{Copper} \longrightarrow \text{Cuprous Sulphide} + \text{Methane}$$
$$CS_2 + 2H_2S + 8Cu \longrightarrow 4Cu_2S + CH_4$$

3. It is also available by passing a mixture of carbon monoxide and hydrogen sulphide over heated copper.

$$\text{Aluminінum Carbide} + \text{Water} \longrightarrow \text{Methane} + \text{Water}$$
$$Al_4C_3 + 12H_2O \longrightarrow CH_4 + H_2O$$

Fairly pure methane is obtained by the action of water on aluminium carbide.

Laboratory Preparation of Methane

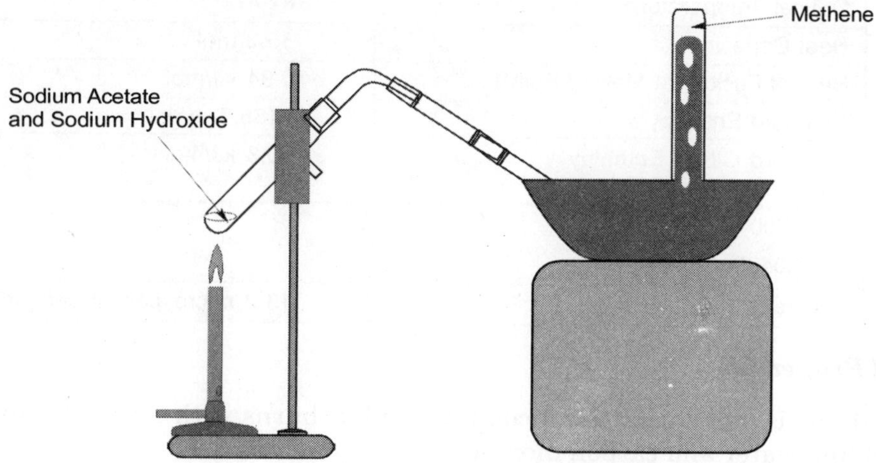

Fig. 14.11. The Lab Preparation of Methane

Theory: Methane in laboratory is prepared by heating sodium acetate and sodium hydroxide mixture. The gas is collected by downward displacement of water. (see Fig. 14.11)

Reaction Equation

Sodium Acetate + Sodium Hyderoxide ⟶ Methane + Sodium Carbonate

$$CH_3COONa + NaOH \longrightarrow CH_4 + NaCO_3$$

Precaution: *The gas should be handled with care because it explodes with oxygen or air mixture (if the mixture is ignited).*

Purification: The obtained methane in this process is not pure and may contain as much as 8% hydrogen and 10% other hydrocarbons such as the chloride (to remove acetylene) and concentrated sulphuric acid (to remove ethylene and moisture). The purified methane still contains hydrogen, which may be removed as water by boiling at 110°C the excess oxygen is absorbed concentrated sulphuric acid.

Heating sodium acetate with barium hydroxide, instead of sodium hydroxide forms nearly pure methane.

Sodium Acetate + Barium Hydroxide ⟶ Methane + Sodium Carbonate + Barium Carbonate

$$CH_3COONa + Ba(OH)_2 \longrightarrow 2CH_4 + NaCO_3 + Na_2CO_3$$

Properties of Methane

Physical Properties

Physical State	Gas
Melting Point	–161.48°C
Boiling Point	–182.46°C
Heat of Vapourization (at boiling point)	8.19 kJ/mol
Critical Temperature	82.47
Heat Capacity	35.3/mol
Heat of Fusion (at Melting Point)	0.94 kJ/mol
Standard Enthalpy	–186.3 J/mol
Standard Gibb's Enthalpy Van der Waal's Constant	50.3 kJ/mol
(*a*) 2.300 bar L^2/mol (*b*) 0.04301 L/mol	
Viscosity	13.2 micro pascal second

Chemical Properties

(*a*) Methane is non-supporter of combustion but burns with a feeble luminous flame yielding water and carbon dioxide.

$$\text{Methane} + \text{Oxygen} \longrightarrow \text{Carbon Dioxide} + \text{Water}$$
$$CH_4 + O_2 \longrightarrow CO_2 + H_2O$$

(*b*) The gas forms an explosive mixture when combines with air or oxygen (if the mixture is ignited)

The uses of open flame type lights are dangerous in the coal mines, as methane is present in the mines, which can cause explosion so to solve this problem. Davy's safety lamp was introduced; it consists of a copper gauge, which by its heat conducting capacity prevents the flame to come contact with the surrounding air.

Methane reacts with chlorine in the presence of light. A mixture of methane and chlorine explodes and methyl chloride is produced along with hydrochloric acid. There is a series of reactions takes place, which at last produces carbon tetrachloride.

$$\text{Methane} + \text{Chlorine} \longrightarrow \text{Hydrogen Chloride} + \text{Water}$$
$$CH_4 + Cl_2 \longrightarrow HCl + H_2O$$
$$\text{Methyl Chloride} + \text{Chlorine} \longrightarrow \text{Hydrogen Chloride} + \text{Methylene Chloride}$$
$$CH_3Cl + Cl_2 \longrightarrow HCl + CH_2Cl_2$$
$$\text{Methylene Chloride} + \text{Chlorine} \longrightarrow \text{Hydrogen Chloride} + \text{Carbon Tetrachloride}$$
$$CH_2Cl_2 + Cl_2 \longrightarrow HCl + CCl_4$$

Methane is saturated hydrocarbon, since all the four valence atoms are satisfied and therefore, does not react with hydrogen. Methane reacts with iodine and fluorine causing an explosion.

$$\text{Methane} + \text{Iodine} \longrightarrow \text{Methyl Iodide} + \text{Hydrogen Iodide}$$
$$CH_4 + Cl_2 \longrightarrow HCl + H_2O$$
$$\text{Methane} + \text{Fluorine} \longrightarrow \text{Carbon Tetrafluoride} + \text{Hydrogen Fluoride}$$
$$CH_4 + 4F_2 \longrightarrow CF_4 + HF$$

Use of Methane

1. As a source of hydrogen, methane is mixed with steam and passed over nickel supported on aluminium at 725°C.

$$\text{Methane} + \text{Steam} \longrightarrow \text{Carbon Monoxide} + \text{Water}$$
$$CH_4 + H_2O \longrightarrow CO + 3H_2O$$

2. For making carbon black methane is decomposed into carbon and hydrogen at 1000°C

$$\text{Methane} \longrightarrow \text{Carbon} + \text{Hydrogen}$$
$$CH_4 \longrightarrow C \quad 2H_2$$

3. Methane is also used for preparation of methyl alcohol and formaldehyde.

ACETYLENE

Invention

Acetylene was discovered by Edmond Davy in 1856, but was carefully studied by Beterworth in the year 1854, who synthesized the gas from the elements by sticking an electrical arc between two carbon electrodes in an atmosphere of hydrogen. Methane exists about 0.6% in coal gas. It is also formed when Bunsen burner 'strikes back'.

Laboratory Preparation of Acetylene

Theory: Acetylene in the laboratory is prepared by the action of water calcium carbide (see Fig. 14.12). The gas is collected by the downward displacement of water.

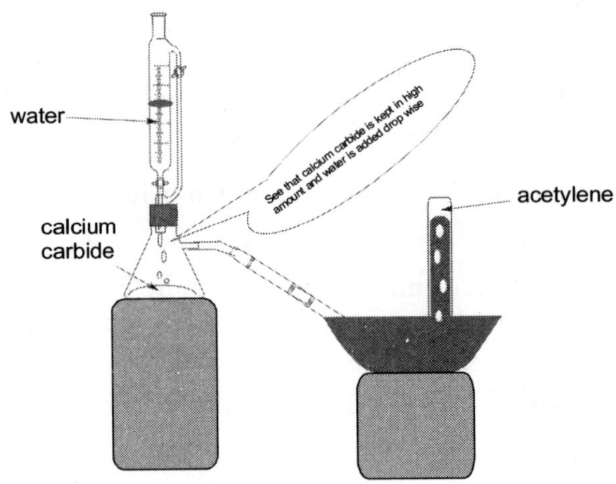

Fig. 14.12. The Lab Preparation of Acetylene

Reaction Equation

$$\text{Calcium Carbide} + \text{Water} \longrightarrow \text{Calcium Hydroxide} + \text{Acetylene}$$

$$CaC_2 + H_2O \qquad Ca(OH)_2 + C_2H_2$$

INDUSTRIAL PREPARATION OF ACETYLENE

Description of The Plant

1. Acetylene Generator {Medium Pressure}

It mainly consists of a twin Hooper that is:

» Agitator and its drive unit

» Carbide Hooper screw feeder and its drive

» Flash back arrestor

» Safety relief valves drain system & the automatic controls system.

Safety: The generator is fitted with five safety valves. One is fitted on the gas outlet frame, one of each of the flash arrestors and one on each of the carbide Hooper. The bottom of the generator is fitted with an air-operated drain valve which discharge slurry in response to the slurry level is high. There is a water inlet valve, which opens in response to the temperature of the slurry.

2. Cooler Condenser: The cooler condenser cools the acetylene gas having the generator, which is hot and laden with moisture and other gaseous impurities.

3. Low Pressure Dryer: This is a pressure vessel charged with solid anhydrous carbide, which absorbs the moisture in the gas.

4. Purifier: This is a large vessel provided with two sectors by wire mesh frame network. Each sections by wire mesh frame work. Each sectors are charged with purifying mass which absorbs gases like hydrogen sulphide, phosgene etc.

5. Ammonia Scrubber: This equipment has a vertical vessel filled with a level gauge and water make-up line. A particular as marked on the level indicator has to be maintained. The ammonia scrubber all the water soluble impurities like ammonia.

6. Acetylene Compressor: The acetylene compressor is a three stage machine which is immersed in water filled tank to avoid any pressurized gas to come in contact with air and also to make sure that all the components are continuously cooled. The compressor is drawn by flame proof motor from outside the tank.

7. High Pressure Dryer: The high pressure dryer consists of three columns. The first column is filled with packing to avoid any void and for proper mixing and the second and third are filled with anhydrous calcium chloride.

8. Filling Manifold: The acetylene gas manifold consists of two headers fitted with non-return valves, one flash back arrestors, pressure gauge and manual isolation valves.

9. Acetone Pump: Acetone pump is supplied along with the plant for charging acetone in cylinders.

10. Automatic controls of acetylene generator:

(*a*) Temperature cum water inlet control.

(*b*) High temperature alarm.

(*c*) Level-cum-slurry discharge control.

(*d*) Level alarm.

(*e*) Pressure controls.

(*f*) Control of high temperature.

(*g*) Control of high level.

11. Level Controls: Level controller maintaining a particular level in the generator. If the level is above normal, water can enter the carbide hop.

12. Level Alarm: The pressure inside the acetylene generator is controlled by a meteoroid switch.

13. Pressure Alarm: This is the important part for the safety of this plant. A pressure switch is connected to the alarm system.

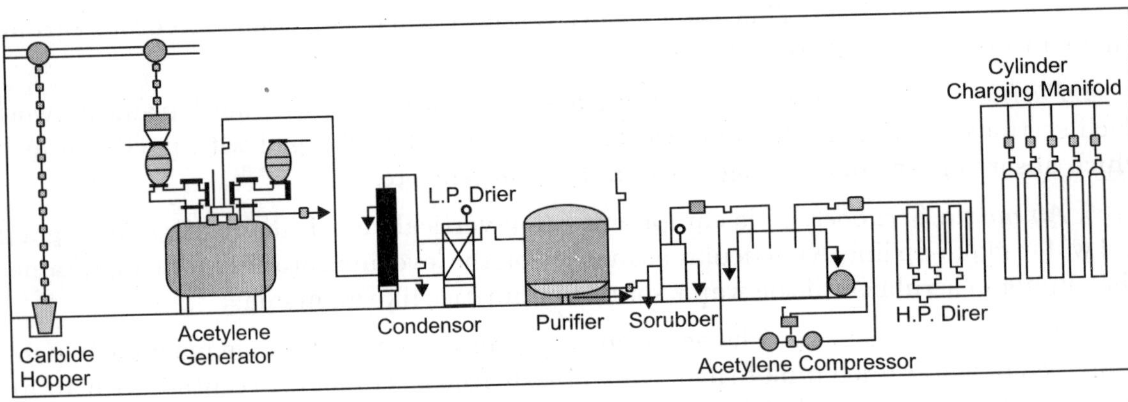

Fig 14.13. The Industrial Preparation of Acetylene

Equipment Schedule for Acetylene Gas Producing Plants

Sl.No.	Description	Sl.No.	Description
1.	Carbide Hopper	11.	Gas Lines
2.	Acetylene Generator	12.	Condenser
3.	L.P. Dryer	13.	Automatic Screw Feed
4.	Purifier	14.	Agitator
5.	Ammonia Scrubber	15.	Temperature Controller
6.	Acetylene Compressor	16.	Pressure Controller
7.	H.P. Dryer	17.	Level Controller
8.	Cylinder Filling Manifold	18.	Slurry Disch Arge Control
9.	Acetone Pump	19.	High Temperature Alarm
10.	Flame Proof Motors	20.	One Charge of Chemicals

Sl.No.	Description	UA - 25	UA - 50	UA-100
1.1	Capacity (acetylene with purity as per IS-30B) Cubic Meter per hour	25	50	100
1.2	Raw material required			
	1.2.1 Process water (litres/kg of calcium carbide processed)	8-10	8-10	8-10
	1.2.2 Calcium carbide (kg/N.cubic metre of gas)	4	4	4
	1.2.3 Calcium chloride (g/N.cubic metre of gas)	50	50	50
	1.2.4 Cooling water make-up (litres/N.cubic metre of gas)	10	10	10
	1.2.5 Power (average requirement)	12kW	15 kW	28kW
	1.2.6 Purifying mass (gas plus acid absorption mix per cum of gas)	15 g/N. Cum + 4 g	15 g/N. Cum + 4 g	15 g/N. Cum + 4 g
	1.2.7 Instrument/Process air	70 litres/ minute	70 litres/ minute	100 litres/ minute
	1.2.8 Acetone	As required	As required	As required
	1.2.9 Hydraulic oil (for screw feed system)	Equivalent to SAE - 30	Equivelent to SAE - 30	Equivalent to SAE - 30
	1.2.10 Lubricating oil (for compressor)	Hylube HDX-40	Hylube HDX-40	Hylube HDX-40
1.3	Space required (abt)			
	1.3.1 Process Area	70 Sq.m.	110 Sq.m.	150 Sq.m.
	1.3.2 Carbide Storage Area	75 Sq.m.	112 Sq.m.	175 Sq.m.
	1.3.3 Cylinder filling area	40 Sq.m.	60 Sq.m.	100 Sq.m.
	1.3.4 Cylinder storage area (depending upon cylinder inventory)	100 Sq.m.	160 Sq.m.	200 Sq.m.
	1.3.5 Lime slurry disposal area	100 Sq.m.	180 Sq.m.	250 Sq.m.
	1.3.6 Free space around the plant (on all sides)	16 m	18.5 m	20 m
	1.3.7 Acetone godown	12 Sq.m.	20 Sq.m.	30 Sq.m.

Purification of Acetylene

Pure Acetylene is obtained from the following processes:

By passing the crude acetylene gas in ammonical cuprous chloride solution, which absorbs acetylene, giving a red cuprous acetylide by heating with strong hydrogen chloride or potassium cyanide solution. The gas may be dried over phosphorous pentoxide and collected over mercury.

Acetylene + Cupric Chloride $\longrightarrow$ Cupric Carbide + Hydrogen Chloride

$$C_2H_2 + 2CuCl \longrightarrow Cu_2C_2 + 2HCl$$

Cuprous Carbide + Hydrogen Chloride $\longrightarrow$ Cupric Chloride + Acetylene

$$Cu_2C_2 + 2HCl \longrightarrow 2CuCl + C_2H_2$$

DEADLY POISON

Cuprous Carbide + Potassium Cyanide + Water $\longrightarrow$ Cuprous Cyanide + Potassium Hydroxide + Acetylene

$$Cu_2C_2 + KCN + H_2O \longrightarrow Cu_2CN + KOH + C_2H_2$$

Properties of Acetylene

Properties of Acetylene

Melting Point	–80.6°C
Boiling Point	–84.57°C
Critical Temperature	35.33°C
Critical Pressure	60.792 atoms
Critical Molar Volume	11 cm^3/mol
Thermal Conductivity at –73°C	21.4 mV/mK (milivolt/mili Kelvin)
Viscosity at 27°C	13.2 micro pascal second
Van der Waal's Constant	(*a*) 4.516 bar/L$_2$ (*b*) 0.05220 L/mol

Chemical Properties

1. Acetylene is colourless gas with a sweet etherical smell, when pure (the unpleasant smell is due to phosphene) acetylene is slightly lighter than air and ver little soluble in water *but very much soluble in acetone.*
2. Non-supporter of combustion, but the gas burns with a smoky flame, separating much carbon and yielding water and carbon dioxide.

   Acetylene Oxygen $\longrightarrow$ Carbon Dioxide + Water

   $$2C_2H_2 + 5O_2 \longrightarrow 4CO_2 + 2H_2O$$

   A mixture of acetylene and air explodes with violence if ignites. On account of its highly endothermic character, compressed acetylene is likely to detonate, therefore the gas is stores in solution (under pressure) in acetone absorbed in a porous material, acetone dissolves, 300 volumes of gas under 12 atmospheric pressure.
3. *Action of chlorine:* Acetylene explodes violently if mixed with chlorine, depositing soot.

   Acetylene + Chlorine $\longrightarrow$ Carbon + Hydrogen Chloride

   $$C_2H_2 + Cl_2 \longrightarrow 2C + 2HCl$$

Exp: When a small piece of calcium carbide is dipped in a saturated solution of chlorine in a jar, there appears a flash of light with simultaneously separation of soot. But by the regulation action of chlorine on acetylene in the presence of a mixture of sulphur chlorine and residual iron, acetylene dichloride $C_2H_2Cl_2$ and acetylene tetrachloride, (westron) $C_2H_2Cl_4$ are produced they are used as solvents.

$$\text{Acetylene} + \text{Chlorine} \longrightarrow \text{Acetylene Dichloride}$$
$$C_2H_2 + Cl_2 \longrightarrow C_2H_2Cl_2$$

The acetylene tetrachloride is produces when trichloroacetylene C_2HCl_3, on heating with lime.

$$\text{Acetylene Tetrachloride} + \text{Calcium Oxide} \longrightarrow \text{Acetylene Trichloride} + \text{Water}$$
$$2C_2H_2Cl_4 + CaO \longrightarrow C_2HCl_3 + H_2O$$

Uses of Acetylene

(*a*) Acetylene is mainly used as a fuel gas

(*b*) Oxygen and acetylene mixed and inflamed produces a flame called Oxy-acetylene flame with a temperature of 3200°C is used for cutting and (gas welding).

(*c*) Acetylene evolved from calcium carbide, which is used for ripening of fruits by putting some carbide on it.

(*d*) It is also used for the preparation of polyethylene and Teflon.

Detection of Acetylene

Bayer's Process: When potassium permanganate is used to dissolve acetylene gas it is decolourized due to presence of unsaturated hydrocarbon.

ETHYLENE, OLEFIENT GAS

Introduction

Bleacher first discovered ethylene gas in the year 1669. If occurs in coal gas about 4 to 10% by volume.

Preparation

Ethylene can be prepared by following processes:

1. Dehydrating ethyl alcohol with sulphuric acid or phosphoric acid.
2. By passing alcohol over heated aluminium oxide.

Laboratory Preparation of Ethylene

Theory: Ethylene is prepared in the laboratory by the dehydrating ethyl alcohol with sulphuric acid.

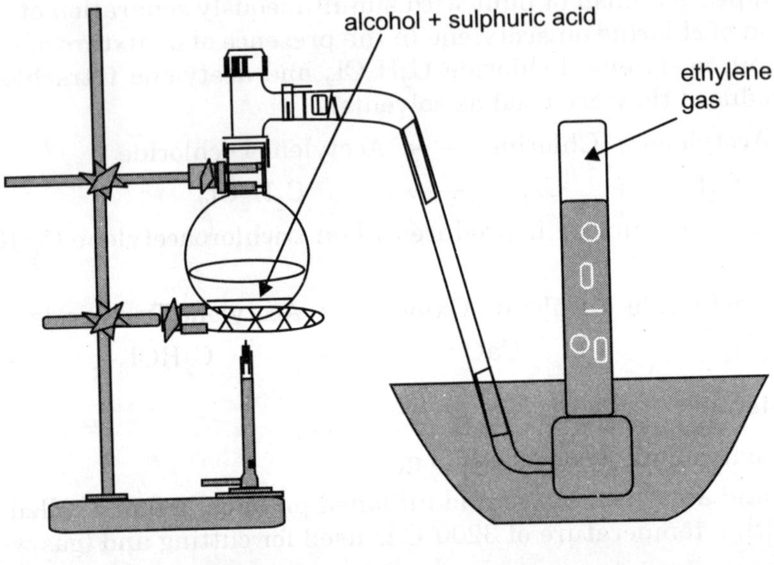

Apparatus	Reagents	Temperature	Pressure
Round bottomed flask, burner, gas jar, delivery tube, etc	30cc ethyl alcohol, 80cc sulphuric acid, sand	160°C	normal

Fig. 14.14. The Lab Preparation of Ethylene

Reaction Equation

Ethyl Alcohol + Sulphuric Acid ⟶ Ehtyl Hydrogen Sulphate + Water

$C_2H_3OH + H_2SO_4 \longrightarrow C_2H_5SO_4 + H_2O$

Before collection of the gas it is washed with potassium hydroxide solution to remove any carbon dioxide and sulphur dioxide and the gas is collected by the downward displacement of water.

Preparation of Pure Ethylene

1. Pure ethylene can be obtained by heating ethylene dibromide with zinc-dust.

   Ethylene Dibromide + Zinc ⟶ Zinc Bromide + Ethylene

   $C_2H_4Br_2 + Zn \longrightarrow ZnBr_2 + C_2H_4$

2. It is also obtained by the action of ethylene iodide and potassium hydroxide

Ethylene Iodide + Potassium Hydroxide ⟶ Ethylene + Water + Potassium Iodide

$2C_2H_4l + 2KOH \longrightarrow 2C_2H_4 + H_2O + 2Kl$

Properties of Ethylene

Physical Properties

Melting Point	103.62°C
Boiling Point	69°C
Critical Temperature	9.34°C
Critical Pressure	5.041 Mpa
Critical Molar	131 cm^3/mol
Volume Viscosity	13.6 m Pascal/second
Thermal Conductivity at 27°C	20 mW/mK
Van der Wall's Constant	(*a*) 4.612 bar L^2/mol (*b*) 0.05821 L/mol

EXPLOSIVE

Ethylene is a colourless and sweet smelling gas and it is non-poisonous and has a little anæsthetic property. It is sparingly soluble in water. The gas is also heavier than air.

Ethylene is a non-supporter of combustion, but it burns in air producing a luminous flame causing an explosion.

$$\text{Methane} + \text{Oxygen} \longrightarrow \text{Carbon Dioxide} + \text{Water}$$

$$C_2H_4 \quad + \quad O_2 \quad \longrightarrow \quad 2CO_2 \quad + H_2O$$

Ethylene reacts with chlorine like methane and acetylene. When ethylene and chlorine mixture burned it causes red flame forming hydrogen chloride and carbon.

$$\text{Methane} + \text{Chlorine} \longrightarrow \text{Hydrogen Chlorine} + \text{Carbon}$$

$$C_2H_4 \quad + \quad O_2 \quad \longrightarrow \quad 2CO_2 \quad + \quad 2C$$

Ethylene reacts with chlorine like methane and acetylene. When ethylene and chlorine mixture burned it causes red flame forming hydrogen chloride and carbon.

But when exposed to light at ordinary temperature an oily liquid is produced, it is also called "Duch Liquid'. Chemically it is called ethylene chloride.

$$\text{Methane} + \text{Chlorine} \longrightarrow \text{Ehylene Chloride}$$

$$C_2H_4 \qquad 2Cl_2 \quad \longrightarrow \quad C_2H_4Cl_2$$

Uses of Ethylene

1. It is used for the manufacture of polymers like polyethylene, plastics.
2. It is also used for the preparation of artificial ripening of fruits.
3. It is also used for the preparation of Ethylene Glycol.

Carbonate and other Componds of Carbon

Carbonates: There are two series of carbonates (1) Carbonates and (2) Bi-carbonates. The carbonates formed basic metallic oxides *e.g.* sodium, potassium calcium etc. The carbonates of sodium potassium are soluble and also stable, but carbonates produced by the calcium are unstable.

The bi carbonates of all the basic hydroxides are available in solid state, those of barium, strontium etc. is found in liquid state and others are unknown.

Structure of Carbonate Ion

This structure is determined by X-Ray study of the material calcite and other metal carbonate, carbonate, is trigonal planner, the three oxygen atoms being symmetrically placed around the carbon atom in the same plane. The bond angles are 120° and the C-O bond distance is 1.32 A. From the planer structure of the iron and the tetra valency of carbon it may be inferred that structure resonates between the three chemical forms stated below:

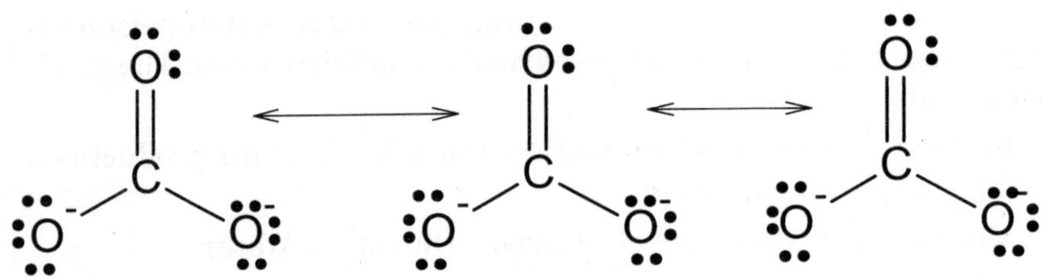

Structure of Carbonate Ion

Para carbonates: When carbon dioxide reacts with permono and perdisulphuric acids two types per carbonic acid are formed they are: Permonocarbonic acid (H_2CO_4) and Perdicarbonic acid ($H_2C_2O_6$).

While Permonocarbonic acid is not known, Perdicarbonic acid is said to form as an unstable solution in potassium per carbonate ($K_2C_2O_6$)

Potassium Percarbonate	+ Phosphoric Acid	$\longrightarrow$	Potassium Hydrogen Phosphate	+	Carbon Dioxide
$2K_2C_6O_6$	+ $2H_3PO_4$		$2K_2HPO_4$	+	$2CO_2$

Potassium Percarbonate reacts with phosphoric acid in an ethereal base.

As many as four different types of per carbonates exist they are as follows:

(*a*) *Perhydrates of Carbonates:* This type consists of hydrogen peroxide. The most important compound in this type is *sodium carbonate perhydrate,* on sodium carbonate followed by preparation with the help of alcohol.

The composition of these compounds is proved (I) by the fact that eather and (II) by the very slow liberation of iodine from neutral solution of potassium iodide.

(*b*) *Permonocarbonates:* Barium permonocarbonate ($BaCO_4$) and sodium permonocarbonate (Na_2CO_4). These compounds are formed when carbon dioxide (one mol) reacts with barium or sodium peroxide. They are of little importance.

(c) *Permonocarbonates:* These are compounds of perdicarbonic acid.

Potassium Perdicarbonate ($K_2C_2O_6$): It is the best known representative of this class. It is obtained by the electrolysis of concentrated solution of potassium bicarbonate at low temperatures and with the high current it stays. The reaction at the positive electrode is exactly same as in the case of perdisulphate.

$$2HCO_3^- \rightarrow H_2C_2O_6 + 2e^-$$

Potassium Percarbonate separates as bluish white, amorphous precipitate. This is washed rapidly with ice. Cold water, alcohol and either and dried over phosphorous pentoxide. When the dry compound is gently heated it gets decomposed as under.

Potassium Percarbonate + Water ⟶ Potassium Carbonate + Carbon Dioxide + Oxygen

$$K_2C_2O_6 + H_2O \longrightarrow 2KHCO_3 + H_2O_2$$

It gets dissolved in ice-cold water without much decomposition. On standing hydrogen peroxide is slowly formed by hydrolysis.

Potassium Percarbonate + Water ⟶ Potassium Carbonate + Hydrogen Peroxide

$$K_2C_2O_6 + H_2O \longrightarrow 2KHCO_3 + H_2O_2$$

Potassium perdicarbonate bleaches indigo and oxidizes lead sulphide to lead sulphate.

Sodium Perdicarbonate ($Na_2C_2O_6$): It is exactly similar to potassium perdicarbonate. It possesses similar chemical by the action carbon dioxide (2 mols) and sodium peroxide in absolute alcohol.

Sodium Peroxide + Carbon Dioxide ⟶ Sodium percarbonate

$$Na_2O + 2CO_2 \longrightarrow Na_2C_2O_6$$

CARBON DISULPHIDE (CS_2)

This important compound of carbon can be prepared by the action of carbon and Sulphuretted hydrogen gas.

Carbon Dioxide + Hydrogen Sulphide ⟶ Carbon Disulphide + Water

$$CO_2 + 2H_2S \longrightarrow CS_2 + 2H_2O$$

But in the industry it is prepared by the action of charcoal and sulphur vapour.

The reaction of these elements takes place in the electrical arc furnace of shaft construction (see Fig. 14.15)

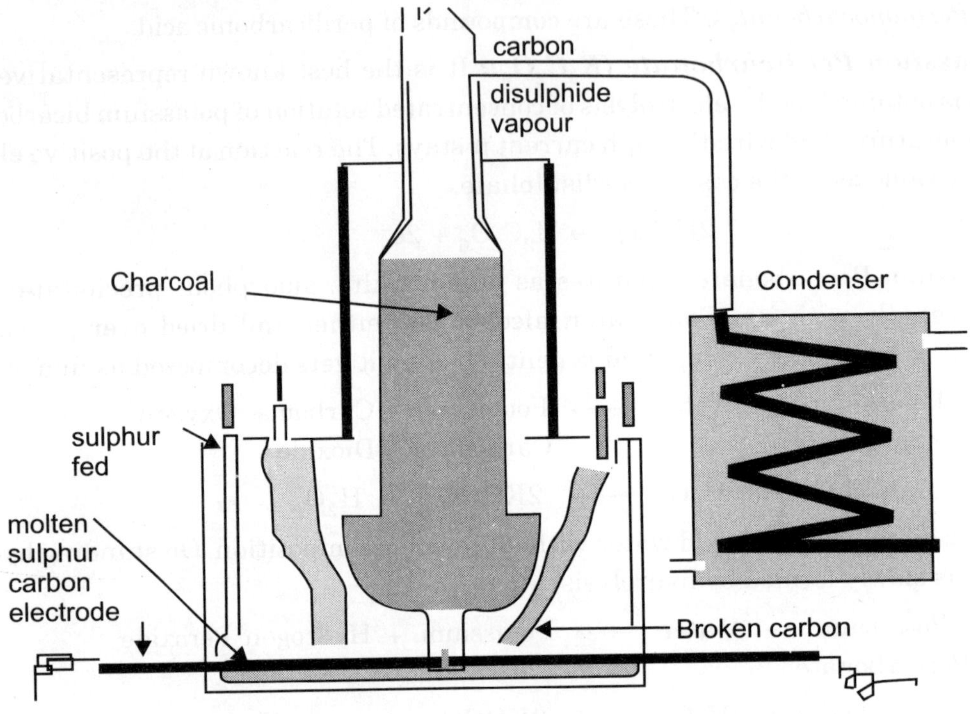

Fig 14.15. The Industrial Preparation of Carbon Disulphide

Reaction Equation

$$\text{Charcoal} + \text{Sulphur} \longrightarrow \text{Carbon Disulphide}$$
$$C \quad 2S \longrightarrow CS_2$$

The reaction is reversible and endothermic. The charcoal is introduced from the top, while sulphur vapour is led through the channels at the sides of the furnace. The electrodes are placed at the base of the furnace and the heat is generated due to the resistance of the broken carbons at the top and is condenses by the means of a water-cooled condenser. The crude carbon disulphide is purified by shaking with mercury and redistilling it.

Properties of Carbon Disulphide

Pysical Properties

Physical State	Liquid
Melting Point	–111.4°C
Boiling Point	46°C
Critical Temperature	279°C
Critical Pressure	7.90 MPa
Critical Molar Volume	173 cm^3/mol
Density	1.3 g/cc

Heat of Capacity	76.4 J/mol K
Standard Enthalpy of Formation	89.0 kJ/mol
Standard Gibb's Enthalpy	64.6 kJ/mol
Standard Entropy	151.3 J/mol K
Enthalpy of Fusion (at Melting Point)	4.40 kJ/mol
Enthalpy of Vapourization (at Boiling Point)	26.7 kJ/mol
Van der Waals constant	(*a*) 11.25 bar L^2/mol (*b*) 0.07262 L/mol

Chemical Properties

1. Carbon disulphide is colourless and very volatile liquid, when pure it possesses an ethereal smell. The commercial product, however, has a yellow colour and a bad smell due to presence of sulphur and hydrogen sulphide. Carbon disulphide has a density of 1.3gm/cc is highly reflective and is insoluble in water.
2. Carbon disulphide vapour is very much inflammable, and on combustion gives, carbon dioxide and sulphur dioxide.

   Carbon Disulphide + Oxygen ⟶ Carbon Dioxide + Sulphur Dioxide

   $$CS_2 + O_2 \longrightarrow CO_2 + 2SO_2$$
3. Carbon disulphide is an acid *thio-anhydride.* Thus when vapour of carbon disulphide is passed over calcium hydroxide calcium carbonate and calcium thio-carbonates are produced

   Calcium Hydroxide + Carbon Disulphide ⟶ Calcium Carbonate + Calcium Thicarbonate + Water

   $$3Ca(OH)_2 + 3CS_2 \longrightarrow CaCO_3 + 2CaCS_3 + 3H_2O$$

   The above reaction is used for removing carbon disulphide from coal gas.
4. It also combines with basic sulphides forming thiocarbonates, e.g. calcium thiocarbonates

   Calcium Sulphide + Carbon Disulphate ⟶ Calcium Thiocarbonate

   $$CaS + CS_2 \longrightarrow CaCS_3$$
5. Carbon Disulphide also reacts with chlorine forming carbon tetrachloride

   Carbon Disulphide + Chlorine ⟶ Carbon Tetrachloride + Sulphur Monochloride

   $$CS_2 + Cl_2 \longrightarrow CCl_4 + S_2Cl_2$$
6. Carbon disulphide is a volatile substance, white phosphorous (2g) dissolved in carbon disulphide (10g) when smeared on a paper catches fire after some times.

Uses of Carbon Disulpide

1. The most important industrial use of the carbon disulphide is in the manufacture of viscose rayon. It is also used in the extraction of oil seeds and for vulcanizing of rubber. All these uses of carbon disulphide are bases on its action as a solvent.

2. The vapour of carbon disulphide is highly poisonous. It is, therefore, used for killing insects also red ants.
3. The carbon disulphide is also used for the preparation of carbon tetrachloride.

Carbides

In its widest sense, the word carbide means that it is a compound of carbon. But it is used to those compounds when carbon combines with any soft element, e.g. silicon, calcium etc. Such compounds are divided into distinct classes; they are again subdivided as shown in the chart below:

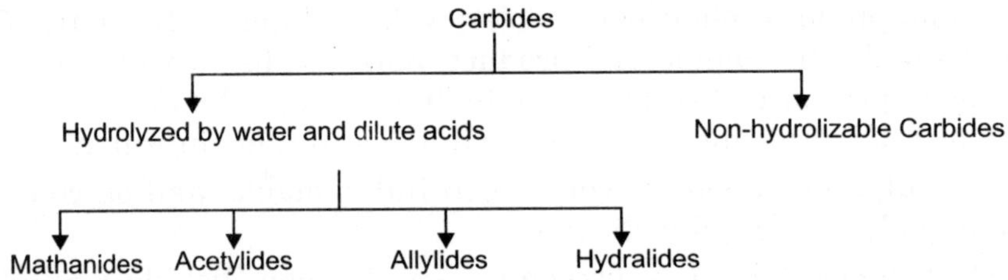

Chart : District Classes of Carbides

(i) Hydrolyzed by Water: These carbides get reached with water, according to the production of gases they are classified:

(a) Methanides: These compounds when reacts with water or dilute acids produce methane they are aluminium carbide, AlC_3 and beryllium carbide Be_2C.

Beryllium Carbide + Water ⟶ Beryllium Hydroxide + Methane

$$Be_2C + 4H_2O \longrightarrow 2Be(OH)_2 + CH_4$$

(b) Acetylide: The carbides of alkali metals (*e.g.* sodium carbide, calcium carbide, etc.), These compounds decompose to form acetylide).

Calcium Carbide + Water ⟶ Calcium Hydroxide + Oxygen

$$Ca_2C + H_2O \longrightarrow Ca(OH)_2 + O_2$$

Such carbides may be taken as derived from acetylene C_2H_2 and are therefore, regards as acrtylides

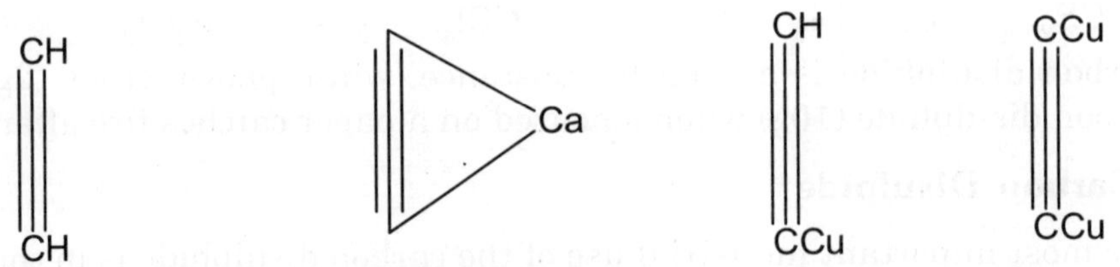

acetylene Calcium Acetylide Cuprous Acetalyde

(c) *Allylides:* When magnesium acetylide (MgC_2) is heated it gives another carbide of magnesium which possesses the formula Mg_2C_3 it is regarded as magnesium allylide as when hydrolyzed. It gives pure acetylene (methyl acetylene).

$$\text{Magnesium Carbide} + \text{Water} \longrightarrow \text{Magnesium Hydroxide} + \text{Methyl Acetylene}$$
$$Mg_2C_3 + 4H_2O \longrightarrow Mg(OH)_2 + C_3H_4$$

(d) *Hydramdes:* Carbides that produces mixture of hydrocarbons (but not acetylene). The carbides of iron group ferric carbide (Fe_3C) manganous carbide (Mg_3C) cobalt carbide (CO_3C). These carbides gets hydrolyzed by water to produce simple hydrocarbons e.g. manganese carbide which gives methane and hydrogen or a mixture of methane ethane, ethylene and even liquid hydrocarbon and free carbon (*e.g.* Ferric carbide).

$$\text{Silicon Carbide} + \text{Sodium Hydroxide} \longrightarrow \text{Sodium Silicate} + \text{Water}$$
$$SiC + NaOH \longrightarrow O_2 + H_2O$$

(e) *Non-hydrollzable Carbide:* The carbides of silicon and boron *e.g.* silicon carbide and boron carbide. They are extremely hard (nearly as hard as diamond) and for that reason, are used as abrasives*. Chemically they are highly resistant and are attached by strong acids, not even by a mixture of hydrofluoric acid and nitric acid (which readily dissolves silicon crystals) they are, however, decomposed by molted sodium hydroxide in the presence of air.

CALCIUM CARBIDE

Invention

Wöhler first invented calcium carbide in 1862. He heated a mixture of sodium hydroxide zinc and charcoal in a furnace and obtained calcium carbide.

Physical Properties

Calcium carbide is a crystalline solid with a typical smell of impure acetylene. It has very high water affinity even it produces acetylene from moisture.

Chemical Properties

(*i*) It is decomposed by water with evolution of acetylene.

$$\text{Calcium Carbide} + \text{Water} \longrightarrow \text{Calcium Oxide} + \text{Acetylene}$$
$$CaCO_2 + H_2O \longrightarrow CaO + C_2H_2$$

(*ii*) When strongly heated in an atmosphere nitrogen calcium cynamide and carbon mixture is produced.

$$\text{Calcium Carbide} + \text{Nitrogen} \longrightarrow \text{Calcium Cynamide} + \text{Carbon}$$
$$CaCO_2 + H_2O \longrightarrow CaNCN + C$$

* Abrasive : Irritants.

Industrial Preparation of Calcium Carbide

Calcium Carbide is obtained by heating a mixture of quick lime and coke at a very heating a mixture of quick lime and coke at a very high temperature (about 3000°C) in an electric furnace (see Fig. 14.16).

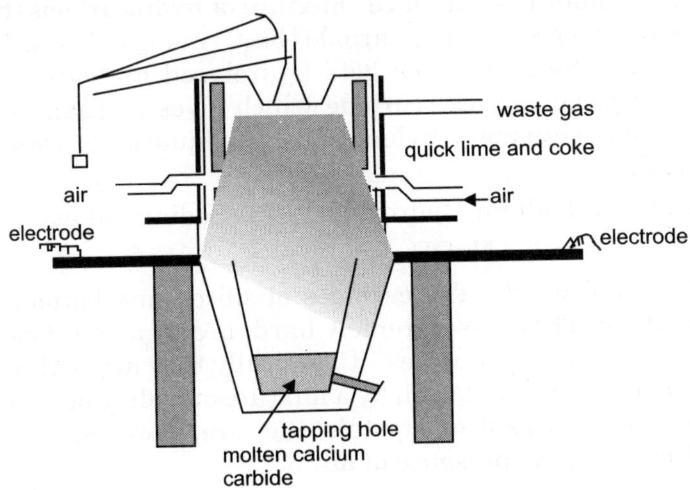

Fig. 14.16. The Industrial Preapration of Calcium Carbide

Reaction Equation

$$\text{Lime} + \text{Coke} \longrightarrow \text{Calcium Carbide} + \text{Carbon Monoxide}$$
$$CaO + 3C \longrightarrow CaC_2 + CO$$

The coke lime mixture is added from time at the top. Before it reaches the region of electrical in the lower part of the furnace, it is pre-heated by carbon monoxide (produced in the reaction burning in air, which enters the furnace through the side tubes, as shown in Fig. 14.16. The calcium carbide is obtained in the fused state and periodically taken from the bottom of the reactor.

Uses of Calcium Carbide

(*i*) *Production of acetylene,* calcium carbide is used mainly for the preparation of acetylene.

(*ii*) It is also used for ripening of fruits.

(*iii*) It is also used as the raw material for the production of calcium cynamide.

(*iv*) It is also used in the synthesis of numerous organic compounds such as acetaldehyde, acetic acid, acetic anhydride, vinyl acetate, polyvinyl acetate etc.

CARBON TETRACHLORIDE

Introduction

Carbon tetrachloride was first produced by Wöllier in the year 1862. He prepared it from carbon disulphide and chlorine.

Physical Properties

Carbon tetrachloride is a liquid and causes much sneeze and unpleasant physical condition, if directly inhaled. Some of the physical properties are given below:

Physical State	Liquid
Melting Point	23°C
Boiling Point	76.9°C
Critical Pressure	4.51 MPa
Critical Temperature	283°C
Critical Molar Volume	273 cm^3/mol
Enthalpy of Formation	–128.2 kJ/mol
Enthalpy of Fusion (at melting Point)	3.28 kJ/mol
Enthalpy of Vapourization (at boiling Point)	29.82 kJ/mol
Heat of Capacity	130.7 J/mol K
Van der Waal's Constant (*a*)	20.10 barL^2/mol
(*b*)	0.1281 L/mol

Chemical Properties

1. Carbon tetrachloride can dissolve carbon in it. It is a colourless volatile liquid of density of 1.5, carbon tetrachloride in non-inflammable, but reacts violently with hot sodium.

   Sodium + Carbon Tetrachloride ⟶ Sodium Chloride + Carbon + Chlorine

   $$4Na + 2CCl_4 \longrightarrow 4NaCl + 2C + 2Cl_2$$

2. Carbon tetrachloride is very much dangerous to health causes lots of sneeze and cough.
3. Carbon tetrachloride is decomposed when exposed to sunlight in the presence of air (oxygen) forming carbonyl chloride and chlorine.

   Carbon Tetrachloride + Oxygen ⟶ Carbonyl Chloride + Chlorine

   $$2CCl_4 + O_2 \longrightarrow 2COCl_2 + 2Cl_2$$

Industrial Preparation of Carbon Tetrachloride

In industry carbon tetrachloride is prepared by the action of carbon disulphide and chlorine, (see Fig. 14.17).

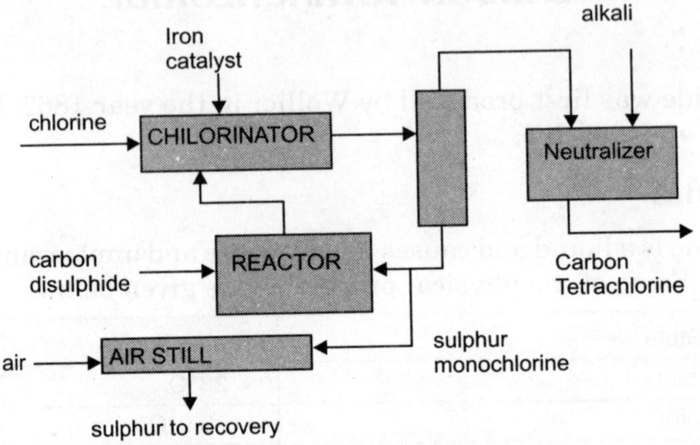

Fig 14.17. The Block Diagram of Industrial Preparation of Carbon Tetrachloride

Reaction Equation

Carbon Disulphide + Chlorine $\xrightarrow{\text{Catalyst}}$ Sulphur Monochloride + Carbon Tetrachloride

$$CS_2 + 3Cl_2 \longrightarrow S_2Cl_2 + CCl_4$$

Carbon Disulphide + Carbon Tetrachloride ⟶ Sulphur + Carbon Tetrachloride

$$CS_2 + CCl_4 \longrightarrow 6S + CCl_4$$

Sulphur + Carbon ⟶ Carbon Disulphide

$$6S + 3C \longrightarrow 3CS_2$$

The yield is 90%

Process: A solution of carbon disulphide in carbon tetrachloride (approx. 40% carbon disulphide and 50% carbon tetrachloride and 10% sulphur monochloride is changes into a lead lined chlorinator fitted with cooling coils. Chlorine is bubbled through the solution, which contains. iron fillings or powder as a catalyst. The iron remains in the chlorinator from batch to batch and is replaced when gets rusted. The temperature is maintained around 30°C by cooling water. An excess of chlorine is used to insure complete conversion of carbon disulphide (about 0.5%, remains unchanged).

The reaction products, consisting carbon tetrachloride (60%) and sulphur monochloride (40%) are passed to a distilling column where they are separated.

The carbon tetrachloride containing 1% sulphur monochloride is send to a neutralizer and dryer where acidic materials are neutralized by agitation with alcoholic sodium hydroxide solution. The moisture content is also decreased. The dried material may be distilled again if wanted to give carbon tetrachloride of 98% purity. This mixture is reacted with carbon disulphide at 60°C in a lead-lined reaction vessel fitted with an agitator. At the end of the reaction, the carbon tetrachloride and most of the sulphur monochloride produced are distilled off. The distillate consisting of carbon disulphide is charged to the collinator to start the cycle.

The residual sulphur, containing to 15% sulphur monochloride, is discharged to an air still. The monochloride is removed sulphur may be converted to carbon disulphide by reaction with coke.

FUEL GASES

Introduction

Carbon is a cheap source of fuel in India. It is easily available and may fuel gas is manufactured from the carbon. The industrial preparation of coal gas, oil gas, producer gas, water gas, are going to be discussed under this subhead.

Definition of Fuel

A substance that can supply energy either alone or by reacting with another substance is known as fuel.

Definition of Calorie

Calorie is defined as the quantity of heat that will raise the temperature of one gram of water through 1°C.

Requisites of Good Fuel

A good fuel must have the following features:

(*a*) It should have a high calorific value per unit weight that is it should evolve a large amount of heart when a measured amount of it is burnt under the conditions in which it is to be used a fuel.

(*b*) It should not give any offensive odour or undesirable product when burnt.

(*c*) It should be economical and easily available.

(*d*) It should yield a little ash as possible.

Type of Fuel

Most of the ordinary fuels contain carbon and they give out heat when burnt in air. The fuels consisting of carbon are three types:

1. **Solid Fuel:** Such as wood, coal, charcoal, coke
2. **Liquid Fuel:** Such as petroleum, kerosene oil, alcohol etc.
3. **Gaseous Fuel:** such as producer gas, water, oil, gas.

Producer Gas

The mixture of nitrogen and carbon monoxide gas prepares producer gas; this producer gas is prepared by blowing air through an incandescent coke. As even a low-grade coal can be used in this process, the gas is used as a cheap industrial fuel. Producer gas has a very low calorific value of only 25.95 Kcal. This is due to presence of huge amount of nitrogen in it.

The furnace known as the producer is shown in the Fig. 14.18. It consists of an airtight cylindrical structure, lined with firebricks having 8 to 10 feet internal diameter and 12 to 15 feet height. Coal is burnt in this furnace while air enters from the bottom through an inlet as, shown in the figure. The following reactions occur in the furnace:

$$C + \tfrac{1}{2}O_2 \rightarrow CO = -27 \text{ K cal}$$

$$CO + \tfrac{1}{2}O_2 \rightarrow CO_2 = +67 \text{ K cal}$$

$$C + O_2 \rightarrow CO_2 = +94 \text{ L cal}$$

Initially, carbon dioxide is the main product but it rises through the bed of coal and combines with the carbon, provided the temperature is about 1000°C forming carbon dioxide.

The over all reaction as seen from the above equation, is endothermic and so the process can go on uninterrupted if the fuel is periodically renewed. The higher the temperature of the carbon the greater the ratio CO: CO_2 in the gas.

$$\text{Carbon Dioxide + Carbon} \longrightarrow \text{Carbon Monoxide}$$

$$CO_2 + C \longrightarrow 2CO$$

A Bunker
B Cast refroctory reactor
C Charcoal fuel
D Metal grate
E Compacted rice hush ash
F Cost refractory dish
G Metal shroud
H Cylinder 1 (reactor)
I Cylinder 2 (settling tank)
J Cylinder 4 (cloth filter)
K Cylinder 5 (cloth filter)
L Cylinder 6 (security filter)
M Outer tank
N Ashport
O Ashport cover
P Cleaning chain
Q Aerodynamic fin

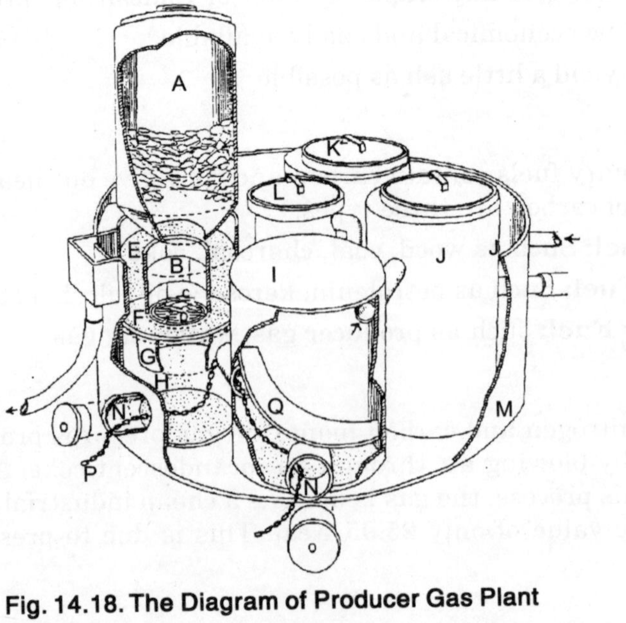

Fig. 14.18. The Diagram of Producer Gas Plant

The gaseous mixture is called producer gas; it consists of mainly carbon dioxide (about 25%) and nitrogen (about 55%) with a small amount of carbon monoxide. It leaves through the exit at the top of the furnace as shown. It is burnt as a fuel where required by admitting more air when carbon dioxide changes into carbon monoxide liberating heat, nitrogen acts as a dilutent* and therefore the calorific value is low.

Water Gas

Water gas is actually a mixture of carbon monoxide and hydrogen. A small amount of carbon dioxide is also present in the mixture. It is manufactured by blowing steam (from this name's 'water gas' has been derived) through layers of glowing coal. The reaction is shown below in the form of equations.

Carbon + Water ⟶ Carbon Monoxide + Hydrogen

$$C + H_2O \longrightarrow CO + H_2$$

Due to endothermic nature of reaction, the coal cools down after a while and the reaction takes a different trend resulting in the formation of carbon dioxide

Carbon Monoxide + Water ⟶ Carbon Dioxide + Water

$$CO + H_2O \longrightarrow CO_2 + H_2O$$

In order to avoid this undesirable trend, the current of steam is however intermittently replaced by a blast of air, which helps in rise of temperature of the coke. When this happens the steam is readmitted. Actually, a small amount of carbon dioxide is formed when carbon monoxide is oxidized by steam. This reaction is reversible and with increase in temperature the equilibrium shifts towards left. Therefore in order to have high value of the ration CO: CO_2 the temperature should be kept as high as possible. It is usually kept between 1350°C to 1450°C. Water gas bums with a blue non-luminous flame. In order to make flame luminous water gas is carbonated by passing it through hot brickwork over which some oil is sprayed. A simple sketch is shown in the Fig. 14.19 to demonstrate the carbonated water generator.

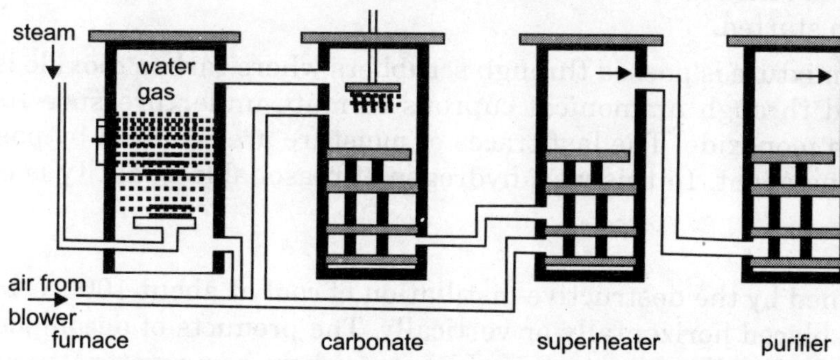

Fig. 14.19. The Industrial Preparation of Water Gas

* Dilutent: A thinging agent.

The furnace consists of a cylindrical vessel made up of stainless steel, 12 to 15 feet height and 8 feet to 10 feet in diameter. The coke is introduced through an opening at the top; there are separate inlets for blow in steam and air (alternatively) from the bottom and the exit near the top for the gases to escape.

The water gas form the furnace passes through carbonator, which is a big tower, packed heated bricks into which a spray of oil is injected. The oil gets volatilized and the vapours mixed with gas, enters the super heater, which is another similarly big tower packed with bricks heated to red ness by the burning of a mixture of producer gas and air. The oil vapour gets cracked at ethylene etc., which bums with a luminous flame.

The carbonated water gas is passed through purifiers, which contains hydrated ferric oxide where hydrogen sulphide (which originates from sulphur present in the coke) is removed by adsorption on ferric hydroxide. When the coke in the water gas furnace has cooled down, the steam is shut off and blast of air is blown through to heat the coke again to required temperature. In this process, some producer gas is formed as well. This is passed through carbonator to heat the breaks and then into super heater where it is made to bum with air so as to heat the bricks to a very high temperature. This again followed by blowing steam into the furnace.

Uses of Water Gas

Water gas has calorific value of 89.5Kcal. It is, therefore, used as a good source of fuel. For this purpose it is sometimes mixed with coal gas in controlled amount. Carbonated water gas is used for heating and as well as for lightning also. It is also used for the manufacture of methyl alcohol. It is an important source for the preparation of commercial hydrogen. For this purpose it is reacted further with steam in the presence of a suitable catalyst (oxides of iron, chromium or cobalt) when the following reactions takes place.

$$\text{Water} + \text{Water Gas} \longrightarrow \text{Carbon Dioxide} + \text{Hydrogen}$$

$$H_2O + H_2CO \longrightarrow CO_2 + 2H_2$$

The heat of the reaction is sufficient to maintain the required temperature once the reaction has been started.

The gaseous mixture is passed through scrubbers where carbon dioxide is eliminated. It is then passed through ammonical cuprous formate under pressure to remove the unreacted carbon monoxide. The last traces of moisture are removed by passing the gas over a dehydrating agent. In this way, hydrogen of reasonable quantity is obtained.

Coal Gas

Coal gas is obtained by the destructive distillation of coal at about 1000°C to 1100°C at a large clay retort placed horizontally or vertically. The products of decomposition are led through the pipe hydraulic main containing water, where some water vapour condenses. The gas then passes through condensers (see Fig. 14.20. block diagram) by air where more coal tar and water vapour condense and most of the ammonia produced dissolves in

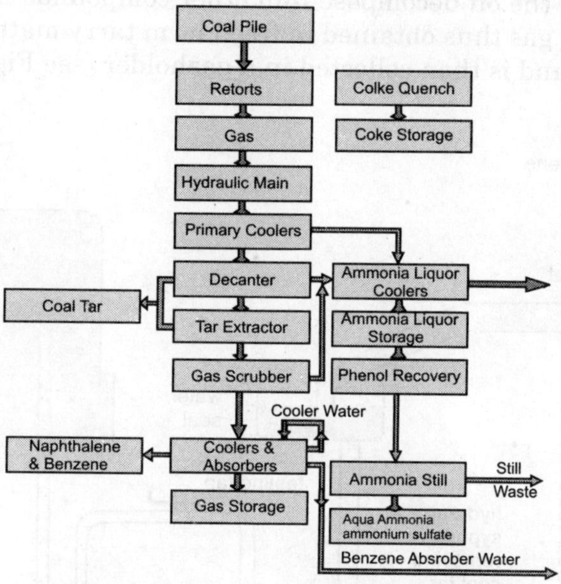

Fig. 14.20. The Block Diagram of Coal Gas

water. The gas is then passed through scrubbers or washing towers to remove the rest of ammonia. The gas still contains harmful impurities like hydrogen sulphide. These are removed by absorption by hydrated ferric amide contained in the purifier. The purified gas is collected over water in gasholders and then distributed to the consumers coal gas is mostly used as a domestic and industrial fuel for the production of heat. Its average composition is stated below:

Hydrogen	56.05% by volume
Methane	22.8% by volume
Ethylene	2.5% by volume
Carbon Monoxide	10.9% by volume
Carbon dioxide	1.3% by volume
Nitrogen	5.0% by volume
Oxygen	0.5% by volume

Its calorific value is about 125Kcal. Besides being a fuel, coal gas is also used to provide reducing atmosphere in metallurgical operation. Important by product of the coal distillation is coke, coal tar and ammonical liquor.

Oil Gas

Oil gas is generally used in laboratories and is obtained by the process of cracking from kerosene oil. A thin stream of kerosene oil is allowed to flow into a red-hot cast iron retort.

Liquid hydrocarbons of the oil decompose into other compounds like methane, ethylene, which are gaseous. The gas thus obtained in freed form tarry matter by washing through a hydrocarbon system and is then collected in a gasholder (see Fig. 14.21).

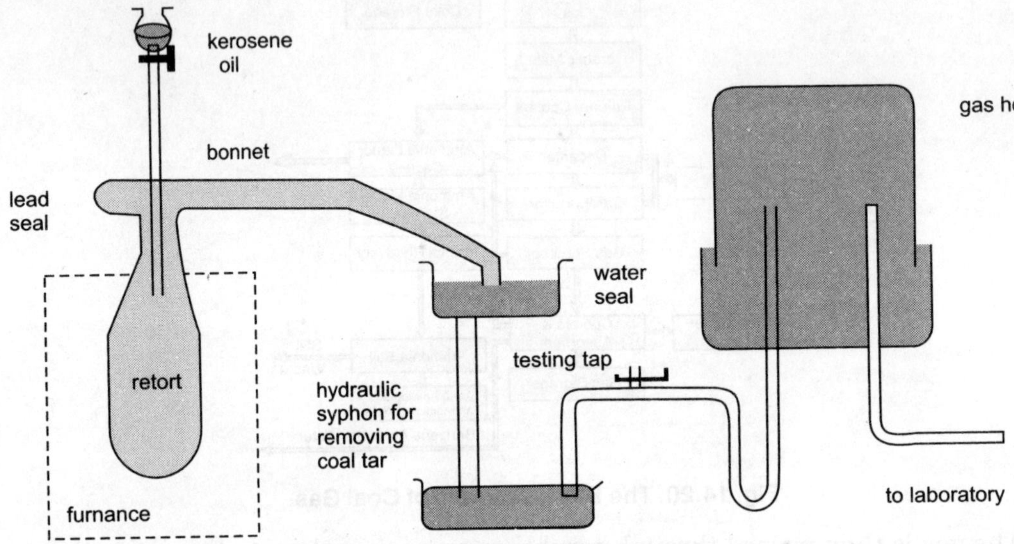

Fig. 14.21. The Oil Gas Manufacturing Plant

CHAPTER

15

Metals of Alkali

Introduction

The metals like sodium, potassium calcium are known as alkali metals because they yield alkali (sodium hydroxide, potassium hydroxide, and calcium hydroxide). Here in this chapter we shall discuss about these metals. This chapter consists the chemistry of lithium, sodium, potassium, magnesium, strontium and beryllium.

We shall discuss about the occurrence, discovery, uses and also the compounds and their properties along with the physical and the chemical tests for identification of these elements and compounds. .

At first we shall start with Lithium.

LITHIUM

Chemical Symbol : Li
Atomic Number : 3
Atomic Weight: 6.941

Melting Point : 179°C
Boiling Point : 1347°C
Lightest Metal

History

The name comes from the Greek word *lithos*. Lithos translates to 'stone'. It was given this name because it was first discovered from a mineral rather than a plant Lithium was first discovered and defined by *J.A. Arfvedson* in 1817 when he did an analysis of a mineral he had found. This mineral, petalite ($LiAl(Si_2O_5)_2$), was first found by Brazilian scientist *José Bonifácio* in 1800. Arfvedson was never able to fully isolate lithium, and it wasn't until 1855 that it was isolated, by W.T. Brande. Lithium was first produced commercially in 1923, by Metallgesellschaft AG.

Occurrence

Lithium is widely distributed in nature in a very small quantity, in number of minerals especially alumionosilicates known as *Lithia Mica.* {$K (Li, Al)_3 (Si, Al)_4O_{10} (OH, F_2)$} Lithium is also found to occur in some mineral springs and in the ashes of many plants.

Properties of The Metal

Physical State	Solid
Density [298K]	534 kg m^{-3}
Crystal Lattice	b.c.c.
Melting Point Heat of	179°C
Fusion	4.6 kJ mol^{-1}
Vapourization	134.7 kJ mol^{-1}
Atomization	161 kJ mol^{-1}
Atomic Mass	6.941
Electronegativity	
Pauling	0.98
Allred	0.97
Absolute	3.01 eV
Electron Affinity	59.6 kJ mol^{-1}
Polarizability	24.3 $Å^3$
Effective Nuclear Charge	
Slater	1.3
Clementi	1.28
Froese Fischer Radius	1.55
Li^{1+}	78 pm
Atomic	52 pm
Covalent	123 pm
Atomic Ionization Energies	
2s	520 kJ mol^{-1}
1s	5596 kJ mol^{-1}
Successive Ionization Energies	
Li →	Li^{+} : 513.3 kJ mol^{-1}
Li^{+1} →	Li^{+2} : 7298 kJ mol^{-1}
Li^{+2} →	Li^{+3} : 11814.8 kJ mol^{-1}
Common Ions	Li^{1+}

	Natural Abun. (%)	Atomic Mass	Half-life	Decay Mode/ Energy (MeV)	Part. En./ Intensity (MeV%)	Cross-sect. Ther. Neut. (b)	Spin (h/2 pi)	Magnetic Dipl. Mom. (nm)	Electric Quad. Mom. (b)	Gamma En./ Intensity (MeV%)
^{5}Li		5.01254	0.3×10^{-21}s				$3/2^-$			
^{6}Li	7.5 %	6.015121				0.039 b	1^+	0.82205 nm	–0.0008	
^{7}Li	92.5 %	7.016003				0.045 b	$3/2^-$	3.25644 nm	–0.041 b	
^{8}Li		8.022485	0.84s	β^-,alpha/ 16.005	12.5/100		2^+	1.6536 nm	0.032 b	
^{9}Li		9.026789	0.177s	β^-/13.6068	13.5/ 75		$3/2^-$	3.439 nm	0.028 b	
^{11}Li			8.7ms	β^-	11/ 25		$1/2^-$			3

It is a silvery white metal harder than sodium and tarnishes on exposure to air and decomposes water with liberation of hydrogen.

Lithium is metal and conducts electricity. The melting point of lithium is 179°C.

Extraction of Lithium

Lithium is extracted from sea water (DOW). It is leached with manganese dioxide in absorbent columns and is then treated with dilute hydrochloric acid; the lithium chloride crystals are obtained (Fig. 15.1).

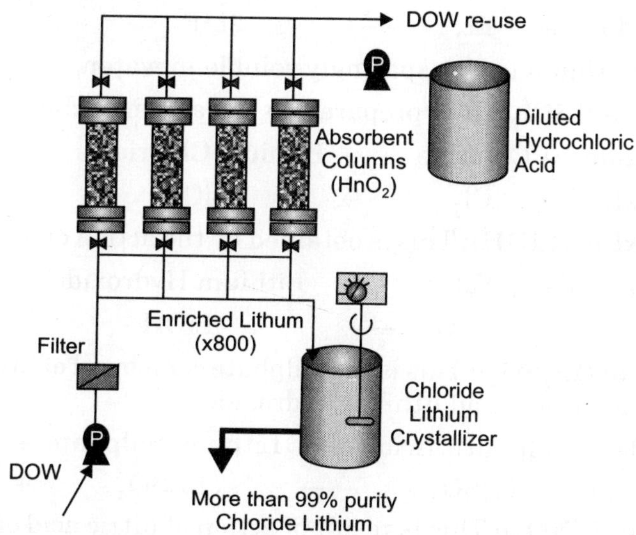

Fig. 15.1. The Lithium Extraction from Lithium Chloride in Seawater

Compounds of Lithium

Lithium Oxide (Li_2O): When lithium directly combines with oxygen to form this oxide.

$$\text{Lithium} + \text{Oxygen} \longrightarrow \text{Lithium Oxide}$$

$$4Li + O_2 \longrightarrow 2Li_2O$$

Lithium Sulphide (Li_2S): Lithium also reacts with sulphur to form lithium sulphide.

$$\text{Lithium} + \text{Sulphur} \longrightarrow \text{Lithium Sulphide}$$

$$2Li + O_2 \longrightarrow 2Li_2O$$

Lithium Nitride (Li_3N): Lithium reacts with nitrogen forming lithium nitride.

$$\text{Lithium} + \text{Nitrogen} \longrightarrow \text{Lithium Nitride}$$

$$6Li + N_2 \longrightarrow 2Li_3N$$

Lithium Carbonate (Li_2CO_3): It is sparingly soluble in water, when lithium oxide reacts with carbon dioxide. This reaction is reversible.

Lithium Oxide + Carbon Dioxide ⇌ Lithium Carbonate

$$Li_2O + CO_2 \rightleftharpoons Li_2CO_3$$

Lithium Phosphate (Li_3PO_4): This compound is produced when lithium directly reacts with phosphoric acid, sparingly soluble in water.

Lithium + Phosphoric Acid ⟶ Lithium Phosphate + Hydrogen

$$3Li + H_2PO_4 \longrightarrow Li_3PO_4 + H_2$$

Lithium Fluoride (LiF): Fluorine reacts with lithium forming lithium fluoride.

Lithium + Fluorine ⟶ Lithium Fluoride

$$2Li + Fl_2 \longrightarrow 2LiF$$

This compound of lithium is also sparingly soluble in water.

Lithium Chloride ($LiCl_2$): It is prepared by the action of chlorine on lithium.

Lithium + Chlorine ⟶ Lithium Chloride

$$2Li + Cl_2 \longrightarrow 2LiCl$$

Lithium Hydroxide (LiOH): This is obtained by the action of lithium oxide on water.

Lithium Oxide Water ⟶ Lithium Hydroxide

$$Li_2O \quad H_2O \longrightarrow LiOH$$

Lithium Sulphate (Li_2SO_4): This is the sulphate compound of lithium and is obtained by the action of sulphuric acid and lithium hydroxide.

Lithium Hydroxide + Sulphuric Acid ⟶ Lithium Sulphate + Water

$$2LiOH + H_2SO_4 \longrightarrow Li_2SO_4 + H_2O$$

Lithium Nitrate ($LiNO_3$): This is result of action of nitric acid on lithium hydroxide.

Lithium Hydroxide + Nitric Acid ⟶ Lithium Sulphate + Water

$$2LiOH + HNO_3 \longrightarrow LiNO_4 + H_2O$$

Lithium Tartrate ($LiC_4H_4O_6$): This is the lithium carbonate and tartaric acid salt, the reaction equation is given below:

Lithium Carbonate + Potassium Hydrogen Tartrate ⟶ Potassium Tartrate + Lithium Tartrate + Carbon Dioxide + Water

$$LiCO_3 + 2K_2C_4H_5O_6 \longrightarrow K_2C_2H_4O_6 + LiC_4H_4O_6 + CO_2 + H_2O$$

Lithium tartrate is also prepared the action of lithium chloride and potassium tartrate.

Lithium Chloride + Potassium Tartarate ⟶ Lithium Tartarate + Potassium Chloride

$$LiCl + K_2C_2H_4O_6 \longrightarrow LiC_4H_4O_6 + 2KCl$$

Lithium Iodide (LiI): This is the compound produced by the action of hydrogen iodide and lithium carbonate.

Lithium Carbonate + Hydrogen Iodide ⟶ Lithium Iodide + Water + Carbon Dioxide

$$LiCO_3 + HI \longrightarrow LiI + H_2O + CO_2$$

Uses of Lithium

Lithium has a number of different uses in different sectors of society. The most common use is in lithium batteries. These are lightweight and are not as toxic as lead and cadmium batteries. These batteries have applications as small as watch batteries and as large as military and space vehicles. In 1949 it was discovered by John Cade, an Australian physician, that lithium carbonate could be used to treat bipolar disorder. It lowers the highs and raises the lows that are symptoms of bipolar disorder. Lithium carbonate now proves effective treatment for 60% of people with bipolar disease. Lithium carbonate is also used in glass and ceramics as a strengthening agent.

SODIUM

Chemical Symbol : Na (Natrium)

Melting Point : 97.5°C

Boiling Point : 880°C

Atomic Number : 11

Atomic Weight : 23

Discovery and Invention

The carbonate and the chloride salt of sodium are known from pre-historic times. But it was isolated in 19th century, when Alexandro Volta introduced his voltaic cell in 1800; Davy used this discovery. Davy isolated the free metal by electrolyzing fused potash and later soda. The electrolysis was carried out by using a platinum spoon, which was connected to the negative pole of the battery and a positive electrode (a platinum wire) to the positive pole of the battery, the other end of which was dipped in the material.

Bright metallic globules of sodium appeared in the spoon and some of these burst into flame with an explosion.

Occurrence

Sodium does not occur in free state as a pure and there are comparatively few naturally occurring compounds. Some of these however are found in enormous quantities, *e.g.* Sodium Chloride (common salt, NaCl), Sodium Nitrate (Chile Saltpetre) ($NaNO_3$), Sodium Sesquicarbonate (Na_2CO_3), $NaHCO_3$ $2H_2O$ and sodium Hexafluoroaluminate (Cryolite) $NaAlF_6$.

Now we shall discuss the extraction of sodium by some industrial processes.

Extraction of Sodium by Down's Cell

In the manufacturing process of sodium the process is electrolysis of sodium chloride (fused). The cell consists of a graphite positive electrode, which is cylindrical and is surrounded by the steel gauge diaphragm and the concentric cylindrical negative electrode (called cathode) also made up of steel. The electrolyte is mainly a mixture of sodium chloride and calcium chloride to approximately 500°C. (Some calcium is therefore liberated with sodium). The gap between anode and cathode is kept as small as possible, to reduce resistance, the heat developed by the current maintains the temperature of the cell. Chlorine is set free and collected as a by-product in the method. The sodium being lighter than fused salt rises upwards and is collected over the cathode from which it is reserved in a separate closed compartment. The diagrammatic representation of the Down's cell in shown in Fig. 15.2.

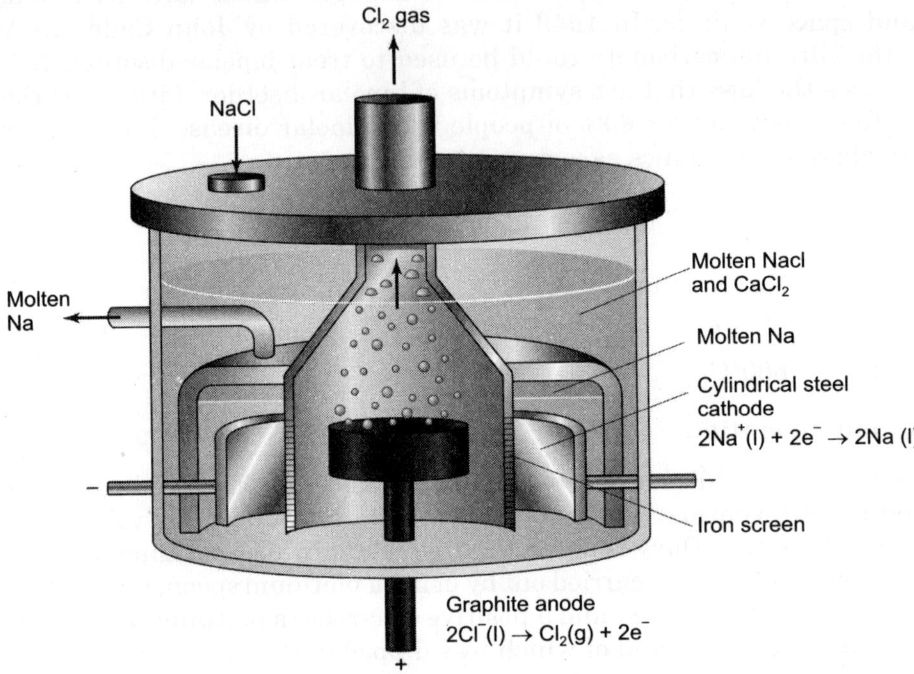

Fig. 15.2. The Downs Cell

Properties of Sodium

Physical Properties

Sodium is a soft silvery white metal. A sharp knife can cut this metal and the fresh cut face shows as the shiny appearance, but the end tarnishes as it reacts with air (moisture), vigorously. The detail physical properties are shown in the table below.

Melting Point	97.5°C
Boiling Point	880°C
Density (at 25°C)	.9674 g/cc
Electrical Resistivity	9.7 ohms/cm^2
Heat of Capacity	6.79 deg^{-1} mol^{-1}
Standard Entropy	12.2 cal deg^{-1} mol^{-1}

Chemical Properties

Reaction of sodium with water: Sodium metal reacts rapidly with water to form a colourless solution of sodium hydroxide (NaOH) and hydrogen gas (H_2). The resulting solution is basic because of the dissolved hydroxide. The reaction is exothermic. During the reaction, the sodium metal may well become so hot that it catches fire and burns with a characteristic orange colour. The reaction is slower than that of potassium (immediately below sodium in the periodic table), but faster than that of lithium (immediately above sodium in the periodic table).

$$2Na(s) + 2H_2O \longrightarrow 2NaOH(aq) + H_2(g)$$

Reaction of sodium with the halogens: Sodium metal reacts vigorously with all the halogens to form sodium halides. So, it reacts with fluorine, F_2, chlorine, Cl_2, bromine, I_2, and iodine, I_2, to form respectively sodium(I) bromide, NaF, sodium(I) chloride, NaCl, sodium(I) bromide, NaBr, and sodium(I) iodide, NaI.

$$2Na(s) + F_2(g) \longrightarrow NaF(s)$$
$$2Na(s) + Cl_2(g) \longrightarrow NaCl(s)$$
$$2Na(s) + Br_2(g) \longrightarrow NaBr(s)$$
$$2Na(s) + I_2(g) \longrightarrow NaI(s)$$

Reaction of sodium with acids: Sodium metal dissolves readily in dilute sulphuric acid to form solutions containing the aquated Na(I) ion together with hydrogen gas, H_2.

$$2Na(s) + H_2SO_4(aq) \longrightarrow 2Na^+(aq) + SO_4^{2-}(aq) + H_2(g)$$

Reaction of sodium with bases: Sodium metal reacts rapidly with water to form a colourless basic solution of sodium hydroxide (NaOH) and hydrogen gas (H_2). The reaction continues even when the solution becomes basic. The resulting solution is basic because of the dissolved hydroxide. The reaction is exothermic. During the reaction, the sodium metal may well become so hot that it catches fire and burns with a characteristic orange colour. The reaction is slower than that of potassium (immediately below sodium in the periodic table), but faster than that of lithium (immediately above sodium in the periodic table). As the reaction continues, the concentration of hydroxide increases.

$$2Na(s) + 2H_2O \longrightarrow 2NaOH(aq) + H_2(g)$$

Reaction with ammonia: When dry ammonia gas is passed over heated sodium. Sodamide ($NaNH_2$) is formed as a white waxy solid

$$2Na + 2NH_3 = 2NaNH_2 + H_2 \uparrow$$

In this reaction, a hydrogen atom of the covalent molecule of NH_3 is replaced by an metal. This reaction also proves that NH_3 contains hydrogen. Sodamide reacts with ordinary temperatures to form-sodium hydroxide and ammonia:

$$NaNH_2 + H_2O = NaOH + NH_3$$

Sodamide is used in many organic syntheses.

Sodium dissolves in liquid ammonia to form a blue solution, which contains ammoniated electrons.

$$(NH_3)_l\ e^- : Na + (NH_3)_l \rightarrow (NH_3)_l\ Na^+ + (NH_3)_l\ e^-$$

Reducing properties: As a highly electropositive metal sodium is a strong reducing agent. It replaces a metal, which is placed below it in the electrochemical series from its salt, generally at elevated temperatures. That is, it reduces the other metal ion into the metallic state. Thus, when aluminium chloride is heated with metallic sodium, sodium chloride is formed and metallic aluminium is set free.

$$AlCl_3 + 3Na = 3NaCl + Al$$

Certain metal-oxides (such as the oxides of uranium and titanium which cannot be effectively reduced to the metal by carbon or other common reducing agents, are easily reduced by heating with metallic sodium. (These reductions are carried out by heating in absence of air, since the presence of oxygen of air readily oxidizes sodium at high temperatures.) An ignited piece of sodium CO_2 gas; as a result, CO_2 is partly reduced to carbon: along with the formation of sodium carbon, thus formed through the reduction of CO_2, is generally in the form of fine black particles or soot.

$$3CO_2 + 4Na = 2NA_2CO_3 + C$$

Uses of Sodium

(*i*) Sodium metal is used in the preparation of many of its important compounds such as sodamide ($NaNH_2$), sodium cyanide (NaCN) etc.

(*ii*) Sodium is used as a reducing agent in the extraction of certain metals.

(*iii*) In the field of organic chemistry, sodium is used in the detection of carbon, SI and halogens in organic compounds and in the drying of organic liquids like ether, a benzene etc.

Compounds of Sodium

While discussing the properties of sodium we will be able to notice many astonishing properties of this metal. The compounds of this metal and their chemistry is stated below.

Sodium Hydride (NaH): This is produced when sodium is allowed to react with the steam of hydrogen gas at a temperature of 400-450°C. It is deposited as a white sublimate of transparent needles in the cooler portions of the reactor. The hydride is readily decomposed by the action of water with the formation of sodium hydroxide and hydrogen.

Sodium hydride, therefore, behaves like a strong reducing agent.

Water		Sodium Hydride	⟶	Sodium Hydroxide		Hydrogen
H_2O	+	NaH	⟶	NaOH	+	H_2

Sodium Monoxide (Na_2O): It can be obtained by heating sodium peroxide with sodium. It is a white amorphous solid substance.

Sodium Peroxide	+	Sodium	⟶	Sodium Monoxide
Na_2O_2		2Na	⟶	$2Na_2O$

It is also formed when *sodium* is burnt in limited supply of air or oxygen freed from moisture and carbon dioxide. It is manufactured by heating sodium in aluminium trays in a current of purified air at 300°C. Pure sodium peroxide is a white solid. The commercial forms a light yellow powder, which can be melted almost undecomposed. But with readily oxidizable substances like cotton, sawdust, straw charcoal, etc also with aluminium powder, it reacts violently with explosion. It reacts with water; the reaction is attended with a considerable amount of heat. It forms a hydrate, Na_2O_2. $8H_2O$, with ice cold water, which can be crystallized out of the aqueous solutions. In dilute aqueous solutions it decomposes with the formation of hydrogen peroxide:

$$\text{Sodium Peroxide} + \text{Water} \longrightarrow \text{Sodium Hydroxide} + \text{Hydrogen Peroxide}$$
$$Na_2O + H_2O \longrightarrow 2NaOH + 2H_2O_2$$

Sodium peroxide is, therefore, used for the preparation of hydrogen peroxide by treatment with cold dilute sulphuric acid.

$$\text{Sodium Peroxide} + \text{Sulphuric Acid} \longrightarrow \text{Sodium Sulphate} + \text{Hydrogen Peroxide}$$
$$Na_2O + H_2SO_4 \longrightarrow Na_2SO_4 + 2H_2O_2$$

With carbon dioxide sodium peroxide forms sodium carbonate with evolution of oxygen.

$$\text{Sodium Peroxide} + \text{Carbon Dioxide} \longrightarrow \text{Sodium Carbonate} + \text{Oxygen}$$
$$2Na_2O_2 + 2CO_2 \longrightarrow 2Na_2CO_3 + O_2$$

Uses of Sodium Peroxide

Sodium peroxide is used in respiration apparatus for fire fighters and divers, also for the purification of air in closed spaces for extremely used for making bleaching baths for silk, wool, straw, feather, hair, bone, tusk, etc. It is also widely used as powerful oxidizing agent in the laboratory for the oxidation of sulphur containing substances to sulphate, of chromic salts to chromates, etc.

Sodium Hydroxide (NaOH): There are two methods for the preparation of sodium hydroxide (I) Electrolytic Methods (II) Chemical Methods. Both of them are going to be discussed in brief.

Electrolytic Method: When sodium chloride is electrolyzed; chlorine is liberated at the anode and at cathode the metallic sodium is also liberated with immediately reacts with water to form sodium hydroxide which can be recognized by performing an experiment. The sodium chloride solution is mixed with phenolphthalein indicator which turns pink when comes in contact with alkali now the electrolysis is done with this solution it will be seen that the solution near the negative electrode is becoming red indicating formation of sodium hydroxide.

In industry sodium hydroxide is manufactured by *Electrolytic Process,* which is being discussed below:

In the Mercury cathode Cell, there are number of graphite positive electrodes, and the mercury is kept as the negative electrode forming a layer of 1/8th inch thick at the bottom of the cell. The mercury is kept flowing continuously throughout the process of electrolysis. The brine (sodium chloride solution) is introduced into cell and makes to flow through it in the same direction as the mercury. Chlorine gas is collected from the positive electrode. As the discharge voltage is high no hydrogen ions are produced on the surface of the negative electrode (mercury). Sodium is liberated and remains dissolved in mercury as an amalgam*. The amalgam flows out of the cell into another vessel containing water in which it comes in contact with metallic iron; here due to lesser discharge potential hydrogen gas is evolved. The anodic (positive electrode corresponding reactions) and cathodic (negative electrode corresponding reactions) occurring inside the cell is stated below.

Anode : $2C = Cl_2 + 2e$

Cathode : $2Na^+ + 2e + xHg = xHg\ Na_2$

Netreaction : $2Na^+ + 2Cl^- + xHg = Hgx\ Na_2 + Cl_2$

In the cell where the amalgam reacts with water in contact with iron.

The following reactions occur between the mercury electrode and the iron electrode where mercury acts as a positive and iron act as cathode. From the amalgam sodium is separated and goes into solution as positive sodium ion and as equivalent quantity of hydrogen is separated at the iron surface from the discharge of the hydrogen ions.

Anode : $xHg\ Na_2 = 2Na^+ + xHg + 2e,$

Cathode : $2H_2O + 2e = H_2 + 2OH^-$

Net Reaction : $Hgx\ Na_2 + 2H_2O = 2Na^+ + 2OH^- + xHg + H_2$

The mercury is recirculated through the cell. The spent brine is again resaturated and sent into the cell for the repetition of the process.

This completes the one of the most important industrial process of the manufacturing of sodium hydroxide.

Chemical Process: Sodium hydroxide can be also is obtained by chemical methods for *e.g.* when a solution of sodium carbonate is boiled with slaked lime the below reaction occurs:

Sodium Carbonate + Calcium Hydroxide ⟶ Calcium Carbonate + Sodium Hydroxide

$$Na_2CO_3 + Ca(OH)_2 \longrightarrow CaCO_3 + 2NaOH$$

Thus, equilibrium is obtained. In order to produce sodium hydroxide commercially a 10% solution of sodium carbonate is used; the idea is that more the solution of sodium carbonate is diluted better will be the percentage yield of the sodium hydroxide. The reaction is carried out in an iron vessel in which lumps of calcium hydroxide in an iron cage is suspended in a 10% solution of sodium carbonate. The solution is wormed by means of steam pipes and agitated by a mechanical stirrer. After the equilibrium point is reached the solution is filtered from sludge+ of calcium carbonate and unreacted calcium hydroxide, still the solution contains some unused sodium carbonate.

* Amalgam: The Alloy of mercury with other metals.
+ Sludge: A muddy deposit

Formerly caustic soda (sodium hydroxide) is also made by heating a mixture of sodium carbonate and ferric oxide to yield sodium ferrite. This was then boiled with water a precipitate of caustic soda and ferric oxide was obtained due to hydrolysis. The detail chemical reaction is given:

Sodium Carbonate + Ferric Oxide $\longrightarrow$ Sodium Ferrite + Carbon Dioxide

$$Na_2CO_3 + Fe_2O \longrightarrow 2NaFeO_2 + CO_2$$

Sodium Ferrite + Water $\longrightarrow$ Sodium Hydroxide + Carbon Dioxide

$$2NaFeO_2 + Fe_2O_3 \longrightarrow NaOH + CO_2$$

The regenerated ferric oxide is turned to the process. The process was known as the ferrite or Lowing Process; but it is now practically obsolete.

Properties of Sodium Hydroxide

Sodium hydroxide forms a white solid; white melting point is 318°C. It is highly soluble in water (also deliquescent*) and the solution evolves quite a good amount of heat. Detail physical properties of sodium hydroxide are given below:

Physical Properties

Boiling Point	1390
Melting Point	319
Density	2.43
Solubility in water	7 g/100cc
Heat of Formation	–101.99 kcal/mol^{-1}
Heat of Capacity	19.2 cal deg^{-1} mol^{-1}

Compounds of Sodium Hydroxide

Sodium hydroxide also produces compounds, of various types with other compounds as well as other elements. They are being stated below:

Sodium Borate ($NaBO_2$): It is formed by the action of sodium hydroxide and boron.

Boron + Sodium Hydroxide + Water $\longrightarrow$ Sodium Borate + Hydrogen

$$2B + 2NaOH + 2H_2O \longrightarrow 2NaBO_2 + 3H_2$$

Sodium Silicate (Water Glass) (Na_2SiO_3): Commercial sodium silicate is manufactured by fusing white sand and washing soda.

Sand + Sodium Carbonate $\longrightarrow$ Sodium Silicate + Carbon Dioxide

$$SiO_2 + Na_2CO_3 \longrightarrow Na_2SiO_3 + CO_2$$

* Deliquescent: Tends to become liquid by absorbing atmospheric moisture.

Sodium Carbonate (Washing Soda) (Na_2CO_3): In crude form it is available in nature. It is manufactured commercially by the action carbon dioxide and sodium hydroxide under pressure in Nelson Cell the reaction equation is

Sodium Hydroxide Carbon Dioxide ⟶ Sodium Carbonate Water

$$2NaOH + CO_2 \longrightarrow Na_2CO_3 + H_2O$$

Sodium carbonate is used as a detergent in the domestic places. Sodium Carbonate is found in a crystalline form as $Na_2CO_3.10H_2O$, which easily crumbles down as $Na_2CO_3.H_2O$ with loss of water in air. The anhydrous salt is very stable to heat and gas a boiling point of 850°C. It is also readily soluble in water, which makes the solution alkaline due to hydrolysis.

Sodium Carbonate + Water ⟶ Sodium Hydroxide + Carbonic Acid

$$Na_2CO_3 + 2H_2O \longrightarrow 2NaOH + H_2CO_3$$

Sodium Carbonate reacts with hydrochloric acid to form sodium chloride water and carbon dioxide.

Sodium Carbonate + Hydrochloric Acid ⟶ Sodium Chloride + Water + Carbon Dioxide

$$Na_2CO_3 + HCl \longrightarrow NaCl + H_2O + CO_2$$

Uses of Sodium Carbonate

Sodium carbonate is used in the manufacture of:

1. Glass.
2. Water glass
3. Soap
4. Caustic Soda
5. Borax
6. For softening hard water, it is also used in the textile industry and as laboratory reagent.

Sodium Bicarbonate ($NaHCO_3$): Commercial sodium bicarbonate is obtained as the intermediate product of the ammonia soda. Normal carbonate may be changed to bicarbonate by passing carbon dioxide through its saturated solution when bicarbonate is precipitated. It is washed with cold water and dried in air.

Sodium Carbonate + Carbon Dioxide + Water ⟶ Sodium Bicarbonate

$$Na_2CO_3 + CO_2 + H_2O \longrightarrow Na_2HCO_3$$

It is gives small white crystals, sparingly soluble in water. The solution is alkaline in nature due to hydrolysis (gives yellow colour to methyl orange and pink to phenolphthalein). It decomposes at 100°C with the evolution of carbon dioxide and leaving normal carbonate behind.

Sodium Bicarbonate ⟶ Sodium Carbonate + Carbon Dioxide + Water

$$Na_2HCO_3 \longrightarrow Na_2CO_3 + CO_2 + H_2O$$

Uses of Sodium Bicarbonate

1. In baking powders

   The evolution carbon dioxide on cooking makes the item spongy.
2. In medicine to neutralize the acidity in the stomach.
3. In effervescent drinks like fruit salt and Coca-Cola etc.
4. In fire extinguisher.

Sodium Chloride, (Common salt) (NaCl): Sodium chloride occurs as a common salt in sea about 3% and also in lakes about 1 %, most of the Indian requirement of salt comes from the sea.

1. *In hot countries:* sea water or lake water or solution of salt is pumped out of mine is allowed to evaporate in sun's heat. Finally the water is drained off and the salt is obtained in free state.

2. *In cold countries:* dilute solution (3% NaCl) is cooled in pits at night when ice separates and is removed. This process is repeated till the limit, *i.e.* 22% of NaCl solution is obtained freezes as such.

Properties of Sodium Chloride

Sodium chloride crystalline in anhydrous cubic crystals, which melts on heating. It's melting point is 820°C and the boiling point is 1440°C. It is a hydrate $NaCl.2H_2O$ at low temperature.

Uses: Common salt is used as an essential constituent of foods, as a flavour also as a preservative of food. In salt glazing the moist salt is thrown in the kiln where the articles are going fired. The salt volatilizes and reacts with silica of the clay to form a fusible silicate or glaze on the surface of the earthenware:

$$\underset{\text{Chloride}}{\text{Sodium}} + \underset{\text{Dioxide}}{\text{Silicon}} + \text{Water} \longrightarrow \underset{\text{Silicate}}{\text{Sodium}} + \underset{\text{Chloride}}{\text{Hydrogen}}$$

$$2NaCl + SiO_3 + H_2O \longrightarrow NaSiO_3 + HCl$$

Sodium chloride serves as the starting material for the preparation of metallic sodium, caustic and for the preparation of many sodium salts, which is going to be discussed now.

Sodium Sulphate (Na_2SO_4): It is prepared in the Leblanc process by heating common salt with sulphuric acid. The salt cake formed is dissolved in water below 32.38°C the decahydrate, $NaSO_4$, $10H_2O$, called *Glauber's salt,* crystallizes out form the aqueous solution; above 32.38°C the anhydrous salt separates out and becomes less soluble with the rise of temperature. The decahydrate effloresces* in air, yielding the anhydrous salt. Sodium sulphate is also prepared by Hargreaves Process by passing sulphur dioxide, air and water vapour over heated common salt.

$$\underset{\text{Chloride}}{\text{Sodium}} + \underset{\text{Dioxide}}{\text{Sulphur}} + \text{Water} + \text{Oxygen} \longrightarrow \underset{\text{Sulphate}}{\text{Sodium}} + \underset{\text{Chloride}}{\text{Hydrogen}}$$

$$4NaCl + 2SO_2 + 2H_2O + O_2 \longrightarrow 2Na_2SO_4 + HCl$$

---

* Efflorescent: To change to a powder, by looses water.

Sodium Bisulphate (Sodium Hydrogen Sulphate) ($NaHSO_4$): This compound of sodium separates out in large crystals from a solution of sodium sulphate in warm concentrated sulphuric acid.

Sodium Sulphate + Sulphuric Acid ⟶ Sodium Hydrogen Sulphate

Na_2SO_4 H_2SO_4 ⟶ $2NaHSO_4$

Its aqueous solution is strongly acidic:

Sodium Sulphate ⟶ Sodium Ion + Hydrosulphate Ion

$2NaHSO_4 \longrightarrow Na^+ + HSO_4^-$

Hydrosulphate Ion ⟶ Hydrogen Ion + Sulphate Ion

$HSO_4^- \longrightarrow H^+ + SO_4^-$

When gently heated it give sodium pyrosulphate:

Sodium Sulphate ⟶ Sodium Pyrosulphate Water

$2NaHSO_4 \longrightarrow Na_2S_2O_7 + H_2O$

Which further decomposes into sodium sulphate and sulphur trioxide at a considerable amount of heat.

Sodium Pyrosulphate + Sodium Sulphate ⟶ Sulphur Trioxide

$Na_2S_2O_7 + Na_2SO_4 \longrightarrow SO_3$

Sodium bisulphate is used in iron and steel industry for cleaning the iron sheets before tinning.

Sodium Thiosulphate ($Na_2S_2O_3$ $5H_2O$): It is the important salt of Thiosulphuric acid, $H_2S_2O_3$. Unlike thiosulphuric acid it is quite stable. The action of sodium sulphite and flowers of sulphur prepare sodium thiosulphate, the mixture is boiled and during boiling continuous stirring is done till the alkaline reaction disappears.

The equation for the reaction is as follows:

Sodium Sulphite + Sulphur ⟶ Sodium Thiosulphate

$Na_2SO_3 + S \longrightarrow Na_2S_2O_3$

The excess sulphur is filtered off and the filtrate is evaporated till it crystallizes.

Sodium Thiosulphate can also be prepared by the action of sulphur and sodium hydroxide.

Sodium Hydroxide + Sulphur ⟶ Sodium Thiosulphate + Sodium Sulphide + Water

$6NaOH + 4S \longrightarrow Na_2S_2O_3 + 2Na_2S + 3H_2O$

Properties of Hypo

1. It is a colourless crystalline white solid soluble in water, and it is also efflorescent substance of the formula $Na_2S_2O_3$. $5H_2O$.
2. On heating it decomposes into sulphur dioxide and sulphur.
3. Dilute acids like sulphuric acid reacts with it to form sulphur. Sodium Thyosulphate is commonly known as *Hypo*.

Sodium Thiosulphate	+ Sulphuric Acid	$\longrightarrow$	Sodium Hydrogen Sulphate	+ Sulphur	+ Water
$Na_2S_2O_3$	+ $2H_2SO_4$	$\longrightarrow$	$2NaHSO_4$	+ S	+ H_2O

The specialty of this reaction is that it produces sulphur dioxide at room temperature, and therefore no extra heat is necessary.

4. Sodium thiosulphate also reacts with silver halides. Silver halides are dissolved in sodium thiosulphate solution (therefore, it is used in photographic developing and printing as a fixer) the reaction is shown below as the example:

Silver Bromide	+ Sodium Thiosulphate	$\longrightarrow$	Sodium Argentothiosulphate	+ Sodium Bromide
AgBr	+ $2Na_2S_2O_3$	$\longrightarrow$	$Na_3[Ag(S_2O_3)_2]$	+ NaBr

5. Silver nitrate also reacts with sodium thiosulphate, a very dilute solution of sodium thiosulphate when reacts gives a precipitate which is white is colour, turns very quickly into yellow finally black due to formation of black silver sulphide.

Silver Nitrate	+ Sodium Thiosulphate	$\longrightarrow$	Silver Thiosulphate	+ Sodium Nitrate
$2AgNO_3$	+ Na_2S_2O		$Ag_2S_2O_3$	+ $NaNO_3$
Silver Thiosulphate	+ Water	$\longrightarrow$	Silver Sulphide	+ Sulphuric Acid
$Ag_2S_2O_3$	+ H_2O	$\longrightarrow$	Ag_2S	+ H_2SO_4

A concentrated solution of sodium sulphate does not react with silver nitrate because the silver thiosulphate gets dissolved in excess sodium thiosulphate.

Silver Thiosulphate	+ Sodium Thiosulphate	$\longrightarrow$	Sodium Argentothiosulphate
$Ag_2S_2O_3$	+ $Na_2S_2O_3$	$\longrightarrow$	$2Na_3[Ag(S_2O_3)_2]$

6. Chlorine reacts as an oxidizing agent to form sodium sulphate:

Chlorine	+ Sodium Thiosulphate	+ Water	$\longrightarrow$	Sodium Sulphate	+ Hydrogen Chloride	+ Sulphur
Cl_2	+ $Na_2S_2O_3$	+ H_2O	$\longrightarrow$	Na_2SO_4	+ 2HCl	+ S

Iodine reacts to decolourize thiosulphate solution to form sodium tetrathionate.

Sodium Thiosulphate	+ Iodine	$\longrightarrow$	Sodium Tetrathionate	+ Sodium Iodide
$2Na_2S_2O_3$	+ I_2	$\longrightarrow$	$Na_2S_4O_6$	+ NaI

7. When barium chloride solution is added to the solution of sodium thiosulphate, barium thiosulphate and sodium chloride is produced.

Sodium Thiosulphate	+ Barium Chloride	$\longrightarrow$	Barium Thiosulphate	+ Sodium Chloride
$2Na_2S_2O_3$	+ $BaCl_2$	$\longrightarrow$	BaS_2O_3	+ 2NaCl

8. When ferric chloride solution is added to sodium thiosulphate a violet coloured precipitate is produced, which is called ferric thiosulphate.

Sodium Thiosulphate	+ Ferric Chloride	$\longrightarrow$	Ferric Thiosulphate	+ Sodium Chloride
$Na_2S_2O_3$	+ $FeCl_3$	$\longrightarrow$	BaS_2O_3	+ 2NaCl

Uses of Sodium Thiosulphate

1. Due to the property of dissolving the silver halides it is used as the fixer under the commercial name hypo.
2. In textile industry sodium thiosulphate solution is used as antichlor* during bleaching.
3. It is also used as an extraction material of gold and silver from their respective ores.
4. It is used as an estimator to determine the strength of iodine in laboratory.
5. Sodium thiosulphate is also used in medicine.

Tests for Sodium Thiosulphate: The only confirmatory test is the iodine test in which it produces sodium tetrathionate.

Sodium Sulphide (Na_2S): It is prepared by the action of sodium sulphate and coke at 1000°C.

$$\text{Sodium Sulphate} + \text{Coke} \longrightarrow \text{Sodium Sulphide} + \text{Carbon Monoxide}$$

$$NaSO_4 + 4C \longrightarrow Na_2S + 4CO$$

Uses: Sodium sulphide is used in textile industry in making dyes and also for extraction of silver.

Sodium Nitrate, (Chile salt petre) ($NaNO_3$): This compound occurs in the Caliche deposits of Chile. The crude caliche contains about 655 of sodium iodate, the rest being mostly sodium chloride and some compounds of potassium. It is lixiviated+ with hot water and sodium nitrate is separated out by crystallization, it contains about 95% $NaNO_3$ and usually also some other compounds but in a very small proportions. It is a white solid deliquescent solid, very much soluble in water. The melting point of the compound is 316°C and decomposes at higher temperature yielding oxygen.

Sodium nitrate is used as a fertilizer in agriculture and for the manufacture of nitric acid, sodium nitrate and also potassium nitrate is needed for the manufacture of gunpowder.

> Generally sodium nitrate not used for the preparation of gunpowder because it hydroscopic.

$$\text{Sodium Nitrate} \longrightarrow \text{Sodium Nitrite} + \text{Oxygen}$$

$$2NaNO_3 \longrightarrow 2NaNO_2 + O_2$$

Sodium Nitrite ($NaNO_2$): It is also known as Nitrate of soda, it is obtained by heating sodium nitrate, or better adding heat pieces of lead in fused sodium nitrate.

$$\text{Sodium Nitrate} + \text{Lead} \longrightarrow \text{Sodium Nitrite} + \text{Lead Oxide}$$

$$NaCO_3 + Pb \longrightarrow 2NaNO_2 + PbO$$

It is now manufactured by passing into caustic soda an equimolecular mixture of nitric oxide and nitrogen dioxide.

---

* Antichlor: It serves as an agent to remove any hypochlorite compound or chlorine after bleaching operation.

* Lixiviated: To extract a soluble content from a solid compound for that purposes it is mixed with a solvent.

Sodium Hydroxide + Nitrogen Sesquioxide ⟶ Sodium Nitrate + Water

$$2NaOH + N_2O_3 \longrightarrow 2NaNO_2 + H_2O$$

It is a pale yellow, crystalline solid soluble in water. It gives a usual reaction of nitrite.

1. All the nitrites are soluble in cold water except silver nitrite; it is soluble in hot water.
2. Dilute hydrochloric acid also decomposes sodium nitrite to form nitrogen dioxide.

Sodium Nitrite + Hydrochloric Acid ⟶ Sodium Chloride + Nitrogen Acid

$$NaNO_2 + HCl \longrightarrow NaCl + HNO_2$$

3. Sodium nitrite is a very good reducing agent; it decomposes potassium permanganate to a colourless compound.

Potassim Permanganate + Sulphuric Acid + Sodium Nitrate ⟶ Potassium Sulphate + Manganese Sulphate + Water + Sodium Nitrate

$$2KMnO_4 + 3H_2SO_4 + 5NaNO_2 \longrightarrow K_2SO_4 + 2MnSO_4 + 3H_2O_3 + 5NaNO_3$$

4. Ferrous nitrososulphate is formed when ferrous sulphate (acidified) reacts with sodium nitrite, which is violet is colour.
5. Meta-phnylenediamine solution in hydrochloric acid, sodium nitrite produces brown colour.

Sodium Phosphate (Na_2HPO_4): It is also known as *normal sodium phosphate* or *disodium phosphate.* This compound of sodium is obtained when a small amount of phosphoric acid reacts with fused sodium carbonate.

Sodium Carbonate + Phosphoric Acid ⟶ Sodium Phosphate + Water + Carbon Dioxide

$$Na_2CO_3 + H_3PO_4 \longrightarrow Na_2HPO_4 + H_2O + CO_2$$

The compound is also prepared by the action of calcium phosphate (bone ash) and sodium carbonate. The solution is further boiled to remove carbon dioxide and exactly neutralize with sodium hydroxide when trisodium phosphate is obtained. It is a white crystalline solid soluble in water. Its aqueous solution is alkaline due to hydrolysis.

Uses: It is also used as detergent.

Monosodium Phosphate (NaH_2PO_4): This compound of sodium is prepared by the reaction of sodium hydrogen phosphate and phosphoric acid in limited proportions. It is a white powder, very much soluble in water; it is an acidic salt, produces sodium pyrophosphate at 230°C and sodium metaphosphate at 360°C. Monosodium Phosphate is used as boiler water treatment, electroplating, dyeing acid cleaners also as a lab reagent.

Phosphoric Acid Sodium Hydrogen Phosphate ⟶ Monosodium Phosphate

$$H_3PO_4 + Na_2HPO_4 \longrightarrow 2Na_2H_2PO_4$$

Monosodium Phosphate + Sodium Meta Phosphate ⟶ Water

$$NaH_2PO_4 + NaPO_3 \longrightarrow H_2O$$

Monosodium Phosphate $\xrightarrow{230°C}$ Sodium Pyrophosphate + Phosphorous Pentoxide + Water

$$4Na_2H_2PO_4 \longrightarrow Na_4P_2O_7 + P_2O_5 + 4H_2O$$

Testing of Sodium

(i) ***Flame test:*** The characteristic colours of the flame becomes golden yellow when a sodium salt is introduced into the flame.

(ii) ***Treatment with Potassium Hydrogen Pyroantimonate*:** When a sodium salt is treated with this compound there is a crystalline precipitate white in colour. (This is generally done with the acidic salt of the sodium and not the basic salt).

Potassium Hydrogen Pyroantimonate + Sodium Chloride ⟶ Sodium Dihydorgen Pyroantimonate + Potassium Chloate

$$K_2H_2Sb_2O_7 + 2NaCl \longrightarrow Na_2H_2Sb_2O_7 + 2KCl$$

(iii) ***Magnesium Urenyl Acetate Test*:** With the addition of this salt sodium compound produces a yellow ppt, called sodium magnesium Uranyl acetate,

$$Na\ Mg\ (UO_2)_3 \cdot (CH_3COO)_2 \cdot 9H_2O$$

This completes the chemistry of sodium and its compounds. We shall now start a new compound is POTASSIUM.

POTASSIUM

Chemical Symbol : K (Kalium)
Atomic Number : 19
Activity : More active than sodium
Atomic Weight : 39.0983
Melting Point : 62°C
Boiling Point : 762°C

Discovery and Isolation

Potassium occurs in wood ash and may be extracted with water. There has been known from earliest times aqueous solution was evaporated in iron pots to yield 'pot ash' hence the present name potassium has occurred. Davy prepared this element.

Occurrence

Potassium is reactive to occur in free state but its compounds are widely spread, important salts are distributed around Stassfort Germany, mainly as *Sylvine, KCl* and *Camellite* $KCl \cdot MgCl_2 \cdot 6H_zO$. Similar deposits elsewhere, probably formed by evaporation of inlet seas. Potassium occurs also in rocks e.g. *feldspar, potassium aluminium silicate,* $(KalSi_3O_8)$ and *mica* $KH_2Al_3(SiO_4)$ potassium is an essential element in plant and animal metabolism. It is therefore found in plant and animal sheep sweat (as a salt of organic acid).

Extraction

Metallic potassium is not prepared commercially in large scale. Generally it is obtained by electrolysis of hydroxide or the cyanide compound of the metal. Metallic potassium is prepared by heating potassium chloride and calcium in vacuum.

Potassium Chloride + Calcium ⟶ Calcium Chloride + Potassium

$$2KCl + Ca \longrightarrow CaCl_2 + 2K$$

Potassium is distilled off and the reaction goes from left to right.

Properties

Potassium resembles sodium in both physical and chemical properties. The metal is rather softer than sodium and exhibits slight radioactivity owing to the presence of isotope 19K40 in very small amounts. Chemical potassium is more reactive that sodium in properties, also it is rather softer than sodium. When it reacts with water the heat of reaction usually causes ignition of the evolved of hydrogen. In excess oxygen potassium bums to give super oxide (KO_2). Now we shall study about the compounds of this metal. The detail physical properties of the potassium is shown in the table below:

Melting Point	62°C
Boiling Point	762°C
Density	0.86 g/cc
Electrical Resistivity	13.6 ohm cm at 64°C
Standard Entropy	15.2 cal deg^{-1} mol^{-1}
Heat of Capacity	6.97 cal deg^{-1} mol^{-1}

Compounds of Potassium

Potassium Monoxide (K_2O): It is formed as a yellow solid when potassium is heated with potassium nitrate.

Potassium Nitrate + Potassium $\longrightarrow$ Potassium Oxide + Nitrogen

$$2KNO_3 + K \longrightarrow 6K_2O + N_2$$

EXPLOSIVE

It dissolves in water with great violence to form potassium hydroxide.

Potassium Tetraoxide (K_2O_4): It is the ultimate compound of potassium when it is burnt in air or oxygen. It forms a chrome-yellow solid, which reacts with normal water, giving potassium hydroxide, oxygen and hydrogen peroxide with an O_2^- anion.

Potassium Hydroxide (KOH): This compound prepared exactly same like the preparation of sodium hydroxide, like sodium very much in physical and chemical properties. It forms a white solid at 360°C and readily absorbs carbon dioxide from atmosphere. Potassium hydroxide is highly soluble in water (97% in 100%). It is a stronger alkali than sodium hydroxide. The fused caustic potassium attacks many metals, (except iron, nickel and silver).

Uses: Potassium hydroxide is used in the laboratory for the adsorption of carbon dioxide, for which sodium hydroxide is unusable. Potassium hydroxide is used for the preparation of soft soap.

Potassium Carbonate (K_2CO_3): Is known from early days as 'potash', being produced from plant ash with water, followed by the evaporation of the solution to dryness and calcination of the residue in the iron pots. This method is still followed in many countries, the pure commercial product is known as 'pearl ash'.

It can be prepared from potassium chloride by a modification of Leblanc's Process; a process similar to the ammonia soda process, is however is extremely soluble in water.

The substance is usually prepared by *Precht's Process.* In this a strong solution of potassium chloride containing hydrated magnesium carbonate in carbonate of potassium and magnesium is precipitated.

Hydrated Magnesium Carbonate + Potassium Chloride + Carbon Dioxide ⟶ Magnesium Potassium Carbonate + Magnesium Chloride

$$3(MgCO_3\ 3H_2O) + 2KCl\ CO_2 \longrightarrow 2Mg(CO_3KHCO_34H_2O) + MgCl_2$$

The precipitate is filtered from the solution and then either heated with water under pressure at 140°C or stirred with a suspension of magnesium oxide in water at 40°C. When potassium carbonate is formed in the solution with the regeneration of the hydrated magnesium carbonate. Which is separated by filtration for use in the process again. From the solution, potassium carbonate is obtained by crystallization as the dihydrate,

$$K_2CO_3 \cdot 2H_2O$$

Magnesium Potassium Carbonate ⟶ Hydrated Magnesium Carbonate + Potassium Carbonate

$$2Mg(CO_3KHCO_3\ 4H_2O) \longrightarrow 3(MgCO_3\ 3H_2O) + K_2CO_3$$

Magnesium Potassium Carbonate ⟶ Magnesium Oxide + Hydrated Mangesium Carbonate + Potassium Carbonate

$$2Mg(CO_3KHCO_3\ 4H_2O) \longrightarrow MgO + 3(MgCO_3\ 3H_2O) + K_2CO_3$$

The anhydrous potassium carbonate is prepared from beet sugar molasses. The molasses are mixed with lime and charred by heating. The residual mass is then extracted with water form the solution of potassium carbonate is obtained by the process of crystallization.

The anhydrous potassium carbonate is white, deliquescent powder with a melting point of 900°C. It is highly soluble in water; the solution is alkaline due to hydrolysis.

Carbonate Ion + Water ⟶ Hydrocarbonate Ion + Hydroxyl Ion

$$CO_3^- + H_2O \longrightarrow HCO_3^- + OH^-$$

It differs with sodium carbonate, being highly soluble and deliquescent but resembles the latter in other properties.

Uses of Potassium Carbonate: It is manly used in glass industry, also for making caustic potash and chromates of potassium.

Potassium Hydrogen Carbonate, (potassium bicarbonate) ($KHCO_3$): When carbon dioxide, is passed over moist potassium carbonate it is converted into potassium bicarbonate due to absorption of carbon dioxide:

Potassium Carbonate + Water + Carbon Dioxide ⇌ Potassium Hydrogen Carbonate

$$K_2CO_3 + H_2O + CO_2 \rightleftharpoons 2KHO_3$$

This is dried on a porous plate over sulphuric acid in an atmosphere of carbon dioxide, when heated to 190°C; it looses carbon dioxide and turns back to normal carbonate. That is why the bi-directional arrows have been shown.

Potassium bi-carbonate is less soluble than the potassium carbonate, but much soluble than sodium bicarbonate. The solution reacts strongly alkaline, due to hydrolysis.

Potassium Chloride (KCl): The main source is Stassfort salt deposit where it occurs as Sylvine and also in combination with magnesium chloride as Carnallite KCl, $MgCl_2 \cdot 6H_2O$.

When a saturated solution of camellite is allowed to crystallize in the cold, potassium chloride separates out leaving the more soluble magnesium chloride in water and its solubility at 0°C and 56.5°C gm at 100°C, dissolve per 100 g of water. Potassium chloride is used as a fertilizer and for the preparation of other potassium compounds.

Potassium Bromide (KBr): The final mother liquor of magnesium chloride after extraction of potassium chloride from camellite contains as appreciable quantity of potassium bromide (KBr). From this liquor bromine is liberated and allowed to react with iron fillings in presence of water to form ferosoferric bromide.

$$\text{Bromine} + \text{Iron} \longrightarrow \text{Ferrosoferric Bromide}$$

$$4Br_2 \quad 3Fe \longrightarrow FeBr_{2.2}FeBr_3$$

The solution of iron bromide is then treated with potassium carbonate, which precipitates the hydrated ferosoferric oxide. The precipitate is filtered off and the potassium bromide remaining in the solution crystallizes out on evaporation.

$$\underset{\text{Bromide}}{\text{Ferrosoferric}} + \underset{\text{Carbonate}}{\text{Potassium}} + \text{Water} \longrightarrow \underset{\text{Bromide}}{\text{Potassium}} + \underset{\text{Dioxide}}{\text{Carbon}}$$

$$FeBr_{2.2}FeBr_3 + 4K_2CO_3 + 4H_2O \longrightarrow 8KBr + 4CO_2$$

It can also be synthesized by passing bromine gas through hot potassium hydroxide solution and dried mixture of KBr and $KBrO_3$ is heated with carbon dioxide to convert $KBrO_2$ also into KBr.

$$\text{Potassium Hydroxide} + \text{Hydrogen Bromide} \longrightarrow \text{Potassium Bromide} + \text{Water}$$

$$KOH + HBr \longrightarrow KBr + H_2O$$

$$\underset{\text{Bicarbonate}}{\text{Potassium}} + \underset{\text{Bromide}}{\text{Hydrogen}} \longrightarrow \text{Water} + \underset{\text{Bromide}}{\text{Potassium}} + \underset{\text{Dioxide}}{\text{Carbon}}$$

$$KHCO_3 + HBr \longrightarrow H_2O + KBr + CO_2$$

Properties and Uses: Potassium bromide is a colourless crystalline solid. Its uses are as follows:

(*i*) In medicine it is used to prevent nervousness and induce sleep.

(*ii*) In photography

(*iii*) In laboratory it is used as a reagent for preparation of bromine gas, with the action of sulphuric acid

Potassium Iodide (KI): It is a compound, which is prepared from iodide through ferosoferric iodide (FeI_2 $2FeI_3$) in the similar manner to that of potassium bromide. It crystallizes out in regular cubes and is readily soluble in water and to some extent alcohol also.

A solution of potassium iodide can dissolve a considerable amount of iodide giving a brown coloured solution. The solution contains potassium tri-iodide, a typical complex salt that can be obtained in a crystalline form:

Potassium Iodine + Iodine ⟶ Potassium Triiodide

$$KI + I_2 \longrightarrow KI_3$$

In aqueous solution the complex tri-iodide anion suffers a considerable dissociation, leading to equilibrium.

Uses: (1) In medicine (2) In photography (3) As a laboratory reagent for dissolving in iodine and forming potassium tri-iodide.

Potassium Iodine + Iodine ⟶ Potassium Triiodide

$$KI + I_2 \longrightarrow KI_3$$

Potassium Sulphate (K_2SO_4): This compound of potassium mainly occurs as Schonite $K_2SOMgSO_46H_2O$ in the Stassfort salt deposits. It is prepared from Schonite by digesting it with potassium chloride solution, when sparingly soluble potassium sulphate crystallizes out first.

Sconite + Potassium Chloride ⟶ Potassium Sulphate + Magnesium Chloride

$$K_2Mg(SO_4)_2 + 2KCl \longrightarrow 2K_2SO_4 + MgCl_2$$

Potassium sulphate is less soluble than sodium sulphate and has a rather high melting point of 1070°C. It is used as fertilizer and in making potash alum.

Potassium Hydrogen Sulphate ($KHSO_4$): When the normal potassium sulphate is heated with concentrated sulphuric acid and then cooled it is formed, it is also formed in the preparation of nitric acid from nitre and sulphuric acid. It melts at 21°C and when heated more strongly, it produces potassium di sulphate or pyrosulphate with loss of water. When the temperature is still increased the compound losses sulphur trioxide with the formation of the normal sulphate.

Potassium Hydrogen Sulphate ⟶ Potassium Pyrosulphate + Water

$$2KHSO_4 \longrightarrow K_2S_2O_7 + H_2O$$

Potassium Pyrosulphate ⟶ Potassium Sulphate + Sulphur Trioxide

$$K_2S_2O_7 \longrightarrow K_2SO_4 + SO_3$$

Potassium hydrogen sulphate is freely soluble in water, giving a strongly acidic which when evaporates deposits K_2SO_4. The acid sulphate and the pyro or the di sulphate are used for analytical purposes in the laboratory, in order to bring many insoluble metallic oxides present in minerals into a soluble form as their sulphates by fusion with them.

Potassium Nitrate, Saltpetre (KNO_3): This compound is better known as Saltpetre or Nitre. It is obtained in Bihar, Punjab, Sri Lanka and Egypt from the efflorescence of the nitre on the soil, which is formed from the oxidation on nitro-fixing bacteria present in the soil, which is formed from the oxidation of nitro-fixing bacteria present in the soil. The nitrate, it is treated with potassium carbonate (from wood-ashes) to precipitate calcium carbonate.

It is however, largely prepared by treating the cheaper, Chile saltpetre or sodium nitrate with a hot concentrated solution of potassium chloride. Sodium chloride crystalline out from the filtrate below 32°C. Potassium nitrate is less soluble than chloride at low temperature. But soluble in higher temperatures.

$$\text{Potassium Nitrate} \longrightarrow \text{Potassium Nitrite} + \text{Oxygen}$$
$$2KNO_3 \longrightarrow 2KNO_2 \quad O_2$$

Potassium nitrate reacts vigorously with sulphur and carbon or piece of charcoal or sulphur comes in comes in contact with molten KNO_3 a violent reaction occurs.

$$\text{Potassium Nitrate} \quad \text{Sulphur} \longrightarrow \text{Potassium Sulphate} \quad \text{Sulphur Dioxide} \quad \text{Nitrogen}$$
$$2KNO_3 + S \longrightarrow K_2SO_4 + SO_2 + N_2$$
$$\text{Potassium Nitrate} + \text{Carbon} \longrightarrow \text{Potassium Carbonate} + \text{Carbon Dioxide} + \text{Nitrogen}$$
$$4KNO_3 + 5C \longrightarrow 2K_2CO_3 + 3CO_2 + 2N_2$$

Uses of Potassium Nitrate: In manufacture of gunpowder and fire-works (Nitre 6 parts, charcoal 1 part and sulphur 1 part). **This ratio of gunpowder is very dangerous and a student should not try, it can cause explosion.**

Potassium Nitrite (KNO_2): It is prepared like sodium nitrite by fusing potassium nitrate and metallic lead. The mass is extracted with water and the nitrite is precipitated from the crystals, slightly soluble in water, the melting point is 441°C. When heated it decomposes like crystalline deliquescent substances.

Potassium Permanganate ($KMnO_4$): It is manufactured by fusing manganese dioxide with caustic potash or potassium carbonate and an oxidizing agent, usually potassium nitrate in free supply of air. The fused mass is dark green due to the formation of potassium manganate.

$$\underset{\text{Dioxide}}{\text{Maganese}} + \underset{\text{Hydroxide}}{\text{Potassium}} + \text{Oxygen} \longrightarrow \underset{\text{Manganate}}{\text{Potassium}} + \text{Water}$$
$$2MnO_2 + 4KOH + O_2 \longrightarrow 2K_2MnO_4 + H_2O$$
$$\underset{\text{Dioxide}}{\text{Maganese}} + \underset{\text{Carbonate}}{\text{Potassium}} + \text{Oxygen} \longrightarrow \underset{\text{Magnate}}{\text{Potassium}} + \underset{\text{Dioxide}}{\text{Carbon}}$$
$$2MnO_2 + 2K_2CO_3 + O_2 \longrightarrow 2K_2MnO_4 + 2CO_2$$

The fused mass is extracted with water and treated with a current of ozone, chlorine or carbon dioxide potassium manganate to permanganate. The purple solution thus obtained is concentrated when it deposits dark purple needle like structured crystals.

$$\underset{\text{Maganate}}{\text{Potassium}} + \text{Ozone} + \text{Water} \longrightarrow \underset{\text{Permanganate}}{\text{Potassium}} + \underset{\text{Hydroxide}}{\text{Potassium}} + \text{Oxygen}$$
$$2K_2MnO_4 + O_3 + H_2O \longrightarrow 2KMnO_4 + KOH \quad O_2$$
$$\underset{\text{Maganate}}{\text{Potassium}} + \text{Chlorine} \longrightarrow \underset{\text{Permanganate}}{\text{Potassium}} + \underset{\text{Chloride}}{\text{Potassium}}$$
$$2K_2MnO_4 + \quad \longrightarrow 2KMnO_4 + 2KCl$$
$$\underset{\text{Maganate}}{\text{Potassium}} + \underset{\text{Dioxide}}{\text{Carbon}} \longrightarrow \underset{\text{Manganate}}{\text{Potassium}} + \underset{\text{Bicarbonate}}{\text{Potassium}} + \text{Manganese}$$
$$3K_2MnO_4 + 4CO_2 \longrightarrow 2KMnO_4 + 4KHCO_3 + MnO_2$$

Properties of Potassium Permanganate: (*i*) Potassium permanganate forms dark purple anhydrous needle like structure with metallic lusture, readily soluble in water and the colour of the solution is deep purple.

(*ii*) Potassium permanganate when heated strongly decomposes into manganese dioxide, potassium manganate and oxygen.

Potassium Permanganate ⟶ Potassium Manganate + Oxygen + Manganese Dioxide

$$2KMnO_4 \longrightarrow K_2MnO_4 + O_2 + MnO_2$$

(*iii*) Oxygen is liberated when heated with concentrated sulphuric acid.

Potassium permanganate can oxidize

1. Halides or halogen acids in halogen.
2. Sulphides or hydrogen sulphide to sulphur.
3. Sulphur dioxide to sulphuric acid.
4. Nitrites and Nitrates.
5. Oxalates and oxalic acids to carbon dioxide.
6. Arsenites to arsenates.
7. Hydrogen peroxide to oxygen.
8. Ferrous salts to ferric and stannous salt to stannic salts.
9. Potassium iodate to iodine.

The Reactions Stated below are very Important must be Studied by the Students also can be Tried in the Lab

(Chlorine)

Potassium Permangante + Hydrochloric Acid ⟶ Potassium Chloride + Chlorine + Water

$$2KMnO_4 + 16HCl \longrightarrow 2KCl + 5Cl_2\Uparrow + 8H_2O$$

(Sulphur)

Potassium Permangante + Sulphuric Acid + Hydrogen Sulphide ⟶ Potassium Sulphate + Water + Sulphur + Manganase Sulphate

$$2KMnO_4 + 3H_2SO_4 + 5H_2S \longrightarrow K_2SO_4 + 8H_2O + 5S\Downarrow + 2MnSO_4$$

(Potassium Nitrate)

Potassium Permanganate + Potassium Nitrite + Sulphuric Acid ⟶ Potassium Sulphate + Manganese Sulphate + Potassium Nitrate + Water

$$2KMnO_4 + 5KNO_2 + 3H_2SO_4 \longrightarrow K_2SO_4 + 2MnSO_4 + 5KNO_3 + 3H_2O$$

(Carbon Dioxide)

Potassium Permanganate + Oxalic Acid + Sulphuric Acid ⟶ Potassium Sulphate + Manganese Sulphate + Carbon Dioxide + Water

$$2KMnO_4 + 5(COOH)_2 + 3H_2SO_4 \longrightarrow K_2SO_4 + 2MnSO_4 + 10CO_2 + 8H_2O$$

The above reaction should be conducted at a temperature of 60°C

(Ferric Sulphate)

Potassium Permanganate + Ferrous Sulphate + Sulphuric Acid ⟶ Potassium Sulphate + Manganese Sulphate + Ferric Sulphate + Water

$$2KMnO_4 + 10FeSO_4 + 8H_2SO_4 \longrightarrow 6K_2SO_4 + 2MnSO_4 + 5Fe_2(SO_4)_3 + 8H_2O$$

(Oxygen)

Potassium Permanganate + Sulphuric Acid + Hydrogen Peroxide ⟶ Potassium Sulphate + Manganese Sulphate + Water + Oxygen

$$2KMnO_4 + 3H_2SO_4 + H_2O_2 \longrightarrow 6K_2SO_4 + 2MnSO_4 + 8H_2O + 5O_2$$

Potassium Dichromate (K_2Cr2O_7): This compound of potassium can be prepared from sodium dichromate by adding potassium chloride to a concentrated solution of the latter. Potassium dichromate being soluble separates out in the form of orange crystals.

Sodium Dichromate + Potassium Chloride ⟶ Potassium Dichromate + Sodium Chloride

$$Na_2Cr_2O_7 + 2KCl \longrightarrow K_2CrO_7 + 2NaCl$$

Properties

1. On heating it decomposes to:

   Potassium Dichromate ⟶ Chromic Acid + Potassium Chromate + Oxygen

   $$K_2CrO_7 \longrightarrow 2Cr_2O_3 + 2K_2CrO_4 + 3O_2$$

2. On heating with alkali it changes to chromate.

   Potassium Dichromate + Potassium Hydroxide ⟶ Potassium Chromate + Water

   $$K_2CrO_7 + 2KOH \longrightarrow 2K_2CrO_4 + H_2O$$

3. On heating with concentrated sulphuric acid it gives oxygen

   Potassium Dichromate + Sulphuric Acid + Potassium Sulphate + Chromic Sulphate + Water + Oxygen

   $$2K_2Cr_2O_7 + H_2SO_4 + 2K_2SO_4 + 2Cr_2(SO_4)_3 + 8H_2O + 3O_2$$

On heating solid dichromate with ammonium chloride is heated in the presence of concentrated Sulphuric Acid, a red vapour of chromyl chloride (CrO_2Cl_2) are given off.

It is a good oxidizing agent in both neutral and acidic solutions.

Potassium Dichromate + Water ⟶ Potassium Hydroxide + Chromic Hydroxide + Oxygen

$$2K_2Cr_2O_7 + H_2O \longrightarrow KOH + 4Cr(OH)_3 + 3O_2$$

Uses of Potassium Dichromate

1. As an oxidizing agent in the laboratory and industry.
2. As a valuable volumetric reagent.
3. In chrome tanning

4. In dyeing and Calico-printing*
5. As starting point for chrome alum, chrome yellow ($PbCrO_4$) chrome red ($PbO \cdot PbCrO_4$) etc.

Potassium Chromate (K_2CrO_4): This compound can be prepared by treating a solution of chromium trioxide or potassium dichromate with the calculated quantity of potassium carbonate or hydroxide and then crystallizing the solution:

$$\text{Potassium Dichromate} + \text{Potassium Hydroxide} \longrightarrow \text{Potassium Chromate} + \text{Water}$$

$$K_2Cr_2O_7 + KOH \longrightarrow K_2CrO_4 + H_2O$$

Potassium chromate reacts with alkaline in solution due to hydrolysis. Hence, potassium chromate is changed to potassium dichromate by requisite treatment of sulphuric acid.

Both potassium chromate and potassium dichromate contain no water and as a very useful reagent in volumetric analysis for oxydimetric titration as a primary standard. The two important oxidation-reduction reactions for which it is frequently used consist in the oxidation of ferrous sulphate and of potassium iodide in acid solutions.

$$\text{Potassium Dichromate} + \text{Sulphuric Acid} + \text{Ferrous Sulphate} \longrightarrow \text{Potassium Sulphate} + \text{Chromic Sulphate} + \text{Ferric Sulphate} + \text{Water}$$

$$K_2Cr_2O_7 + H_2SO_4 + FeSO_4 \longrightarrow K_2SO_4 + Cr(SO_4)_3 + 3F_2(SO_4)_3 + 7H_2O$$

$$\text{Potassium Iodide} + \text{Sulphuric Acid} + \text{Potassium Dichromate} \longrightarrow \text{Potassium Sulphate} + \text{Chromic Sulphate} + \text{Water} + \text{Iodine}$$

$$6KI + 7H_2SO_4 + 2K_2Cr_2O_7 \longrightarrow 4K_2SO_4 + Cr_2(SO_4)_3 + 7H_2O + 3I_2$$

As potassium dichromate has no action on dilute hydrochloric acid it can be used for titration of solutions containing chlorides, for which potassium permanganate becomes unstable.

Potassium dichromate in acid solution on reduction with sulphur dioxide gives equimolecular proportions chromium sulphate and potassium sulphate which deposit crystals of chrome alum, $K_2SO_4.Cr_2(SO_4)_3 \cdot 24H_2O$.

Potassium Ferrocyanide ($K[Fe(CN)_6]$): It is prepared from. (1) Nitrogenous refuse like horns, hoofs, leather, scrap etc, are heated with potassium carbonate and scrap iron. The mass is extracted with water and concentrated lemon yellow crystals are obtained.

2. *From Spent oxides:* Hydrated iron oxide used in the purification of coal gas reacts with Hydrogen cyanide, present to extract of about 0.3% in coal gas and Prussian blue ($Fe_4[Fe(CN_6]_3$). This is boiled with lime and calcium ferrocyanide formed is extracted with water. This is converted into potassium ferrocyanide by treating with potassium carbonate.

$$\text{Ferrous Hydroxide} + \text{Hydrogen Cyanide} \longrightarrow \text{Ferrous Cyanide} + \text{Water}$$

$$Fe(OH)_2 + 2HCN \longrightarrow Fe(CN)_2 + 2H_2O$$

$$\text{Ferric Oxide} + \text{Hydrogen Cyanide} \longrightarrow \text{Ferric Cyanide} + \text{Water}$$

$$Fe(OH)_2 + 3HCN \longrightarrow Fe(CN)_3 + 3H_2O$$

---

* Calcio Printing: The printing on cotton fabric.

Ferrous Cyanide + Ferric Cyanide ⟶ Ferric-ferrocynide

$$Fe(CN)_2 + Fe(CN)_3 \longrightarrow Fe_4[Fe(CN_6)]_3$$

Calcium Ferrocyanide + Potassium Carbonate ⟶ Potassium Ferrocyanide + Calcium Carbonate

$$Ca_2[Fe(CN)_6] + 2K_2CO_3 \longrightarrow K_4[Fe(CN)_6] + 2CaCO_3$$

Potassium Ferrocyanide ($K_3[Fe(CN)_6]$): It is obtained as the result of oxidation of potassium ferrocyanide, with chorine and $KMnO_4$, also nitric acid or hydrogen peroxide. It is separated from potassium chloride by fractional crystallization. It forms anhydrous dark red crystals, which gives yellowish brown solution in water.

An alkaline solution of potassium ferricyanide is a good oxidizing agent. It is converts lead monoxide to dioxide, chromic compounds to chromates and hydrogen sulphide to sulphur.

Potassium Ferricyanide + Potassium Hydroxide + Chromic Oxide ⟶ Potassium Ferrocyanide + Water + Potassium Chromate

$$6K_3[Fe(FN)_6] + 10KOH + Cr_2O_3 \longrightarrow 6K_4[Fe(CN)_6] + 5H_2O + 2K_2CrO_4$$

Uses: Potassium ferricyanide is used as an oxidizing agent in the laboratory and in the preparation of blue prints. It is also used for the analysis of ferric ions in samples.

Potassium Cyanide (KCN): It is prepared by the action of potassium ferrocyanide and potassium metal.

POISON

Potassium Ferrocyanide + Potassium ⟶ Potassium Cyanide + Iron

$$K_4[Fe(CN)_6] + 2K \longrightarrow 6KCN + Fe$$

It is a white solid soluble in water. It is treated as in lab it is used as both organic and inorganic compound, also used for the extraction of gold.

(Students are requested not to try this above reaction in **any circumstances because it is a deadly poison, and also needs a poison license)**

Potassium Chlorate ($KClO_3$): It may be prepared by passing chlorine into hot caustic potash, solution.

Chlorine + Potassium Hydroxide ⟶ Potassium Chloride + Potassium Chlorate + Water

$$3Cl_2 + 6KOH \longrightarrow 5KCl + KClO_3 + H_2O$$

As the potassium chlorate is less soluble crystallizes out first. Potassium chlorate is nowadays manufactured by the process of electrolysis. A hot and concentrated sodium chloride is electrolyzed in an undivided cell.

$$\text{Sodium Chloride} \xrightarrow{\text{Electrolysis}} \text{Sodium (cathode)} + \text{Chlorine (anode)}$$

$$NaCl \xrightarrow{\text{Electrolysis}} Na + Cl$$

$$\text{Sodium} + \text{Water} \xrightarrow{\text{at cathode}} \text{Sodium Hydroxide} + \text{Hydrogen}$$

$$2Na + H_2O \longrightarrow 2NaOH + H_2$$

Chlorine evolved at anode reacts with NaOH formed at cathode.

Sodium Hydroxide + Chlorine ⟶ Sodium Chlorate + Water

$$NaOH + 3Cl_2 \longrightarrow 5NaClO_3 + H_2O$$

Now this sodium chlorate thus obtained is treated with required quantity of potassium chloride. When potassium chlorate is obtained by double decomposition.

Potassium Chloride + Sodium Chlorate ⟶ Potassium Chlorate + Sodium Chloride

$$KCl + NaClO_3 \longrightarrow KClO_3 + NaCl$$

On cooling the less soluble potassium chlorate crystallizes out. Remaining solution of sodium 8 chloride is electrolyzed.

Note: In place of sodium chloride solution a solution of potassium chloride directly is used but the difficulty is that potassium chlorate being far less soluble than $NaClO_3$, crystallizes out during electrolysis and increases the resistance of the cell and wastes energy.

Properties of Potassium Chlorate

1. Potassium Chlorate is a white solid crystalline powder, soluble in water.
2. Potassium chlorate decomposes on heating in two stages producing oxygen:
   (*i*) $2KClO_3 + KClO_4 = KCl + O_2$ (380°C)
   (*ii*) $KClO_4 = KCl + 2O_2$ (610°C) (here a catalyst is used to reduce the temperature)
3. On heating with concentrated sulphuric acid chlorine dioxide (ClO_2) is first produced, then explodes into chlorine and oxygen.

Potassium Chlorate + Sulphuric Acid ⟶ Potassium Sulphate + Chlorine Dioxide + Water + Chloric Acid

$$2KClO_3 + 2H_2SO_4 \longrightarrow 3KHSO_4 + 2ClO_2 + H_2O + HClO_4$$

Chlorine Dioxide ⟶ Chlorine + Oxygen

$$2ClO_2 \longrightarrow Cl_2 + 2O_2$$

Potassium chlorate when heated with concentrated hydrochloric adds it gives a mixture of chlorine and chlorine dioxide called *euchlorine.*

Potassium Chlorate + Hydrochloric Acid ⟶ Potassium Cloride + Chlorine + Chlorine Dioxide + Water

$$2KClO_3 + HCl \longrightarrow 2KCl + 2ClO_2 + Cl_2 + H_2O$$

It is best known as potassium chlorate is a very powerful oxidizing agent a mixture of potassium chlorate and phosphorous is highly explosive, also a mixture of sugar and potassium chlorate catches fire when a few drops of sulphuric acid is added. All of these reactions take place because the potassium chlorate is an oxidizing agent.

Uses: Because of its oxidizing properties it is used:

1. In the manufacture of explosives and fire works also matches.
2. In medicine in the form of lozenges for sore throat treatment.
3. For the preparation of oxygen in the laboratory.
4. In Calico-printing also.

Test for Potassium

1. Flame test: When the nitrous compounds of this metal is used it produces a lilac* coloured flame due to the production of nitrogen dioxide. If the metal it self or hydroxide compound is used it produces an orangish flame.

2. Sodium tetraphenylborate: This test is quantitative, when sodium tetraphenylborate is added an immediate precipitate of potassium tetraphenylborate is produced.

CALCIUM

Chemical Symbol : Ca
Atomic Number : 20
Atomic Weight : 40.078
Melting Point : 842°C
Boiling Point : 1494°C

Discovery and Isolation

In the year 1808 Davy prepared calcium amalgam by electrolysis of a moist mixture of lime and mercuric oxide, using mercury as the negative electrode. When mercury was distilled off impure calcium was left behind.

Occurrence

Calcium is too reactive to occur in free state like sodium and potassium but its compounds occur widely and abundantly. Calcium carbonate ($CaCO_3$), is found as *marble, limestone, chalk, calcite, aragonite and Iceland spar*. Calcium sulphate ($CaSO_4$) occurs in the anhydrous state as anhydrite and as the dihydrates gypsum and alabaster. Calcium phosphate $Ca_3(PO_4)_2$ and calcium fluoride CaF are also found as minerals. Calcium is an essential constituent of many living organism, occurring mainly as skeletal material in bones, teeth sea shell and egg shells.

Extraction

Metallic calcium is obtained by the electrolysis of a mixture of calcium chloride and calcium fluoride; the compound calcium fluoride is used because it reduces melting point of the reaction mixture below that of calcium. The mixture is melted in a graphite pot, which also acts as an anode electrode, and the cathode is the iron rod kept dipping in the molten mass and is gradually raised as electrolysis proceeds. ill this way a rod of calcium metal is formed which is protected by a film of the electrolyte round it to protect it from corrosion by air or attack by the chlorine at the anode.

Properties

Calcium is a greenish white hard metal, with a silvery luster. On the exposure to the air the metallic calcium gets covered with a white powder consisting of the hydroxide radical and a trace of nitrate. The metal will burn in air to give the oxide, CaO and some nitride also. It dissolves rapidly, but violently in water evolving hydrogen. The detail physical properties are shown below:

---

* Lilac: Mauve

Physcial Properties

Melting Point	842°C
Boiling Point	1494°C
Density	1.57cc
Electrical Resitivity (at 0°C)	3.43 ohm cm
Heat of Capacity	6.43 cal deg^{-1} mol^{-1}
Standard Entropy	9.95 cal deg^{-1} mol^{-1}

Chemical Properties

Calcium violently reacts with water producing calcium hydroxide

$$\text{Calcium} + \text{Water} \longrightarrow \text{Calcium Hydroxide} + \text{Hydrogen}$$

$$Ca + 2H_2O \longrightarrow Ca(OH)_2 + H_2$$

$$\text{Calcium} + \text{Oxygen} \longrightarrow \text{Calcium Oxide (quick lime)}$$

$$2Ca + O_2 \longrightarrow 2CaO$$

$$\text{Calcium Oxide} + \text{Water} \longrightarrow \text{Calcium Hydroxide (slaked lime)}$$

$$CaO + H_2O \longrightarrow Ca(OH)_2$$

Calcium also reacts with chlorine forming calcium chloride and with ammonia it forms calcium ammoniate.

$$\text{Calcium} + \text{Chlorine} \longrightarrow \text{Calcium Chlorine}$$

$$Ca + Cl_2 \longrightarrow CaCl_2$$

Calcium Oxide, Quick Lime (CaO): It is generally prepared by heating calcium carbonate in kilns.

$$\text{Calcium Carbonate} \xrightarrow{800°C} \text{Calcium Oxide} + \text{Carbon Dioxide}$$

$$CaCO_3 \longrightarrow CaO + CO_2$$

The carbon dioxide is removed from the kiln because the reaction is reversible so to favour the forward movement of the reaction this is done. The temperature is kept steady because if it becomes high it may cause the formation of fusible silicates of calcium due to interaction of calcium and clay in the kiln.*

Properties of Calcium Oxide: Lime is a white porous solid, which is infusible and gives brilliant white limelight when heated in an oxy-acetylene flame, when added to water produces calcium hydroxide and with it produces calcium chloride and bleaching powder

Uses of Calcium Oxide

(I) It is used for manufacturing calcium carbide.

(II) For drying gases and alcohol.

*Kiln: Over or a furnace.

(III) It is also used for preparation of ammonia and soda lime (lime soaked in caustic soda).

(IV) It is used for basic lining of furnaces.

(V) It is used for making calcium hydroxide, which is used, for making caustic soda, sodium carbonate and sodium carbonate, bleaching powder and basic calcium nitrate.

(VI) It is used for the manufacture of limelight.

(VII) As milk of lime in refining sugar and white washing.

(VIII) Limewater is used as a reagent in laboratory for detection of carbon and also used in medicine.

Mortar: It is the mixture of slaked lime with three to four times its bulk of sand and enough water producing a thick paste, it is used in building as materials for bricks. It hardens on exposure to air due to loss of water, by evaporation. Also the slaked lime absorbs carbon dioxide from air and produces calcium carbonate which is very hard mass.

$$\text{Calcium Hydroxide} + \text{Carbon Dioxide} \longrightarrow \text{Calcium Carbonate}$$
$$Ca(OH)_2 + CO_2 \longrightarrow CaCO_3$$

Sand keeps the mass porous and allows evaporation, hence checks cracks.

Portland Cement: It is manufactured from limestone and clay. Powdered limestone and clay is mixed in a suitable ratio, and then heated to 1400°C to 1600°C in a rotary cement kiln fired by coal. The mass fuses to give clinkers*. The clinkers are mixed with 2 to 3% of gypsum and ground to a very fine powder, which is known as cement. This is sometimes called hydraulic cement because it has the property of setting or hardening under water. It is one of our most important building materials used today. When mixed water and sand it becomes an inter-locked crystal of calcium and aluminum silicates.

Mixed with sand gravel, cement is used for flooring and concrete and a cement construction is also imbedded with steel frame-reinforced cement.

Calcium Carbonate ($CaCO_3$): Calcium carbonate is found in Limestone, Marble, Dolomite and seashells in a large amount. Pure calcium carbonate (precipitated rock) carbon dioxide into calcium hydroxide solution or by adding a solution of calcium chloride to a solution of sodium carbonate.

It is insoluble in water but dissolves in presence of excess of carbon dioxide due to the formation of soluble calcium bicarbonate.

$$\text{Calcium Carbonate} + \text{Carbon Dioxide} + \text{Water} \longrightarrow \text{Calcium Bicarbonate}$$
$$CaCO_3 + CO_2 + H_2O \longrightarrow Ca(HCO_3)_2$$

Uses

(*a*) It is used for manufacture of lime, cement, washing soda and glass.

(*b*) As a building material.

(*c*) As chalk in paints toothpaste and powders.

---

* Clinkers: A stony matte fused together.

Calcium Carbid (CaC_2): In the year 1862 Wöhler, heated a mixture of carbon calcium and zinc he got calcium carbide as the result. Now in modem times calcium carbide is generally prepared by the Moissan's Process. In this process a mixture of calcium oxide and powdered coke is heated strongly in an electric furnace at a temperature of 3000°C

$$\text{Calcium Oxide} + \text{Carbon} \longrightarrow \text{Calcium Carbide} + \text{Carbon Monoxide}$$

$$CaO + 3C \longrightarrow CaC_2 + C$$

Calcium Sulphate ($CaSO_4$): It is also known as *Gypsum* $CaSO_4.2H_2O$. it is sparingly soluble in water and on heating its solubility decreases further. It dissolves in ammonium sulphate solution forming $CaSO_4$ $(CNH_4)SO_4$ H_2O.

Gypsum looses water on heating at a temperature of 120°C to 130°C to form a semi hydrate $CaSO_4$. $1H_2O$ or $2CaSO_4.H_2O$ which is called Plaster of Paris. When made into paste with little water it sets into a hard mass, which expands on setting and hence gives rise to fine castings. The final product of setting is $CaSO_4 \cdot 2H_2O$.

Gypsum when heated strongly looses the entire water molecule and does not get solidified like Plaster of Paris, then it is said to be 'dead burnt'.

Plaster of Paris is used for making statues and castings and immobilizing broken bones in surgery. Gypsum and important product used in paper industry.

Calcium Chloride ($CaCl_2$): It is the by product of Solvay Ammonia Soda process $CaCl_2 \cdot 6H_2O$ is a deliquescent compound, which when strongly heated at 200°C yields its anhydrous salt and it fuses at 773°C. The fused calcium chloride contains a little free lime or basic salt due to hydrolysis.

Calcium chloride is highly soluble in water and is extremely hydroscopic fused *calcium chloride also absorbs ammonia gas to yield $CaCl_2 \cdot 6NH_3$ so it is not used to dry ammonia gas, except ammonia gas all other gases are dried with help of calcium chloride.*

Uses of Calcium Chloride: (1) It is used as a reagent for drying gases (except ammonia). (2) It is also used is refrigeration plants and for and for making freezing mixtures.

Calcium Nitrate ($Ca(NO_3)_2$): It is present in the soil and is assimilated by plants. It is manufactured by neutralizing dilute nitric acid with limestone and evaporating the solution when calcium nitrate separates out.

$$\text{Calcium Carbonate} + \text{Nitric Acid} \longrightarrow \text{Calcium Nitrate} + \text{Water} + \text{Carbon Dioxide}$$

$$CaCO_3 + 2HNO_3 \longrightarrow Ca(NO_3)_2 + H_2O + CO_3$$

Calcium nitrate decomposes on heating

$$\text{Calcium Nitrate} \longrightarrow \text{Calcium Oxide} + \text{Nitrogen Dioxide} + \text{Oxygen}$$

$$2Ca(NO_3)_2 \longrightarrow 2CaO + 4NO_2 + O_2$$

When mixed with lime it gives Basic Calcium Nitrate ($Ca(NO_3)_2$). CaO also known as Norwegian saltpetre, which is extremely important chemical agent as a fertilizer in agriculture.

Bleaching Powder ($CaOCl_2$): When chlorine gas is passed over dry slaked lime, it is rapidly absorbed giving rise to what is called bleaching powder, or chloride of lime. To some extent the reaction the reaction equation comes as follows:

Calcium Hydroxide + Chlorine $\longrightarrow$ Bleaching Power + Water

$$Ca(OH)_2 + Cl_2 \longrightarrow CaOCl_2 + H_2O$$

It is manufactured by spreading slaked lime over the floors of closed lead chambers and raking the mass into furrows to expose a large surface area; dilute chlorine is passed into the chambers connected in series. The powder is turned over within wooden rakes and the absorption allowed continuing till the product contains 36% of available chlorine.

In continuous mechanical process slaked lime is made to move through a system of rotating load or iron pipes against a counter current of chlorine. The absorption of chlorine continues and a product containing 35-37% of available chlorine is obtained.

Bleaching powder is white amorphous substance and smell strongly as chlorine. It contains a good deal of unchanged slaked lime. It is not water soluble, but when boiled with water it is found that the compound contains calcium hypochlorite. It is not a definite compound and id now regarded as a mixture of calcium hypochlorite and a basic chloride of calcium. The reaction leading to its formation should therefore like below:

Calcium Hydroxide + Chlorine $\longrightarrow$ Bleaching Power + Water

$$3Ca(OH)_2 + 2Cl_2 \longrightarrow CaOCl_2, CaCl_2, Ca(OH)_2, H_2O + H_2O$$

The presence of hypochlorite in the product is shown by the fact that chlorine is liberated from bleaching powder by the action carbon dioxide or by the action of dilute acids. The powder is not hydroscopic hence it does not contain any calcium chloride and no calcium chloride can be extracted from bleaching power by the action of dilute acid corresponds to the composition of the mixture representing bleaching power, as given in the equation.

Bleaching Power + Sulphuric Acid $\longrightarrow$ Calcium Sulphate + Chlorine + Water

$$CaOCl_2, CaCl_2, Ca(OH)_2, H_2O + 3H_2SO_4 \longrightarrow 3CaSO_4 + 2Cl_2 + H_2O$$

The chlorine, which is set free by the action of dilute sulphuric acid, is better known as *available chlorine.* Bleaching powder in acid solution; therefore acts as an oxidizing agent, due to the liberation of chlorine and its value is determined by the percentage of available chlorine present in it. In the practice the percentage of available chlorine is estimated by titration of the iodine liberated from the solution of potassium iodide, acidified with dilute acid, by the addition of a given quantity of bleaching powder.

Bleaching Powder + Sulphuric Acid $\longrightarrow$ Calcium Sulphate + Chlorine + Water

$$CaOCl_2,CaCl_2,Ca(OH)_2,H_2O + 3H_2SO_4 \longrightarrow 3CaSO_4 + 2Cl_2 + H_2O$$

Bleaching powder also oxidizes sodium arsenite to sodium arsenate. In the laboratory bleaching powder is used for the preparation of chlorine, nitrogen and oxygen. Chlorine is produced by the action of sulphuric acid on bleaching powder.

Nitrogen is formed by the action of ammonia and oxygen is produced by the action of hydrogen peroxide and bleaching powder.

Bleaching Power + Ammonia ⟶ Calcium Chloride + Calcium Hydroxide + Nitrogen + Water

$$3\{CaOCl_2, CaCl_2, Ca(OH)_2, H_2O\} + 2NH_3 \longrightarrow 6CaCl_2 + Ca(OH)_2 + N_2 + 6H_2O$$

The above reaction equation is for the preparation of oxygen.

Uses of Bleaching Powder

1. The bleaching powder is largely used as an oxidizing agent;
2. It is also used for the purification and disinfecting.
3. Due to this bleaching powder it bleaches the colour of the clothes also.

Calcium Nitride (Ca_3N_2): Calcium nitride is a brownish-yellow solid obtained by passing nitrogen over fused metallic calcium. The melting point of this compound is 1195°C. The nitride is decomposed by steam with the formation of ammonia and calcium hydroxide.

Calcium + Nitrogen ⟶ Calcium Nitride

$$3Ca + Na_2 \longrightarrow Ca_3N_2$$

Calcium Nitride + Water ⟶ Calcium Hydroxide + Ammonia

$$Ca_3N_2 + 6H_2O \longrightarrow 3Ca(OH)_2 + 2NH_3$$

Calcium Superphosphate ($Ca(H_2PO_4)_2 \cdot 2CaSO_4 \cdot 2H_2O$) (Super Phosphate of lime): It is manufactured by treating normal phosphate (powdered phosphate rock, $Ca_3(PO_4)_2$ with calculated quantity of commercial sulphate acid, which reacts as:

Calcium Phosphate (normal) + Sulphuric Acid + Water ⟶ Calcium Phosphate (soluble primary phosphate) + Calcium Sulphate (Gypsum)

$$Ca_3(PO)_2 + H_2SO_4 + H_2O \longrightarrow Ca(H_2PO_4)_2 \cdot H_2O + 2CaSO_{4.2}H_2O$$

The reactants are mixed mechanically for a few minutes and then allowed to stand for 24 hours, the reaction is exothermic. Some carbon dioxide (limestone in rock) and hydrogen fluoride (from calcium fluoride) escapes. The residue is the super phosphate of lime.

Uses: As the compound is water soluble compound and is easily assimilable* by plants it is used as a fertilizer in agriculture.

Double and Triple Superphosphate of Calcium: These days double and triple superphosphates are being produced in which the amount of available phosphorous is much more as compared to that in superphosphate of lime. To manufacture triple phosphate, a weighted quantity of rock phosphate $Ca_3(PO_4)_2$ is treated with 54% phosphoric acid in a mixture. The reaction product is allowed to flow on a belt conveyer when it sets into a solid mass of triple supper phosphate.

Calcium Rock Phosphate + Phosphoric Acid ⟶ Triple Phosphate

$$Ca_3(PO_4)_2 + 4H_3PO_4 \longrightarrow 3CaH_4(PO_4)_2$$

---

* Assimilable: Taking in.

Calcium Fluoride (CaF_2): It occurs in nature as fluorspar and is issued as a flux in metallurgical operations and for the preparation of fluorine. In the laboratory it is obtained as a white sparingly soluble precipitate by the addition of a solution of sodium fluoride to that of calcium chloride or calcium nitrate.

Calcium Chloride + Sodium Fluoride ⟶ Calcium Fluoride + Sodium Chloride

$$CaCl_2 + 2NaF \longrightarrow CaF_2 + 2NaCl$$

Uses of Metallic Calcium: Metallic calcium is used as a drying agent, *e.g.* to dehydrate alcohol (which unlike sodium is does not react) as deoxidizers for metals to remove oxygen from castings as a 'getter'. That is to remove the last traces of air from high vacuum and as additive to lead to harden it for used for cable covering or in bearing metal. Calcium is also used to reduce the compounds like uranium and plutonium tetrafluoride to their respective metals, this reduction being the final stage in their extraction. The reaction refractory, once started the strong exothermic reaction maintains the reaction temperature for it to continue. The reaction s carried out in the atmosphere of argon.

Analytical Test for Calcium

(*a*) ***Flame Test:*** When a portion of Metallic calcium is introduced in a flame it produces the brick red colour.

(*b*) Ammonium Oxalate produces a white precipitate of calcium oxalate.

(*c*) Also Ammonium Carbonate produces a white precipitate of calcium carbonate.

This completes the chemistry of calcium now we shall start the chemistry of Barium, the element of alkaline earth metals.

BARIUM

Chemical Symbol : Ba
Atomic number : 56
Atomic Weight : 137.327

Melting Point : 725°C
Boiling Point : 1640°C

Alkaline Earth Metal

Occurrence

It was first prepared by Scheele at the year 1774 as an oxide from *heavy spar.* Latter in the year 1808 Berzzelius extracted the metallic element in a very impure form; Davy and Bunsen prepared late in somewhat pure form by electrolysis of the fused chloride of the element.

Barium is less available than calcium but more than strontium. Barium in nature is mostly found in two forms the heavy spar $BaSO_4$ and Whiterite $BaCO_3$.

Extraction of Barium

Industrially barium is prepared by the process of electrolysis of hot barium chloride, using the mercury as the negative electrode. The barium amalgam, which forms, is heated in the vacuum to separate the metallic barium from mercury.

Barium can also be prepared by heating barium oxide with aluminium powder or silicon at a temperature of 1300°C in an evacuated steel tube.

Properties

Barium is a shiny white, soft metal and has a lusture, which inflames continuously in air. It resembles heavier alkaline metals. Barium hydroxide is formed when barium gets quickly dissolved in water. Barium also forms barium nitride, BaN_2 with nitrogen and hydride of barium BaH_2 with hydrogen, when it is little warmed the above reactions takes place. Here the metal resembles calcium in nature. The detail physical properties are stated in the table below:

Physical Properties

Melting Point	725°C
Boiling Point	1640°C
Desnity	3.6 g/cc
Electrical Resistivity (at 0°C)	60 ohm cm
Heat of Capacity	6.30 cal deg^{-1} mol^{-1}
Standard Entropy	16 cal deg^{-1} mol^{-1}

Now we shall study the different compounds of barium.

Compounds of Barium

Barium Oxide (BaO): This compound of barium is obtained by heating barium carbonate with carbon at a temperature of 800°C.

Barium Carbonate + Carbon ⟶ Barium Oxide + Carbon Monoxide

$BaCO_3$ C ⟶ BaO 2CO

Avery high temperature is necessary to decompose barium carbonate without addition of carbon. Barium oxide forms a white infusible powder. It absorbs water readily and often used as a dehydrating agent.

Barium Peroxide (BaO_2): Barium Peroxide is technically prepared by heating the monoxide in a current of air, freed from carbon dioxide and water, at 500°C - 600°C under pressure of 2 atmosphere. Barium peroxide is formed with evolution of much heat.

Barium Oxide + Oxygen ⟶ Barium Peroxide

$2BaO + O_2 \longrightarrow 2BaO_2$

Above 800°C it decomposes back into barium monoxide and oxygen as the back reaction preponderates with the rise of temperature the reaction being exothermic. The reduction of pressure will also accelerate the decomposition of the peroxide (according to Le Chatelier's Principal) Brin's process for the preparation oxygen from air with the use of barium oxide was based upon this principle. The process has been, however, becomes obsolete now.

On treatment with dilute acids barium peroxide yields hydrogen peroxide.

Barium Peroxide + Nitric Acid $\longrightarrow$ Barium Nitrate + Hydrogen Peroxide

$$BaO_2 + HNO_3 \longrightarrow Ba(NO_3)_2 + H_2O_2$$

Barium Peroxide + Sulphuric Acid $\longrightarrow$ Barium Sulphate + Hydrogen Peroxide

$$BaO_2 + H_2SO_4 \longrightarrow BaSO_4 + H_2O_2$$

Barium Peroxide + Hydrochloric Acid $\longrightarrow$ Barium Chloride + Hydrogen Peroxide

$$BaO_2 + HCl \longrightarrow BaCl_2 + H_2O_2$$

A hydrated variety, $BaO_2 \cdot 8H_2O$, is obtained as a white precipitate by adding a solution of hydrogen peroxide to a solution of barium hydroxide, and for igniting Thermit powder, which is the mixture of aluminium powder and iron oxide powder when ignited evolves a great amount of heat, used in explosives.

Barium Hydroxide ($Ba(OH)_2$): The hydroxide of barium is made by the treatment of barium oxide and hot water:

Barium Oxide + Water $\longrightarrow$ Barium Hydroxide

$$BaO + H_2O \longrightarrow Ba(OH)_2$$

It dissolves in water sufficiently to give a strongly alkaline solution. The solution of barium hydroxide is commonly known as 'Baryta' and is used as the volumetric and is used as the volumetric agent in the laboratory. *Baryta solution is kept in such a way that it does not come in contact with atmospheric carbon dioxide; otherwise it will react with it and form barium carbonate.*

Baryta water is, therefore, used for the preparation of pure hydrogen by electrolysis.

Barium Carbonate ($BaCO_3$): This compound of barium is obtained as a white insoluble precipitate by the addition of ammonia and ammonium carbonate to a solution of barium chloride. It occurs in nature as the mineral 'Whiterite'. It is also prepared by the reduction of barytes with carbon and treating the barium sulphide obtained with carbon dioxide.

Ammonium Carbonate + Barium Chloride $\longrightarrow$ Barium Carbonate + Ammonium Chloride

$$(NH_4)_2CO_3 + BaCl_2 \longrightarrow BaCO_3 + 2NH_4Cl$$

Barium Carbonate is a white powder; it decomposes on heating at a much higher temperature.

Barium Carbonate $\longrightarrow$ Barium Oxide + Carbon Dioxide

$$BaCO_3 \longrightarrow BaO + CO_2$$

Barium Chloride ($BaCl_2$): This compound of barium is prepared as a dihydrate by dissolving whitherite in hydrochloric acid. It crystallizes from aqueous solution as $BaCh_2H_2O$. It is soluble in water, but looses water and forms the anhydrous chloride.

Barium Nitrate [$Ba(NO_3)_2$]: Barium nitrate is prepared by dissolving barium carbonate in nitric acid. It is less soluble than barium chloride and sodium nitrate:

Barium Carbonate	+	Nitric Acid	$\longrightarrow$	Barium Nitrate	+	Carbonic Acid
$BaCO_3$	+	$2HNO_3$	$\longrightarrow$	$Ba(NO_3)_2$	+	H_2CO_3
Barium Chloride	+	Sodium Nitrate	$\longrightarrow$	Barium Nitrate	+	Sodium Chloride
$BaCl_2$	+	$2NaNO_3$	$\longrightarrow$	$Ba(NO_3)_2$	+	$2NaCl$

All aqueous soluble barium salts are poisonous.

Barium nitrate is used in green fireworks.

Barium Sulphate ($BaSO_4$): This compound of barium is obtained as a fine white precipitate, insoluble in water and dilute acids, by the addition of dilute sulphuric acid or the solution of a sulphate that of barium salt.

Barium Chloride	+	Sulphuric Acid	$\longrightarrow$	Barium Sulphate	+	Hydrochloric Acid
$BaCl_2$	+	$2NaNO_3$	$\longrightarrow$	$BaSO_4$	+	$2HCl$

The precipitate is, however, somewhat soluble in concentrated hydrochloric acid or sulphuric acids. Barium sulphate naturally occurs as heavy spars or barytes.

From a solution of the substance in hot concentrated sulphuric acid, an acid sulphate, and called barium bisulphate [$Ba(HSO_4)_2$] separates out. Barium sulphate is insoluble in concentrated ammonium sulphate solution unlike calcium sulphate. Barium sulphate, being a very fine precipitate, possesses the power of carrying down with many dissolved salts by adsorption for the solution in which it is precipitated. Hence during the estimation of barium or sulphate care must be taken to minimize this adsorption. By fusion of potassium carbonate in excess, or by boiling with a strong solution of the same, barium sulphate is changed into barium carbonate.

Barium Sulphate		Potassium Carbonate	$\longrightarrow$	Barium Carbonate		Potassium Sulphate
$BaSO_4$	+	K_2CO_3	$\longrightarrow$	$BaCO_3$	+	K_2SO_4

Barium sulphate is used as a pigment, an also as permanent white the glaze for paper.

Barium Sulphide (BaS): Barium sulphate when heated with carbon barium sulphide is produced.

Barium Sulphate	+	Carbon	$\longrightarrow$	Barium Carbonate	+	Carbon Monoxide
$BaSO_4$	+	$4C$	$\longrightarrow$	BaS	+	$4CO$

In similar manner it reacts with acids to form barium salts. Barium sulphide is used to manufacture the white pigment lithopone, which is a mixture of sulphide of zinc and barium sulphate. The pigment is prepared from a solution of barium sulphate and zinc sulphide.

Analytical Test of Barium

1. *Flame test:* Barium salts impart an apple green colour in the flame
2. *Chromate Test:* Potassium chromate solution gives a yellow precipitate of barium chromate with any soluble barium salt.
3. *Sulphate Test:* Barium salts gives a white precipitate of sulphate when reacts with sulphuric acid, or any other sulphate, which is insoluble in hydrochloric acid.

This ends the chemistry of the barium and we will now go to the next alkaline earth metal the magnesium.

MAGNESIUM

Chemical Symbol : Mg
Atomic Number : 12
Alkaline Earth Metal

Melting Point : 651°C
Boiling Point : 1 107°C
Atomic Weight : 24.3050

Discovery and Isolation

Magnesium sulphate (Epsom salt) ($MgSO_4.7H_2O$) was discovered in spring water at Epsom in 1695. Impure metallic magnesium was obtained electrolytically by Davy in 1808.

Occurrence

Magnesium is not found free but it's compounds are widely distributed and plentiful of Dolomite, [$MgCa(CO_3)_2$] constitutes the Dolomite and other metals like calcium is also present in it. Carnallite ($KC1.MgCl_6H_2O$) occurs in deposits in Germany; other minerals like Manesite ($MgCO_3$) and asbestos [$CaMg(SiO_3)_4$]. Magnesium halides are found in sea waters and magnesium is an essential constituent of chlorophyll, the green coloured pigment present in plants.

Extraction

Magnesium is obtained commercially by the electrolysis of the fused anhydrous magnesium chloride, mixed with some sodium chloride to lower melting point (or potassium chloride if the source is carnellite. Small amounts of alkaline earth fluorides are added to improve the yield of magnesium. The first problem is heated four molecules of water are lost fairly readily but attempts to remove the other two molecules result in the formation by hydrolysis of basic salt.

The anhydrous chloride may therefore be made by heating crushed magnesite with coke and passing chlorine over it the magnesite is first decomposed to give the oxide, which is then reduced.

$$\text{Magnesium Carbonate} \rightleftharpoons \text{Magnesium Oxide} + \text{Carbon Dioxide}$$
$$MgCO_3 \rightleftharpoons MgO \quad CO_2$$
$$\text{Magnesium Oxide} + \text{Carbon} + \text{Chlorine} \longrightarrow \text{Magnesium Chloride} + \text{Carbon Monoxide}$$
$$MgO + C + Cl_2 \longrightarrow MgCl_2 + CO$$

In the electrolysis cell a carbon anode and iron or steel rod is used as cathode. The magnesium is collected as drops on the cathode surface, these drops grow together and rise to the surface of the melt from which a layer of magnesium can be collected.

Since molten magnesium reacts with oxygen and nitrogen coal gas is used as active atmosphere.

Magnesium is also extracted from seawater in a large plant in the Gulf of Mexico. The first stage is to precipitate magnesium hydroxide by addition of lime.

Magnesium Chloride + Calcium Hydroxide ⟶ Magnesium Hydroxide + Calcium Chloride

$$MgCl_2 + Ca(OH)_2 \longrightarrow Mg(OH)_2 + CaCl_2$$

The hydride is then converts to the to the chloride and the latter crystallizes as far as possible without hydrolysis and this in completely dehydrated chloride is then added to the fixed chloride in the electrolytic cell. Each ton of magnesium requires the treatment of about 849.5055cubic metre of seawater.

Another method for the large-scale production of magnesium uses the reversible reaction of magnesium oxide carbon.

Magnesium Oxide + Carbon ⟶ Magnesium + Carbon Monoxide

$$MgO + C \longrightarrow Mg + CO$$

At 2000°C both products of the forward reaction are gaseous and would normally recombine but if the products are cooled rapidly, magnesium vapour is frozen out and the forward reaction goes to completion.

Propertiers

Magnesium is a white lustrous soft ductile (*e.g.* magnesium ribbon). It tarnishes somewhat in moist air, a film of the oxide, gets formed. On heating of the ribbon in the air magnesium burns with a brilliant white light forming both the oxide and the nitride (MgO and Mg_3N_2). These compounds appear as a white smoke, which can be collected as deposits on any flat polished surface as thick white powder. The later has been used as standard of whiteness and also to study the size of liquid droplets formed when a jet of liquid breaks up; the droplets are allowed to full on the white layer where they produce circular depressions produced are proportional to their sizes. Magnesium will burn also in steam and in carbon dioxide.

Magnesium + Water ⟶ Magnesium Oxide + Hydrogen

$$Mg + H_2O \longrightarrow MgO + H_2$$

Form the electrochemical series, it would be expected that magnesium would react with cold, however a layer of magnesium hydroxide [$Mg(OH)_2$] reduces the rate of attack of water on the metal, which therefore dissolves only at a very slow rate. Dilute acids including nitric acid and solutions of salts, which are acids by hydrolysis, readily, attack magnesium to give hydrogen. The detailed physical properties of the magnesium are shown in the table below:

Physical Properties

Melting Point	651°C
Boiling Point	1107°C
Density	1.74 g/cc
Electrical Resistivity	3.9 ohm cm
Heat of Capacity	5.71 cal deg^{-1} mol^{-1}
Standard Entropy	7.77 cal deg^{-1} mol^{-1}

Compounds of Magnesium

Now we shall study about some of the compounds of magnesium.

Magnesium Oxide (MgO): This is also called magnesia; it is prepared by heating magnesite in air.

$$\text{Magnesite} \longrightarrow \text{Magnesium Oxide} + \text{Carbon Dioxide}$$

$$MgCO_3 \longrightarrow MgO + CO_2$$

It is also formed when magnesium burn is oxygen. Magnesium hydroxide, magnesium nitrate and basic carbonate of magnesium also leave a residue of magnesium oxide on heating.

Magnesium oxide is a white powder, very refractory, which a high melting point of 2600°C. It is a basic oxide and reacts with acid to form salts and water. It is slowly changed by water to magnesium hydroxide, which is sparingly soluble, through suspension is alkaline to litmus. At very high temperature in the electric furnace it reacts with carbon dioxide from air.

Magnesium is used for making basic refractory bricks for furnace linings. It is moistened with water and mixed with magnesium chloride to form a hard cement (when sets in solid) it is known as Sorrel's cement it is used as the stopping of tooth.

With water it forms the hydroxide and with acids it forms basic salts.

$$\text{Magnesium Oxide} + \text{Water} \longrightarrow \text{Magnesium Hydroxide}$$

$$MgO + H_2O \longrightarrow Mg(OH)_2$$

$$\text{Magnesium Oxide} + \text{Hydrochloric Acid} \longrightarrow \text{Magnesium Chloride} + \text{Water}$$

$$MgO + 2HCl \longrightarrow MgCl_2 + H_2O$$

Magnesium is also used as an insulating material for lagging gas pipes for boilers.

Magnesium Hydroxide, Mg(OH)₂: When a solution of sodium hydroxide is added to that of magnesium chloride or sulphate, magnesium hydroxide is produced as a white precipitate.

$$\text{Magnesium Chloride} + \text{Sodium Hydroxide} \longrightarrow \text{Magnesium Hydroxide} + \text{Sodium Chloride}$$

$$MgCl_2 + 2NaOH \longrightarrow Mg(OH)_2 + 2NaCl$$

The precipitate is insoluble in alkali, but dissolves slowly in ammonium chloride solution. This is probably due to the continuous removal of hydroxyl ions in the form of slightly dissociated ammonium hydroxide. Magnesium hydroxide is sparingly soluble in water, the water is saturated with magnesium hydroxide, which is being a strong base dissociates more or less completely into magnesium and hydroxyl ions. The following equilibrium, therefore occurs in the system:

$$\underset{\text{(solid)}}{Ma(OH)_2} \rightleftharpoons \underset{\text{(liquid)}}{Mg(OH)_2} \rightleftharpoons \underset{\text{(ions in the solution)}}{Mg^{++} + 2OH^-}$$

$$NH_4Cl \rightleftharpoons NH_4^+ + Cl^-$$

$$NH_4^+ + OH \rightleftharpoons NH_4OH^-$$

As hydroxyl ions are removed as ammonium hydroxide solution, more magnesium hydroxide passes into the solution to keep up the equilibrium represented by the first equation. The following equilibrium, therefore, occurs in the system.

$$\underset{\text{Hydroxide}}{\text{Magnesium}} \quad \underset{\text{Chloride}}{\text{Ammonium}} \rightleftharpoons \underset{\text{Chloride}}{\text{Magnesium}} + \underset{\text{Hydroxide}}{\text{Ammonium}}$$

$$Mg(OH)_2 + NH_4Cl \rightleftharpoons MgCl_2 + 2NH_4OH$$

The process continues till all the magnesium hydroxide dissolves. Magnesium hydroxide is therefore completely precipitated by ammonia and not at all in presence of ammonium chloride.

An aqueous solution of magnesium hydroxide reacts strongly alkaline to litmus and dissolves in acids to form salts. It slowly absorbs carbon dioxide from air. When heated it looses water and forms magnesium oxide.

Magnesium Carbonate ($MgCO_3$): The normal carbonate occurs in nature as magnesite; which is isomorphous with calcite ($CaCO_3$).

The white precipitates formed by the addition of a solution of sodium or potassium carbonate to a solution of magnesium hydroxide. Magnesium hydroxide and carbonate are known as magnesia Alva Levis and magnesia Alva ponderosa respectively. Both have same approximate composition represented by $MgCO_3 \cdot Mg(OH)_2$ The basic carbonates dissolves when carbon dioxide is passed into their suspension in water, forming the bicarbonate.

$$\text{Magnesium Hydroxide + Carbon Dioxide + Water} \longrightarrow \text{Magnesium Carbonate}$$
$$\text{Magnesium Carbonate}$$

$$Mg(OH)_2\ MgCO_3 + 3CO_2 + H_2O \longrightarrow 2Mg(HCO_3)_2$$

If ammonium carbonate is added to a solution of magnesium salt containing ammonium chloride, no precipitation occurs; for, ammonium carbonate in aqueous solution is more or less completely hydrolyzed into ammonium hydroxide and ammonium bicarbonate.

Magnesite and the basic carbonate, when heated, decompose into magnesium oxide and carbon dioxide.

Magnesium Sulphate (Epsom salt) ($MgSO_47H_2O$): It occurs in mineral springs at Epsom and hence the name. It is prepared from Kieserite $MgSO_4 \cdot H_2O$ (of Stassfort deposits), which is dissolved in water and crystallized.

$$\text{Magnesium Oxide} \quad \text{Sulphuric Acid} \longrightarrow \text{Magnesium Sulphate} \quad \text{Water}$$

$$MgO + H_2SO_4 \longrightarrow MgCO_3 + H_2O$$

$$\underset{\text{Carbonate}}{\text{Magnesium}} + \underset{\text{Acid}}{\text{Sulphuric}} \longrightarrow \underset{\text{Sulphate}}{\text{Magnesium}} + \text{Water} + \underset{\text{Dioxide}}{\text{Carbon}}$$

$$MgCO_3 + H_2SO_4 \longrightarrow MgSO_4 + H_2O + CO_2$$

It is colourless crystalline compound, soluble in water. When heated, it looses water and becomes anhydrous.

Uses of Epsom Salt

1. In dressing cotton goods, in dyeing, tanning and in medicine as purgative.
2. As a catalyst in the contact process of sulphuric acid manufacture.
3. As magnesia mixture (solution of magnesium sulphate + ammonium chloride + Hydrochloric acid) in laboratory for the detection of phosphate radical.

Magnesium Chloride ($MgCl_2$ $6H_2O$): When a solution of magnesite in dilute solution of hydrochloric acid is cooled, colourless prismatic crystals of the hexahydrate, $MgCl_2.6H_2O$ are deposited. Hydrates contain 12, 8 and 4 molecules of water have also been prepared. The hexahydrate is very deliquescent and when heated it is hydrolyzed and converted into oxychloride with loss of water and hydrochloric acid. The other hydrates also behave also in the same manner of heating.

The anhydrous chloride, $MgCl_2$, is obtained by heating the hydrated chloride in a steam of hydrogen chloride gas or by heating the double salt on drying, then the ammonium chloride sublimes away, leaving behind the anhydrous magnesium chloride. The anhydrous chloride is also formed when the hexahydrate is carefully heated in vacuum at 175°C. The technical method of preparing the anhydrous chloride consists of the metal and carbon, in a current of chlorine

$$\text{Magnesium Oxide} + \text{Chlorine} + \text{Carbon} \longrightarrow \text{Magnesium Chloride} + \text{Carbon Monoxide}$$
$$MgO + Cl_2 + C \longrightarrow MgCl_2 + CO$$

A paste of magnesium oxide and concentrated solution of magnesium chloride sets on standing to a hard mass, known as Sorrel cement. The actual chemical composition of this compound is $MgCl_2.5H_2O.17H_2O$.

Magnesium Perchlorate [$Mg(ClO_4)_2$]: This is prepared from magnesium hydroxide and perchloric acid. It is used as the most efficient drying agent, under the name of *Anhydrone*. On absorbing the moisture it forms a hexahydrate, which becomes anhydrous again when heated to 250°C. Thus the product can be regenerated easily after use. A trihydrate is also formed as an intermediate product, which is also used a desiccating agent.

Magnesium Nitride (Mg_3N_2): This compound of magnesium is produced as a colourless crystalline compound by passing nitrogen or ammonia over metallic ammonia over metallic magnesium, heated to redness.

$$\text{Magnesium} + \text{Ammonia} \longrightarrow \text{Magnesium Nitride} + \text{Hydrogen}$$
$$3Mg + 2NH_3 \longrightarrow Mg_3N_2 + 3H_2$$

It is hydrolyzed to magnesium oxide and ammonia with reaction to water.

$$\text{Magnesium Nitride} + \text{Water} \longrightarrow \text{Magensium Hydroxide} + \text{Ammonium Hydroxide}$$
$$Mg_3N_2 + H_2O \longrightarrow Mg(OH)_2 + 2NH_4OH$$

Magnesium Phosphate ($MgHPO_4$,$7H_2O$): A white precipitate of magnesium hydrogen phosphate is obtained by adding a solution of disodium hydrogen phosphate, Na_2HPO_4, to a solution of magnesium chloride in the cold. On heating the mixture, the precipitate changes into the insoluble normal phosphate and a soluble acid salt.

Magnesium Phosphate ⟶ Magnesium Phosphate (tribasic) + Magnesium Phosphate (Monobasic)

$$4MgHPO_4 \longrightarrow Mg_3(PO_4)_2 + MgH_4(PO_4)_2$$

STRONTIUM

Chemical Symbol : Sr

Atomic Number : 38

Alkaline Earth Metal

MeltingPoint : 769°C

Boiling Pint : 1390°C

Atomic Weight : 87.62

History and Occurrence

Strontium was rather prepared in the impure state in the year 1808. It occurs in small quantities in many calcium minerals. There are very few strontium minerals, they are *Strontianite* ($SrCO_3$); *Celestine* ($SrSO_4$).

Preparation and Properties

The process of electrolysis of a fused mixture strontium chloride and potassium chloride can prepare strontium metal. It can also be prepared by heating strontium oxide with finely crushed aluminium in vacuum at a temperature of 1000°C

Strontium is a lustrous, silver white metal, which in air readily turns yellow; it is almost active as calcium. The detail physical properties of strontium are given below.

Physical Properties

Melting Point	769°C
Boiling Point	1390°C
Density	2.54 g/cc
Electrical Resistivity (at 0°C)	23 ohm cc
Heat of Capacity	6.0 cal deg^{-1} mol^{-1}
Standard Entropy	13.0 cal deg^{-1} mol^{-1}

We shall now study the different compounds of strontium.

Strontium Oxide (SrO): Strontium oxide is prepared by heating strontium nitrate at a temperature of 700°C.

It forms a white powder, which absorbs moisture and carbon dioxide and moisture very easily.

Strontium Hydroxide [$Sr(OH)_2$]: When a strontium salt is treated with a solution of sodium hydroxide, the octahydrate $Sr(OH)_2 8H_2O$, crystallizes out from the solution. It is a strong base and is employed for the recovery of sugar from the molasses.

Strontium Peroxide (SrO_2): This compound of strontium is formed when strontium oxide is heated at 400°C in a current of oxygen under a pressure of 90 atms. It can also be obtained as a precipitate by the addition of ammonical hydrogen peroxide solution to that of strontium at a temperature of 50°C. It is thus obtained as a white powder of the composition of $SrO_2.8H_2O$.

Strontium Hydride (SrH_2): This compound of strontium is prepared from the oxide salt of strontium, the oxide is heated at the temperature of 700°C in a current of hydrogen, which is converted to hydride of the strontium.

Strontium Nitrate [$Sr(NO_3)_2$ $4H_2O$]: It is obtained by dissolving the carbonate in dilute nitric acid and is used in fire works for producing crimson coloured flame.

Strontium Chloride ($SrCl_2$ $6H_2O$): It is prepared by dissolving the carbonate in dilute hydrochloric acid and evaporating the solution. It is readily dissolved in water but less deliquescent than calcium chloride.

Strontium Carbonate ($SrCO_3$): It occurs in nature as Strontianite and also can be prepared commercially from Celestine $SrSO_4$ by fusion with sodium carbonate.

Strontium Sulphate + Sodium Carbonate ⟶ Sodium Sulphate + Strotium Carbonate

$$SrSO_4 + Na_2CO_3 \longrightarrow Na_2SO_4 + SrCO_3$$

It resembles calcium carbonate but decomposes on heating to a still higher temperature.

Analytical Tests for Strontium

1. *Flame Test:* Strontium compounds impart crimson colour to the flame.
2. *Sulphate Test:* It is done adding ammonium sulphate solution to a strontium salt solution.

This completes the chemistry of strontium now we shall start about the Beryllium compound.

BERYLLIUM

Chemical Symbol : Be
AtomicNumber : 4
Atomic Weight : 9.012
Alkaline Earth Metal

Melting Point : 1283°C
Boiling Point : 2970°C

Occurrence

Vauquelin discovered beryllium in 1707. It was prepared by the action of potassium on beryllium chloride by Wöhler in 1828.

The various minerals of beryllium are as follows:

(*i*) Beryl, 3BeO, $Al_2O_36SiO_2$; it contains about 11 % to 15% Beryllium oxide.

(*ii*) Crysoberyl, BeO, Al_2O_3; It contains about 7% Beryllium oxide.

(*iii*) Phenacite, Be_2SiO_4 it contains about 45% of Beryllium oxide.

(*iv*) Brommelite, it also contains 45% of beryllium oxide.

Extraction

Beryllium is extracted from the mineral beryl by two processes such as

1. Electrolysis Reduction Method: This method includes following steps:

***(A) Conversion of Fluorides*:** The process is known as *Capaixis* method. In this method the finely powdered beryl is heated with sodium fluosilicate, (Na_2SiF_6) to a temperature of 800 - 850°C when the double fluoride of sodium and beryllium Na_2BeF_4 is obtained.

$$\text{Beryllium Oxide} + \text{Silicon Tetrafluoride} \longrightarrow \text{Silicon Dioxide} + \text{Beryllium Fluoride}$$

$$BeO + SiF_4 \longrightarrow SiO_2 + 2BeF_2$$

$$\text{Sodium Fluoride} + \text{Beryllium Fluoride} \longrightarrow \text{Double Fluoride of sodium and Beryllium}$$

$$NaF + 2BeF_2 \longrightarrow Na_2BeF_4$$

The aluminium present in the beryl ore is covered into insoluble sodium aluminium fluoride (Na_3AlF_6).

***(B) Removal of Aluminium*:** The reaction product obtained from the step (*A*) is mixture of sodium beryllium fluoride and sodium aluminium fluoride. When the reaction mixture is extracted with boiling water, sodium beryllium fluoride goes into the solution leaving the bulk of sodium aluminium fluoride (Na_3AIF_6) as the insoluble residue.

***(C) Electrolysis of Beryllium Fluoride*:** The aqueous extract obtained from the step (*B*) is evaporated to dryness. The solid mass is mixed with sodium fluoride and beryllium fluoride and fused to give a mixture, which is less volatile at higher temperature and is able to conduct the electric current reasonably well.

The fused mixture is electrolyzed in a graphite crucible which acts as anode.

The cathode consists of a water-cooled iron pipe dipping in the melt. When the electricity is passed and the cathode is continuously raised slowly, a thick rod of beryllium is obtained. At regular intervals, double fluoride of sodium and beryllium is added to maintain the concentration of beryllium in the bath constant.

2. Chemical Reduction Method: This method is specially used for manufacturing alloys of beryllium. This method involves the simultaneous reduction of beryllium and metallic oxides by carbon in an electric furnace. This process involves the following steps.

(*i*) Conversion to Beryl Oxide: Beryl is made to fuse in an electrical arc furnace at about 1550°C and poured into cold water. A crystalline mass is formed which is crushed and digested with concentrated sulphuric acid when a mixture of silica, beryllium and aluminum sulphate is obtained.

$$BeO{\cdot}Al_2O_{3.6}SiO_2 + 6H_2SO_4 \longrightarrow 3BeSO_4 + Al_2(SO_4)_3 + 6SiO_2 + H_2O$$

The sulphates obtained above are first separated out with water and then made to treat with ammonium sulphate when a sparingly soluble double salt (soda alum) is obtained from the filtrate, which is purified. It is ignited at 1350°C to give beryllium. (See Fig. 15.3).

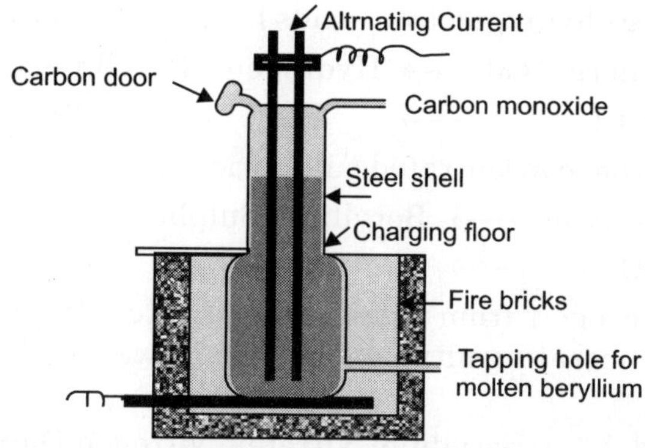

Fig. 15.3. The Process of Beryllium Oxide Redution in an Electrical Arc Furnace

***(ii)* Reduction of Beryllium Oxide:** Beryllium oxide by the above method is mixed with coke. This furnace is enclosed in firebricks and is provided with a tapping hole (as shown in the figure). It serves as one of the electrodes; Beryllium oxide gets reduced to beryllium.

$$\text{Beryllium Oxide} + \text{Carbon} \longrightarrow \text{Beryllium} + \text{Carbon Monoxide}$$

$$BeO + C \longrightarrow Be + CO$$

Properties of Beryllium

The properties of beryllium are divided into (*i*) physical property and (*ii*) chemical property.

Physical Properties

1. It is a silvery white metal.
2. It is as hard as quartz it takes up polish very easily.
3. It is one of the light metals having the conductivity 1/12th of the copper. The detail properties are shown in the table below:

Melting Point	1283°C
Boiling Point	2970°C
Density	1.85 gm/cc
Specific Heat	
Heat of Capacity	4.28 cal deg^{-1} mol^{-1}
Standard of Entropy	2.28 cal deg^{-1} mol^{-1}

Chemical Properties

1. Air when dry do not attract beryllium, but moist air slowly oxidizes beryllium slowly gets oxidized. When heated in oxygen a layer of oxide forms on the elements surface as a protective surface. Beryllium bums in air at a temperature of 1250°C brilliantly to form the oxide of beryllium and nitride of beryllium.

2. Water does not react with beryllium even on strong heating.
3. Beryllium displaces hydrogen from acids like hydrochloric and sulphuric.

Beryllium + Hydrochloric Acid ⟶ Hydrogen + Beryllium Chloride

$$Be + HCl \longrightarrow H_2 + BeCl_2$$

It liberates sulphur from concentrated sulphuric acid.

Beryllium + Sulphuric Acid ⟶ Beryllium Sulphate + Sulphur Dioxide + Water

$$Be + H_2SO_4 \longrightarrow BeSO_4 + SO_4 + H_2O$$

It also displaces hydrogen from nitric acid (dilute). It reacts violently with hot concentrated nitric acid, liberating nitrogen dioxide. However, all concentrated nitric acid makes it passive.

Beryllium + Nitric Acid ⟶ Beryllium Nitrate + Nitrogen Dioxide + Water

$$Be + 4HNO_3 \longrightarrow Be(NO_3)_2 + 2NO_2 + 2H_2O$$

Beryllium is slowly attracted by alkalis forming fused beryllates and liberating hydrogen.

Beryllium + Sodium Hydroxide ⟶ Sodium Beryllate + Hydrogen

$$Be + 2NaOH \longrightarrow Na_2BeO_2 + H_2$$

It reacts with halogen also halides. Now we shall about the compounds of beryllium.

Compounds of Beryllium

Beryllium Monoxide (BeO): This compound is obtained by igniting beryllium hydroxide or carbonate. It melts at a temperature of 1400°C. The oxide is weakly basic and possesses some acidic properties as well, since it is soluble in aqueous solution of alkalis. It is however, reprecipitated from its alkaline solution on boiling.

Beryllium Hydroxide [Be(OH)$_2$]: Beryllium hydroxide is obtained as a white gelatinous substance solid precipitate from solutions of beryllium salts by alkalis or ammonia. The precipitate is soluble in excess of caustic alkalis forming beryllates.

Beryllium Hydroxide + Sodium Hydroxide ⟶ Sodium Beryllite + Water

$$Be(OH)_2 + NaOH \longrightarrow Na_2BeO_2 + 2H_2O$$

In this it resembles aluminum hydroxide, but unlike the latter, beryllium hydroxide is reprecipitated from the alkaline solution on boiling.

Beryllium Fluoride (BeF$_2$): This compound of beryllium is prepared by heating the double fluoride $BeF_2.2NH_4F$ in the absence of air. It is a glassy hydroscopic substance, readily soluble in water.

Double Fluoride ⟶ Ammonia + Beryllium Fluoride + Hydrogen Fluoride

$$BeF_2 \cdot 2NH_4F \longrightarrow 2NH_3 + BeF_2 + 2HF$$

Beryllium Chloride, BeCl$_2$: This compound of beryllium is prepared by the action of hydrochloric acid on the metal.

Beryllium + Hydrochloric Acid ⟶ Beryllium Chloride + Hydrogen

$$Be + HCl \longrightarrow BeCl_2 + H_2$$

It forms an extremely hydroscopic white solid, which is readily hydrolyzed in aqueous solution. The melting temperature of this compound is 400°C.

Beryllium Sulphate [$BeSO_4$ $4H_2O$]: A solution of beryllium oxide in hot concentrated sulphuric acid deposits on cooling, crystals of $BeSO_4$ is obtained. It crystallizes from water as $BeSO_4.4H_2O$. In aqueous solution the salt is hydrolyzed.

Beryllium Nitrate [$Be(NO_3)$]: The compound beryllium nitrate is prepared by the action of nitric acid of the element itself where the element replaces the hydrogen from the acid:

Beryllium + Nitric Acid ⟶ Beryllium Nitrate + Hydrogen

$$Be + HCl \longrightarrow Be(NO_3)_2 + H_2$$

Beryllium nitrate is extremely hydroscopic. Beryllium nitrate also shows a tendency to form complexes with organic compounds.

Basic Beryllium Acetate [$Be_4O(CH_3CO_2)_6$]: This compound is obtained by evaporating a solution of beryllium hydroxide in acetic acid, is ready volatile. It is also known as *AnI.*

Beryllium Carbide (Be_2C): This compound of beryllium is prepared by direct interaction of beryllium and carbon.

Beryllium Carbon ⟶ Beryllium Carbide

$$Be + C \longrightarrow BeC_2$$

This compound is found in hard crystalline forms. It is attacked vigorously by strong alkali solutions forming methane gas and alkline beryllate. The compound decomposes at 1500°C.1t is used in nuclear reactor cores.

Beryllium-Copper Alloy: Alloy of copper and beryllium the actual ratio is

copper: beryllium: nickel :: 97.4 : 2.25 : 0.35.

It is used in electrical parts, for making contact points corrosion free.

Beryllium Hydride (BeH_2): This also prepared by direct action of the elements.

Beryllium + Hydrogen ⟶ Beryllium Hydride

$$Be + H_2 \longrightarrow BeH_2$$

It is white solid reacts violently with water. It also reacts with the dilute acids and methanol forming hydrogen. It also decomposes at 220°C. It is used in as fuels in experimental rockets.

Beryllium Metaphosphates [$Be(PO_3)_2$]: This compound is prepared by the action of beryllium oxide and phosphoric acid.

Beryllium Oxide + Phosphoric Acid ⟶ Beryllium Phosphate + Water

$$BeO + 2H_3PO_4 \longrightarrow Be(PO_3)_2 + 3H_2O$$

This is very toxic compound of beryllium. It has a high melting point, it is also insoluble in water. It is used for special ceramic compositions, also as catalyst carrier.

Beryllium Nitride (Be_3N_2): This compound of beryllium is also prepared by the direct reaction of nitrogen and beryllium.

$$\text{Beryllium} + \text{Nitrogen} \longrightarrow \text{Beryllium Nitride}$$

$$3Be + N_2 \longrightarrow Be_3N_2$$

There are also some double salts of beryllium they are beryllium potassium sulphate ($BeSO_4.K_2SO_4$); beryllium sodium fluoride ($BeF_2.2NaF$). These are also called associated compounds of beryllium.

CHAPTER

16

Heavy Metals

Introduction

In this chapter we are going to discuss the chemistry of those metals, which are having higher atomic weights. We shall study their physical and chemical properties also their uses and industrial preparations. This chapter also includes the different compounds of these elements.

ALUMINIUM

Atomic Weight : 26.88
Atomic Number : 13

Melting Point : 659°C
Boiling Point : 2450°C

Occurrence

Aluminium occurs in its natural ore called Bauxite ($AI_2O_3.H_2O$). It is also available from Cryolite ($NaAlF_6$) and Corundum (Al_2O_3). Next to silica aluminum is the widely distributed element present in India.

Physical Properties

Aluminium is bluish white metal having a brilliant lusture, which however is soon destroyed due to the formation of an oxide layer. It is malleable and ductile and can be rolled into sheets, foils and wires. It is a very good conductor of electricity and heat next to copper. It is a very lightweight metal with a specific gravity of 2. 7. It melts at 659°C and boils at 2450°C.

Extraction of Aluminium

Aluminium is produced from bauxite in two stages normaly:

1. Extraction of Alumina from Bauxite and
2. Reduction of Alumina into Aluminium.

1. Extraction of Alumina from Bauxite: The Bayer's Process of extraction of alumina from bauxite is adopted (See Fig. 16.1).

Bauxite and Alumina

Alumina (aluminium oxide Al_2O_3) is a fine white material similar in appearance to salt. While alumina is also used in abrasive, ceramics and refractory industries, Queensland Alumina Limited produces only smelter-grade alumina for the owners reduction plants in Australia and overseas.

The QAL process was designed to refine bauxite located at Weipa in North Queensland. The extensive deposits of this ore were formed by weathering of sands and rocks millions of years ago, increasing the alumina content as other more soluble elements were removed.

Bauxite occurs close to the surface in seams varying from one metre to nine metres, formed as small reddish pebbles (pisolites). The ore is shipped to Gladstone following "beneficiation" to remove low-grade material, and blending to provide a consistent grade.

pisolites

aluminium oxide

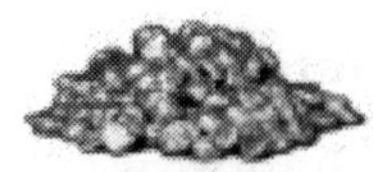
aluminium

The Bayer Process

An economical method of producing aluminium oxide - was discovered by an Austrian chemist Karl Bayer and patented in 1887.

The process dissolves the aluminium component of bauxite ore in sodium hydroxide (caustic soda); removes impurities from the solution; and precipitates alumina trihydrate, which is then calcined to aluminium oxide.

A Bayer Process plant is principally a device for heating and cooling a large recirculating stream of caustic soda solution. Bauxite is added at the high temperature point, red mud is separated at an intermediate temperature, and alumina is precipitated at the low temperature point in the cycle.

Bauxite usually consist of two forms of alumina - a monhydrate form Boehmite ($Al_2O_3 \cdot H_2O$) and a trihydrate form Gibbsite ($Al_2O_3 \cdot 3H_2O$).

The QAL uses the Bayer Process to refine two grades of Weipa bauxite, the bulk of which is "monohydrate" grade bauxite.

Average analyses of Weipa bauxites

Constituent	Monohydrate Grade %	Trihydrate Grade %
Al_2O_3	55	50
Available Al_2O_3	50*	44#
Fe_2O_3	12	17
SiO_2	5	4
TiO_2	3	3
Other (mainly H_2O)	25	26

*40% from Gibbsite and 10% from Boehmite.

Gibbsite only is extractable in sweetening.

Boehmite requires elevated temperatures (above 200°C) to dissolve readily in 10% sodium hydroxide solution. The trihydrate grade bauxite is mainly Gibbsite, which dissolves readily in 10% sodium hydroxide solution at temperatures below 150°C. Consequently, monohydrate bauxite undergoes high temperature extraction under pressure in digesters, while trihydrate grade material is added as a "sweetening bauxite" to the flash tanks where temperatures are less than 200°C. The design of the plant meets the requirement of smelters of coarse or sandy alumina for reduction to aluminium. The recovery rate is about one tonne of alumina per 2.2 tonnes of bauxite. From the plants, million-tonne bauxite stockpile to the A-frame alumina storage sheds is a processing journey of about 2.5 days. The QAL plant circulates some 550 million litres of caustic soda solution through four distinct stages, the functions of which are detailed in this process description.

PROCESS

Digestion of Bauxite

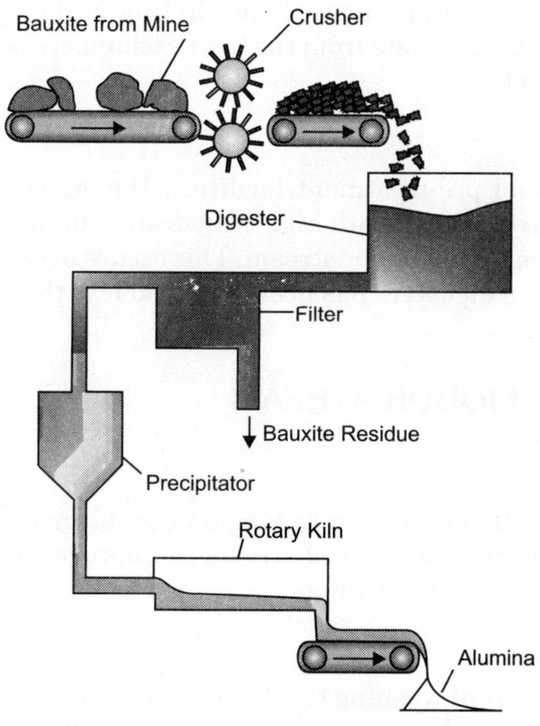

Fig. 16.1. The Bayer's Process

Grinding

Pisolitic, monohydrate-grade bauxite sized to a maximum of 20mm, is ground in 10 mills (each with one compartment of rods and one of balls) to allow better solid liquid contact during digestion. Recycled caustic soda solution is added to produce a pumpable slurry, and lime is introduced for phosphate control and mud conditioning.

Desilication

The silica component of the bauxite is chemically attacked by caustic soda, causing alumina and soda losses by combining to form solid desilication products. To desilicate the slurry prior to digestion, it is heated and held at atmospheric pressure in pre-treatment tanks, reducing the build-up of scale in tanks and pipes. Most desilication products pass out with the mud waste as sodium aluminium silicate compounds.

Digestion

The plant has three digestion units. The monohydrate slurry is pumped by high pressure pumps through two agitated, vertical digester vessels operating in series. Mixed with steam and caustic solution, alumina in the bauxite forms a concentrated sodium aluminate solution leaving undissolved impurities, principally inert iron and titanium oxides and silica compounds. Reaction conditions to extract the monohydrate alumina are about 250°C and a pressure about 3500 kPa, achieved by steam generated at 5000 kPa in coal-fired boilers. Under these conditions, the chemical reactions are rapid:

$$2NaOH + Al_2O_3 \cdot 3H_2O \longrightarrow 2NaAlO_2 + 4H_2O$$

$$2NaOH + Al_2O_3 \cdot H_2O \longrightarrow 2NaAlO_2 + 2H_2O$$

By sizing the vessel to optimum holding time, about 97% of the total available alumina is extracted and the silica content of liquor is reduced.

Heat Recovery

After digestion about 30% of the bauxite mass remains in suspension as a thin red mud slurry of silicates, and oxides of iron and titanium. The mud-laden liquor leaving the digestion vessel is flash-cooled to atmospheric boiling point by flowing through a series of flash vessels, which operate at successively lower pressures.

The flash steam generated is used to preheat incoming caustic liquor in tubular heat exchangers located parallel to the flash tank line. Condensate from the heat exchangers is used for boiler feed water and washing waste mud.

Sweetening

The trihydrate bauxite has separate grinding and pre-treatment facilities. During the pass through the flash tanks, this additional bauxite slurry with high trihydrate alumina content is injected to maximise the alumina content of the liquor stream. This occurs in the appropriate flash vessels when the slurry from the digesters has been cooled to less than 200°C.

CLARIFICATION OF THE LIQUOR STREAM

Settlers

Most red mud waste solids are settled from the liquor stream in single deck 40 metre diameter settling tanks. Flocculants are added to the settler feed stream to improve the rate of mud settling and achieve good clarity in the overflow liquor.

Washers

The mud is washed with fresh water in counter-current washing trains to recover the soda and alumina content in the mud before being pumped to large disposal dams on Boyne Island.

Slaked lime is added to dilute caustic liquor in the washing process to remove carbonate (Na_2CO_3) which forms by reaction with compounds in bauxite and also from the atmosphere and which reduces the effectiveness of liquor to dissolve alumina. Lime regenerates caustic soda, allowing the insoluble calcium carbonate to be removed with the waste mud.

$$Na_2CO_3 + Ca(OH)_2 \longrightarrow CaCO_3 + 2NaOH$$

Filters

Settlers overflow liquor containing traces of fine mud is filtered in Kelly-type constant pressure filters using polypropylene filter cloth. Slaked lime slurry is used to produce a filter cake. Mud particles are held on the filter leaves for removal and treatment in the mud washers when filters are sequentially taken off line.

Heat Interchange

With all solids removed, the pregnant liquor leaving the filter area, contains alumina in clear supersaturated solution. It is cooled by flash evaporation, the steam given off being used to heat spent liquor returning to digestion.

PRECIPITATION OF ALUMINA HYDRATE

Crystallisation

Dissolved alumina is recovered from the liquor by precipitation of crystals. Alumina precipitates as the trihydrate $Al_2O_3 .3H_2O$ in a reaction which is the reverse of the digestion of trihydrate.

$$2NaAlO_2 + 4H_2O \longrightarrow Al_2O_3 \cdot 3H_2O + 2NaOH$$

The cooled pregnant liquor flows to rows of precipitation tanks, which are seeded with crystalline trihydrate alumina, usually of an intermediate or fine particle size to promote crystal growth. Each precipitation tank is agitated, with a holding time of about three hours. During the 25-30 hours pass through precipitation, alumina of various crystal sizes is produced. The entry temperature and the temperature gradient across the row, seed rate and caustic concentration are control variables used to achieve the required particle size distribution in the product.

As correct particle size is important to smelter operations, sizing is carefully controlled. The QAL precipitation plant was designed to operate on a continuous basis to produce "sandy" or coarse alumina.

Classification

The finished mix of crystal sizes is settled from the liquor stream and separated into three size ranges in three stages "gravity" classification tanks. The primary classifiers collect the coarse fraction which becomes the product hydrate. The intermediate and fine crystals from the secondary and tertiary classifiers are washed and returned to the precipitation tanks as seed.

Spent Liquor

Spent caustic liquor essentially free from solid overflows from the tertiary classifiers and is returned through an evaporation stage where it is reconcentrated, heated and recycled to dissolve more alumina in the digesters. Fresh caustic soda is added to the stream to make up for process losses.

Washing

A slurry of coarse hydrate ($Al_2O_3 \cdot 3H_2O$) from the primary thickeners is pumped to hydrate storage tanks and is filtered and washed on horizontal-table vacuum filters to remove process liquor.

Calcining

The resulting filter cake is fed to a series of calcining units - an 1800 tonnes a day circulating fluidised bed calciner or one of nine rotary kilns each 100m long and 4m in diameter. The feed material is calcined to remove both free moisture and chemically-combined water. Firing-zone temperatures above 1100°C are used, achieved by firing with natural gas. The circulating fluidised bed calciner is more energy efficient than the older rotary kilns. Product sandy alumina particles are 90%+ 45 μm (microns) in size.

Cooling

Rotary or satellite coolers are used to cool the calcined alumina from the rotary kilns, and to pre-heat secondary combustion air for the kilns. Fluidised-bed coolers further reduce alumina temperature to less than 90°C before it is discharged on to conveyor belts which carry it to storage buildings where it is stockpiled for shipment.

Description of Process of Conversion of Alumna into Aluminium

Alumina received our alumina plant is reduced to aluminium in electrolytic cells by fused ash electrolysis. Alumina, which is an oxide of aluminium, gets dissociated to aluminium and oxygen ions in the cells. Aluminium having positive charge deposits at the cathode (which is the negative electrode) and oxygen being negatively charged collects at the anode (which is the positive electrode). The cathode is placed at the bottom of the cell. The oxygen goes to the anode and there it reacts with the carbon part of the anode (calcined with petroleum coke plus hard pith after baking at 1 150°C) and from CO_2 and Co depending on the operating conditions *e.g.* bath temperature, electrolytic composition, etc. (see Fig. 16.2).

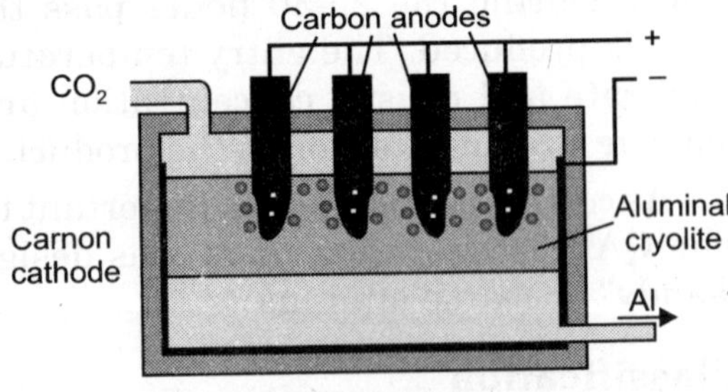

Fig. 16.2. The Electrolysis Cell for Alumina conversion

The electrolytic bath consists of cryolite and 8% calcium fluoride, which dissolve alumina up to 12% at operating temperature (970°C). Cryolite is chemical mixture of sodium fluoride (NaF) and aluminium fluoride (AlF_3) and for maximum current efficiencies the ratio NaF: AlF3 is maintained to desirable limits (1.36-1.40). To maintain this ratio which normally goes up if no additions are made, aluminium fluoride is added which compensates for any loss of fluoride due to volatilization during operating and to neutralize the Na_2O coming with the alumna. This Na_2O in alumina is of the order of 0.5% average. Calcium fluoride, which is maintained at round 8% of the cryolite in bath, reduces the melting point of cryolite (1008°C) enabling to maintain lower operating temperature, without any loss in the conductivity of the electrolyte. As far as cryolite is concerned, it is required to compensate for loss of electrolyte in the used linings (cathode); bath skimming and mechanical losses wit used anode butts coming out of the cells.

Soda ash needed is only while starting a new cell because sodium of NaF gets absorbed by new linings (cathodes) and sodium of soda ash compensates for the same while CO_2 goes to atmosphere.

The two polarities (anode and cathode) of the cells where the alumina is reduced to aluminium by this electrolytic process are made of carbon materials under special conditions. The quality of carbon materials required after a definite interval since it gets consumed due to formation of CO_2 and CO during continuous reduction of alumina.

Negative polarity, which is cathode, is also changed but the frequency is less and depends on quality old materials used for its construction and operating conditions. Old linings (cathode), which have been hanged, cannot be used again.

Reaction Equations

Electrolysis of the alumina/cryolite solution gives aluminium at the cathode and oxygen at the anode.

$$4Al^{3+} + 12e^- \longrightarrow 4Al \text{ (aluminium metal at the (–) cathode).}$$

$$6O^{2-} - 12e^- \longrightarrow 3O_2 \text{ (oxygen gas at the (+) anode).}$$

Aluminium is more dense than the alumina/cryolite solution, and so falls to the bottom of the cell where it can be tapped off as pure liquid metal. The overall reaction is

$$\text{Aluminium oxide} \longrightarrow \text{Aluminium + Oxygen.}$$

$$2Al_2{}^{3+}O_3{}^{2-}{}_{(l)} \longrightarrow 4Al_{(l)} + 3O_{2(g)}$$

Oxygen is discharged at the positive carbon (graphite) anode.

Oxygen reacts with the carbon anode to form carbon dioxide gas.

$$\text{Carbon + Oxygen} \longrightarrow \text{Carbon Dioxide.}$$

$$C_{(s)} + O_{2(g)} \longrightarrow CO_{2(g)}$$

Anodes: Anode is the positive electrode of the system, is made of the calcinated petroleum coke which is crushed ground and sized before it mixed with hard pitch (16-17%) in steam heated mixtures at 150%-160%. It comes out in the form of a paste and is pressed to confirm to our anode size in hydraulic presses available for the purpose at 914.667 kg. The pressed block thus received is called green anode and is sent for baking to improve conductivity and crushing strength by removing volatilize of pitch and baking the remaining binder coke in the blocks to 1150°C. This is done in special baking furnaces, which are pit type, and are oil or gas fired. Fuel oil is used for the firing of baking furnaces which gets consumed during baking it self. While baking anodes are packed are in the furnace with calculated petroleum coke and green petroleum coke mixture (75% plus 25%) which forms a blanket and keeps the necessary reducing atmosphere, in burnt and damaged. The other advantage of green coke mixing in the blanket material present in the green coke, temperatures can be maintained well, even at the top layer of anodes.

The backed anodes coming out of the furnaces are sent for fixing of copper bars and thereafter they are ready to go into the cells.

Cathodes: These are the negative electrodes of the system. These are made from the pot linings, which are result of ramming* a mixture of coal and 12% low softening point pitch. Anthracite coal is the only material commercially used for giving reasonable life of the cathode since it can stand the stresses developed at the operating conditions due to the bath and metal seepage in the cathode because of frost heave phenomena.

To construct the cathodes, calcinated anthracite coal is graded and sized (crushed and screened) and then mixed in steam heat mixtures with low softening pitch where it forms paste. This paste is taken to potrooms in special boxes and rammed into position with pneumatic rammers and necessary shape is given. Passing current then packs this. The other advantages of anthracite use are that it absorbs fewer baths so the cathode drops are less long life, because of list higher graficiticy.

Properties of Aluminium

The detail physical properties of aluminium are given below in the table below:

Melting Point	659°C
Boiling Point	2450°C
Density	2.708 g/cc
Electrical Resistivity	1.14×10^{-6}
Tensile Strength	318813.56368 Mpa* Annealed 110.332112 Mpa Cold Rolled Sheets
Heat of Capacity	5.817 cal deg^{-1} mol^{-1}
Standard Entropy	6.789 cal deg^{-1} mol^{-1}

* ramming: hammering

Chemical Properties

1. Action of Air: Aluminium is not affected by pure dry air, but in moist air a film of oxide is formed at the surface of the metal which protects the metal from further corrosion. It burns readily in air or oxygen giving a blue light.

Oxygen + Aluminium ⟶ Aluminium Oxide

$$O_2 + 4Al \longrightarrow 2Al_2O_3$$

The heat evolved in the oxidation of aluminium is used in the termite process for the reduction of chromium iron, manganese etc.

2. Action of Alkali: Aluminium readily dissolves in sodium and potassium hydroxide and forming the corresponding aluminates.

Aluminium + Sodium Hydroxide ⟶ Water + Sodium Aluminate + Hydrogen

$$2Al + 2NaOH \longrightarrow 2H_2O + 2NaAlO_2 + 4H_2$$

Aluminium + Potassium Hydroxide + Water ⟶ Potassium Aluminate + Hydrogen

$$2Al + 2KOH + 2H_2O \longrightarrow 2KAlO_2 + 4H_2O$$

It also dissolves in sodium carbonate.

3. Action of Water: Pure aluminium is not affected by pure water. However, the impure from is readily corroded by water containing salts. It dissolves boiling water and liberates hydrogen.

Aluminium + Water ⟶ Aluminium Hydroxide + Hydrogen

$$2Al + H_2O \longrightarrow 2Al(OH)_2 + 3H_2$$

Aluminium mercury couple is therefore used as a reducing agent.

4. Action of Acids: Aluminium dissolves in dilute hydrochloric and sulphuric acid evolving hydrogen. But concentrated nitric acid as no reaction on aluminium.

Aluminium + Hydrochloric Acid ⟶ Aluminium Chloride + Hydrogen

$$2Al + 6HCl \longrightarrow 2AlCl_3 + 3H_2$$

Aluminium + Sulphuric Acid ⟶ Aluminium Sulphate + SulphurDioxide + Water

$$2Al + H_2SO_4 \longrightarrow Al_2(SO_4)_3 + 3SO_2 + H_2O$$

5. Action of Halogen Hydrogen and Nitrogen

Aluminium + Chlorine ⟶ Aluminium Chloride

$$2Al + 3Cl_2 \longrightarrow 2AlCl_3$$

Aluminium + Nitrogen ⟶ Aluminium Nitride

$$2Al + N_2 \longrightarrow 2AlN$$

Aluminium + Hydrogen ⟶ Aluminium Hydride

$$2Al + H_2 \longrightarrow 2AlH$$

6. Displacement of Other Metals: Aluminium Displaces copper zinc and lead from the solutions of their salts.

$$\text{Aluminium + Zinc Sulphate} \longrightarrow \text{Aluminium Sulphate + Zinc}$$

$$2Al + 2ZnSO_4 \longrightarrow Al_2(SO_4)_3 + 2Zn$$

Uses of Aluminium

- Aluminium, being very light, is used as the material for making household airplanes parts; precision parts also some surgical instruments.
- Since it is unattached by nitric acid it is used as the container for carrying nitric acid.
- Mixed with oil, it is used for steam pipes and other metal pipes for protection.
- It is used as the reducing agent for extraction of metals such as chromium, iron manganese.
- Oxide of aluminium known an alumina is used for making refractory brick and ukramarine.
- In certain forms like feldspar and mica it is used in the construction buildings.
- Aluminium is also used in paper industry.

THERMIT WIELDING

Aluminium is used in thermit wielding of metals. This process is also called Goldschimidts Aluminothermic Process (see Fig. 16.3).

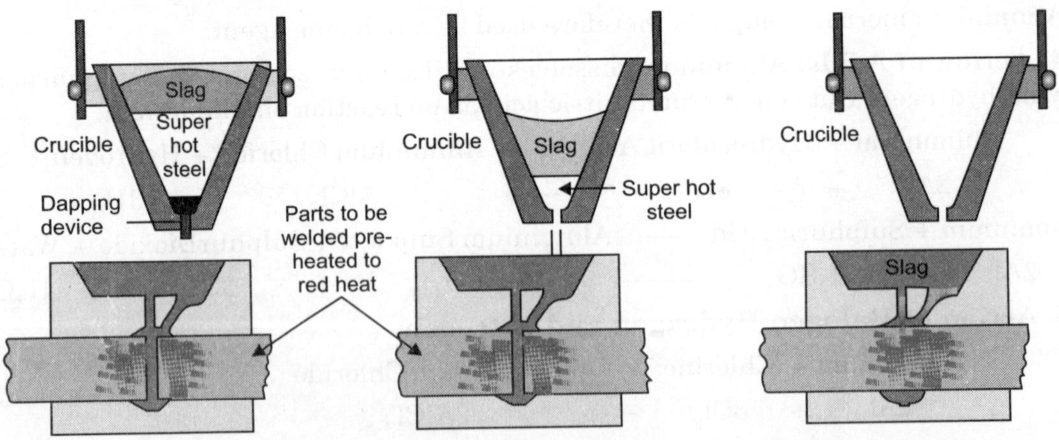

Fig. 16.3. The Process of Thermit Wielding

A mixture containing three parts of ferric oxide and one part of aluminium powder is placed in a crucible lined with a layer of mixture of magnesium powder and barium peroxide, with a magnesium ribbon interested into it to acts as a fuse. The broken ends of a rail or grinder, etc. are brought nearer and thoroughly cleaned and surrounded by a fire clay mould when the magnesium ribbon fuse is ignited the reaction occurs as shown below:

$$\text{Aluminium} + \text{Ferrous Oxide} \longrightarrow \text{Aluminium Oxide} + \text{Iron}$$

$$2Al + Fe_2O_3 \longrightarrow Al_2O_3 + 2Fe$$

Starts producing a tremendous amount heat as a result of which the iron melts. These heated ends of the iron rods (to be wielded) actually melt and mix with this molten metal added, giving a firm and strong wield.

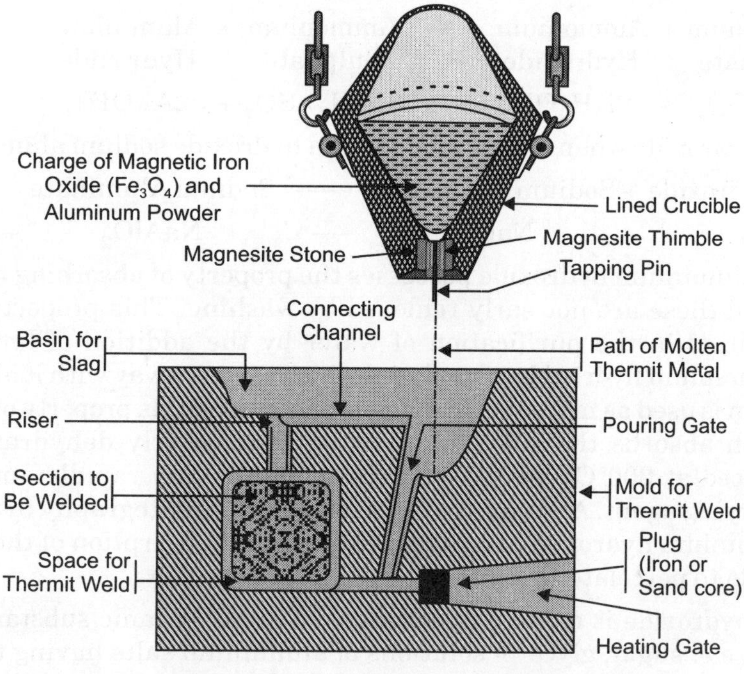

Fig. 16.4. The Thermit Crucible and the Detail System

ALLOYS OF ALUMINIUM

Aluminium forms a number of important allows which are given in the table below:

Alloy	Appropriate Composition	Uses
1. Aluminium Bronze	Aluminium 10% Copper 90%	For hard non-corrosive vessels Airplanes and Non-corrosive vessels.
2. Duralimin	Aluminium 95% Copper 3% Manganese 1% Magnesium 1%	
3. Y-Alloy	Aluminium 92.5% copper 4% Nickel 12%	
4. Magnelium	Magnesium 10% Aluminium 10%	

Its melting point is 2050°C and boiling point is 2980°C. It is very stable and extremely inactive. It is reduced with great difficulty. Alumina in the form of bauxite is used in making bauxite bricks. Fused alumina is used as a refractory material; alumina is now extensively used in chromatography for the separation of mixture. Alumina in the form of bauxite is used for making refractory bricks fused alumina is used as a refractory material. When heated in an electrical arc furnace at a temperature of 3000°C, cooled and pulverized*, alumina gives a very hard powder known as aluminium, which is also used as abrasive+.

Aluminium Hydroxide [Al(OH)$_3$]: It is a white gelatinous precipitate when potassium hydroxide or sodium hydroxide or ammonia is added to any aluminium salt.

Aluminium Sulphate + Ammonium Hydroxide ⟶ Ammonium Sulphate + Aluminium Hydroxide

$$Al_2(SO_4)_3 + 6NH_4OH \longrightarrow 3(NH_4)_2SO_4 + 2Al(OH)_3$$

Aluminium hydroxide when reacts with sodium hydroxide sodium aluminate is formed.

Aluminium Hydroxide + Sodium Hydroxide ⟶ Sodium Aluminate Water

$$Al(OH)_3 + NaOH \longrightarrow NaAlO_2 + H_2O$$

Amorphous aluminium hydroxide possesses the property of absorbing many substance form solution and these are not early removed by washing. This property of aluminium hydroxide is utilized in the purification of water by the addition of lime and alum. A precipitate of aluminium hydroxide is produced which carries away with it all the impurities of water. The alum is used as mordent+ in dying is based upon this property of the aluminium hydroxide, which absorbs the colouring matter. Particularly dehydrated aluminium hydroxide produced at 200°C - 250°C, absorbs moisture very rapidly and is used as an absorbent and drying agent. Aluminium gel is used in chromatography in which a column of hydrated aluminium hydroxide is used for preferential absorption of the components of the solution made to percolate through it.

Aluminium hydroxide is soluble in solutions of many organic substances containing hydroxyl groups, *e.g.* sugar, glycerol solutions of aluminium salts having these substance in sufficient quantities do not give any precipitate of aluminium hydroxide by the addition of any, alkali. A complex compound of aluminium with this hydrolysis organic substance is believed to be formed in these cases.

Aluminium Sulphate [Al$_2$(SO$_4$)$_3$]: Aluminium sulphate is prepared on large scale by dissolving the pure hydrate aluminium oxide in hot concentrated sulphuric acid. The crystals obtained on cooling are dissolved again in water and precipitated, by adding alcohol. The oily supersaturated solution first, separating readily solidifies to lustrous scaly crystals of the hydrated aluminium sulphate variety prepared by heating bauxite or kaolin with sulphuric acid, contains ferric sulphate as an impurity, which cannot be removed by crystallization.

* Pulverized: Converted into fine powder.

+ Abrasive: Used for polishing.

+ Mordant: Fixation of dye.

Kaolin + Sulphuric Acid ⟶ Aluminium Sulphate + Silicon Dioxide + Water

$$Al_2O_3 \cdot 2SiO_2 2H_2O + 3H_2SO_4 \longrightarrow Al_2(SO_4)_3 + 2SiO_2 + 5H_2O$$

A crude mixture of aluminium and ferric sulphate prepared from bauxite known as alumio-ferric is used as sewage purifier (also known as drain ex).

Aluminium Chloride ($AlCl_3$): Passing a current produces the anhydrous chloride over heated of aluminium oxide and charcoal in the presence of chlorine. The whole apparatus set is shown in the Fig. 16.5.

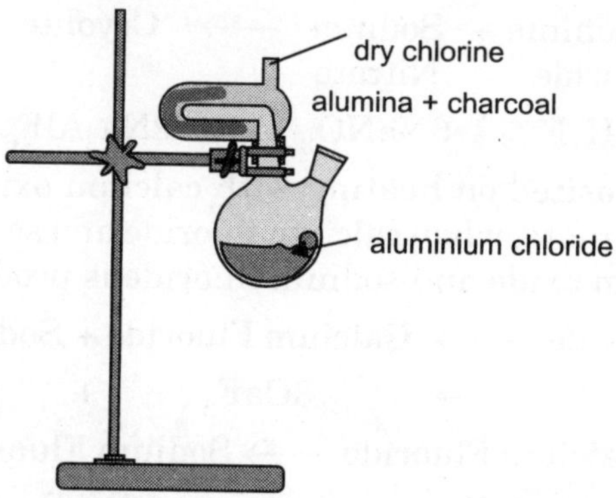

Fig. 16.5. The Lab Preparation of Aluminium Chlorde

Reaction Equation

Aluminium Oxide + Charcoal + Chlorine ⟶ Aluminium Chloride + Carbon Monoxide

$$Al_2O_3 + 3C + 3Cl_2 \longrightarrow 2Cl_2 + 3CO$$

The aluminium chloride hydrate, $AlCl_3 6H_2O$ is obtained as crystals by evaporation of the solution of metallic aluminium and hydrochloric acid. When heated the hydrate decomposes with the formation of aluminium oxide.

Therefore it is most possible to prepare anhydrous aluminium chloride by heating hydrated salt.

Anhydrous aluminium chloride is a white crystal, extremely hydroscopic salt. It sublimes at 183°C. The vapour density corresponds to the molecular formula Al_2Cl_6 This indicates the covalent character of the binding between aluminium chloride in organic solvents and by its chloride is readily soluble in water but it is extremely hydrolyzed.

Aluminium Chloride + Water ⟶ Aluminium Hydroxide + Hydrogen Chloride

$$AlCl_3 + H_2O \longrightarrow Al(OH)_3 + 3HCl$$

In laboratory aluminium chloride is mainly used as the considering agent in the Friedel-Crafts reaction in organic chemistry.

Aluminium Fluoride (AlF_3): This usually prepared by the reaction of aluminium sulphate and sodium fluoride.

Aluminium Sulphate + Sodium Fluoride ⟶ Aluminium Fluoride + Sodium Sulphate

$$Al_2(SO_4)_3 + 6NaF \longrightarrow 2AlF_3 + 3Na_2SO_4$$

Aluminium is scarcely soluble in water and is neutral to acids and basic hydroxides. It is however, hydrolyzed when heated in a steam.

Aluminium Fluoride + Water ⟶ Hydrogen Fluoride + Aluminium Hydroxide

$$AlF_3 + H_2O \longrightarrow 3HF + Al(OH)_3$$

Aluminium Fluoride in the form of Na_3AlF_6 is found in nature. This compound can be prepared in the lab by the action of aluminium fluoride ammonium fluoride and sodium nitrate.

Aluminium Fluoride + Ammonium Fluoride + Sodium Nitrate ⟶ Cryolite + Ammonium Nitrate

$$2AlF_3 + 6NH_4F + 6NaNO_3 \longrightarrow 2Na_3AlF_6 + 6NH_4NO_3$$

These crystals gets deposited on heating with calcium oxide forming at first calcium fluoride and sodium trialuminate, when calcium fluoride and sodium trialuminate is heated aluminium fluoride, calcium oxide and sodium fluoride is produced.

Cryolite + Calcium Oxide ⟶ Calcium Fluoride + Sodium Trialuminate

$$Na_3AlF_6 + 3CaO \longrightarrow 3CaF_2 + Na_3AlO_3$$

Sodium Trialuminate + Calcium Fluoride ⟶ Sodium Fluoride + Calcium Oxide

$$Na_3AlO_3 + CaF_2 \longrightarrow 6NaF_2 + 6CaO$$

Aluminium Bromide ($AlBr_3$): It is prepared by the action of aluminium chloride carbon and bromine; it forms a crystal of $AlBr_36H_2O$.

Aluminium Oxide + Carbon + Bromine ⟶ Aluminium Bromide + Carbon Monoxide

$$Al_2O_3 + 3C + 3Br_2 \longrightarrow 2AlBr_3 + 3CO$$

Aluminium bromide has a melting point of 93 C and boiling point 263 C.

Aluminium Iodide (AlI_3): It is formed by heating aluminium and iodine in a shielded enclosure or by acting on aluminium with iodine dissolved in carbon tetrachloride.

Aluminium + Iodine ⟶ Aluminium Iodide

$$2Al + 3l_2 \longrightarrow 2AlI_2$$

It has a melting point of 185°C and boiling point 350°C aluminium iodide reacts with carbon tetrachloride forming carbon tetraiodide and aluminium chloride.

Aluminium Iodide + Carbon Tetrachloride ⟶ Aluminium Chloride+ Carbon Tetraiodide

$$4AlI_3 + 3CCl_4 \longrightarrow 4AlCl_3 + 3CI_4$$

Aluminium iodide and also aluminium bromide have the formula Al_2I_6 and Al_2Br_6 respectively in their vapour state.

Aluminium Nitride (AIN): This aluminium compound is formed by heating aluminium powder in nitrogen at 740°C.

Aluminium + Nitrogen ⟶ Aluminium Nitride

$$2Al + N_2 \longrightarrow 2AlN$$

It is decomposed by hot water.

Aluminium Nitride + Water ⟶ Aluminium Hydroxide + Ammonia

2AlN 3H_2O ⟶ $Al(OH)_3$ NH_3

Aluminium Nitrate [$Al(NO_3)_3$ $9H_2O$]: It is deliquescent salt prepared by dissolving precipitated aluminium hydroxide in dilute nitric acid evaporating and crystallizing. It decomposes on heating, aluminium being formed. It is used as a mordent and for adding to oxide in making glass mantles*.

Aluminium Phosphide (AlP): It is formed by heating aluminium powder and red phosphorous, it is decomposed by water with evolution of phophene.

Aluminium + Phosphorous ⟶ Aluminium Phosphide

2AlP + P_2 ⟶ 2AlP

Aluminium Phosphide + Water ⟶ Aluminium Oxide + Phosphene

2AlP 3H_2O ⟶ Al_2O_3 2PH_3

Aluminium Phosphate ($AlPO_4$): This compound is produced when a solution of aluminium chloride is added to a solution of sodium phosphate, a gelatinous mass is precipitated.

Aluminium Chloride + Sodium Phosphate ⟶ Aluminium Chloride + Sodium Chloride + Chlorine

$AlCl_3$ + $NaPO_4$ ⟶ $AlPO_4$ + NaCl + Cl_2

It is soluble in mineral acids and alkalis and also ammonia.

Aluminium Sulphide (Al_2S_3): It is formed by adding a calculated amount of sulphur to fused aluminium, or by passing sulphur or carbon disulphide vapour over a strongly heated mixture of alumina and carbon.

Aluminium Oxide + Carbon + Sulphur ⟶ Aluminium Sulphide + Carbon Monoxide

Al_2O_3 + 3C + 3S ⟶ Al_2S_3 + 3CO

In pure state it is yellow in colour. It is completely hydrolyzed by water.

Aluminium Sulphide + Water ⟶ Aluminium Hydroxide + Hydrogen Sulphide

Al_2S_3 + H_2O ⟶ 2$Al(OH)_3$ + 3H_2S

Aluminium Sulphate [$Al_2(SO_4)_3$]: This salt of aluminium occur as *hair salt or feather alum* basic aluminium sulphate is called websterite, $Al(OH)_4.7H_2O$ a crystalline mass is precipitated known as hydrated alummium sulphate $Al_2(SO_4)_3$ $16H_2O$.

Aluminium Oxide + Sulphuric Acid ⟶ Aluminium Sulphate + Water

Al_2O_3 + 3H_2SO_4 $Al(SO_3)_3$ + 3H_2O

This is purified by redissolving in a little water and adding alcohol when an oily supersaturated solution separates, which soon solidifies into a lustrous scaly crystals. On heating, the crystals shrink and loose water leaving a white mass of anhydrous aluminium sulphate, which decomposes at 800°C.

---

* Mantes: Covers.

Aluminium Sulphate $\longrightarrow$ Aluminium Oxide + Sulphur Trixide

$$Al(SO_4)_3 \quad Al_2O_3 \quad + \quad 3SO_3$$

Impure aluminium solution is evaporated and the crystals processed. The product may contain a considerable amount of ferric sulphate (especially if bauxite) which, although it is does not form mixed crystals with aluminium sulphate, cannot be separated from it; used for the precipitation of the colloidal matter from sewage. If the ferric salt is reduced to a ferrous salt, say hydrogen sulphide, the aluminium sulphate may be crystallized at alone.

Aluminium Acetate [Al($CH_3COO)_3$]: A solution of this salt is prepared by mixing solutions of aluminium sulphate and lead acetate and filtering off the precipitated lead sulphate.

Aluminium Sulphate + Lead Acetate $\longrightarrow$ Aluminium Acetate + Lead Sulphate

$$Al(SO_4)_3 \quad + 3Pb(CH_3COOH) \longrightarrow \quad Al(CH_3COO_2)_3 \quad + \quad 3PbSO_4$$

Aluminium Carbide (A_4C_3): It is prepared by heating aluminium or aluminium oxide with carbon in an electrical furnace.

Aluminium + Carbon $\longrightarrow$ Aluminium Carbide

$$4Al \quad + \quad 3C \quad \longrightarrow \quad Al_4C_3$$

Alum

Is that double compound which is found when two compounds one consisting of monovalent or radical and other consisting of trivalent unite, *e.g.* M_2SO_4, $Mr_2(SO_4)_3 24H_2O$,

where

M = Monovalent element or radical such as lithium, sodium potassium ammonium etc.

M = Aluminium, iron, chromium.

The most common alum is potash alum, $K_2SO_4.Al_2\ (SO_4)\ 24H_2O$. It is prepared as follows:

$$M_2SO_4,\ M^1{}_2(SO_4)_3\ 24H_2O$$

where

M = monovalent element or radical such as lithium , sodium potassium or ammonium etc.

M^1 = iron, aluminium, chromium.

(i) From alunite or alum stone: Alunite is boiled with dilute sulphuric acid and filtered. The filtrate is mixed with a requisite quaintly of potassium sulphate and crystallized.

Alunite + Aluminium Hydroxide + Sulphuric Acid $\longrightarrow$ Potassium Sulphate + Aluminium Sulphate + Water

$$K_2SO_4Al_2(SO_4)_3 + 4Al(OH)_3 + 6H_2SO_4 \longrightarrow K_2SO_4 + 3Al_2(SO_4)_3 + 12H_2O$$

Potassium Sulphate + Aluminium Sulphate + Water $\longrightarrow$ Potassium Alum

$$K_2SO_4 \quad + \quad 3Al_2(SO_4)_3 \quad + 12H_2O \longrightarrow K_2SO_4\ Al_2(SO_4)_3$$

(ii) From Alum Shale: It is actually aluminium silicate consisting of iron pyrites. It is roasted to get the aluminium sulphate. This is extracted with water, mixed with potassium sulphate and crystallized.

(iii) From Bauxite: Bauxite is dissolved in sulphuric acid treated with calculated amount of potassium sulphate and crystallized. It is a white crystalline substance soluble in water. On heating looses its water of crystallization and swells up.

Potash alum is used for purification of water, a mordant in dyeing for tanning of leather, for sizing of paper and as an antiseptic to arrest bleeding. Some other alums are:

1. Ammonium Alum, $(NH_4)_2SO_4 . Al_2(SO_4)_3 . 24H_2O$
2. Sodium Alum, $Na_2SO_4 . Al_2(SO_4)_3\ 24H_2O$
3. Ferric Alum, $(NH_4)2SO_4Fe_2(SO_4)_3\ 24H_2O$
4. Chromic Alum, $K_2SO_4Cr(SO_4)_3\ 24H_2O$

SCANDIUM

Atomic Number : 21
Atomic Weight : 44.9559

Melting Point : 1539°C
Boiling Point : 2727°C

History and Occurrence

This element was discovered by Nilson in the year 1880. Its properties were found to correspond closely with those attributed by Mendeleef to an element undiscovered, at the time of his description about it. Although vividly distributed in small quantities, scandium is a source element and usually occurs in association with the rare earths. A mineral *thortveitite*, $Sc_2Si_2O_7$ is, however known that consists essentially of scandium silicate.

Extraction

Scandium is generally extracted from scandium fluoride, by heating it with calcium or titanium in an inert atmosphere using tantalum crucible. Unfortunately, molten scandium dissolves up to 5wt % tantalum.

The reduced metal is first purified by vacuum induction melting to remove the more volatile impurities. Then the scandium is distilled away from the residual tantalum crucible contamination and other less volatile impurities. The scandium vapour in a closed end tantalum tube inverted over the crucible that holds the metal to be purified and as long as the condensed scandium is not melted, it will not pick up any tantalum. Scandium cannot be achieved by electro transport purification of distilled scandium metal. A high DC is passed through a scandium rod of 3mm diameter and 150 mm long, which, causes most of the impurities to migrate the anode end of the rod thus purifying the rod near the cathode end.

SOME NOTES ON INDUCTION FURNACE

The coreless induction furnace is composed of a refractory container, capable of holding the molten bath, which is surrounded by a water cooled helical coil connected to a source of alternating current. Figure shown is a simplified cross section of a coreless induction furnace below:

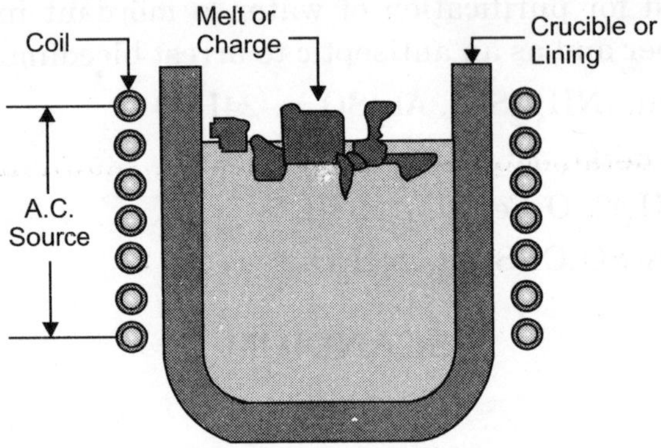

Induction Melting Furnace

The principle of operation of the induction furnace is the phenomenon of electro-magnetic induction.

The induction (generation) of the electrical current in a conductive metal (charge) placed within a coil of conductor carrying electrical current is known as electromagnetic induction of secondary current.

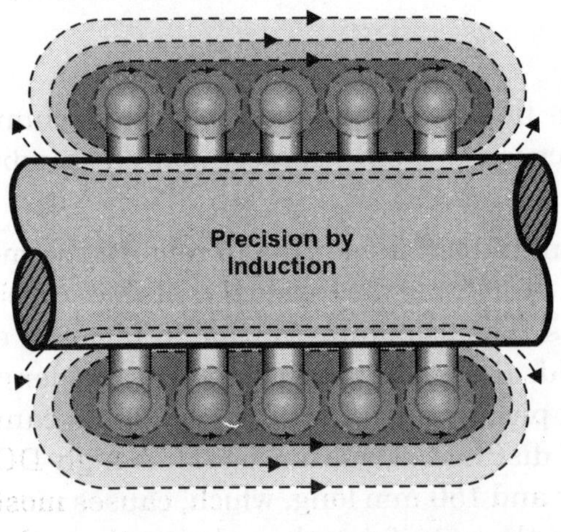

Figure of the Coil of the Furnace

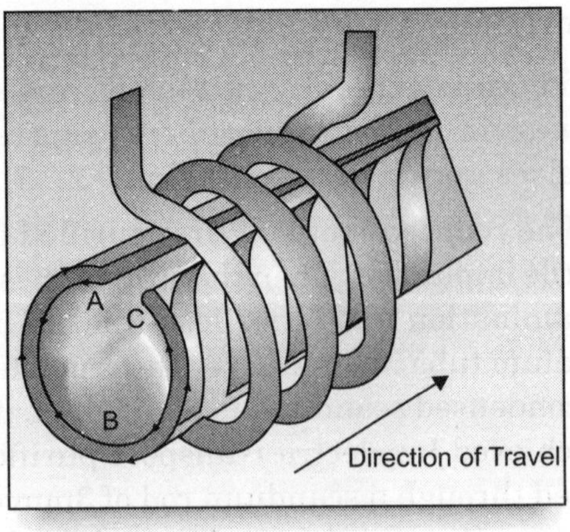

Figure of the Magnets

The alternating current applied to the coil produces a varying magnetic field, which is concentrated within the helical coil. This magnetic field passing through the charge induces secondary current in the charge piece. The current circulating in the charge produces electrical (I^2R) losses, which heat the charge and eventually melt it.

INDUCTION HEATING - A BETTER WAY

Why Induction Heating?

Induction heating is a sub-field of industrially used electric heating. All electrically - conductive materials can be heated quickly and cleanly with pollution free induction heating. Features include:

- ➢ Selected parts of the work piece can be heated.
- ➢ Heating time and temperature can be precisely controlled.
- ➢ There is no smoke or soot pollution;
- ➢ Heating operation can be integrated in semi-automatic production sequences;
- ➢ Induction equipment can usually be operated by unskilled personal.

Key advantages of induction heating

- ➢ Very fast partial hardening;
- ➢ High production rates;
- ➢ Significant reduction in pollution, distortion, forging scale, energy and space requirements;
- ➢ High degree of reproduction and automation;

What is Induction Heating ?

Placing a metal body in an alternating magnetic field creates eddy currents, causing losses through which the metal is heated. Skin effects concentrate these currents in the outer layers. The inductor traversed by an alternating current creates a magnetic field, which should be optimally adapted to the workpiece. The depth of heating can be influenced by varying the AC frequency, but it also depends on the concentration of flux capacity, on the length of treatment and the material of the workpiece, *i.e.* its heat conduction properties. Medium frequency is usually used for melting , forging heat treatment and annealing. For hardening and soldering, medium or high frequency may be used, depending on the treatment required.

Major Uses of Induction Heat Treatment

- ➢ Melting ferrous and non- ferrous metals with temperatures up to 1800°C.
- ➢ Heating for forging up to 1250°C.
- ➢ Stress relieving and normalizing steels after cold forming between 7500c and 950°C.
- ➢ Surface hardening of steel or cast iron work pieces at 8500c to 950°C.
- ➢ Soft soldering and brazing up to 1100°C.

➢ Recrystallisation.

➢ Tube and Pipe Welding.

Tube Welding Process: In the HFI (High Frequency Induction) tube welding process, high frequency current at approximately 400 kHz is induced in the open seam tube, an induction coil located ahead of (upstream from) the weld point.

Properties

It is a silvery white solid, does not tarnish in air.

Physical Properties

The detail of physical properties are shown in the table.

Melting Point	1539°C
Boiling Point	2727°C
Transformation Point	1537°C
Density	2.99 g/cc
Electrical Resistivity	70.9 ohm cm
Heat of Capacity	25.51 J/mol
Standard Entropy	34.78 J/mol K
Heat of Transformation	4.01 kJ/mol
Heat of Fusion	14.10 kJ/mol
Heat of Sublimation	377.8 kJ/mol
Debye Temperature	345.3°K
Electronic Specific Heat Constant	10.334 mJ/mol K^2
Atomic Volume	15.041 cm^3/mol
Atomic Radius	1.44 Å
Ionic Radius	0.41 Å
Magnetic Susceptibility X_A^{298} (A) X_A^{298} (C)	297.6 × 10^{-6} emu/mol 2.88.6 × 10^{-6} emu/mol
Isothermal Compressibility	1.78 × 10^{-12} m^2/N
Thermal Expansion	7.55 × 10^{-6}
Bulk Modulus	5.67 × 10^{10} N/m^2
Yong's Modulus	7.52 × 10^{10} N/m^2
Stress Modulus	2.94 × 10^{10} N/m^2
Poisson's Ratio	0.279

Chemical Properties

In its compounds somewhat resembles beryllium, but it is uniformly trivalent. The hydroxide is a stronger base than aluminium hydroxide, aluminium hydroxide, weaker than the hydroxides of rare earths. It is distinguished from aluminium hydroxide by its lack of amphoteric properties, shown by its insolubility in caustic alkalis, but unlike the hydroxides of rare earths can be precipitated by the addition of sodium thiosulphate solution, which are weakly alkaline through hydrolysis. This made use in one process for the separation of scandium from these elements. Scandium hydroxide is not so weak base that scandium carbonate cannot be precipitated from the solution; another distinction from aluminium, but scandium sulphide like aluminium sulphide is decomposed by water, at any rate on boiling. The halides of scandium are more salt like than there of aluminium, they are less hydrolyzed in solution and are much less volatile. All these properties predicted from position in the periodic table.

Compounds of Scandium

Scandium Oxide (S_2O_3): This is the oxide of scandium, which is prepared from thortveitite by heating it.

$$\text{Trotveitite} \longrightarrow \text{Silicon Dioxide} + \text{Scandium Oxide}$$

$$Sc_2Si_2O_7 \longrightarrow 2SiO_2 + Sc_2O_3$$

Scandium Fluoride (ScF_3): It is prepared by the action of scandium hydroxide and hydrogen fluoride.

$$\text{Scandium Hydroxide} + \text{Hydrogen Fluoride} \longrightarrow \text{Scandium Fluoride} + \text{Water}$$

$$Sc(OH)_2 + 3HF \longrightarrow ScF_3 + 3H_2O$$

It is also prepared by the direct action of scandium and fluorine.

$$\text{Scandium} + \text{Fluorine} \longrightarrow \text{Scandium Fluoride}$$

$$2Sc + 3F_2 \longrightarrow 2ScF_3$$

Scandium fluoride resembles aluminium very much in its properties. It is a very variable compound only slightly soluble in water, but freely soluble in fluoride solutions, provided that the fluoride ion concentration has not been reduced by the addition of acids. These solutions contain fluoscandates, yielding the ion ScF_6 and stable in the solid state. The complex is fairly stable, since ammonia will not precipitate the scandium hydroxide form fluoscandates solutions. The fluoscandates by their much greater stability in water.

Scandium Chloride ($ScCl_2$): It is obtained by the action of a mixture of chlorine and sulphur monochloride on hot scandium oxide.

$$\text{Scandium Oxide} + \text{Chlorine} + \text{Sulphur Monochloride} \longrightarrow \text{Scandium Monochloride} + \text{Sulphur Dioxide}$$

$$2SCO \quad Cl_2 \quad S_2Cl_2 \quad 2ScCl_3 \quad SO_2$$

It is a colourless solid, which melts at 930°C and begins to sublime at about 800°C so it is much less volatile than aluminium chloride. Its solutions are less hydrolyzed than these of aluminium chloride are and it will not dissolve in alcohol.

Scandium Sulphate [$SC_2(SO_4)_3$]: This compound can be prepared by the action of dilute sulphuric acid on scandium or scandium hydroxide.

Scandium Hydroxide + Sulphuric Acid ⟶ Scandium Sulphate + Water

$$2Sc(OH)_3 + 3H_2SO_4 \longrightarrow Sc_2(SO_4)_3 + 3H_2O$$

And crystallized from these solutions as the hexahydrate $Sc_2(SO_4)_3 6H_2O$.

TITANIUM

Atomic Number : 22
Atomic Weight : 47.88

Melting Point : 1725°C
Boiling Point : 3260°C

Occurrence

Titanium was discovered by Georger (1791) and independently by Kalporth in the year (1794), but the metal itself in a state of reasonable purity has only recently be prepared. It is one of the most abundant of the metals, but though it is widely diffused ores rich in titanium are common and until the recent increased demand for titanium they have not been worked on a very large scale.

The principal ores are *rutile,* TiO_2 *ilmenite*, $FeO.TiO_2$. Ilmenite found in South India and also in U.S.A. The rutile is available in Norway and also in U.S.A.

Extraction

Ferro-titanium is made by two processes. If an alloy rich in carbon is required, ilmenite is reduced with carbon in the electrical furnace. Ferro-titanium free form carbon is prepared by *Goldschimidts Process*. Many processes have been devised for the preparation of titanium compounds. In the *Monk-Irwin* Process ilmenite is ground up and heated in water gas. The product is a mixture of iron and the dioxide of titanium, which is freed from iron by treatment with dilute sulphuric acid. The dioxide is then converted to the sulphate by fusion with sodium hydrogen sulphate and the product is dissolved in water to form a solution from which fairly pure titanium dioxide can be precipitated by boiling. In the American process ilmenite is dissolved in sulphuric acid and the ferric salts are then reduced to the ferrous conditions. Titanium dioxide can then be precipitated from the solution by boiling it.

Uses of Titanium

In the form of ferro-titanium and alloy containing some 80% of iron, titanium has long being used for the cleansing of steel, that is, for the removal of oxygen and nitrogen. This demand is satisfied by quite small quantities of titanium. The resent developments have greatly increased the output. Titanium paint is made either from the dioxide or from barium titanate and has a greater covering power than white lead, it is non-toxic, unaffected by light and also remains unchanged by impurities in the atmosphere.

Compounds of titanium are also used as mordant in dyeing and in staining leather. The tetravalent salts, the titanious compounds are also used in the laboratory. As a reducing agent it is used on the dyeing industries for stripping out the colours. It is also used for the smoke screen.

Metallic Titanium

The hard very brittle metal is obtained by reducing the tetrachloride compound of titanium with finely powdered sodium and mixed with iodine, is placed in an evacuated chamber heated to about 2000°C and transversed by tungsten filaments. As the rods thicken the heating current is increased to maintain the temperature.

The process amounts to a distillation of titanium so prepared is freely malleable and ductile. When heated to a fairly high temperature in the presence of air, it burns to form the dioxide,

TiO_2 and at 800°C it will also combine with nitrogen to form titanium nitride, TiN. Titanium reacts with the nitrogen but in hydrochloride or sulphuric acid it liberates hydiogen and dissolves to form a violet solution, which contains the tetravalent ion **Ti**$^{\cdots}$ and is slowly oxidized by the air.

Physical Properties

Melting point	1725°C
Boiling Point	3260°C
Density	4.54 g/cc
Crystalline Form	Hexagonal closed packed, transforming at 800°C to body centre cubic
Specific Heat	0.13 cal/(g) (°C)
Latent Heat of Fusion	96 cal/g
Leaner Coefficient of Expansion	8.5×10^{-6} per °C
Electrical Resistivity	56×10^{-6}
Modulus of Elasticity	1034.3655 Mpa
Metallic Radius	147 pm
Heat of Fusion	18.8 kJ/mol
Heat of Vapourization	42.4 kJ/mol
Electronegativity	1.5

Compounds of Titanium

Titanium Monoxide (TiO): It is made by heating titanium dioxide with carbon in the electric arc furnace.

$$\text{Titanium Dioxide} + \text{Carbon} \longrightarrow \text{Titanium Monoxide} + \text{Carbon Monoxide}$$
$$TiO_2 + C \longrightarrow TiO + CO$$

It is weakly basic.

Titanium Sesquioxide (Ti_2O_3): It is obtained by reducing titanium dioxide with hydrogen.

$$\text{Titanium Dioxide} + \text{Hydrogen} \longrightarrow \text{Titanium Sesquioxide} + \text{Water}$$
$$2TiO_2 + H_2 \longrightarrow Ti_2O_3 + H_2O$$

When caustic alkali is added to a titanious salt such as titanium trichloride, a reddish brown precipitate of the titanium hydroxide is produced.

Titanium Dichloride + Sodium Hydroxide ⟶ Sodium Chloride + Titanium Hydroxide

$$TiCl_2 + 2NaOH \longrightarrow 2NaCl + Ti(OH)_2$$

It dissolves in sulphuric acid producing a violet precipitate of titanium sulphate.

Titaniuim Sesquioxide + Sulphuric Acid ⟶ Titanium Sulphate + Water

$$Ti_2O_3 + H_2SO_4 \longrightarrow Ti_2(SO_4)_3 + H_2O$$

Titanium Dioxide (TiO_2): It is the most stable oxide of titanium, occurs in an impure form called rutile. It may be precipitated from titanic solutions of alkalines, or from solutions of titanates simply by boiling them (hydrolysis). At low temperature the precipitate may consist of tetrahydroxide of titanium $Ti(OH)_4$, sometimes called orthotitanic acid, but X-Ray diffraction fails to reveal any definite hydroxides of titanium. It can be converted to the dioxide by strong by strong heating.

Titanium dioxide is produced when titanium burns in oxygen and can reduced to the metal only with the greatest difficulty, in which it resembles silica. It is amphoteric and when in hydrated form it will dissolve in acid to form titanic compounds, probably giving rise to the colourless titanyl ion $TiO^{\bullet\bullet}$ and in alkalis to produce the titanides, where characteristics ion is TiO_3'' or some more complex form. Titanic acid is so weak that the solubility of titanium dioxide in aqueous alkalis is very limited and the titanates are best prepared from the dioxide and fused alkalis or carbonates like silicic acid. Titanic acid can exist as solutions in form of colloid and precipitated from solutions of titanates or boiling.

The titanates are often associated with the silicates, calcium titanate occurs in the Ural Mountains and ferrous titanates; $FeTiO_3$ is the principal constituent of the ilmenite. As a typical soluble titanate we may mention potassium metatitanate, $K_2TiO_3 \cdot 4H_2O$. This compound is prepared by action of titanic acid and potassium hydroxide.

Tinanic Acid + Potassium Hydroxide ⟶ Potassium Metatitanate + Water

$$Ti(OH)_4 + 2KOH \longrightarrow K_2TiO_3 + H_2O$$

It is a colourless hydroxide solid, where solutions have strongly alkaline reactions from hydrolysis.

Titanium Trioxide, Titanium Peroxide (TiO_3): In the hydrated form it is also called pertitanic acid. It is also obtained as a yellow precipitate by adding hydrogen peroxide to a solution of titanium disulphate in the presence of sulphuric acid neutralized by ammonia. Before the ammonia is added the solution is orange and possibly contains a per-acid. Since the titanium in sulphuric acid, the next change may be represented in the equation shown below:

Titanium Dioxide + Hydrogen Peroxide ⟶ Titanium Peroxide + Water

$$TiO_2 + H_2O_2 \longrightarrow TiO_3 + H_2O$$

The orange coloured solution is used as a test for both titanium and hydrogen peroxide. The colour is destroyed by reducing agent, which will liberate chlorine from hydrochloric acid.

Titanium Nitrides (TiN): At high temperature titanium will burn in nitrogen to produce a nitride to produce a nitride if titanium dioxide is reduced in ammonia gas the oxygen can actually be replaced by nitrogen.

Titanium Dioxide + Nitrogen ⟶ Titanium Nitride + Oxygen

$$2TiO_2 + N_2 \longrightarrow 2TiN + 2O_2$$

At lower temperature the nitride of titanium is hydrolyzed by water with the formation of titanium oxides and ammonia.

Not only diamond but also titanium nitride is as hard as diamond

It is golden yellow in colour and as hard as diamond.

Titanium Halides, Tetrahalides (TiX_4): These are the most stable halides of titanium. The tetrafluoride $TiF_4 \cdot 2H_2O$ is a colourless hydroscopic solid of deficiency salt like character, which readily forms complex called fluorotitanates, with other fluorides.

Titanium Tetrafluoride + Fluorine ⟶ Titanium Hexafluoride

$$TiF_4 + F_2 \longrightarrow TiF_6$$

The fluorotitanate are stable compounds among them the only compound which is slightly soluble in water is potassium fluorotitanate.

The tetrachloride of titanium (TiC_4) is one of the most important compounds of titanium and is obtained by the action of chlorine on the metal.

Titanium + Chlorine ⟶ Titanium Tetrachloride

$$Ti + 2Cl_2 \longrightarrow TiCl_4$$

Or, by the reaction of titanium dioxide and carbon tetrachloride.

Titanium Dioxide + Carbon Tetrachloride ⟶ Titanium Tetrachloride + Carbon Dioxide

$$TiO_2 + CCl_4 \longrightarrow TiCl_4 \quad CO_2$$

It is a liquid boiling point at 135°C and it is used in the determination of atomic weight of titanium in the thoroughly purified state. When titanium chloride reacts with water, the reaction is exothermic, but titanyl ions TiO and the products of hydrolysis of which the final stage hydrolysis is the formation of TiO_4'''', the orthotitanate ion. On boiling the hydrolysis is complete, unless there is an excess of titanium dioxide and iodine.

Titanium Iodate + Oxygen ⟶ Titanium Dioxide + Iodine

$$TiI_4 + O_2 \longrightarrow TiO_2 + 2I_2$$

Trihalides are also stable like other halides and are also the most definitely salt like compounds of titanium. When hydrochloric acid of titanium tetrachloride is reduced with aluminium tetrachloride or electrolytically, a violet solution is produced which is titanium trichloride ($TiCl_3$). The anhydrous salt forms violet crystals, which are hydroscopic in nature and are also like their solutions gets readily oxidized and decolourized by air. Titanious chloride solutions are powerful reducing agents. The oxidation potentials of $Ti^{\cdot\cdot\cdot}/TiO''$ is 0.10 volt. They will precipitate copper, silver gold or mercury form solutions of their salts, reduce nitrates to ammonia, perchlorate and to reduce ferrous salts with ammonium thiocyanate as internal indicator. They are also used in titration of organic nitro-compounds, which they reduce to amines. As the solutions are oxidized by air they are kept in the environment of hydrogen as shown in the Fig. 16.6.

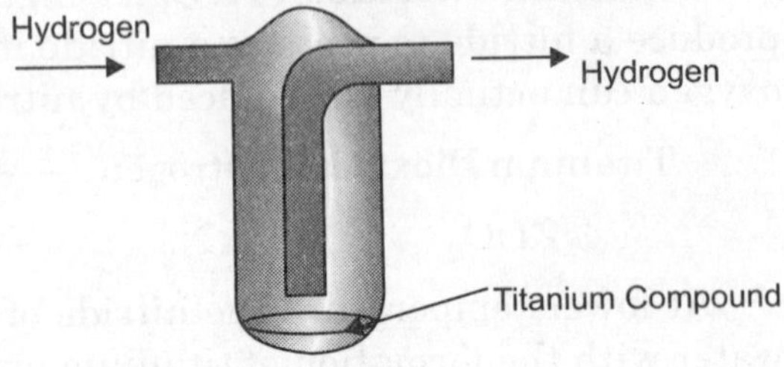

Fig. 16.6. The preservation of Titanium compounds

Now-a-days titanium is replaced by titanious sulphate, obtained by electrolytic reduction of a solution of titanium dioxide in sulphuric acid, in either solution excess of acids must be present to stop the hydrolysis.

Titanious Bromide ($TiBr_3$): This compound of bromine is produced together with the dibromide ($TiBr_2$) when titanium is carefully heated n hydrogen bromide gas.

Titanium + Hydrogen Bromide ⟶ Titanious Bromide + Hydrogen

$$4Ti + 6HBr_2 \longrightarrow 4TiBr_3 + 3H_2$$

Titanious bromide decomposes in at higher temperature producing tetrabromides.

Titanious Bromide ⟶ Titanious Tetrabromide + Titanious Dibromide

$$4TiBr_3 \longrightarrow TiBr_4 + TiBr_2$$

Titanious Dibromide ⟶ Titanium + Titanious Tetrabromide

$$TiBr_2 \longrightarrow 2Ti + TiBr_4$$

The dihalides are very unstable compounds obtained by reducing the trihalides with hydrogen from it. Titanium dichloride produces a hissing sound when brought in contact with water and liberates hydrogen from it.

Titanium Dichloride Water ⟶ Titanic Acid + Hydrogen Chloride + Hydrogen

$$2TiCl_2 + 4H_2O \longrightarrow Ti(OH)_4 + HCl + H_2$$

It also burns in air forming the dibromide and the tetrachloride.

Titanium Dichloride + Oxygen ⟶ Titanium Dioxide + Titanium Tetrachloride

$$2TiCl_2 + O_2 \longrightarrow TiO_2 + TiCl_2$$

Titanium Sulphates [$Ti(SO_4)_2$]: The disulphate of titanium is obtained in solution by boiling the hydrated dioxide with sulphuric acid. When electrically reduced it yields violet solution of titanious sulphate [$Ti_2(SO_4)_3$] in sulphuric acid and is used in volumetric analysis.

VANADIUM

Atomic Number : 23 | Melting Point : 1900°C
Atomic Weight : 50.9415 | Boiling Point : 3000°C

History

Vanadium was discovered in 1830 by Sefström in a Swedish iron ore and was given the name after the mythological goddess of Scandinavia, Vandis. It was subsequently shown that the same element was discovered some thirty years ago by Del Reo).

Occurrence and Extraction

The quantities of vanadium extracted its ores are still abundant comparatively more abundant than copper.

The principal ore of vanadium is *patronite,* (VS_4)*;* it is basically vanadium sulphate found in Peru. This is mixed with a flux and roasted in a reverberatory furnace, when the vanadium passes into the slag, which is then reduced to ferro-vanadium in blast furnace. The acid extraction process has also been widely used, particularly for the poorer deposits. In this process the ore is crushed and heated with dilute sulphuric acid at 200°C under a pressure of fifteen atmospheres. The solution is then filtered and evaporated to dryness, the product contains the vanadium, which is heated strongly to convert it to the oxide. It is mixed with sodium carbonate, roasted in the air and dissolved in hot water from which carbon dioxide will precipitate any aluminium as oxide. Sodium vandate can be obtained from the solution by concentration.

Elementary Vanadium

In the early researches of Berzzelius the oxychloride of vanadium ($VOCl_3$) was taken for the chloride and oxide VO for the element. The metal forms oxides and nitrides which are very much stable from which the metal is not possible to extract. A 95% pure vanadium can be obtained from the pentoxide and aluminium powder by the Goldschimidts Process and a somewhat purer substance by reducing the chloride with hydrogen, sodium hydroxide. Attempts to prepare the metal by the process of electrolysis is unsuccessful.

Vanadium is a gray lustrous metal therefore it is regarded as the hardest metal. The melting point of vanadium is 1900°C. It is oxidized by heating in the air and burns if heated in oxygen, but it is quite unaffected by water. Among the acids only hydrofluoric acid and oxidizing acids are able to dissolve vanadium. It combines with chlorine, nitrogen, silicon, and other elements.

> Vanadium is the hardest metal. Its density is 6 g/cc.

About 90% of world's vanadium comes from ferro-vanadium. The addition of vanadium to steel increases the toughness and the tensile strength and much steel is used in high speed tools and the parts of the machinery for these purpose is only 2% of vanadium is necessary in steel. The detail physical properties are shown in the table below.

Physical Properties

Melting Point	1900°C
Boiling Point	3000°C
Density	6.11 g/cc
Specific Heat (20-100°C)	0.12 cal/(g) (°C)
Linear Co-efficient of Thermal Expansion	$9.7 \pm 3 \times 10^{-6}$/°C
Thermal Conductivity	0.074 cal/(sec) (sqcm) (°C/cm)
Electrical Resistivity	24.8 µ ohm cm
Hardness (annealed)	76
Rockwell	B

Compounds of Vanadium

Vanadium has number of compounds which are discussed below:

Vanadium Dioxide (VO): It is prepared by heating a trioxide and vanadium mixture.

Vanadium Trioxide + Vanadium ⟶ Vanadium Oxide

$$V_2O_3 + V \longrightarrow 3VO$$

At high temperature it is insoluble in water but soluble in acids, with which it yields violet solutions of hypovandate salts, containing the bivalent cation $V^{\bullet\bullet}$. These solutions can also be prepared from solutions of other vanadium compounds by reduction with sodium amalgam or at cathode during electrolysis and the hydroxide $V(OH)_2$ can be precipitated from them by the addition of caustic alkalis. The oxidation potential of the charge $V^{\bullet\bullet\bullet}$ ...$V^{\bullet\bullet}$ is –0.2 volt and the hypovandous compound in which the element is bivalent are among the strongest reducing agents. In acid solutions they rapidly evolve hydrogen and even the insoluble hydroxide slowly decomposes water. They have been used in the dearsenification* of hydrochloric acid, as they will reduce arsenic chloride, which can be filtered off.

Vanadium Trioxide (V_2O_3): It is obtained by reducing the pentoxide with hydrogen at a moderate temperature.

Vanadium Pentoxide + Hydrogen ⟶ Vanadium Trioxide + Water

$$V_2O_5 + 2H_2 \longrightarrow V_2O_3 + 2H_2O$$

It is a black substance with a very high melting point and burns when heated in the air. It will dissolve only in the oxyacids. Solutions of vanadeous salts, in which vanadium is tetravalent, must therefore be obtained by reduction of solutions of other vanadium compounds and this can be carried out either electrochemically or with magnetism or zinc. The resulting solutions; which are green in colour, are powerful reducing agents, which slowly decompose with the evolution of hydrogen and are rapidly oxidized on the atmospheric exposure. The insoluble hydroxide $V(OH)_3$ is precipitated form them by caustic alkalis.

---

* Dearsenification: to remove arsenic.

Vanadium Tetraoxide, Vanadyl Oxide [V_2O_4, VO_2, V(O)O]: This compound of vanadium is prepared by the gentle reduction of vanadium pentoxide, in this case vanadium pentoxide is fused with oxalic acid where vanadium tetraoxide is produced.

Vanadium Pentoxide + Oxallic Acid ⟶ Vanadium Tetraoxide + Carbonic Acid + Carbon Dioxide

$$V_2O_5 + H_2C_2O_4 \longrightarrow V_2O_4 + H_2CO_3 + CO_2$$

It is dark blue solid of very high melting point, which absorbs oxygen from the air. It dissolves readily in acids to form solution of vanadyl salts; containing the negative VO^- and from these reactions derives its alternative name and formula, vanadyl oxide V(O)O, salts of quadrivalent vanadium usually decompose on heating into salts in which the metal is tetravalent. Vanadyl solutions are blue and can be prepared by gentle reduction of solutions Vanadium compounds for e.g. with hydrogen sulphide, sulphur dioxide or even hydrogen. They are weak reducing agents. Vanadium tetraoxide is amphoteric and has weak acidic properties, thus it gets dissolved in alkaline solutions to form yellow solutions of hypovandate where anion appears to be V_4O_9''.

Vanadium Tetraoxide + Sodium Hydroxide ⟶ Sodium Hypovandate + Water

$$V_2O_4 + 2NaOH \longrightarrow Na_2V_4O_9 + H_2O$$

These yellow solutions are easily oxidized by air to colourless solutions called vandates.

Vanadium Pentoxide (V_2O_5): It is the most important compound of vanadium. It is usually prepared by the ignition of ammonium metavandate in a platinum crucible.

Ammonium Metavandate ⟶ Vanadium Petoxide + Ammonia + Water

$$2NH_4VO_3 \longrightarrow V_2O_5 + 2NH_3 + H_2O$$

In the pure state it is obtained by hydrolyzing the oxychloride of vanadium.

Vanadium Oxychloride + Water ⟶ Vanadium Pentoxide + Hydrogen Chloride

$$2VCl_3 + 3H_2O \longrightarrow V_2O_5 + 6HCl$$

The pentoxide red or yellow substance, which, can exist in various physical form. It melts at 595°C and is stable towards the heat, but at much higher temperature it decomposes to trioxide and oxygen.

Vanadium Pentoxide ⟶ Vanadium Trioxide + Oxygen

$$V_2O_5 \longrightarrow V_2O_3 + O_2$$

In the electrical furnace it can be reduced to dioxide by reducing it with hydrogen.

Vanadium Pentoxide + Hydrogen ⟶ Vanadium Dioxide + Water

$$V_2O_5 + 3H_2 \longrightarrow V_2O_2 + 3H_2O$$

It is insoluble in water, but a number of vanadic acids can be precipitated by acidifying vandate solutions. The oxide is amphoteric and will dissolve either in acids or in alkalis. Vanadic salts are oxidizing agents and when the pentoxide dissolves as a vanadyl salt, hydrochloric acid. Chlorine and vanadyl chloride is obtained.

Vanadium pentoxide is used as catalyst in the preparation of sulphur trioxide.

Vanadates: These compounds of vanadium are obtained by dissolving the vanadium pentoxide in potassium or sodium potassium hydroxide.

Vanadium Petoxide ⟶ Sodium Hydroxide + Sodium Orthovandate + Water

$$V_2O_5 \longrightarrow 6NaOH + 2Na_3VO_4 + 3H_2O$$

Vanadium Petoxide ⟶ Potassium Hydroxide + Potassium Orthovandate + Water

$$V_2O_5 \longrightarrow 6NaOH + 2K_3VO_4 + 3H_2O$$

These are regarded as the orthovanadates, but these exist only in the strong alkaline solution.

Metavandates: Among them ammonium metavanadate is well-known compound of vanadium. This compound is prepared by the addition of ammonium chloride on orthovanadic acid.

Ammonium Chloride + Orthovanadic Acid ⟶ Ammonium Metavanadate + Hydrogen Chloride

$$NH_4Cl + HVO_3 \longrightarrow NH_4VO_3 + HCl$$

Vanadic Acid Sodium Peroxide

Pervanadates: When vanadic acid reacts with different peroxide, respective pervandates are produced.

Vanadic Acid Sodium Peroxide ⟶ Sodium Pervanadate + Hydrogen Peroxide

$$2HVO_3 + 2Na_2O_3 \longrightarrow 2Na_2VO_4 + H_2O_2$$

But potassium pervanadate is produced by a different reaction.

Potassium Metavanadate + Hydorgen Peroxide ⟶ Potassium Pervanadate + Potassium Hydroxide

$$K_3VO_4 + H_2O_4 \longrightarrow KVO_4 + 2KOH$$

In addition to these many oxychlorides are also known, of which few are soluble in water. The fluorides and chlorides of vanadium are being briefly stated below:

Halides : The halides of vanadium are shown in the list below:

Fluorides	Difluorides ———	Trifluorides VF_3	Tetrafluorides VF_4	Pentafluorides VF_5
Chloride	Dichloride VCl_2	Trichloride VCl_3	Tetrachloride VCl_4	Pentachloride VCl_5
Bromide	Dibromide VBr_2	Tribromide VBr_3	Tetrabromide ———	Pentabromide ———
Iodides	Diiodide VI_2	Triiodide VI_3	———	———

Fluorides: The action of fluorine on vanadium produces a mixture of substances and the fluorides are more easily obtained by the action of hydrogen fluoride on chlorides of vanadium.

It is a greenish yellow refractory* substances insoluble in water.

Vanadium Tetrafluoride (VF_4): It is obtained in a similar way by the action of hydrogen fluoride on the cold tetrachloride.

Vanadium Pentafluoride (VF_5): This compound of vanadium is obtained together with the trifluoride and nitrogen at 650°C and collects as a sublimate in the cooler parts of the apparatus in the form of a white powder. It sublimes at 111°C and is soluble in water, but the solution suffers a hydrolysis with the formation of oxyfluoride. It is the only petahalide of vanadium.

Vanadium Pentafluoride is the only petahalide of vanadium.

CHLORIDES

Vanadium Dichloride, Hypovandous Chloride, VCl_2: It is the best known dichloride or dihalide of vanadium, a high temperature in a stream of nitrogen.

Vanadium Trichloride ⟶ Vanadium Dichloride + Chloride

$$2VCl_2 \longrightarrow 2VCl_2 + Cl_2$$

or by passing a mixture of the tetrachloride and hydrogen through a red hot tube. It forms pale green crystals, which volatilize unchanged at a high temperature (about 1000°C) and readily absorb oxygen or water. The violet solutions have the properties already described.

Vanadium Trichloride, Vanadous Chloride, VCl_3: It is obtained by decomposition of tetrachloride at moderate temperature.

Vanadium Tetrachloride ⟶ Vanadium Trichloride Chlorine

$$2VCl_4 \longrightarrow 2VCl_3 + Cl_2$$

The tetrachloride may be heated nearly to boiling while a current of dry carbon dioxide is blown through it to remove the chlorine. The trichloride remains as violet, it is very much hydroscopic and crystalline in structure also very much prepared by dissolving the trihydroxide in hydrochloric acid.

When strongly heated in an inert gas the anhydrous substance yields the dichloride and the tetrachloride and at a high temperature it can be reduced to the metal in a current of hydrogen.

Vanadium Tetrachloride (VCl_4): It is obtained by the action of chlorine on vanadium or ferrovanadium, or by the action of a mixture of chlorine and sulphur monochloride on the hot patronite. The product of these reacts can be purified by distillation and is a dark brown viscous liquid freezing at –26°C and boiling at 154°C, it has a normal vapour density. It dissolves in water to form a blue solution of oxychloride, $VOCl_2$ and decomposes into trichloride and chloride, slowly at ordinary temperature, but more rapidly on heating.

---

* Refractory: Difficult to fuse or burn or corrode generally bricks used in furnace.

Sulphides: The trisulphide (V_2S_3) and a tetrasulphide (VS_4) are known. Neither of them is very much soluble in acids, but can be brought into solutions by strong oxidizing agents such as nitric acid or hot concentrated sulphuric acid and they form oxides when heated in air.

Vanadium Trisulphide (V_2S_3): It is greatly obtained by heating the tetrasulphide in an inert atmosphere.

$$\text{Vanadium Tetrasulphide} \longrightarrow \text{Vanadium Trisulphide} + \text{Sulphur}$$

$$4VS_4 \longrightarrow 2V_2S_3 + 5S_2$$

Like pyrites (FeS_2) the tetrasulphide, probably contains S_2 group. It is the chief constituent of the patronite.

Hypovandous Sulphate ($VSO_{47}H_2O$): This compound is prepared by the electrolytic reduction of a solution of vandyl sulphate in the absence in of the air followed by concentration in vacuum. It is similar to ferrous sulphate.

Vanadeous Hydrogen Sulphate [$VH(SO_4)_{24}H_2O$]: It is obtained by the electrolytic reduction of sulphuric acid and vanadium pentoxide mixture followed by the precipitation of concentrated sulphuric acid. Its solutions are reducing agents, which are used in the place of titanious solutions in volumetric analysis. They appear to be less sensitive to the atmospheric oxidation.

Vanadyl Sulphuric ($VOSO_4 . 2H_2O$): This compound of vanadium is prepared by the reaction of vanadium pentoxide and sulphur dioxide in dilute sulphuric acid.

$$\text{Vanadium Pentoxide} + \text{Sulphur Dioxide} \xrightarrow{\text{Sulphuric Acid}} \text{Vanadyl Sulphate}$$

$$V_2O_5 + 2SO_2 \longrightarrow 2VOSO_4$$

And then followed by evaporation. It is a hydroscopic sky blue salt.

Nitrates: Nitric acid oxidizes vanadium compounds and nitrates are produced as yellow compounds.

$$\text{Vanadium Pentoxide} + \text{Nitric Acid} \longrightarrow \text{Vanadoxyl Nitrate} + \text{Water}$$

$$V_2O_5 + HNO_3 \longrightarrow 2VO_2(NO_3)_2 + H_2O$$

CHROMIUM

Atomic Number : 24
Atomic Weight : 51.9961
Melting Point : 1900°C
Boiling Point : 4470°C

History

In 1762-1765 Lehman described a new mineral called Crocorite or Crocosite. In the year 1798, Vauquelin showed that it consisted the oxide of lead and the oxide of new metal called chromium, because it formed colourful compounds (Greek, Chrome = colour). He was able to isolate the metallic chromium in an impure state by the fusion of chromic acid and carbon. Pure chromium was obtained by Deville in the year 1857, but it was not till 1895 that the reasonably pure metal was prepared by Moissan.

Occurrence and Extraction

Chromium is fairly abundant element, which occurs principally as Chromite (Cr_2O_3 FeO) sometimes called Ferro-chrome ore, which is found in South Africa, United States and else where. This mineral is washed finally ground with lime and potassium carbonate and finally roasted, potassium chromate is there by produced.

Potassium Carbonate + Chromite + Oxygen ⟶ Potassium Chromate + Carbon Dioxide

$$4K_2CO_3 + 2Cr_2O_3 + 3O_2 \longrightarrow 4K_2CrO_4 + 4CO_2$$

This is water soluble compound extracted after washing by water then it is converted to potassium dichromate ($K_2Cr_2O_7$) by adding sulphuric acid and re-crystallizing. Potassium dichromate is then converted to chromite ($K_2Cr_2O_4$) by reacting with starch or some other reagents and chromium sesquioxide is produced which is washed with water. This chromium sesquioxide is the source of commercial chromium. Conversion of metallic chromium from its metallic sesquioxide is generally done with the use of carbon or alkali metals like sodium etc, but the process is very difficult. This is done commercially by Goldschimidts Process. In this process one or two hundred weights of chromium is produced (see Fig. 16.7).

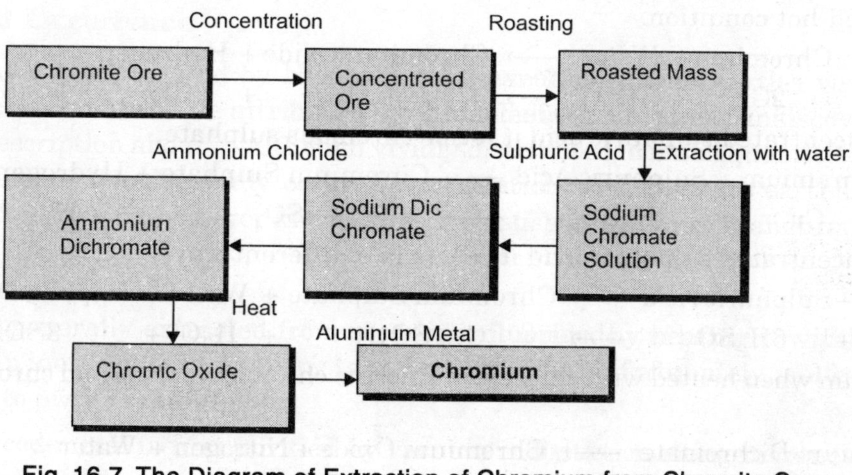

Fig. 16.7. The Diagram of Extraction of Chromium from Chromite Ore.

Properties

Chromium is a bluish-white lustrous metal with a melting point of 1900°C. It takes a good polish which lasts long because of the formation of an oxide coating, which also protects the metal from corrosion. Pure metal is soft and malleable but as usually obtained, chromium is of the strongest of the common metals. This is due to the presence of a small amount of carbon in it as impurity. It is also resistant to the corrosive materials. The compounds are all coloured. The detail physical properties are tabulated below:

Chemical Properties

Melting Point	1900°C
Boiling Point	4470°C
Density	7.43 g/cc
Metallic Radius	127 pm
Electrical Restivity	185 × 10^6 ohm cm
Electronegativity	1.5
Heat of Capacity	5.58 cal deg^{-1} mol^{-1}
Standard Entropy	5.68 cal deg^{-1} mol^{-1}

Chemical Properties

Chromium is not affected by air or atmosphere in normal conditions and therefore does not tarnish in air. However, in the presence of oxygen or air when it is heated it is converted to chromic oxide, which is green in colour.

Chromium + Oxygen ⟶ Chromic Oxide

$$4Cr + 3O_2 \longrightarrow 2Cr_2O_3$$

Chromium does not react with water at ordinary temperature, but it decomposes steam when in a red-hot condition.

Chromium + Water ⟶ Chromium Oxide + Hydrogen

$$2Cr + 3H_2O \longrightarrow Cr_2O_3 + 3H_2$$

With concentrated sulphuric acid it forms chromous sulphate.

Chromium + Sulphuric Acid ⟶ Chromium Sulphate + Hydrogen

$$Cr + H_2SO_4 \longrightarrow CrSO_4 + H_2$$

With concentrated sulphuric acid it reacts in a different way.

Chromium + Sulphuric Acid ⟶ Chromium Sulphate + Water + Sulphur Dioxide

$$2Cr + 6H_2SO_4 \longrightarrow Cr_2(SO_4)_3 + H_2O + 3SO_2$$

Chromium when heated with chlorine or fluorine chromic chloride and chromic fluoride are formed.

Ammonium Dichromate ⟶ Chromium Oxide + Nitrogen + Water

$$(NH_4)_2Cr_2O_7 \longrightarrow Cr_2O_3 + N_2 + 4H_2O$$

Industrial Application of Chromium

The most important use of chromium is the production of special steels for this purpose it is mixed with iron to form an alloy called ferro-chrome.

The chromium is now being used for plating metals, because it is a very good corrosion protective metal. Alloys of nickel and chromium, called nichrome using 20% of chromium are used for making the heating coils and elements in electrical irons and hot plates since it resists oxidation when red-hot.

Chromium is also a continuant of acid resistant alloy. The most important of these alloys are stainless steels.

Chrome Plating: As mentioned above, a larger amount of chromium is used in industry for electroplating steel parts to improve their appearance and to protect them from rusting. For this purpose steel is first plated with copper or nickel or both since the chromium does not made negative and the positive is the lead and the bath contains the solution of chromium trioxide. The solution contains about 250 g of chromium trioxide per litre of water and 3 to 5 g of chromium trisulphate. The process is carried out at 40°C using a high current of 11 amps per sq dm of cathode. Chromium deposits on the metal at the cathode in the form of layer with a bright bluish white lusture, which is not easily lost.

Compounds of Chromium

Chromium produces many compounds, which we are now going to discuss.

Oxides of Chromium

Chromium Sesquioxide (Cr_2O_3): The preparation of this compound is done from chromite and form ammonium dichromate.

If once started the reaction goes on itself and this is the basis of the well-known volcano experiment in which a heap of ammonium dichromate is ignited by a red-hot fire thrust into it.

Chromium sesquioxide is thus prepared in the form of a voluminous green powder insoluble in water; it has been used as a green pigment. It is very difficult to reduce but it is easy to oxidize by potassium permanganate and is converted to potassium chromate. Its rate of solution in acids depends upon the temperature to which it has been heated and if it is strongly ignited is very small.

Associated with varying amounts of water, it is precipitated as a jelly from chromite (which cannot exist in aqueous solution) as can be shown by dialysis, which separates the whole of the sesquioxide. It is also precipitated from such alkaline solutions by boiling or even on long such alkaline solutions by boiling even on long standing. If ammonia is added to chromic solutions same precipitate settles, but it dissolves in excess of ammonia with a violet colour due to a complex chromium ammonia ion.

Chromium Dioxide (CrO_2): This compound is prepared by heating the hydroxide in oxygen, on strong heating the hydroxide, which decomposes on heating into sesquioxide and oxygen. It an also be prepared in the cold condition from the reaction of chromite and chromic salt. It is also produced by heating the chromic nitrate.

$$\text{Chromium Nitrate} \longrightarrow \text{Chromic Acid} \longrightarrow \text{Nitrogen Dioxide}$$

$$Cr(NO_3)_2 \longrightarrow CrO_3 \longrightarrow 3NO_2$$

$$\text{Chromic Acid} \longrightarrow \text{Chromium Dioxide} + \text{Oxygen}$$

$$2CrO_3 \longrightarrow 2CrO_2 + O_2$$

Chromium Trioxide, Chromic Acid (CrO_3): This is prepared by the reaction of concentrated sulphuric acid and concentrated potassium dichromate.

Potassium Dichromate + Sulphuric Acid ⟶ Potassium Sulphate + Chromic Acid + Water

$$K_2Cr_2O_7 + 2H_2SO_4 \longrightarrow 2KHSO_4 \quad 2CrO_3 + H_2O$$

Chromium trioxide looks like scarlet needles filtered off through the asbestos filter. The crystals are very hydroscopic. The chromium trioxide dissolves in water producing dichromic acid.

Chromic Acid + Water ⟶ Dichromic Acid

$$2CrO_3 \quad H_2O \quad H_2CrO_7$$

It is powerful oxidizing agent in acid solution it is usually reduced to chromic state. It oxidizes sulphur dioxide, hydrogen sulphide, ferrous sulphate, arsenious oxide, stannous chloride, stannous iodide, etc. Usually a solution of potassium dichromate acidified with sulphuric acid for this purpose.

Chromium trioxide in aqueous solution acid and compounds such as Potassium Chromate (K_2CrO_4) and Potassium Dichromate ($K_2Cr_2O_7$) corresponding to sulphates and disulphates. Acid chromates of type ($KHCrO_4$) are not yet known. The free chromic acid and dichromic acid are not known in pure state.

Chromous Compounds

Chromous Oxide (CrO): It is best prepared by warming chromium alum with dilute nitric acid which dissolves the mercury and oxidized the chromium. It is a black powder, which may inflame in the air forming the sesquioxide and the acidic trioxide - $Cr_2O_3.CrO_3$. It is decomposed by boiling water into the sesquioxide and chromic acid.

Chromic Acidoxide + Water ⟶ Chromic Acid + Chromium Sesquioxide

$$Cr_2O_3.CrO_3 + H_2O \longrightarrow H_2CrO_4 + \quad Cr_2O_3$$

With excess of alkali carbonate it forms complex salts of Sodium Chromocarbonate $Na_2Cr(CO_3)_2.10H_2O$.

Chromous Sulphate ($CrSO_4$ $7H_2O$): This compound is blue needle shaped crystal, isomorphous with ferrous sulphate. It is prepared by dissolving metallic chromium in dilute sulphuric acid, followed by cooling.

Chromium + Sulphuric Acid ⟶ Chromium Sulphate + Hydrogen

$$Cr + H_2SO_4 \longrightarrow CrSO_4 + H_2$$

Chromous Chloride ($CrCl_2$): It is obtained in colourless deliquescent crystals by passing gaseous hydrogen chloride over heated metallic chromium or by heating anhydrous chromic chloride in dry hydrogen.

Hydrogen Chroride + Chromium ⟶ Chromous Chloride + Hydrogen

$$2HCl + Cr \longrightarrow CrCl_2 + H_2$$

Chromic Chloride + Hydrogen ⟶ Chromous Chloride + Hydrogen Chloride

$$2CrCl_3 + H_2 \longrightarrow 2CrCl_2 + 2HCl$$

It is readily dissolve in water producing a blue solution.

Chromic Compounds

Chromic Hydroxide [$Cr(OH)_3$]: It is formed as a green gelatinous precipitate, by the addition of alkali hydroxide, ammonia or ammonium sulphide to the solution of chromic salt.

Chromic Chloride + Ammonium Sulphide + Water ⟶ Chromic Hydroxide + Ammonium Chloride + Hydrogen Sulphide

$$2CrCl_3 + 3(NH_4)_2S + 6H_2O \longrightarrow 2Cr(OH)_3 + 6NH_4Cl + 3H_2S$$

Freshly prepared chromic hydroxide is soluble in alkali forming a deep green solution, which contains chromite, *e.g.* sodium chromite (when treated with sodium hydroxide). Chromic hydroxide is weak base. When chromic chloride is treated with hot concentrated ammonia in the presence of ammonium chloride a considerable amount of chromium remains in solution giving a pink colour to the latter, due to the formation of chromoamines *e.g.* $Cl_3.Cr(NH)_3H_2O$.

Chromic Chloride ($CrCl_3$): It is obtained as red violet peach-blossom coloured sublimate of scaly crystals by passing a current of chlorine over red hot chromium sesquioxide and carbon.

Chromium Sesquioxide + Carbon + Chlorine ⟶ Chromic Chloride + Carbon Monoxide

$$Cr_2O_3 + 3C + 3Cl_2 \longrightarrow 2CrCl_3 + 3CO$$

It can also be prepared by heating the hydrated chloride ($CrCh_6H_2O$) in a current of hydrogen chloride about 250°C. The anhydrous chromic chloride is insoluble in water. Hydrated chloride ($CrCl_2O$) is obtained as green crystals by evaporating of a solution of chromium trioxide (CrO_3) in boiling concentrated hydrochloride acid. There are there different modifications of the crystalline hexahydrate of chromic chloride, having different structure and properties:

1. **Grayish blue [$Cr(H_2O)_6Cl_3$]** All the chlorine are present as the chloride ion are precipitated by silver nitrate from solution.
2. **Light-Green [$Cr(H_2O)_5Cl.\ Cl_2H_2O$]** Only one of the chlorine atoms is precipitated by silver nitrate from solutions and two molecules of water are easily removed.
3. **Dark-Green [$Cr(H_2O)_4Cl_2CI.2H_2O$]** Only one of the chrome atoms is precipitated by silver nitrate from solutions and two molecules of water are easily removed.

Chromic Sulphate [$Cr(SO_4)_3$]: This anhydrous salt is red. The sulphates are obtained as violet grey or green coloured crystals depending upon the number of molecules of water. The violet crystals, with 16 or $18H_2O$ are obtained by dissolving the chromium hydroxide in hot strong sulphuric acid and allowing the green solution to stand for some weeks when it becomes violet and then precipitating the solution with a little alcohol. The anhydrous sulphate is obtained by heating the hydrate in a current of carbon dioxide at 280°C - 400°C. The green variety is formed by heating the violet crystals at 90°C until they have the composition $Cr_2(SO_4)_36H_2O$.

A freshly prepared solution of the crystals does not give any precipitate with alkalis on barium chloride and their molecular formula is represented by $[Cr(H_2O)_6]_2(SO_4)_36H_2O$.

Chromic sulphate forms with the sulphates of alkalis some important compounds called alums, chrome alum; $K_2SO_4.Cr(SO_4)_324H_2O$ or $KCr(SO_4)_210H_2O$. Potassium chrome alum is also known as chrome alum is obtained as deep purple octahedral crystals by reducing a solution of potassium dichromate in dilute sulphuric acid with sulphur dioxide, alcohol, or oxalic acid.

Chrome alum is used m calico printing, dye and chrome tanning.

Chromates and Dichromates

When chromium sesquioxide is made to react with sodium hydroxide m the presence of oxygen sodium chromate is produced.

Chromium Sesquioxide + Sodium Hydroxide + Oxygen ⟶ Sodium Chromate + Water

$$Cr_2O_3 + 4NaOH + O_2 \longrightarrow 2Na_2CrO_4 + H_2O$$

But the industrial process is quite a different one, this process includes the heating of finely powdered chromite sodium carbonate and quicklime mixture to red-hot state. The reaction is carried out m a reverberatory furnace, with free access of air. The chromium is slowly oxidized to chromate and the iron to ferric oxide.

Ferric Chromate + Sodium Carbonate + Calcium Oxide + Oxygen ⟶ Ferric Oxide + Sodium Chromate + Calcium Carbonate

$$4FeCrO_4 + 8N_2CO_3 + 8CaO + 7O_2 \longrightarrow 2Fe_2O_3 + 8Na_2CrO_4 + 8CaCO_3$$

The sodium chromate is extracted with water and is converted into sodium dichromate by treatment concentrated sulphuric acid.

Sodium Chromate + Sulphuric Acid ⟶ Sodium Dichromate + Sodium Sulphate + Water

$$2Na_2CrO_4 + H_2SO_4 \longrightarrow Na_2Cr_2O_7 + Na_2SO_4 + H_2O$$

sodium sulphate separates out and is removed. The solution on concentration deposits deliquescent red crystals of sodium dichromate $Na_2Cr_2O_7$.

Potassium Dichromate ($K_2Cr_2O_7$): This compound can be prepared by adding potassium chloride to sodium dichromate.

Potassium dichromate being less soluble separates out in the form of dark red crystals

Sodium Dichromate + Potassium Chloride ⟶ Potassium Dichromate + Sodium Chloride

$$Na_2Cr_2O_7 + 2KCl \longrightarrow K_2Cr_2O_7 + 2NaCl$$

Chromates and dichromates are used as oxidizing agents as mordents in dying and in making pigments.

Potassium Chromate ($K_2Cr_2O_4$): It is obtained is lemon yellow crystals. By treating a solution of chromium trioxide or potassium dichromate with calculated amount of potassium hydroxide and then crystallizing the solution.

Potassium Dichromate + Potassium Hydroxide ⟶ Potassium Chromate + Water

$$K_2Cr_2O_7 + KOH \longrightarrow 2K_2CrO_4 + H_2$$

Potassium Chromate reacts like alkali in water due to hydrolysis and potassium dichromate reacts acid in water. Both potassium chromate and potassium dichromate contains no water molecule and therefore they are not deliquescent salts. Potassium dichromate is used in laboratory for volumetric analysis and also as an oxidizing agent for many organic preparations. The two important oxidation-reduction reactions for which it is frequently used consist in the oxidation of ferrous sulphate and potassium iodide in the acid solutions.

Ferrous Sulphate + Sulphuric Acid + Potassium Dichromate ⟶ Potassium Sulphate + Chromium Sulphate + Ferric Sulphate + Water

$$6FeSO_4 + 7H_2SO + K_2Cr_2O_7 \longrightarrow K_2SO_4 + Cr(SO_4)_3 + 3Fe_2(SO_4)_3 + 7H_2O$$

Potassium Iodide + Sulphuric Acid + Potassium Dichromate ⟶ Potassium Sulphate + Chromium Sulphate + Iodine + Water

$$6Kl + 7H_2SO_4 + K_2Cr_2O_7 \longrightarrow K_2SO_4 + Cr(SO_4)_3 + 3l_2 + 7H_2O$$

As potassium dichromate has no action on dilute hydrochloric acid it can he used for titration of solutions containing chlorides for which the potassium permanganate is unstable.

Potassium dichromate in acid solution on reduction with sulphur dioxide gives equimolecular portions of chromium sulphate and potassium sulphate, which deposits crystals of chrome-alum $K_2SO_4.Cr(SO_4)_3.24H_2O$.

Ammonium Chromate [$(NH_4)_2CrO_4$]: This compound is unstable and changes from chromate to dichromate by loosing ammonia. The dichromate is prepared by adding a calculated amount of ammonia in chromium trioxide.

Chromium Trioxide + Ammonia + Water ⟶ Ammonium Dichromate

$$2CrO_3 + 2NH_3 + H_2O \longrightarrow (NH_4)_2Cr_2O_7$$

It decomposes violently on heating yielding nitrogen and chromium sesquioxide (volcano experiment).

Barium Chromate ($BaCrO_4$): This compound is a lemon yellow substance produced by the reaction of barium hydroxide and potassium chromate.

Potassium Chromate + Barium Hydroxide ⟶ Potassium Hydroxide + Barium Chromate

$$K_2CrO_4 + Ba(OH)_2 \longrightarrow 2KOH + BaCrO_4$$

It is almost insoluble in water but freely soluble in strong acids. As the chromate of calcium is soluble in water. This substance is used in the separation of barium in analysis.

Lead Chromate ($PbCrO_4$): This compound occurs naturally in the form of crocosite and is prepared in the form of bright yellow substance named "chrome yellow" and it is used as pigment. It is highly insoluble in water, but soluble in acids and it sometimes used in the analysis of lead compounds.

Silver Chromate ($AgCrO_4$): It is a brick red coloured compound prepared by the reaction of silver nitrate and sodium chromate.

Silver Nitrage + Sodium Chromate ⟶ Sodium Nitrate + Silver Chromate

$$2AgNO_3 + Na_2CrO_4 \longrightarrow 2NaNO_3 + Ag_2CrO_4$$

If a drop of a silver nitrate solution is placed as a layer of gelatine containing sodium chromate in solution and spread on a glass plate, the precipitate from in concentric rings separated by clear spaces. They are called *Lisengang's Rings* after the name of the discoverer the exact explanation of this experiment is not known.

Potassium Chlorochromate ($KCrO_3Cl$): This substance is deposited in red brown crystals, form potassium dichromate when it reacts with worm hydrochloric acid.

Potassium Dichromate + Hydrochloric Acid ⟶ Potassium Chlorochromate + Water

$$K_2Cr_2O_7 + HCl \longrightarrow 2KCrO_3Cl + H_2O$$

Chromyl Chloride (CrO_2Cl_2): This substance is prepared by distillation of a dichromate and a chloride, with concentrated sulphuric acid. The product contains hydrogen chloride and chlorine from which it may also be prepared by the addition of concentrated sulphuric acid to a well cooled mixture of chromium chloride, forms a layer and may be separated.

Chromic Trioxide + Hydrogen Chloride ⟶ Chromyl Chloride + Water

$$CrO_3 + 2HCl \longrightarrow CrO_2Cl_2 + H_2O$$

Chromyl chloride is a heavy liquid, density 1.96, with a beautiful deep red colour, freezing at –96°C and boiling at 117°C. It is a pure covalent compound, soluble in organic liquid such as carbon tetrachloride. It is immediately decomposed into chromic acid and hydrochloric acid.

Chromyl Chloride + Water ⟶ Chromic Acid + Hydrogen Chloride

$$CrO_2Cl_2 + 2H_2O \longrightarrow H_2CrO_4 + 2HCl$$

For this reason it fumes in moist air. It is a powerful oxidizing agent and also chlorinating agent. If added to alcohol or phosphorous it immediately explodes.

Chromyl Chloride explodes when added to phosphorous.

Chromyl Fluoride (CrO_2F_2): It is similar to chromyl chloride. It is obtained by heating lead chromate, calcium fluoride and sulphuric acid.

Calcium Fluoride + Lead Chromate + Sulphuric Acid ⟶ Chromyl Fluoride + Lead Sulphate + Calcium Hydroxide

$$CaF_2 \quad PbCrO_4 \quad H_2SO_4 \longrightarrow CrO_2F_2 \quad PbSO_4 \quad Ca(OH)_2$$

Detection and Estimation

Chromium compounds produce an intense green colour in a borax bead, both in the oxidizing flame and the reducing flame. Also the dichromates and chromates are detected in the blue colour in the acid hydrogen peroxide solution. Chromium in other compounds can be oxidized to chromate, by titration against ferrous disulphate using an internal indicator.

MANGANESE

Atomic Number : 25 Melting Point : 1245°C

Atomic Weight : 54.93 Boiling Point : 2097°C

History and Occurrence

Manganese is obtained mainly from *Pyrolusite* (MnO_2) the most important ore; it is actually the black manganese dioxide. In those early days pyrolusite was confused with magnesite, (Fe_2O_4). Gahn first obtained the metal in an impure state in 1775 by reducing pyrolusite with carbon.

Manganese though not a very abundant element it is available easily. It occurs in the form of oxides and less frequently as sulphide, the most important or is pyrolusite, as said above is also known as manganese dioxide, which is found mainly in India. Other common ores of manganese are *Braunite* (Mn_2O_3), *Manganite* (Mn_2O_3), (H_2O) *Manganese Blende, MnS*. Manganese compounds occur in traces in plants and animals and play an important part in biological process.

Extraction

Impure manganese can be obtained by strong heating of the oxide Mn_3O_4 with carbon.

Manganese Tetraoxide + Carbon ⟶ Manganese + Carbon Monoxide

$$Mn_3O_4 + 4C \longrightarrow 3Mn + 4CO$$

Pure metal is obtained by reducing the oxide Mn_3O_4 with aluminium by Thermit process.

Manganese Tetraoxide + Aluminium ⟶ Manganese + Aluminium Oxide

$$Mn_3O_4 + 8Al \longrightarrow 3Mn + 4Al_2O_3$$

Pure manganese can also obtained by electrode composition from an acidified solution of manganese ammonium sulphate.

Properties

Manganese is a grey hard brittle metal; the physical properties are given in the table below:

Physical Properties

Melting Point	1245°C
Boiling Point	2097°C
Density	7.44 g/cc
Moh's Hardness	5
Standard Free Energy	.33 kcal/mol^{-1}
Standard Entropy	7.59 cal deg^{-1} mol^{-1}
Heat of Capacity	6.59 cal deg^{-1} mol^{-1}

Manganese in pure state is stable in air but if it contains carbide and other impurities, it is rapidly corroded. When quite pure, it does not react with pure water; sometimes it reacts with warm water forming manganous hydroxide and hydrogen is evolved. Manganese reacts with dilute acids to form hydrogen.

$$\text{Manganese + Sulphuric Acid} \longrightarrow \text{Manganese Sulphate + Hydrogen}$$
$$Mn + H_2SO_4 \longrightarrow MnSO_4 \quad H_2$$

With chlorine or hydrogen, chloride it forms manganous chloride.

$$\text{Manganese + Hydrogen Chloride} \longrightarrow \text{Manganese Chloride}$$
$$Mn + HCl \longrightarrow MnCl_2$$

At much high temperature it reacts with sulphur, phosphorous an also nitrogen.

$$\text{Manganese + Nitrogen} \longrightarrow \text{Manganese Nitride}$$
$$Mn + N_2 \longrightarrow MnN_2$$
$$\text{Manganese + Phosphorous} \longrightarrow \text{Manganese Phosphide}$$
$$Mn + P_2 \longrightarrow MnP_2$$

It also forms carbide when it combines in an electric furnace at very high temperature with carbon.

Uses of Manganese

1. About 80% of manganese produced is used in the manufacture of manganese steel. Manganese when added to steel performs two functions if added in small proportions it acts as a scavenger (cleaning agent) in removing oxygen and sulphur forming manganese oxide and manganese sulphide, which can be easily eliminated. This improves the strength of the final product.
2. When added upto 14% it produces a special hardness and toughness to the steel. Ordinary steel contains from 0.6 to 1.5% of manganese, the toughness in iron ore increases with the addition of manganese.
3. The steel containing about 14% of manganese known as manganese steel is very hard and for making crushers and steel helmets etc.
4. Some other alloys of manganese is manganin (Cu = 84%, Ni = 3%, Fe = .4% and Mn = 12%) which has a very low temperature co-efficient of resistance.

5. Among the other useful alloys of manganese with copper and zinc. Its approximate composition is (Cu = 64%, Zn = 35%, Fe = 0.5%, Mn = 0.5%). It is resistance towards water and is therefore used in ship building industry.
6. Manganese compounds like Manganese dioxide; potassium permanganate is used as laboratory reagent.

Compounds of Manganese

Manganic Oxide (Mn_2O_3): This compound of manganese is available in nature in hydrated form. It can also be obtained by passing a brisk current of air for some hours through an ammonical solution of manganous salt and ammonium chloride. Manganic oxide is a black solid which can be heated in the air without change to nearly 1000°C above that the oxide will lose oxygen and become trimanganese tetraoxide. It is easily reduced to the monoxide, by heating in hydrogen. With concentrated acids it yields solutions of magnetic salts. When heated acids these compounds produce manganous salts and precipitates hydrated manganous dioxide.

Trimanganese Tetraoxide (Mn_3O_4): It is stable oxide of manganese at high temperature and can be prepared by heating any other oxide to 1000°C or over. The colour depends upon the method of preparation and the state of subdivision. In the hydrated form it can also be prepared by caustic alkalis mixed with manganous and manganic solutions. With acids it forms manganese dioxide.

Manganese Dioxide (MnO_2): It occurs in nature as pyrolusite, the principal ore of manganese. In the hydrated form it can be prepared from solutions of manganous compounds by many different reactions. Manganous salts can be reacted in alkaline medium with sodium peroxide to form manganese dioxide.

Manganous Carbonate + Sodium Peroxide ⟶ Manganese Dioxide + Sodium Carbonate

$$MnCO_3 + Na_2O_2 \longrightarrow MnO_2 + Na_2CO_3$$

Manganese dioxide is a black grainy compound, nearly insoluble in water. When strongly heated it yields sesquioxide and tetraoxide at about 600°C.

Manganese Dioxide ⟶ Manganese Sesquioxide + Manganese Tetraoxide

$$MnO_2 \longrightarrow MnO_2O_3 + Mn_3O_4$$

Manganese also reacts with cold dilute hydrochloric acid forming an insoluble solution of manganese tetrachloride.

Manganese Dioxide + Hydrochloric Acid ⟶ Manganese Tetrachloride + Water

$$MnO_2 + 4HCl \longrightarrow MnCl_4 + H_2O$$

This solution evolves chlorine on heating.

Manganese Tetrachloride ⟶ Manganese Dichloride + Chlorine

$$MnCl_4 \longrightarrow MnCl_2 \quad Cl_2$$

When heated with concentrated sulphuric acid below 140°C the dioxide, yields oxygen and manganic sulphate, the manganic sulphate changes to manganous sulphate at higher temperature.

Manganese Dioxide + Sulphuric Acid ⟶ Manganese Sulphate + Water + Oxygen

$$4MnO_2 + 6H_2SO_4 \longrightarrow 2Mn_2(SO_4)_3 + 6H_2O + O_2$$

Manganese Sulphate + Water ⟶ Manganese Sulphate + Sulphuric Acid + Oxygen

$$2Mn_2(SO_4)_3 + 2H_2O \longrightarrow 4MnSO_4 + H_2SO_4 + O_2$$

Manganese Heptoxide (Mn_2O_7): This compound is the anhydrous permanganic acid. It is obtained by adding powdered potassium permanganate to well cooled concentrated sulphuric acid, and then treating the dark green solution with ice cold water.

Potassium Permanganate + Sulphuric Acid ⟶ Potassium Sulphate + Manganese Heptoxide + Water

$$2KMnO_4 + 2H_2SO_4 \longrightarrow 2KHSO_4 + Mn_2O_7 + H_2O$$

It forms an opaque brown heavy oily liquid with specific gravity 2.4. When warmed to 40°C-50°C, it gives a violet vapour, in presence of organic matter it can cause explosion. The heptoxide decomposes on heating by liberating ozonized oxygen and separation of manganese dioxide.

Manganese Heptoxide ⟶ Manganese Dioxide + Ozone

$$Mn_2O_7 \longrightarrow MnO_2 + O_3$$

The heptoxide when dissolved in cold water produces permanganic acid.

Manganese Heptoxide + Water ⟶ Permanganic Acid

$$Mn_2O_7 \quad H_2O \longrightarrow 2HMnO_4$$

Permanganic acid is generally prepared by the reaction of potassium permanganate and hydroflurosilicic acid in pure state.

Potassium Permanganate + Hydroflurosilicic Acid ⟶ Potassium Fluorosilicate + Permanganic Acid

$$2KMnO_4 + H_2SiF_6 \longrightarrow K_2SiF_6 + 2HMnO_4$$

or a solution of barium permanganate with dilute sulphuric acid.

Barium Permanganate + Sulphuric Acid ⟶ Permanganic Acid + Barium Sulphate

$$Ba(MnO_4) + H_2SO_4 \longrightarrow 2HMnO_4 + BaSO_4$$

The salts of permanganic acid consist of a dark violet colour and they dissolve in water to form a purple solution. The most important of these is potassium salt, which is described in the next sub head.

Potassium Permanganate ($KMnO_4$): Potassium permanganate is one of the most important compounds of manganese. It is prepared on a commercial scale from pyrolusite and potassium hydroxide in the presence of potassium chlorate.

A mixture of potassium chlorate and potassium hydroxide is heated in an iron dish; finely powdered manganese dioxide is added very little at a time with constant stirring and the mixture is continuously heated to keep the mixture in molten state. The mixture is then cooled and boiled in distil water during this time carbon dioxide is passed through the solution till a small amount of permanganic acid is left. The solid matter is allowed to settle and the clear liquid is taken out, and then allowed to crystalline the final product is thus obtained.

Properties of Potassium Permanganate

- Potassium permanganate forms dark purple crystals with a metallic lusture. It dissolves in water, giving a dark purple solution.
- When heated strongly, decomposes giving potassium manganate and manganese dioxide, with oxygen.

Potassium Permanganate	$\longrightarrow$	Potassium Mangnate	+ Manganese Dioxide	+ Oxygen

$$2KMnO_4 \longrightarrow H_2SO_4 + K_2SO_4 + O_2$$

- With concentrated sulphuric acid it gives manganese heptoxide.

Potassium Permanganate	+ Sulphuric Acid	$\longrightarrow$	Potassium Sulphate	+ Manganese Heptoxide	+ Water

$$2KMnO_4 + 2H_2SO_4 \longrightarrow 2KHSO_4 + Mn_2O_7 + H_2O$$

- Potassium permanganate is a powerful oxidizing agent both in solid and in solution.
- When solid potassium permanganate is heated in hydrogen it burns giving potassium hydroxide and manganous oxide and steam.

Potassium Permanganate	+ Hydrogen	$\longrightarrow$	Potassium Hydroxide	+ Water	+ Manganous Oxide

$$2KMnO_4 + 5H_2 \longrightarrow 2KOH + 4H_2O + 2MnO$$

- A mixture of potassium permanganate with sulphur phosphorous or organic matter explodes when rubbed or heated.
- In neutral solution brown hydrated manganese dioxide and potassium hydroxide are produced and three atoms of oxygen becomes available for oxidation per two molecules of permanganate.

Potassium Permanganate	+ Water	$\longrightarrow$	Potassium Hydroxide	+ Manganese Dioxide	+ Oxygen

$$2KMnO_4 + H_2O \longrightarrow 2KOH + MnO_2 + 3O$$

- In alkaline solution, potassium permanganate reacts to form potassium manganate.

Potassium Permanganate	+ Potassium + Hydroxide	$\longrightarrow$	Potassium Manganate	+ Water	+ Oxygen

$$2KMnO_4 + 2KOH \longrightarrow 2K_2MnO_4 + H_2O + O$$

- Acid solution (sulphuric acid) reacts with potassium permanganate according to the equation below:

Potassium Permanganate +	+ Sulphuric Acid	$\longrightarrow$	Potassium Sulphate	+ Manganese Sulphate	+ Water	+ Oxygen

$$2KMnO_4 + 3H_2SO_4 \longrightarrow 2K_2MnO_4 + 2MnSO_4 + 3H_2O + 5O$$

Sodium Permanganate ($NaMnO_4$): Sodium permanganate is prepared by the same process as potassium permanganate, but instead of potassium hydroxide and potassium chlorate, sodium hydroxide and sodium chlorate is used. But is very difficult to crystallize and therefore it is used in aqueous state under the name of *Condy's Fluid.*

Manganous Chloride ($MnCl_2$): This is prepared by heating manganese dioxide and ammonium chloride.

$$\underset{\text{Dioxide}}{\text{Manganese}} + \underset{\text{Chloride}}{\text{Ammonium}} \longrightarrow \underset{\text{Chloride}}{\text{Manganese}} + \text{Ammonia} + \text{Water} + \text{Nitrogen}$$

$$3MnO_2 + NH_4Cl \longrightarrow 3MnCl_2 + 4NH_3 + 6H_2O + N_2$$

The residue is extracted with boiling water and the solution evaporated when the pink crystals of ($MnCl_2.2H_2O$) separates at lower temperature the hexahydrate ($MnCh_6H_2O$) is formed. A hydrate ($MnCl_22H_2O$) is obtained at higher temperature. The anhydrous salt (MnCh) is generally prepared by heating the dihydrate at 98°C, or by heating the hydrated salts with concentrated hydrochloric acid. The anhydrous salt forms rose red coloured deliquescent salts.

Manganous Carbonate ($MnCO_3$): This occurs in nature as precipitate by adding a solution of sodium carbonate and chloride.

$$\text{Mangnous Chloride} + \text{Sodium Carbonate} \longrightarrow \text{Mangnous Carbonate} + \text{Sodium Chloride}$$

$$MnCl_2 + Na_2CO_3 \longrightarrow MnCO_3 + 2NaCl$$

Manganous chloride slowly hydrolyzed by water and is changed to manganous oxide on decomposition.

$$\text{Manganese Carbonate} \longrightarrow \text{Manganous Oxide} + \text{Carbon Dioxide}$$

$$MnCO_3 \longrightarrow MnO + CO_2$$

Manganous Nitrate [$Mn(NO_3)_2$]: It forms many hydrated nitrates, including the pink coloured deliquescent crystals of hexahydrated nitrate [$Mn(NO_3)_26H_2O$] are obtained by evaporating a solution of manganese carbonate in slight excess of dilute nitric acid. The salt decomposes on heating.

$$\text{Manganese Nitrate} \longrightarrow \text{Manganous Dioxide} + \text{Nitrogen Dioxide}$$

$$Mn(NO_3)_2 \longrightarrow MnO_2 + NO_2$$

The anhydrous salt can be obtained by warming the hydrated salt with concentrated nitric acid.

Manganous Phosphate [$Mn_3(PO_4)_2.7H_2O$]: It is obtained as a white precipitate by the addition of an excess of disodium hydrogen phosphate to a manganous salt, in the presence of ammonia and ammonium chloride. A pinkish coloured manganous ammonium phosphate is formed as a crystalline precipitate. The precipitate, when dried at 110°C, has the molecular formula $MnNH_4PO_4.H_2O$. On ignition, it forms the manganous pyrophosphate.

Manganese is estimated in this form in the gravimetric process.

Manganous Ammonium Phosphate ⟶ Manganous Pyrophosphate Ammonia Water

$$2MnNH_4PO_4 \longrightarrow Mn_2P_2O_7 + 2NH_3 + H_2O$$

Manganous Sulphide (MnS): This compound is produced when hydrogen sulphide is passed through a solution of manganous salt.

Manganous Carbonate + Hydrogen Sulphide ⟶ Manganous Sulphide + Carbonic Acid

$$MnCO_3 + H_2S \longrightarrow MnS + H_2CO_3$$

This from of manganese sulphide is a metastable form; it changes to stable green coloured crystalline form, more hydrogen sulphide is passed at a temperature of 250°C. Manganous sulphate gets dissolved in acid.

Manganous Sulphate ($MnSO_4.H_2O$): This forms as crystals when manganese dioxide reacts with sulphuric acid.

Manganese Dioxide + Sulphuric Acid ⟶ Manganous Sulphate + Water

$$MnO_2 + H_2SO_4 \qquad MnSO_4 + H_2O$$

Then evaporating the solution and gently igniting the residue to decompose the ferric sulphate formed form the oxides of iron present in the manganese dioxide. The undecomposed manganous sulphate is extracted with water and the solution evaporated, when the pink efflorescent crystals of the tetrahydrate ($MnSO_4.4H_2O$) separate. At a temperature below 9°C the heptahydrate ($MnSO_47H_2O$) (similar to FeSO47H2O) between 9°C - 27°C, the pentahydrate ($MnSO_4.5H_2O$) (similar to $CuSO_4.5H_2O$) is obtained. The anhydrous sulphate (MnSO4) is formed by heating the hydrated salts at 280°C and is almost colourless.

Manganic Fluoride (MnF 33H_2O): Manganic Fluoride is produced when manganous sesquioxide reacts with hydrofluoric acid.

Manganous Sesquioxide + Hydrogen Fluoride ⟶ Manganic Floride + Water

$$Mn_2O_3 + 6HF \longrightarrow 2MnF_3 \quad 3H_2O$$

Manganic fluoride can also be obtained by the action of fluorine on manganous iodine.

Manganous Iodide + Fluorine ⟶ Manganous Fluoride + Iodide

$$2Mgl + 3F_2 \longrightarrow 2MnF_3 + I_2$$

when heated it decomposes into manganous fluoride and fluoride it forms complex fluoride of the composition $K_2MnF_5.H_2O$.

Manganic Chloride ($MnCl_3$): It is found in the solution of manganese dioxide and hydrochloric acid.

Manganese Dioxide + Hydrogen Acid ⟶ Manganic Chloride + Water + Chlorine

$$2MnO_2 + 8HCl \longrightarrow 2MnCl_3 + H_2O + Cl_2$$

When manganic chloride is heated chlorine is produced.

Follow the statement below.

Manganic Chloride ⟶ Manganous Chloride + Chlorine

$$2MnCl_3 \longrightarrow 2MnCl_2 + Cl_2$$

Pure manganic chloride is obtained by the action of dry hydrogen chloride and manganous acetate, but this is obtained only under the temperature of –26°C above this temperature the compound is decomposed into chlorine and manganous chlorine (as shown in the equation above).

Manganic Phosphate, Manganic Orthophosphate ($MnPO_4$ H_2O): This compound of manganese is prepared by the action of manganous sesquioxide and orthophosphoric acid.

Manganous Sesquioxide + Orthophosphoric Acid ⟶ Manganic Phosphate + Water

$$Mn_2O_3 + 2H_3PO_4 \longrightarrow 2MnPO_4 + 3H_2O$$

The metaphosphate is only red in colour and others are all violet in colour. All the phosphate compounds of manganese are insoluble in water.

Manganic Acetate [$Mn(C_2H_3O_2)_3$ $2H_2O$]: This compound is obtained in the form of brown crystals, when metallic manganese is dissolved in hot acetic acid.

Manganese + Acetic Acid ⟶ Manganic Acetate

$$Mn + 3CH_3COOH \longrightarrow Mn(C_2H_3O_2)_3$$

Complex Cyanides of Manganese

No simple cyanide of manganese has been isolated. When a solution of manganous salt is treated with potassium cyanide, a yellowish green precipitate is formed. This dissolve in excess of cyanide and dark violet crystals of potassium manganocyanide [$K_4Mn(CN)_6H_2O$] are deposit from the solution. When a solution of manganocyanide is exposed to air it occurs a deep-red colour and manganocyanide is formed. The potassium manganicyanide, [$K_3Mn(CN)_66H_2O$] are deposited from the solution. The potassium manganicyanide, [$K_3Mn(CN)_6$] is obtained in dark red crystals which are very unstable.

Potassium Manganocyanide + Water + Oxygen ⟶ Potassium Manganicyanide + Potassium Hydroxide

$$4K_4Mn(CN)_6 + 2H_2O + O_2 \longrightarrow 4K_3Mn(CN)_6 + 4KOH$$

Sodium and potassium salts of complex cyanides with univalent manganese of *e.g.* sodium manganous cyanide [$Na_5Mn(CN)_6$] and potassium manganous cyanide [$K_5Mn(CN)_6$]. These salts are also been prepared from the reduction of corresponding manganocyanides either by metallic aluminium in alkaline solution (for the sodium salt) or by the electrolytic method (potassium salt).

Detection and Estimation of Manganese

The presence of manganese in a substance can be detected easily by fusing it with sodium carbonate and nitre, when a green mass of manganese is produced. This green mass when acidified turns violet due to the formation of permanganate with separation of manganese dioxide. A very sensitive test is carried out with the reaction of the substance with sodium bismuthate or lead dioxide, when the solution obtains a violet-red colour of manganese, indicated due to the presence of permanganic acid. The reaction with sodium bismuthate is used for the volumetric analysis of manganese.

From solution, manganese can be gravimetrically estimated from the precipitation of manganese ammonium phosphate ($MnNH_4PO_4H_2O$) and igniting the precipitate to pyrophosphate ($Mn_2P_2O_7$).

IRON

Atomic Number : 26 — Melting Point : 1536°C

Atomic Weight : 55.847 — Boiling Point : 3000°C

History and Occurrence

Iron was known to man form the early days, during the time of Egyptians (3400-3000 B.C.) but extremely scarce and used only as beads for jewellery. It may have been obtained from meteoric form of iron, since it contained nickel. Iron of this early period is also known for Mesopotamia. The use of iron seems to have spread from the Hittites of Asia Minor. It was very much used by the Assyrians about 600B.C. In the Mycenean (the preclassic Greek) period described by Homer iron was still a rare metal at that time because Achilles was gifted a lump of iron as prize. But the Greeks brought with them the use of iron. Iron was known in India from 900B.C. and in China from 550 B.C.

Iron occurs very rare in the free state. Some ballistic rocks are said to contain it in small quantities in free state. Many meteorites contain iron in the form of alloy with nickel about 2 to 3% of nickel. It is believed that the inner core of the earth largely consists of iron in the 4.7% of the earth's crust. It is the fourth in abundance amongst the elements; the first three are being oxygen silicon and aluminium.

The most important ores of iron are the oxides, carbonates, and the sulphides, listed below:

1. **Magnesite (Fe_3O_4):** Which is usually black in colour and is found in India, Sweden, Germany and England.
2. **Alaemattite (Fe_2O):** Which is usually crystalline in nature and red in colour and also found in Indies, England, France and Germany.
3. **Limonite, or Hydrated Ferric Oxide (Fe_2O_3, $3H_2O$):** Which is found in England, France, Germany, Spain and Canada.
4. **Iron Pyrites (FeS_2):** Which occurs abundantly but are used more as a source of sulphur dioxide, for the manufacture of sulphuric acid and than a source of Iron.

Varieties of Iron

There are three commercial varieties of iron, which are put to different industrial uses, they are *Cast iron, Wrought Iron, Pig iron and steel.* The differences in their properties depends upon their carbon content and also their phosphorous content. *Cast Iron* is the least pure from of the iron, it consists of only 2.5 to 4.5% of carbon together with some sulphur, phosphorous and manganese. *Wrought Iron* is the purest form of iron consisting not more that 0.5% of carbon and other impurities. Steel consisting not more that 0.5% of carbon and other impurities. Steel comes in between cast iron and wrought iron, its carbon content lies between 0.4% to 1.5%. It usually contains some manganese as well.

Impurities in the Iron Ore their Effects: The principal impurities found in the iron ore are silica and aluminium oxide. The presence of these devalues the iron ore because they not only reduce the iron content in the ore but also increase the cost of production as larger quantity of flux and fuel are required for production of iron and steel. The presence of calcium carbonate or magnesium carbonate would decrease the amount of flux and consequently of silicon, aluminium oxide and limestone, is such that the ore is more cheaply. The famous *Minettite* ore is the example of this type.

Sulphur and phosphorous are always found in iron ores in varying proportions in the form of ferrous sulphate, calcium sulphate and calcium phosphate. Both sulphur and phosphorous are not desired in iron also in steel. Sulphur may vary from 0.01 to 0.05%. Sulphur and phosphorous are also contributed by the fuel used. The steel normally should not contain more than 0.05% sulphur and 0.05% phosphorous.

Sulphur can be removed in the blast furnace slag but the cost of production goes high due to the high flux and fuel consumption. Sometimes sulphur is removed by introducing calcium carbonate, phosphorous cannot be removed from the slag and thus goes into the pig iron which gets converted into steel in the converter; due to this the cost of production goes high. Small amount of manganese is generally present in the good steel, as it helps in the reduction of sulphur by forming manganous sulphide. So if manganese is absent in the ore it is added externally.

Fuel: This is used mainly for two purposes, firstly to provide energy to the plant and secondly it reduces the iron ore to metallic iron by combining with oxygen of the ore. Oxygen supplied in the furnace should be regulated because a part of the oxygen is obtained from the ore itself.

The fuel is coke, which is prepared by burning coal in an oxygen deficient atmosphere to remove the volatile matter. The volatile matter is removed, as a fuel slag which goes to the lower part of the furnace where the heat require is maximum. During coke production the volatile matter goes off and the coke becomes porous in nature due to swelling. The hardness of the coke produced is also important, as it has to bear a great load in the furnace and still should not get-disintegrated. The passage of oxygen will be choked which is very dangerous. The coke should be very hard enough and should not react to form carbon monoxide. Porosity produces great surface area, which increases the rate of reaction, also the sulphur and the ash contents should be as low as possible. The composition of coke is given in the list below:

Analysis of Coke

Fixed Carbon	85% - 89%
Sulphur	7 - 1.3%
Volatile Matter	0.9% - 1.5%
Ash	5.5 - 13.0%

India is having a good reserve of coal but a good quality of coaking coal is not available.

For 1 tonne of iron production 900kg of coke is required. For making coke all types of coal are not suitable, because some produces soft coke and others in powdered form, although after prolonged heating.

Flux: It is the chemical substance, which is used in the furnace to remove the silicates and ferrous sulphate and convert them in the form of slag. Here limestone is used as flux. The reactions of flux are shown below.

$$\text{Calcium Carbonate} \longrightarrow \text{Calcium Oxide} + \text{Carbon Dioxide}$$

$$CaCO_3 \longrightarrow CaO + CO_2$$

$$\text{Calcium Oxide} + \text{Silicon Dioxide} \longrightarrow \text{Calcium Silicate}$$

$$CaO + SiO_2 \longrightarrow CaSiO_3$$

$$\text{Ferrous Sulphide} + \text{Calcium Hydroxide} + \text{Carbon Sulphate} \longrightarrow \text{Calcium Monoxide} + \text{Carbon} + \text{Iron}$$

$$FeS + CaO + C \longrightarrow CaS + CO + Fe$$

Then flux is used in the central or the middle region of the furnace where the temperature is about 850°C - 950°C. The composition of flux are shown below:

Flux (Lime Stone)

Lime	30 - 54%
Silica	0.6% - 4.5%
Phosphorous	0.001 - 0.025%
Silica	0.004% - 0.132%
Aluminium Oxide	0.4 - 3.0
Magnesia	22%

Manufacture of Pig Iron and the Description of Blast Furnace

Pig Iron is produced by melting iron ore with flux in blast furnace. Oxygen is blown through a number of symmetrically placed pipes called tuyers. The molten iron and slag is collected at the bottom of the blast furnace and the gases are removed from the top. The slag layers floats on the heavy iron, therefore the iron is tapped from the bottom and the slag is tapped from the top of the mixture. The normal capacity of the blast furnace is about 105 tonnes a day. Fig. 16.8 shows the diagram of the blast furnace.

The furnace is about 30.5m tall and 10m in diameter. There are two frustums* of cones the larger one is called stack and the smaller one is called *bosh*. The upper one is supported on iron structure so that the bottom can be opened for repairs etc. The bosh rests on a cylindrical section called *hearth*. The stack gradually widens from top to the bottom because to accommodate the explanation of charge, because of heating aid is always kept full with charge.

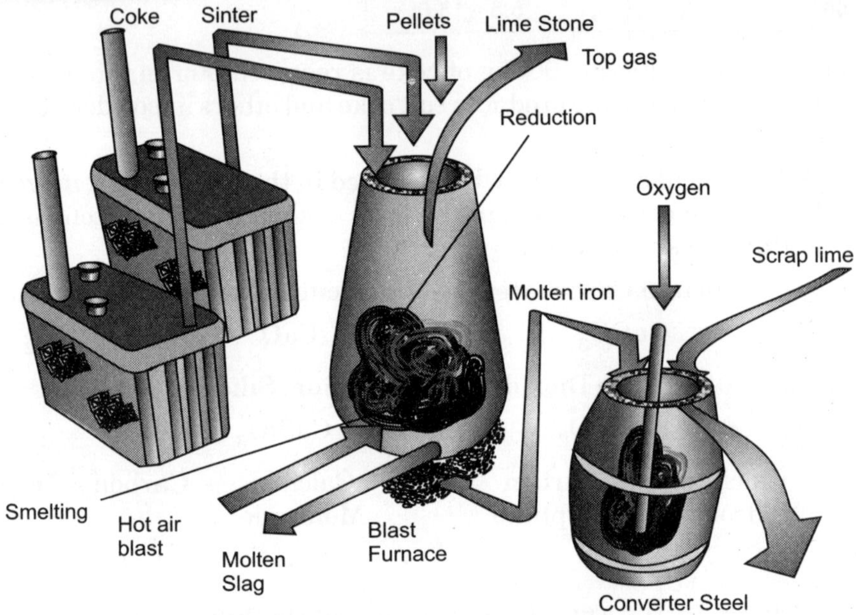

Fig. 16.8. The Operation of Blast Furnace

The iron oxide reacts with carbon monoxide and carbon and also flux producing iron. Calcium carbonate decomposes to give calcium oxide and carbon dioxide. The temperature in the stack less and therefore, the charges always remain in the solid form. Fusion takes place in the bosh, which is having a diameter of 6m and 4m height. The bosh angle is 75 to 80°. The hearth acts as a crucible where molten iron and slag are collected (the detail blast furnace is shown in the Fig. 16.9).

The saft of the furnace is made up of steel plates lined with firebricks of thickness of 1.2 to l.5m. The walls are water-cooled and stream is raised out of the heat. The walls have 0.5 to 1.0m lining of firebreaks and crucible is about 1 m thick. From the bottom of the crucible the molten iron is taken from the bottom of the crucible and the slag from the top through a tap hole, which is about 1.5 m high. The out going gas mixture is passed to the dust catcher and then to wash lined and finally stored to be used for heating oxygen apart.

The slag is either mixed with flowing water quenches or it is sent to a settling tank where water is mixed with slag gets settled and then the clear water is re used for different purpose. The slag is used for cement manufacture, railway ballast or floor tiles.

---

* Frustums: -a cone cut into "equal halves, the lower on

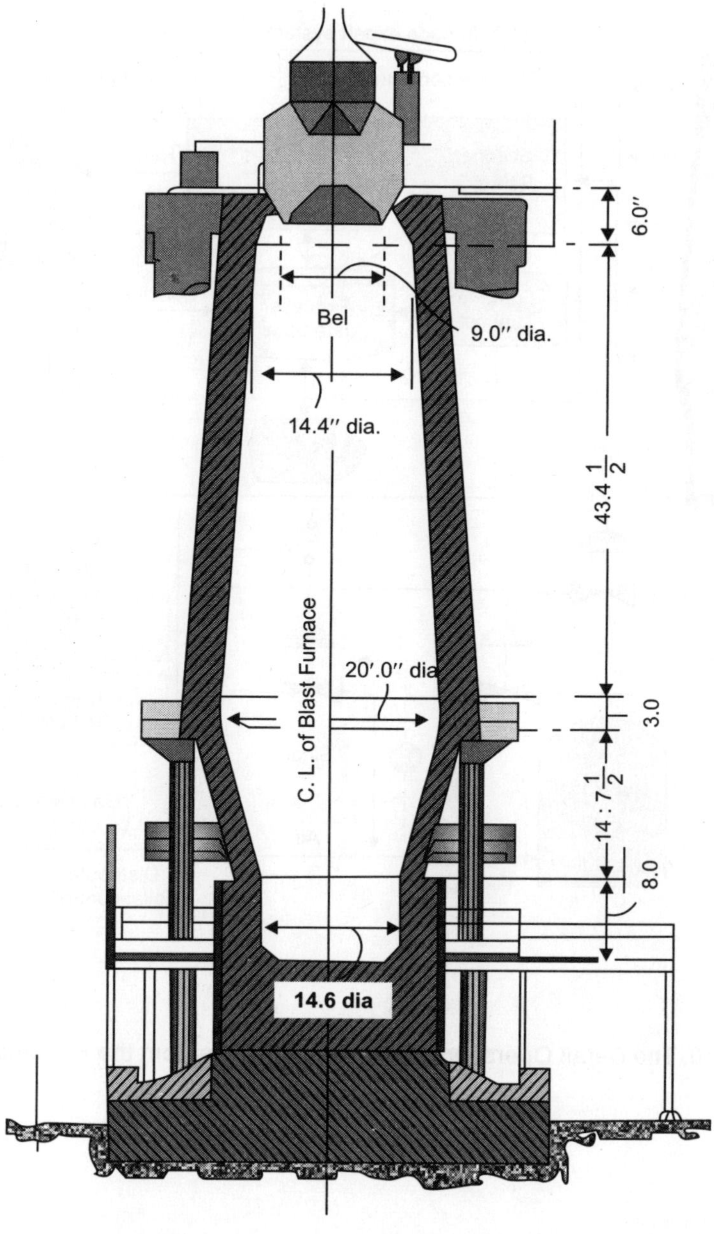

Fig. 16.9. The Blast Furnace

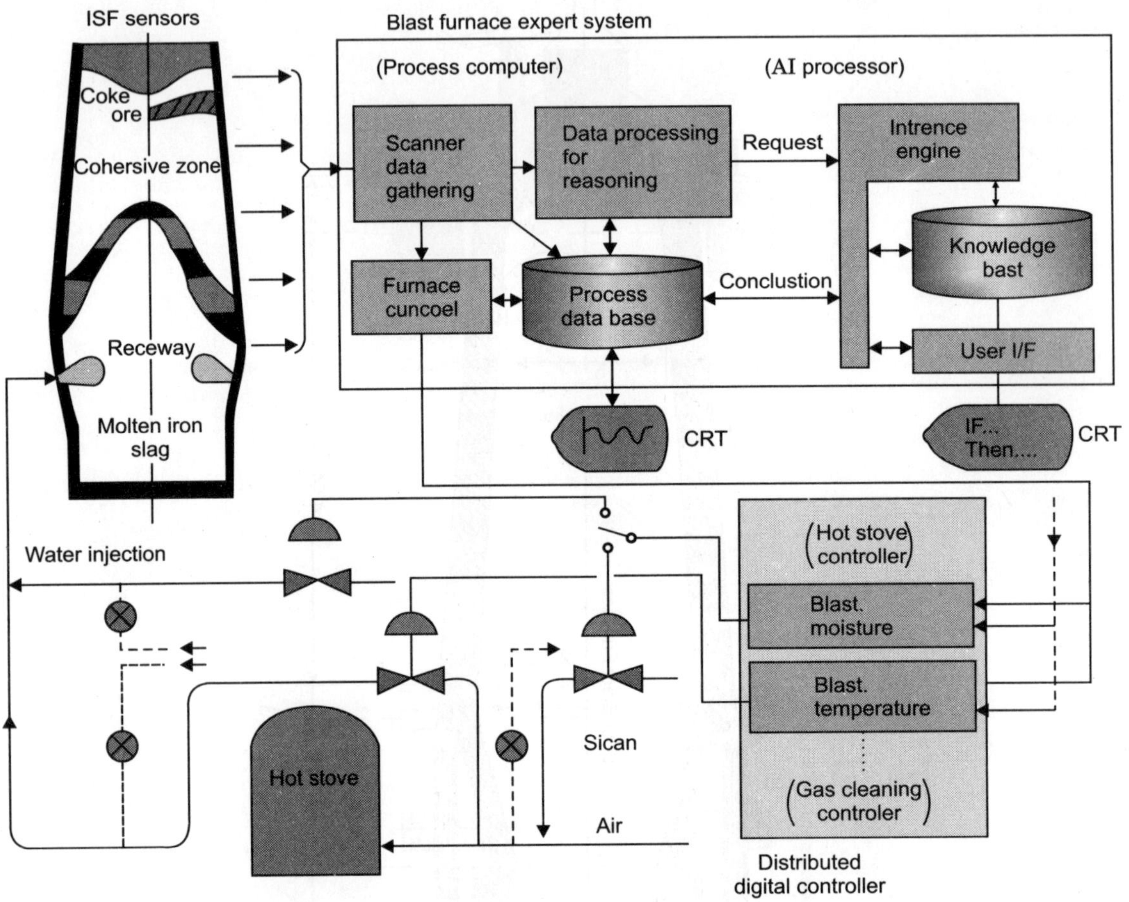

Fig. 16.10. The Detail Operation of the Blast Furnace from the Furnace Control

The Blast Furnace Temperatures and Reactions

Depth Below Stack line	Temperature	Reaction Equation
3.04 m	450°C	$2Fe_2O_3 + 6CO \longrightarrow 4Fe + 6CO_2$ $Fe + CO_2 \longrightarrow FeO + CO$
6.09 m	575°C	$Fe + CO \longrightarrow FeO + C$ $3Fe_2O_3 + C \longrightarrow 2Fe_3O_2 + CO$
9.144 m	650°C	$Fe_3O_4 + CO \longrightarrow 3FeO + CO_2$
12.1 m	755°C	$FeO + CO \longrightarrow FeO + CO_2$
13.7 m	900°C	$MgCO_3 \longrightarrow MgO + CO_2$
15.2 m	1050°C	$Fe_3O_4 + 5C \longrightarrow Fe_3C + CO_2$ $MnO + C \longrightarrow Mn + CO$ $CaO + Al_2O_3 + 2SiO_2 \xrightarrow{Mn+CO} CaSiO_3 + Al_2SiO_3 + O_2$
18.2 m	1200°C	$MnO + C \longrightarrow Mn + CO$
21.3 m	1350°C	$3Fe + C \longrightarrow Fe_3C$ $CaO + Al_2O_3 + 2SiO_2 \xrightarrow{Fe_3C} CaSiO_3 + Al_2SiO_3 + O_2$
24.3 m	1800°C	$Ca_3P_2O_8 + 3SiO_2 + 5C + 2Fe \longrightarrow 3CaO.SiO_2 + 5CO + 2FeP$
27.5 m	1900°C	$SiO_2 + 2C + 2Fe \longrightarrow FeSi + 2CO$

Cast Iron

Cast iron is produced by passing the metal into moulds of desired shape and size. The metal gets cooled and solidified and takes the desired shape. The pig iron and scrap is melted in an open furnace which is also known as cupola. actually it is a cylindrical shell of boiler plates lined with fire bricks and resting on a square bed plate. The opening of the bedplates is closed by a two semicircular cast iron hinged doors opening downwards. These support the change and help in cleaning etc. From the bottom of the cupola, air for combustion is blown in. At the bottom are iron tap holes and slag tap holes. Cupolas are cleaned and repaired from time to time, these are lightened with wood and waste oil and then the charge of coke is poured from the top of the wood etc. when the bed charge is well ignited iron is put in. The charge of coke with every charge of iron is about 6 to 7kg iron is used per kg of coke. The blast is turned on; iron melted into moulds sometimes for longer operation cupola flux is added.

For small scale production cupola furnace is commonly used it has certain disadvantages for e.g. alignment and maintenance of temperature.

These disadvantages have led to the development of rotary furnace for special work. The furnace consists of a cylindrical drum with conical ends the heating is done from one end by burning some fuel the products goes out from the other end of the operator, higher production and better temperature control. The sands for main moulds are also of special type. The main requirement of a good mould is that it should withstand high temperature without damage to the surface of contact. There should be no softening and change of shape in the mould. There must be adequate number of pores for the escape of gases, steams.

Composition of Cast Iron

Cast iron normally consists of 3 to 5% of carbon also nitrogen, phosphorous sulphur and manganese in small proportions. The presence of iron carbide also known as cementite is present with graphite, which change in composition depending on the rate of cooling. If the rate of cooling is slow the formation of cementite is less which decreases the hardness property of the cast iron and if the rate of cooling is high the cementite is precipitated and give the hardness to the cast iron.

Hardness of the cast iron depends upon the cementite deposition. which on the other hand depends upon the rate of cooling.

Silica, when present in large quantity (about 2 to 3%) fabours the decomposition of cementite to graphite which results in the formation of grey cast iron.

Manganese removes sulphur and provides an increment in tensile strength and hardness. Manganese content varies from 0.5% to 1 % white cast iron is the cheapest and is not of much used, except for making weights. Most of the cast iron used is the gray cast iron because due to the presence of graphite. This type of iron is less hard in nature, with good working properties, specially enough resistance to rupture. It is therefore, used for making base plates bodies etc.

Malleable Cast Iron

The total carbon content is low but not less that about 2% the freezing point temperature would be high below this, and would cause difficulties, sulphur should b quite low as its presence reduces the conversion of cementite to graphite and ferrite. Phosphorous has no appreciable effect but its presence is restricted to a maximum of 20%.

It is prepared by slowly cooling white cast iron of special composition so that the hardness is less. The usual composition of malleable cast iron is carbon = 2 to 2.55, silicon = 0.8 to 1.1 %, sulphur = 0.8%, Phosphorous = 0.1 % to 0.2% and manganese = 3%.

Wrought Iron

It is nearly free from carbon and other impurities. This is very soft and fibrous structure and resistant to vibration and fatigue. This is used for the manufacture of chains draw bars and making wrenches and spanners.

Wrought iron is generally prepared by melting pig iron in a reverberatory furnace mixed with iron ore. The iron ore takes part in an important role in the process of manufacture, as the oxygen of the ore reacts with the carbon present in the pig iron and forms the monoxide of carbon and also dioxide of carbon. Then the carbon is and the iron left is wrought iron, which is first made in a shape of balls and then with hammers it is given a proper shapes, of billet, strip and rods etc.

Steel

Steel is an alloy of carbon and iron and has carbon less than wrought iron. Normally steel have carbon content of 3%. A part from carbon it has sulphur and phosphorous in small quantities which is not desirable, manganese and silicon are desirable.

Between very soft wrought iron and very hard cast iron likes steel, which has all the desirable properties of strength, hardness and care of fabrication.

Steel is generally manufactured by four processes they are:

1. Bessemer's Process
2. Lintz-Donwtz Process
3. Open-Hearth Process
4. Electric Arc Furnace.

Bessemer's Process: This process was discovered by an English engineer whose name was Henry Bessemer (1815-1899). The process consists of blowing air through iron oxidizing the impurities and simultaneously converting pig iron to steel. The converter is made of steel with brick lining and is pear shaped vessel with a perforated bottom as shown in the Fig. 16.11. The molten pig iron is mixed in mixtures and then charged into converters. About 15 to 16 tones of iron is charged at a time. The converter may be rotated on a horizontal axis.

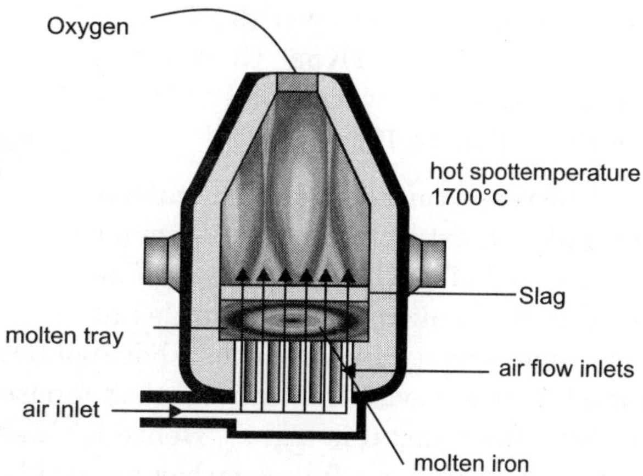

Fig. 16.11. The Bessemer's Converter

After charging, cold air is blasted through tuyers at a pressure of about 2.5kg/cm^2 of about 10 to 15 minutes during which the impurities are oxidized after this the blast is shut off and the recarburizers are added to get steel.

The converter actually has three parts : (1) detachable bottom, (2) a cylindrical portion and (3) upper conical portion.

This converter is lined with highly refractory material. The thickness of the material is about 40cm.

During the operation oxygen of the blast combines with silica and produces a dull red flame. After this the flame turns yellower because of burning of carbon monoxide. The carbon content that is 0.04% as has been indicated earlier here, therefore carbon must be added. Apart from this the metal must be deoxidized otherwise large number of blow holes would be produced. For this purpose recarburizers and deoxidizers are added. Coke is added as decarburizer and materials like silicon, manganese is added as deoxidizer. The finished steel is then transferred to a bucked refractory lined steel shell with a hole at the bottom through which the molten metal is allowed to flow into moulds. When the ingots have cooled successfully, these are reheated and rolled to various shapes and sizes.

Lintz-Donawitz Process (L-D Process)

It was invented in the year 1932 in Australia and is now in use at many places through Asia, India, Europe and other places. The Hindustan Steel Ltd at Rourkela uses the same process Fig. 16.12.

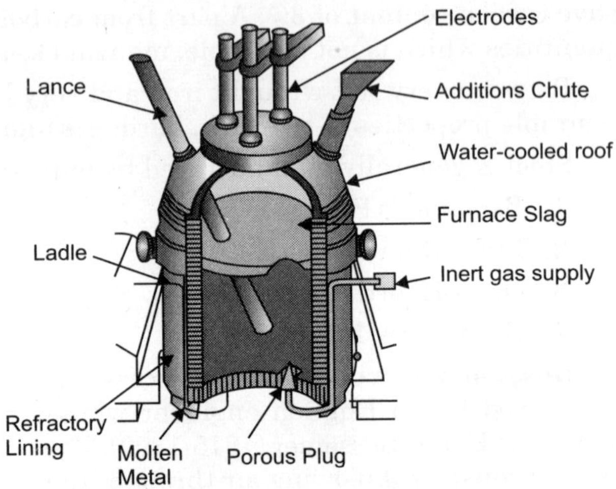

Fig. 16.12. The Ladle Furnace

The process is preferable for flats and plates. The process consists of introducing pure oxygen (99%) at a pressure of .6894Kpa through a specially designed water-cooled lane as shown in Fig. 16.12.

The vessel containing iron metal has a similar construction as a Bessemer process, but of much improved way. The oxygen is supplied through nozzles at very high velocity and produces a hot spot (temperature about 25,000°C) striding with the metal. It reacts with iron, silicon, manganese and carbon, refining the metal. The density of the refined metal is higher, Hence it sinks to the bottom, allowing unrefined metal to come to the top for refining carbon monoxide is produced in the reaction.

The decomposition is automatically achieved. The entire process takes an hour to complete. In the L-D Process the unwanted sulphur can be effectively reduced by oxidation. Silicon is removed and manganese is controlled. The L-D steel has low oxygen gas and contents and low phosphorous and also low sulphur, nitrogen content is also low as pure oxygen is used instead of air for blasting L-D steels has superior properties as compared to open hearth steel. It has good resistance to cleavage* fracture, better ductility.

The L-D Process has been found to be suitable for India; raw materials, the Indian pig iron contains higher phosphorous and therefore, is not suitable for steel conversion by open-hearth process. The basic open-hearth process is slow duplexing is often done which is the combination of acid Bessemer and basic: open-hearth furnace

* Cleavage: -Divided.

Open Hearth Furnace Process

The process has an open-hearth furnace. The gaseous fuel are burned with controlled amount of oxygen enriched and to produce high-temperature required for steel making the construction of the open hearth furnace is shown in Fig. 16.13. The furnace is a regenerative type with brick checker works to utilize heat of the outgoing gases by heated by' the fuel gas. The hearth formerly contained acid lining but later on for the removal of the phosphorous hearth was constructed of magnesite birth, covered with a layer of burnt dolomite or magnesite to replace the sand of open hearth furnace.

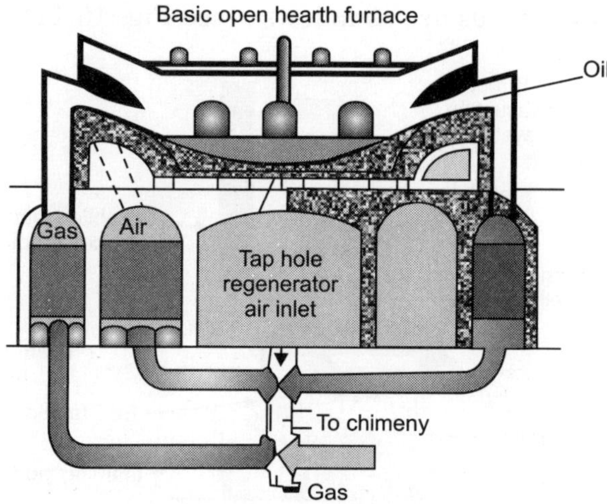

Fig. 16.13. The Open Hearth Furnace

The arched roof is made of silica, the advantage of this type of furnaces are: (1) Better temperature control, (2) acts as the Oxidizing agent and (3) easy elimination of phosphorous.

The normal capacity of an open-hearth furnace is 200-300 tonnes but bigger furnace is also in operation; sometimes liquid fuels are used instead of gaseous fuel. (The roofs are lined with chrome-magnesite bricks). The hearth is about 12.5m long and 4.8m width and about 50cm deep. The hearth is made of firebricks lined with, magnesite bricks. On the top of which a lining of calcinated magnesite and open-hearth slag is pit. Ports are provided for admission of fuel and air and are made of silica bricks. The airport is kept above the gas first to avoid to unnecessary oxidation of the metal produced. A suitable hole is provided at the centre of the hearth to tap out the metal. The slag is removed from the far end of the hearth.

The furnace charges consist of pig iron ore and limestone. Manganese is prepared in higher amount as it helps in removal of sulphur should as low as possible.

Acid Hearth Process

The principal difference is the difference in the composition of the hearth. In acid furnace it is made of sand. The charge consists of pig iron and scrap iron in the order of pig-scrap-pig-scrappig-scrap.

The pig at the bottom protects the lining and the pig at the top protects the scrap from getting oxidized. The pig iron has to be low in phosphorous and sulphur content, as both cannot be eliminated.

During melting all the silicon, manganese and part of carbon are oxidized. The silicon dioxide and manganese oxide from the slag are to be removed. Sometimes small amount of lime is added to increase the fluidity. Carbon is removed by adding ore. After this ferro-silicon is added as oxidizer.

Electric Arc Furnace

The heating is done in an electric arc furnace to facilitate rapid steel production. The advantages are as follows : (see Fig. 16.14).

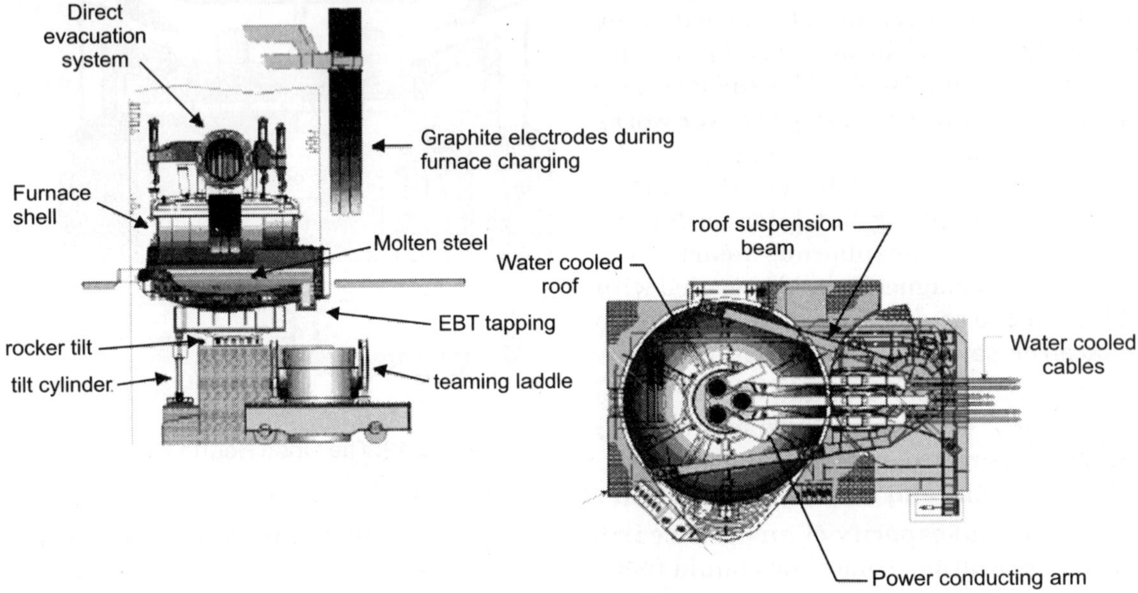

Fig. 16.14. The Diagram of Electric Arc Furnace

1. Clean plant
2. Rapid production and maintenance of the desired composition.
3. Phosphorous, sulphur and oxygen can be totally removed as oxidizing, reducing and neutralizing conditions can be made as will.

Advantages and disadvantages of the electric arc furnace process.

The disadvantage is that the electricity used in the furnace is costly which increases the cost of production of steel.

The features of this furnace are (1) Arc and (2) Induction.

The arc can be produced between either two carbon electrodes or one carbon electrode and the charge. In case of induction the charge is heated induction current.

Charging: Plate scrap and limestone is charged to the hearth to prevent from reaching with silica produced from silicon in the charge. It also delays the decomposition of limestone to the reaction below.

$$\begin{array}{ccccc} \text{Calcium Carbonate} & \longrightarrow & \text{Calcium Oxide} & & \text{Carbon Dioxide} \\ CaCO_3 & \longrightarrow & CaO & + & CO_2 \end{array}$$

Lime stone being at the bottom acts as insulator to heat transfer from the fuel charge. The limestone reacts with oxidized phosphorous and removes the same as slag. Scrap and pig iron are charged to the furnace because scrap is purer and if directly comes in contact with the form. It would get oxidized.

It requires about 2 hours to get the charge molten, scrap iron get oxidized during the process, being at the top the silicon and manganese present get oxidized by the oxygen of the ore and go to the slag. Phosphorous is oxidized to phosphorous pentoxide, which combines with iron oxide, and ore forming iron-manganese phosphate goes into the slag. Later this reacts with calcium oxide of limestone to produce calcium phosphate, which goes to the slag, oxidation of carbon proceeds slowly, during this carbon monoxide formed goes out bubbling through the metal and thus assisting in purification.

After this is over limestone decomposes quickly ($CaCO_3 - CaO + CO_2$) and carbon dioxide produced reacts with carbon in the bath to form carbon monoxide which goes out and calcium oxide thus formed reacts with manganese and iron phosphate to produce calcium phosphate which goes into the slag.

After silicon, manganese and phosphorous are removed the carbon concentration is adjusted. One method would be to remove significant part of carbon in the furnace itself and then recarburize the metal in the ladles.* The metals temperature at which it is poured to ladles must be high so that the metal is in the fluid state, during the recarburization. The other procedure is to control the concentration in the furnace by controlled addition of pig iron consisting of 3.5% of carbon (when carbon removal rate in furnace is high because of oxidation) or ore (when carbon removal rate in the furnace is low). To remove phosphorous also by oxidation sometimes pig iron is added to compensate for the high rate of carbon removal by oxidation reaction taken place as shown below:

Ferric Phosphide + Oxygen ⟶ Ferric Oxide + Phosphorous Pentoxide

$$3FeO + 11O \longrightarrow 6FeO + P_2O_5$$

Ferric Oxide + Phosphorous Pentoxide ⟶ Ferric Oxide·Phosphorous Pentoxide

$$3FeO \quad P_2O_5 \longrightarrow (FeO_3)\, P_2O_5$$

Phosphorous Pentoxide

Ferric Oxide + Calcium Oxide ⟶ Calcium Oxide Phosphorous Pentoxide + Ferric Oxide

$$(FeO)_3\, P_2O_5 + 3CaO \longrightarrow (CaO)_3P_2O_5 + 3FeP$$

Significant fraction of world production is by open Hearth Furnace.

Description of Electrical Arc Furnace

It consists of round dome shaped furnace with concave hearth made up of magnesite and the roof is made up of silica bricks. The graphite electrodes are fitted at the roof projected towards the hearth and the distance between the electrodes are water-cooled and have precise smoother movements due to threaded joints. Electric current is passed, arc is produced and the charge is heated to refine the steel. Normally complete (see Fig. 16.15) operation of the conversion of pig iron to steel is not economical due to expensive electrical power the steel obtained from the open hearth furnace is refined here to get better quality.

* Ladles: Enormous spoon.

Steel scrap is the usual charge but charges also contain pig iron from the open-hearth process. When the furnace has been charged, current is passed and ore produced. The charge gets melted and forms a molten metal. Some lime and iron ore are fed to facilitate formation of slag. Firstly phosphorous and sulphur etc., are removed then some coke is introduced to adjust the carbon content of the steel. Coke also reacts with lime to form calcium carbide, samples of slag are taken out will liberate acetylene gas when comes in contact with water. Small amount of ferro-manganese, ferro-silicon are added as deoxidizers and then the metal tapped to prepare ingots*:

(a) ***Hammering:*** As the name suggests the ingots are hammered with hand or mechanical hammer to give the material the desired shape and size.

(b) ***Drop Forging:*** This consists of hammering the steel between two hardened pieces to give desired shape and size. These are two portions, the upper is attached with hainmer called punch and the lower is called die. Because of forging the material is given the desired shape. It gives quite uniform properties unlike in hammering and forging.

(c) ***Rolling:*** Ingots are passed through rollers to give the round shape to the metal. The practice is very common for the production of rounds of various shapes and sizes for structural purpose. Continuous seamless+ steel can be produced by rolling the ingots and then forcing the round through a mandrel+ to get the desired tube section.

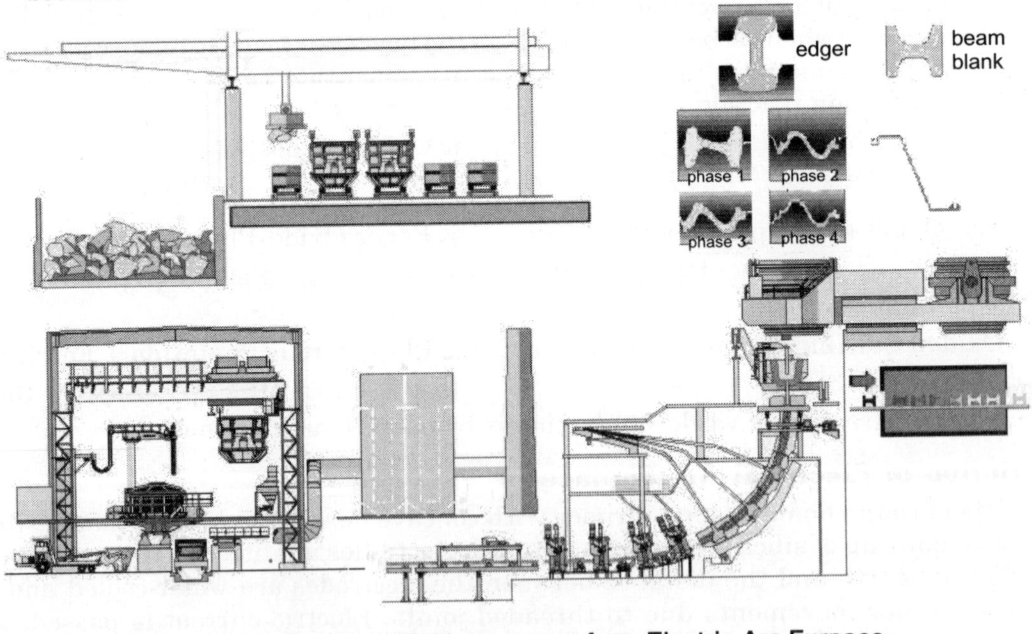

Fig. 16.15. The Steel Making process from Electric Arc Furnace

---

* Ingots: Raw metal before finishing.
+ Mandrel: Bar of iron in the shape of spindle used for holding pices of iron for further machining.
+Seamless: A continuous sheet without joining.

Physical Properties of Iron

Pure is silver-white metal. Physical properties of different types of iron stated in the next table.

Name	Melting Point	Structure	Hardness	Ductile	Mailleable	Tenuous	Tempered
1. Wrought	1500°C	Fibrous	Soft	Yes	Yes	Yes	No
2. Mild Stell	1450°C	Granular	Moderately Soft	"	"	"	"
3. Hard Steel	1375°C	Granular	Hard to very hard	Not much	Not much	Not much	Yes
4. Cast Iron	1200°C	Crystalline	very hard	No	No	"	No

Melting Point	1536°C
Boiling Point	3000°C
Density	7.87 g/cc
Thermal Conductivity	0.9 cal/sec/sqcm/°C/cm
Electrical Conductivity	100, 000 micro ohm/cm
Young's Modulus for Elasticity	2.0684271^{8} Pascal
Tensile Strength	275789 Pascal
Yield Strength	82735.084 Pascal
Elongation in 5.08 cm at 20°C	20 - 45%
Reduction of Area	70 - 90%
Binell Hardness	50 - 70%
Impact Strength (Izod notch Bar)	.762 m/Kg
Grip Strength at Flowing Temperature	
(I) 300°C	117210 Pascal
(II) 400°C	82737 Pascal
(III) 600°C	1736 Pascal
Thermal Expansion Per Degree Centigrade	
100°C	0.0000126 cm
500°C	0.0000146 cm
600°C	0.000016 cm
Specific heat	
100°C	0.12 cal/g
500°C	0.16 cal/g
750°C	0.32 cal/g
900°C	18.5 cal/g
Resistivity	2800000
Maximum Permitivity	6.03 cal deg^{-1} mol^{-1}
Heat of Capacity Standard Entropy	6.49 cal deg^{-1} mol^{-1}

CHEMICAL PROPERTIES

Rustion of Iron

The Corrosion Process

The following will treat corrosion as a process, which cannot occur without the presence of water and therefore excludes other types of attack, such as those associated with high temperature oxidation or sulphidation.

Corrosion is an electrochemical process in which a metal reacts with its environment to form an oxide or other compound. The cell which causes this corrosion process has three essential constituents: an anode, a cathode and an electrolyte (electrically conducting solution). The anode is the site at which the metal is corroded; the electrolyte is the corrosive medium; and the cathode (part of the same metal surface, or of another metal surface in contact with it) forms the other electrode in the cell and is not consumed in the corrosion process. At the anode the corroding metal passes into the electrolyte as positively charged ions, releasing electrons electrons, which participate in the cathodic reaction. Hence the corrosion current between anode and the cathode consists of electrons flowing within the metal and ions flowing within the electrolyte.

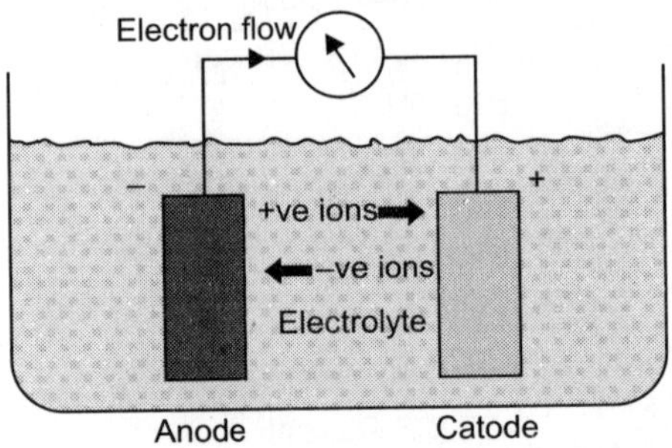

Electrolytic Cellular Arrangement of the Rusting Process

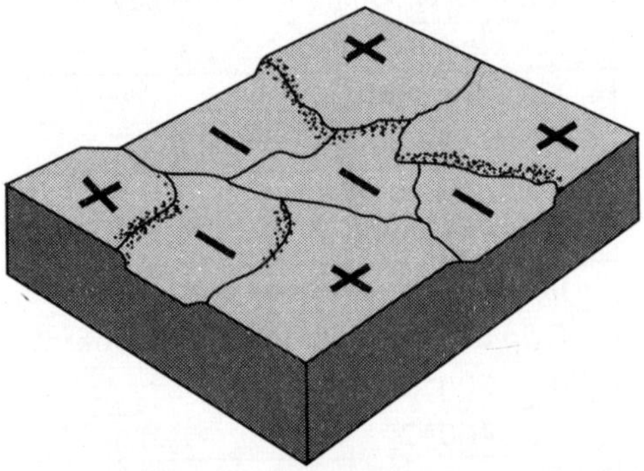
Cellular arrangement on the metal block

The surface of one component may become the anode and the surface of another component in contact with it the cathode. Usually, corrosion cells will be much smaller and more numerous, occurring at different points on the surface of the same component. Anodes and cathodes may arise from differences in the constituent phases of the metal itself, from variations in surface deposits or coatings on the metal, or from variations in the electrolyte. The metal may be immersed in an electrolyte or the electrolyte may be present only as a thin condensed or adsorbed film on the metal surface. The rate of corrosion

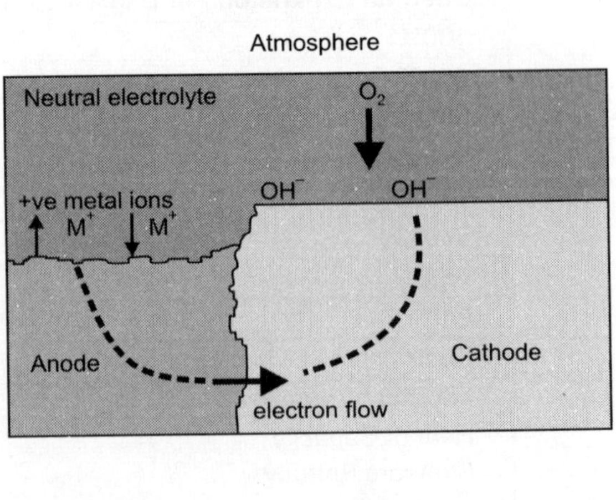

is influenced considerably by the electrical conductivity of the electrolyte. Pure water has poor electrical conductivity and the corrosion rate will be much lower than say an acid solution of high conductivity.

The ability of metals to resist corrosion is to some extent dependent upon their position in the electrochemical series.

The Reactions Related with Rusting

$$\text{Carbonic Acid} \rightleftharpoons \text{Carbonic Ion} + \text{Hydrogen Ion}$$

$$\left.\begin{array}{l} H_2CO_3 \rightleftharpoons HCO_3^- + H^+ \\ \text{Water} \longrightarrow \text{Hydroxyl Ion} + \text{Hydrogen Ion} \\ H_2O \longrightarrow OH^- + H^+ \end{array}\right\} \text{in the water film}$$

$$4Fe^{2+} + O_2 + 8OH^- + (2x - 4)\, H_2O \longrightarrow 2Fe_2O_3xH_2O \text{ ... At the anode area}$$

or,

$$\left.\begin{array}{l} 4Fe^{2+} + 2H_2O + O_2 \longrightarrow 4Fe^{3+} + 4OH^- \\ Fe^{3+} + 3OH^- \longrightarrow Fe(OH)^3 \end{array}\right\} \text{Rust}$$

$$\left.\begin{array}{l} 2H^+ + 2e \longrightarrow H_2 \\ 2H_2 + 2O_2 \longrightarrow 2H_2O \end{array}\right\} \text{At the cathodic area}$$

Electrochemical Series

Element	Ion	Electrode Potential (Volts)	Hydrogen Overvoltage (Volts)
Magnesium	Mg^{2+}	–1.87 (**Base End**)	0.7
Aluminium	Al^{3+}	–1.35	0.5
Zinc	Zn^{2+}	–0.76	0.7
Chromium	Cr^{2+}	–0.6	0.32
Iron	Fe^{2+}	–0.44	0.18
Cadmium	Cd^{2+}	–0.4	0.5
Cobalt	Co^{2+}	–0.29	
Nickel	Ni^{2+}	–0.22	0.15
Tin	Sn^{2+}	–0.14	0.45
Lead	Pb	–0.13	0.45
Hydrogen	H^+	0.00	–
Antimony	Sb^{3+}	+0.11	0.42
Copper	Cu^{2+}	+0.34	0.25
Silver	Ag^+	+0.8	0.1
Gold	Au^{3+}	+1.3 (**Noble End**)	0.35
Oxygen	OH^-	+0.4	
Chlorine	Cl^-	+1.36	

The farther two metals are separated from one another in the electrochemical series, the more powerful is the electric current produced by their contact in the presence of an electrolyte. Also the more rapidly the metal towards the top of the table is attacked and the more will the metal towards the bottom of the table be protected. It must be remembered, however, that the order in the above series may vary under special corrosive conditions, and the galvanic series in service media, *e.g.* sea water, are often more useful from the corrosion aspect.

GALVANIC SERIES IN SEA WATER

Corroded End (Anodic)

Magnesium
Zinc
Aluminium
Cadmium
Mild Steel
Cast Iron
Stainless Steels 18/8 (Active)
Lead
Tin
Nickel (Active)
Brass
Copper
Aluminium Bronze
Cupro nickel
Silver Solders
Nickel (Passive)
Stainless Steel 18/8 (Passive) Silver
Titanium
Graphite
Gold
Platinum

Protected End (Cathodic)

An example of a corrosion cell is provided by an imperfect coating of copper on steel immersed in dilute sulphuric acid. The current generated passes from the copper to the steel by the path of lowest resistance and and returns to the copper through the solution by the passage of ions. The steel, which has the greatest negative potential, dissolves and is called the anode; whilst the copper is called the cathode. In such acid attack the hydrogen, which is freed as the iron dissolves, is deposited on the surface of the copper cathode and as it increases in amount two things may occur. The corrosion of the steel is either reduced because of the formation of an opposing hydrogen electrode, *i.e.* the cell is polarised; or the hydrogen may be evolved as bubbles which stream away, with the result that the corrosion will occur continuously. In the first case the corrosion will be accelerated by exposure to oxidising agents (*e.g.* air), which remove the hydrogen from the cathode. The size of the cathode relative to the anode is important, e.g. copper rivet in a large steel plate, is quickly polarised and corrosion on the plate is small. On the other hand, a large cathode coupled to a small anode has the opposite effect, with rapid attack of the anode.

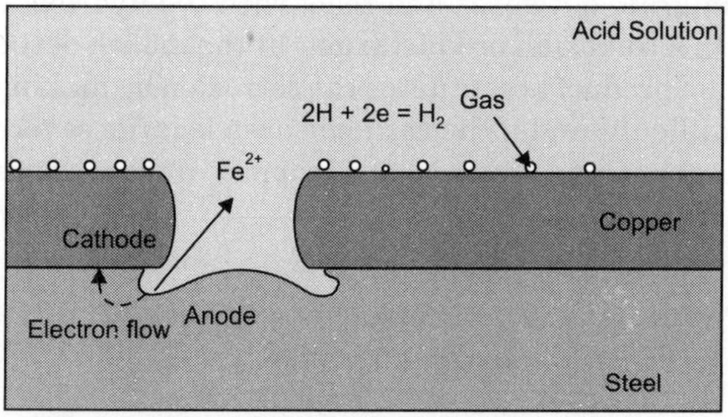

The reaction effect of the steel in the acid solution

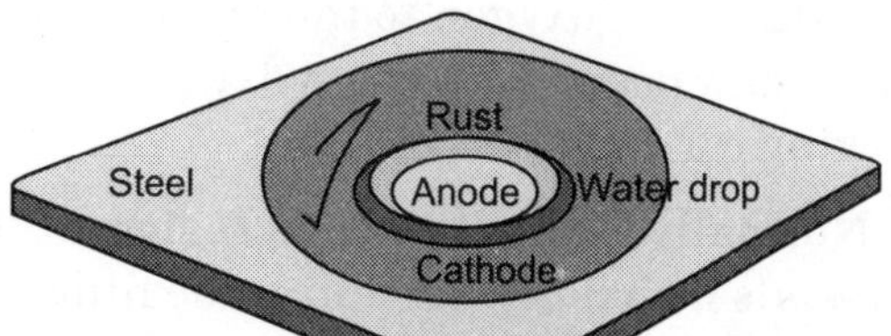

The arrangement of water droplet on iron sheet

Iron and steel are the most common materials of construction, and their corrosion characteristics in neutral waters is important. When steel corrodes, the corrosion rate is usually governed by the cathodic reaction of the corrosion process, and oxygen is an important factor. In neutral waters free from dissolved oxygen, corrosion is usually negligible. The presence of dissolve oxygen in the water accelerates the cathodic reaction; and consequently the corrosion rate increases in proportion to the amount of oxygen available for diffusion to the cathode. Where oxygen diffusion is the controlling factor, the corrosion rate tends to increase also with rise in temperature. In acid waters (pH < 4), corrosion can occur even without the presence of oxygen.

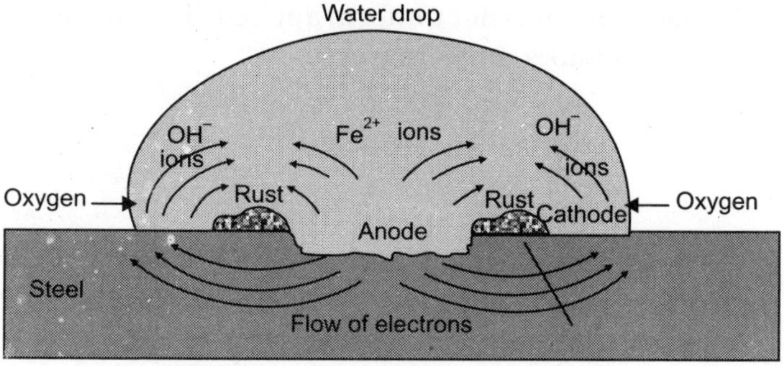

The Effect of water droplet on iron sheet

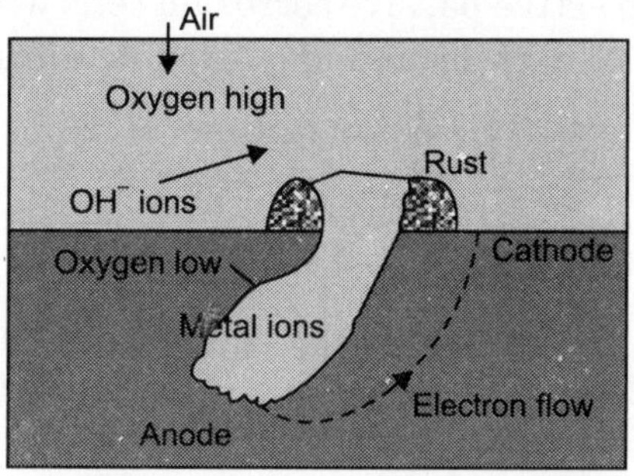

The involvement of oxygen in rusting

Pitting and Crevice Corrosion

Electrochemical corrosion can be stimulated from not only differences in the metal surface, but also from variations in the electrolyte. The above is effected to some degree by this mechanism, as oxygen diffuses into the water drop a concentration gradient is set up, where the oxygen content at the extremities is the highest and the lowest being at the centre where the anode forms. Cavities in metal surfaces and metal surfaces partially covered by another material are prone to this type of attack. The diffusion of oxygen into cavities or crevices is impeded and results in these areas becoming anodic to the surrounding metal to which oxygen can easily reach (oxidation-concentration cell or differential aeration cell). The metal ions formed in the cavity migrate outwards and react with the hydroxide ions flowing in the opposite direction to form a corrosion product (rust) at the mouth of the cavity or crevice. This position of the corrosion product accentuates the corrosion by making the diffusion of oxygen to the anode more difficult, and if the cathodic area is large severe pitting may occur. Also when dry conditions prevail moisture can be trapped in the cavities allowing corrosion to continue.

Normally very corrosion resistant materials which rely on thin oxide films for protection, such as stainless steel, can suffer from this type of corrosion attack. These materials rely on oxygen being present, so that they can maintain their oxide films (passive state). When oxygen is excluded and the oxide films break down, the material surface becomes active and corrodes readily.(See galvanic series).

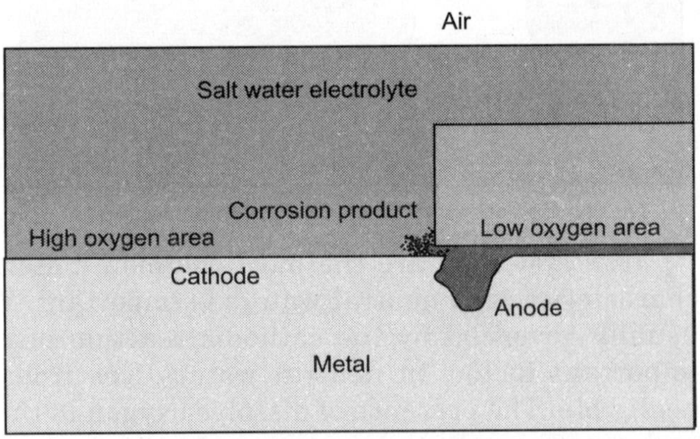

The effect of salt in the corrosion of the metal

The effects of corrosion can be accelerated or induced when operating in conjunction with stress and various wear mechanisms. Usually the mechanisms work by not allowing the corroded metal to become passive by continually removing protective films and setting up active/passive corrosion cells where the mechanism is not uniform applied. The corrosion products formed may provide abrasive debris to make matters worse.

- Corrosion fatigue
- Cavitation corrosion
- Erosion corrosion
- Fretting corrosion
- Stress corrosion cracking

Protecting Metals from Corrosion

Ideally, a material which is inherently resistant to its service environment, meets with the mechanical, formability and economic requirements would be the first choice for selection. Unfortunately, this is not often the case. Many materials will need a method of corrosion control and there are three main approaches:

- ➢ Modification of the environment to which the material is exposed
- ➢ Electrical methods of control
- ➢ Use of protective coatings.

Thermal Spray Coatings for Corrosion Protection

Thermal spray coatings are widely used in preventing corrosion of many materials, with very often, additional benefits of properties such as wear resistance etc. due to the very wide selection of coatings that can be sprayed. Broadly, thermal spray coatings fall into three main groups:

- ➢ Anodic Coatings
- ➢ Cathodic Coatings
- ➢ Neutral Coatings.

Anodic Coatings

Anodic coatings for the protection of iron and steel substrates are almost entirely limited to zinc and aluminium coatings or their alloys. Where coatings anodic to the substrate are applied, the corrosion protection is referred to as cathodic protection or sacrificial protection. The substrate is made to be the cathode and the coating the sacrificial corroding anode. The mechanisms of corrosion protection inferred by these coatings fall into two classes:

- ➢ Cathodic or sacrificial protection and
- ➢ A barrier to the environment.

An ordinary sprayed coating of zinc or aluminium although somewhat porous, to a large extent excludes the environment and provides cathodic protection. Where desired the porosity can be sealed with organic sealers, or the coating painted, which can in some cases prolong the life of the protective system by increasing the barrier effect. It is argued that sealing or painting these coatings reduce the cathodic protection effect and reduce the overall effectiveness. Sealing and painting certainly reduces the cathodic protection by decreasing the area of contact of the coating with the environment, but in many cases where the substrate becomes exposed there is more than enough coating exposed to keep the substrate as the cathode. Also, the barrier effect prolongs the life of the coating. It could be argued that this is not applicable to all situations. Depending on choice of coating system and environment the life expectancy can be well in excess of 20 years with no maintenance. This method is generally regarded as providing superior corrosion protection than galvanizing, plating and painting without excessive cost penalties.

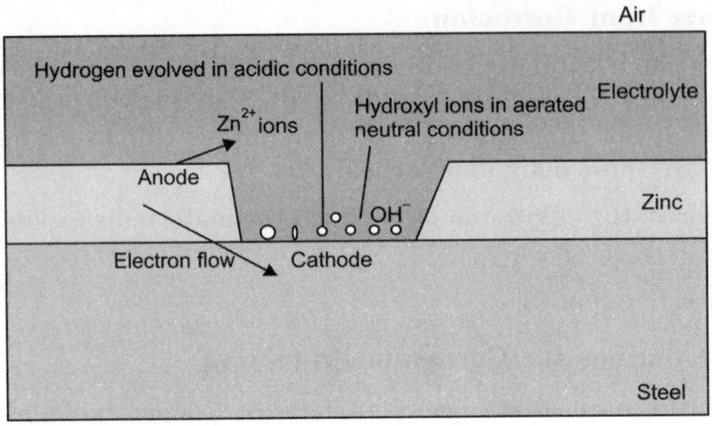

Effect of Zinc on the Rusting of Iron

Cathodic Coatings

Cathodic coatings are those which comprise a coating metal which is cathodic with respect to the substrate. A stainless steel or nickel alloy coating would be cathodic to a steel base. Cathodic coatings can provide excellent corrosion protection. There is a very wide choice particularly for steel base materials ranging from stainless steel to more exotic materials like tantalum to cater for the more extreme corrosive environments. However, an outstanding limitation of such coatings is that they must provide a complete barrier to the substrate from the environment. If the substrate is exposed to the corrosive environment, the substrate will become the anode and corrosion will be dramatically accelerated resulting in spalling of the coating.

Generally, sealing of these coatings is always recommended. Processes which provide the densest coatings are preferred (HVOF, plasma and fused coatings). Thick coatings will provide better protection than thin coatings.

Neutral Coatings

Neutral materials such as alumina or chromium oxide ceramics provide excellent corrosion resistance to most corrosive environments by exclusion of the environment from the substrate. Generally a neutral material will not accelerate the corrosion of the substrate even if the coating is somewhat permeable (An exception to this is with stainless steel type materials where the exclusion of oxygen can cause crevice corrosion, nickel chromium bond coats are required to stop this), although any corrosion of the substrate interface with the coating should be avoided to prevent coating separation. Again sealing of the coatings is recommended. The densest and thickest plasma sprayed coatings are recommended.

Protective Action of Tin: Tin is resistant to corrosion hence a tin coating of this metal protects the iron form rusting. Tin coating is however, is bit unreliable because if the protective coating is broken the area which is exposed starts getting rusted and therefore, more electropositive than iron. Iron, therefore looses iron and thus rusting is enhanced. In this case hydrogen ion (It) originating from the electrolytes are discharged at the depolarizer. Electrons pass from iron (anodic area) to tin (cathode area).

Thus at the exposed iron surface (anodic area) ferrous ions passes into solution. These Fe^{2+} ions oxidized to Fe^{3+} ions by dissolved oxygen and produced rust as explained before. The protective action of tin is shown in the Fig. 16.16.

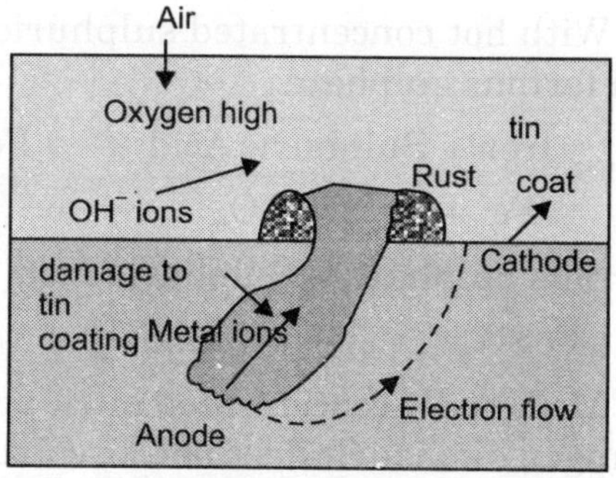

Fig. 16.16. The Break of tin Coating

Passivity: Place a piece of iron in dilute sulphuric acid a vigorous reaction take place and hydrogen gas is evolved. The same piece of iron will displace copper from a solution of copper sulphate. It is noticed, however, that if the iron piece were first kept in concentrated nitric acid and then washed thoroughly it can neither produce hydrogen nor it can separate copper from copper sulphate.

In other words, by treatment with concentrated nitric acid iron has so changed that it would not behave like the ordinary iron would do. This similar behavior is also noticed with chromium nickel and some other metals.

This inertness exhibited by metals under conditions in which a definite chemical activity is expected is called Passivity.

Not only nitric acid other oxidizing agents such as chromic acid acidifies potassium permanganate, etc causes the metals to become passive by forming a thin layer of oxide on the surface of the metal and this oxide film protects the metal from further corrosion of the metal.

> Passivity is the artificial chemical interness cause to some metals to arrest their chemical reactions.

According to Evans this oxide film is very thin and transparent so that it is quite invisible normally. If this is removed mechanically or chemically the metal becomes active again, thus when passive iron is heated in nitric acid at 75°C it becomes reactive and starts getting dissolved in nitric acid.

In electroplating or electrolysis or electrolysis sometimes the anode becomes passive if a high current density is employed, or a strong oxidizing agent is present in the present in the solution.

1. Action of Oxygen: Red hot iron burns in oxygen giving sparks.

$$\text{Iron} \quad \text{Oxygen} \longrightarrow \text{Ferric Oxide}$$

$$Fe + O_2 \longrightarrow Fe_3O_4$$

Iron + Sulphuric Acid ⟶ Ferrous Sulphate + Hydrogen

$$4Fe + H_2SO_4 \longrightarrow FeSO_4 + H_2$$

2. Action of Acids: From dilute hydrochloric and sulphuric acid it displays hydrogen. With nitric acid it produces ammonium nitrate

Iorn + Nitric Acid ⟶ Ferrous Nitrate + Ammonium Nitrate

$$4Fe + 10HNO_3 \longrightarrow 4Fe(NO_3)_2 + 3H_2O$$

With hot concentrated sulphuric acid it gives sulphur dioxide and a mixture of ferric and ferrous sulphate.

Iron + Sulphuric Acid ⟶ Ferrous Sulphate + Sulphur Dioxide + Water

$$Fe + 2H_2SO_4 \longrightarrow FeSO_4 + SO_2 + H_2O$$

Ferrous Sulphate + Sulphuric Acid ⟶ Ferrous Sulphate + Sulphur Dioxide + Water

$$FeSO_4 + H_2SO_4 \longrightarrow Fe_2(SO_4)_3 + SO_2 + H_2O$$

Moderately concentrated nitric acid dissolve iron giving ferric nitrate. Sand also liberates nitrogen dioxide. It does not dissolve in fuming nitric acid because it surface is changed in such a way that it fails to exhibit the properties of normal iron. It is said to be passive.

3. Action of Chlorine, Sulphur: (*a*) Chlorine reacts with iron forming ferric chloride;

Iron + Chlorine ⟶ Ferric Chloride

$$2Fe + 3Cl_2 \longrightarrow 2FeCl_3$$

(*b*) Sulphur reacts with iron directly on heating strongly formic ferrous sulphide.

Iron + Sulphur ⟶ Ferrous Sulphide

$$2Fe + 2S \longrightarrow 2FeS$$

Compounds of Iron

Generally iron from two types of compounds the ferrous compounds which are divalent and other one the ferric compounds which are trivalent. The properties of these two types of compounds are given in the table below:

Property	Ferrous Compounds	Ferric Compounds
1. Colour	Anhydrous: White Hydrate: Light green	White, yellow or reddish brown
2. Hydroxide	Weak base, light green in colour and easily oxidized	Much weaker base forms amptoteric salts
3. Reducing or oxidizing property	Reducing Property *e.g.*, it reduces acidified $KMnO_4$, $K_2Cr_2O_7$, $HgCl_2$	Oxidizing property *e.g.* it oxidizes potassium iodid.
4. White Nitric oxide	Gives dark brown nitroso-compounds	No action

Ferrous Sulphate ($FeSO_4$ $6H_2O$): This also known as *Green Vitrium rather Viridis Vitrium* and may be obtained in the chemical laboratory from the Kipp's Waste, it is also prepared in the lab by the reaction of scrap iron and dilute sulphuric acid.

The liquid is concentrated in the presence of scrap iron, which reacts with excess of sulphuric acid to give ferrous sulphate. On filtration and sedimentation $FeSO_4.7H_2O$ crystallized out. Ferrous sulphate containing 6, 5, 3, 2, 1, or no molecules of water of crystallization is also known.

In commerce ferrous sulphate is obtained by exposing big heaps of moist iron pyrites to air when slow oxidation takes place.

Iron Pyrites + Water + Oxygen ⟶ Ferrous Sulphate + Sulphuric Acid

$$2FeS_2 + H_2O + 7O_2 \longrightarrow FeSO_4 + 2H_2SO_4$$

From the solution obtained crystals of ferrous sulphate are obtained in the laboratory method.

Properties and Uses: (I) Light green crystals of ferrous sulphate loose water and turn brown with exposed to air, due to oxidation.

Ferrous Sulphate + Water + Oxygen ⟶ Ferrous Hydrogen Sulphate

$$4FeSO_4 + 2H_2O + O_2 \longrightarrow 4Fe(OH)SO_4$$

(II) On heating it decomposes as follows:

(III) With nitric acid ferrous sulphate turns black with the formation of nitro so ferrous sulphate $FeNOSO_4$ (Ring Test).

(a) Potassium Nitrate + Sulphuric Acid ⟶ Potassium Sulphate + Nitric Acid

$$KNO_3 + H_2SO_4 \longrightarrow KHSO_4 + HNO_3$$

(*b*) Ferrous Sulphate + Nitric Acid + Sulphuric Acid ⟶ Ferric Sulphate + Nitric Oxide + Water

$$6FeSO_4 + 2HNO_3 + 3H_2SO_4 \longrightarrow 2Fe_2(SO_4)_3 + 2NO + 4H_2O$$

(*c*) Ferrous Sulphate + Nitric Oxide ⟶ Ferrous Nitrososulphate (violet ring)

$$FeSO_4 + NO \longrightarrow FeNOSO_4$$

reactions of ring test

(IV) Reducing property, Ferrous sulphate converts acidifies potassium dichromate to a green solution.

Ferrous Sulphate + Potassium Dichromate + Sulphuric Acid ⟶ Potassium Sulphate + Chromic Sulphate + Ferric Sulphate + Water

$$6FeSO_4 + K_2Cr_2O_7 + 7H_2SO_4 \longrightarrow K_2SO_4 + Cr_2(SO_4)_3 + 3Fe_2(SO_4)_3 + H_2O$$

(V) Ferrous sulphate also forms alums *e.g.* $K_2SO_4.FeSO_4.6H_2O$.

(VI) With ammonium sulphate it forms ferrous ammonium sulphate (Moh's salt), $FeSO_4(NH_4)2SO_4.6H_2O$. It is not oxidized so easily as ferrous sulphate therefore it is used in volumetric analysis.

Ferrous sulphate is largely used in the dyeing industry as mordent also in tanning industry. It is also used in the manufacture of Mohr's salt and ferric oxide.

Ferric Sulphate [$Fe_2(SO_4)_3$]: It is prepared by the oxidation of acidified ferrous sulphate solution with nitric acid.

Ferrous Sulphate + Nitric Acid + Sulphuric Acid ⟶ Nitric Oxide + Water

$$6FeSO_4 + 2HNO_3 + 3H_2SO_4 \longrightarrow 2NO + 4H_2O$$

The yellowish brown solution on concentration deposits a whitish mass having the composition of $Fe_2(SO_4)9H_2O$.

The anhydrous salt can be obtained by heating the hydrate.

With sulphates of alkali metals, it also gives alums that is ferrous alums; $(NH)_2SO_4 \cdot Fe_2(SO_4)_3 \cdot 24H_2O$ also used in volumetric analysis.

Ferric Chloride ($FeCl_2$): Anhydrous ferric chloride is prepared in the lab be passing dry chlorine over heated iron (see Fig. 16.17). When the salt sublimes over as deep yellow peletes.

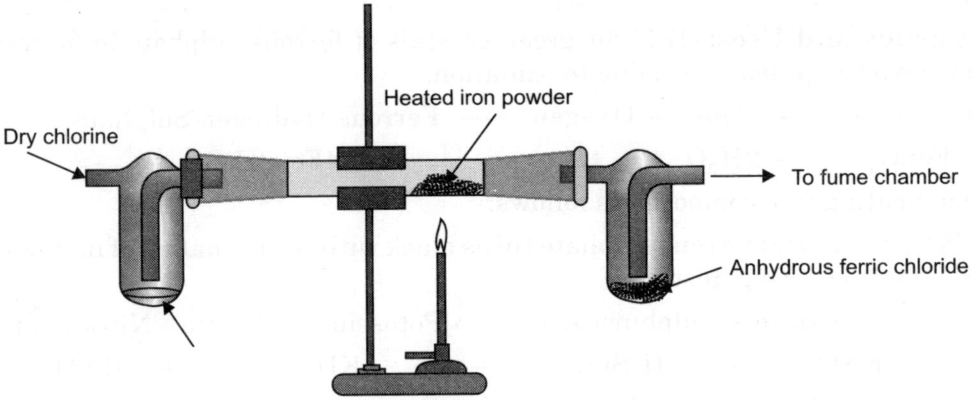

Fig. 16.17. The Diagram of Lab Preparation of Ferric Chloride

Anhydrous solution of the salt is obtained by dissolving ferric hydroxide in dilute hydrochloric acid.

Ferric Hydroxide + Hydrochloric Acid ⇌ Ferric Chloride + Water

$$Fe(OH)_3 + 3HCl \rightleftharpoons FeC_3 + H_2O$$

The anhydrous salt is deliquescent and soluble in water at high level.

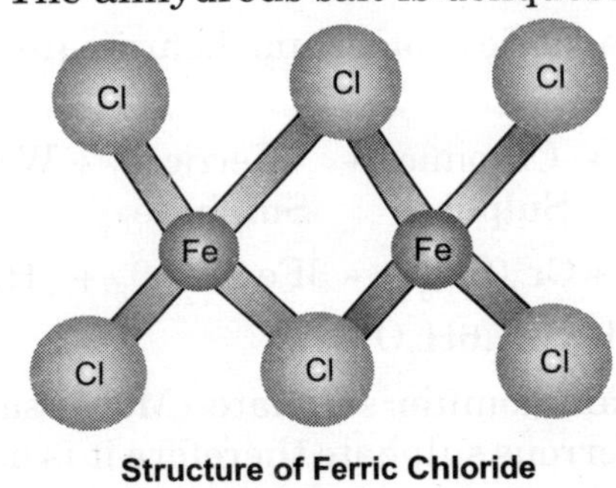

Structure of Ferric Chloride

Its vapour density correspond to the formula $FeCl_2$ at 440°C and $FeCl_3$ at 750°C. Hydrate of ferric chloride containing 12, 5, 7 and 4 molecules of water of crystallization are known. It forms a number of double salt *e.g.* $2KCl.FeCl_3H_2O$ ferric chloride is most probably covalent.

It is used as reagent in the laboratory and in medicine.

Ferrous Oxide [FeO]: This oxide is different to prepare in the pure state. The usual method of preparation it consists in heating ferrous oxalate in absence of air.

Ferrous Oxalate ⟶ Ferrous Oxide + Carbon Tetrachloride + Carbon Dioxide

$$FeC_2O_4 \longrightarrow FeO + CO + CO_2$$

Another method is to reduce ferric oxide be means of hydrogen. When hydrogen is passed over hot ferric oxide, it is converted to ferrous oxide.

Ferric Oxide + Hydrogen ⟶ Ferrous Oxide + Water

$$Fe_3O_2 + H_2 \longrightarrow FeO + H_2O$$

But in the both cases metallic iron may be present as the impurity in the compound.

Ferrous oxide is a black powder. It usually burns in air, after spontaneously, giving ferrous salts.

Ferrous oxide is basic alkaline in nature and when treated with sulphuric acid, give ferrous salts.

Ferrous Hydroxide [Fe(OH)$_2$]: This compound is obtained as white precipitate by mixing air free solutions of sodium hydroxide and a ferrous to ferric.

When quite pure, ferrous hydroxide forms a white powder. But it immediately changed by oxidation in air to a green producer possibly a basic *ferrous ferric hydroxide.* (OH)Fe.O.Fe(OH)$_2$. Ultimately it is oxidized to brown ferric hydroxide [Fe(OH)$_3$]. Ferrous hydroxide dissolves in dilute acids yielding ferrous salts. It is a powerful reducing agent.

Ferrous Carbonate (FeCO$_3$): It is obtained naturally from siderite similar to cacite. It is formed as a white precipitate by the action of alkali carbonate to a ferrous salt.

Ferrous Sulphate + Sodium Carbonate ⟶ Ferrous Carbonate + Sodium Sulphate

$$FeSO_4 + Na_2CO_3 \longrightarrow FeCO_3 + Na_2SO_4$$

But the precipitate rapidly turns green and finally brown on exposure to air, owing to oxidation to ferric hydroxide. Ferrous carbonate is moderately soluble in water to form ferrous by carbonate, Fe(HCO$_3$)$_2$.

Ferrous Carbonate + Water ⟶ Ferrous Bicarbonate

$$2FeCO_3 + H_2O \longrightarrow 2Fe(HCO_3)_2$$

But the solution on exposure to air yields a deposit of ferrous hydroxide.

Ferrous Carbonate + Water ⟶ Ferrous Hydroxide + Carbon Dioxide

$$FeCO_3 + H_2O \longrightarrow Fe(OH)_2 \quad CO_2$$

Ferrous Fluoride (FeF$_2$): It is produced as a white powder when iron or anhydrous ferric chloride is heated in a steam of hydrogen fluoride.

Hydrogen Fluoride + Ferrous Chloride ⟶ Ferrous Fluoride+ Hydrogen Chloride

$$2HF + FeCl_2 \longrightarrow FeF_2 + 2HCl$$

It is at first obtained as hydrate salt, which is green in colour from which anhydrous salt is obtained.

Ferrous Chloride ($FeCl_2$): It is obtained in anhydrous white lustrous scales by passing a stream of hydrogen chloride over iron.

$$\text{Iron + Hydrogen Chloride} \longrightarrow \text{Ferrous Chloride + Hydrogen}$$
$$Fe + 2HCl \longrightarrow FeCl_2 + H_2$$

Crystals of hydrate ($FeCl_2H_2O$) with pale blackish-green colour may be obtained from the solution of iron in dilute hydrochloric acid. The anhydrous chloride melts at a red heat and volatilizes at about 1000°C, its vapour density shows a partial association into double molecules.

$$\text{Ferrous Chloride} \rightleftharpoons \text{Ferrous Chloride}$$
$$Fe_2Cl_4 \rightleftharpoons 2FeCl_2$$

But at higher temperature there is a complete dissociation of the double molecules.

Ferrous Bromide ($FeBr_2$): This compound is formed as a yellow crystalline substance by the action of bromine on iron to temperature of 600°C.

$$\text{Iron} \quad \text{Bromine} \longrightarrow \text{Ferrous Bromide}$$
$$Fe + Br_2 \longrightarrow FeBr_2$$

In aqueous solution the compound is formed when iron fillings is dropped in a strong solution of bromine water. From the solutions the hydrates with $6H_2O$ can be obtained other hydrates are known.

Ferrous Iodide (FeI_2): It is obtained by heating iron with iron. The union of the elements obtains an aqueous solution in presence of water. The anhydrous salt is a grey solid. Crystals of the hexahydrate (FeI_26H_2O) separate on evaporation, of the aqueous solution. Its solution is readily oxidized in air with separation of iodine.

Ferrous Nitrate [$Fe(NO_3)_2$]: This compound of iron is obtained when lead nitrate reacts with ferrous sulphate.

$$\text{Ferrous Sulphate + Lead Nitrate} \longrightarrow \text{Ferrous Nitrate + Lead Sulphate}$$
$$FeSO_4 + Pb(NO_3)_2 \longrightarrow Fe(NO_3)_2 + PbSO_4$$

On filtering and evaporating the filtrate in the form of hexahydrate $Fe(NO_3)26H_2O$ are formed. It is formed when iron dissolves in hot nitric acid. Ferrous nitrate is highly unstable and decomposes rapidly.

Ferrous Phosphate [$Fe_3(PO)_48H_2O$]: This is obtained by the action of phosphoric acid and ferrous sulphate.

$$\text{Ferrous Sulphate + Phosphoric Acid} \longrightarrow \text{Ferrous Phosphate + Sulphuric Acid}$$
$$3FeSO_4 + 2H_3PO_4 \longrightarrow Fe_2(PO_4)_2 + 3H_2SO_4$$

Ferrous Sulphide (FeS): It is formed when red-hot iron reacts with sulphur.

$$\text{Iron} \quad \text{Sulphur} \longrightarrow \text{Ferrous Sulphate}$$
$$Fe \quad S \longrightarrow FeS$$

It is also produced when hydrogen sulphide is passed through a solution of ferrous sulphate in the presence of sodium acetate.

$$\text{Ferrous Sulphate} + \text{Hydrogen Sulphate} \xrightarrow{\text{Sodium Acetate}} \text{Sulphuric Acid} + \text{Ferrous Sulphide}$$
$$FeSO_4 + H_2S \longrightarrow H_2SO_4 + FeS$$

Sodium sulphide when added to ferrous sulphate also produces ferrous sulphate.

$$\text{Sodium Sulphide} + \text{Ferrous Sulphate} \longrightarrow \text{Sodium Sulphate} + \text{Ferrous Sulphide}$$
$$NaS + FeSO_4 \longrightarrow Na_2SO_4 + FeS$$

It is a black solid that melts at 1200°C and can be easily cast in sticks.

Ferrous Disulphide (Fe_2S_2): It occurs abundantly as the mineral known as iron pyrites. It can be prepared in the laboratory by boiling a suspension of ferrous sulphide and sulphur in water.

$$\text{Ferrous Sulphide} + \text{Sulphur} \longrightarrow \text{Ferrous Disulphide}$$
$$FeS + S \longrightarrow FeS_2$$

The disulphide forms as yellow crystals. It is insoluble in water and also decomposes into ferrous sulphide and sulphur when heated strongly in the absence of air.

Iron pyrites is generally used in the preparation of sulphur dioxide for the manufacture of hydrochloric acid in the presence of moist air and sulphuric acid.

$$\text{Ferrous Disulphide} + \text{Water} + \text{Oxygen} \longrightarrow \text{Ferrous Sulphate} + \text{Sulphuric Acid}$$
$$2FeS_2 + H_2O + O_2 \longrightarrow 2FeSO_4 + 2H_2SO_4$$

Ferrosoferric Oxide (F_2O_4): This oxide occurs in nature as magnesite, which forms shining black masses. Because of the action of the earth's magnetism, the mineral acquires north and south polarity. Hence, pieces of magnesite (loadstone) behave as permanent magnets.

The compound is artificially obtained by burning the metal in oxygen it in steam

$$\text{Iron} + \text{Oxygen} \longrightarrow \text{Ferosoferic Oxide}$$
$$2Fe + 2O_2 \longrightarrow Fe_2O_4$$
$$\text{Iron} + \text{Water} \longrightarrow \text{Ferrosoferic Oxide} + \text{Hydrogen}$$
$$2Fe + H_2O \longrightarrow Fe_2O_4 + H_2$$

Ferrosoferric oxide is stable up to 3000°C when heated above this temperature in the presence of oxygen, it yields ferric oxide.

$$\text{Ferosoferic Oxide} + \text{Hydrogen} \longrightarrow \text{Iron} + \text{Water}$$
$$4Fe_2O_4 \quad 4H_2 \longrightarrow 8Fe \quad H_2O$$

When heated with hydrochloric acid, it forms a mixture of ferrous and ferric chloride.

$$\text{Ferosoferric Oxide} + \text{Hydrogen Chloride} \longrightarrow \text{Ferric Chloride} + \text{Ferrous Chloride} + \text{Water}$$
$$Fe_2O_4 + 8HCl \longrightarrow FeCl_3 + FeCl_2 + 4H_2O$$

When heated in a current of hydrogen, ferrosoferric oxide is reduced to metal.

Ferosoferic Oxide + Hydrogen ⟶ Iron + Water

$$Fe_2O_4 + H_2 \longrightarrow 2Fe + 4H_2O$$

Other acids do not appreciably attack the oxide.

Ferrosoferric Oxide [Fe(OH)$_3$]: It is formed as a black precipitate by adding sodium hydroxide to mixed solutions containing calculated proportions of ferrous and ferric salts. It dissolves in hydrochloric acid to give a green coloured solution, from which crystals of ferosoferic chloride (Fe3C1s.18HzO) may by separated out by evaporation ferosoferric bromide (Fe3Brs) is formed as a byproduct in the manufacture of bromine and is employed in preparing potassium bromide.

Potassium Ferrocyanide [K$_4$[Fe(CN)$_6$] 3H$_2$O]: This compound is prepared in the industry by heating nitrogenous animal refuse, such as blood, horns, hooves leather etc, with iron fillings and potassium carbonate and then lixiviating* the product with water. On concentrating the solution, lemon-yellow crystals of the salt separate out.

In the laboratory potassium ferrocyanide is prepared by the addition of potassium cyanide to ferrous cyanide is left.

Ferrous Sulphate + Potassium Cyanide ⟶ Ferrous Cyanide + Potassium Sulphate

$$FeSO_4 + 2KCN \longrightarrow Fe(CN)_2 + K_2SO_4$$

Ferrous Cyanide + Potassium Cyanide ⟶ Potassium Ferrocyanide

$$Fe(CN)_2 + 4KCN \longrightarrow K_4[Fe(CN)_6]$$

POISON

Potassium ferrocyanide forms lemon-yellow crystals, which are readily soluble in water. The solution when acidified with hydrochloric acid gives a white precipitate of hydroferrocyanic acid, $H_4[Fe(CN)_6]$.

Potassium ferrocyanide is also a reducing agent it is oxidized by potassium permanganate to potassium ferrocyanide.

Uses

(*i*) Potassium ferrocyanide is used as reagent in the laboratory, for the detection of ferric, cupric and zinc ions.

(*iii*) It is also used for the production of potassium ferricyanide.

(*iii*) The oxidation may also be effected by the means of potassium permanganate in the in the presence of hydrochloric acid.

Potassium Permanganate + Hydrochloric Acid + Potassium Ferrocyanide ⟶ Potassium Chloride + Manganese Chloride + Potassium Ferricyanide + Water

$$2KMnO_4 + 16HCl + 10K_4Fe(CN)_6 \longrightarrow 12KCl + 2MnCl_2 + 10K_3Fe(CN)_6 + 8H_2O$$

* Lixiviating: The mixture is percolated with water and then it is separated in the form of lye.

Potassium ferricyanide forms dark red anhydrous crystals readily soluble in water. It is a mild oxidizing agent. The most important reaction of potassium ferrocyanide is with ferrous salt with which it gives a deep blue precipitate of Turnbill's blue. Ferric salts give only a brownish colour with the salt. The salt is used as an external indicator in the titration of ferrous salt by means of potassium dichromate.

Sodium Nitroprusside [$Na_2Fe(CN)_5NO\ 2H_2O$]: Nitroprusside of sodium is prepared by the addition of dilute nitric acid, dilute with its own volume of water to a concentrated solution of potassium ferrocyanide, potassium nitrate separates out on cooling and the salt of potassium nitrate is deposited can then be removed. The residual solution is neutralization with sodium carbonate and then -evaporated to crystallization. Sodium nitroprusside forms beautiful rhombic prisms, which are readily soluble in water. It is used as a presence of the soluble sulphide and of acetone.

Iron Pentachloride [$Fe(CO)_5$]: Iron forms three carbonyls of which the pentacarbonyl is the most important. When finely divided iron combines with carbon monoxide, it is the pentacarbonyl that is produced. The reaction is slow at the ordinary temperature but proceeds more rapidly at higher temperature and pressures.

Manufacture: The commercial method for the preparation of the pentacarbonyl consists in heating finely divided iron with carbon. The gaseous product, consisting of the pentacarbonyl and unchanged carbon monoxide at about 200°C and a pressure of 15198Pa. The gaseous product, consisting of the pentacarbonyl is condensed out, the uncondensed carbon monoxide is returned to the reaction vessel.

Physical Properties: Iron pentacarbonyl is a colourless highly refractive only liquid having a density of about 1.5 g/cc and the boiling point is at 101.9°C and freezes at about –20°C. The commercial product may have a pale yellow to dark red colour due to the presence of traces of non-carbonyl. The vapour density and the freezing point depression in benzene corresponds to the formula; $Fe(CO)_5$.

Iron pentacarbonyl is readily soluble in benzene petroleum, ether amyl and higher alcohols and many other organic solvents.

Inflammability in Air: Iron pentacarbonyl is highly inflammable and its flashing point is below –15°C, when burnt in air it produces ferric oxide, in a form of most suitable for use as a pigment and polishing agent.

Thermal Decomposition of the Vapour: When the pentacarbonyl is heated in the absence of air, it gets decomposed as follows:

$$\text{Iron Pentacarbonyl} \longrightarrow \text{Carbon Monoxide} + \text{Iron}$$

$$Fe(CO)_5 \longrightarrow 5CO + Fe$$

The iron powder thus obtained consists of almost spherical particles.

Uses

1. The first use of iron carbonyl as an antiknock. This application is not much use because due to mechanical problem, as some ferric oxide is deposited in the engine.
2. As already mentioned, the use of carbonyl iron has been found to be great commercial importance in the electronic industry.

3. The carbonyl iron being the purest form of the metal, is an important catalyst.
4. The ferric oxide obtained by the burning of the pentacarbonyl it is used as a pigment and a polishing agent.

Precaution

(*i*) The presence of iron carbonyl in the gas used in the incandescent lamps is undesirable because the deposition of ferric oxide on the gas mantles will decrease the luminosity.

(*ii*) Because of formation of iron pentacarbonyl, carbon monoxide may penetrate through heated iron flues stoves burning with insufficient air.

Structure of Iron Pentacarbonyl [Fe(CO)$_5$]: The modern views of the structure of the compound mainly based on its electrons diffractions study is that the five CO groups are arranged around the metal atom at the apices of a rectangular trigonal bipyramid as shown below:

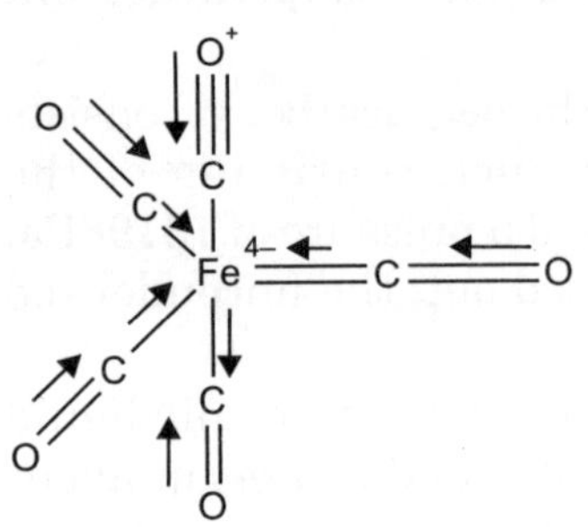

The bond lengths have been flamed to the M – C 1.85 Å, C – O 1.10 Å. The dipole moment though finite is small. From the trigonal pyramid structure of the compound it is concluded that the zero valent iron atoms(iron ions) has taken up an electronic configuration that allows 3dz^2 orbital to take as shown in the following diagram.

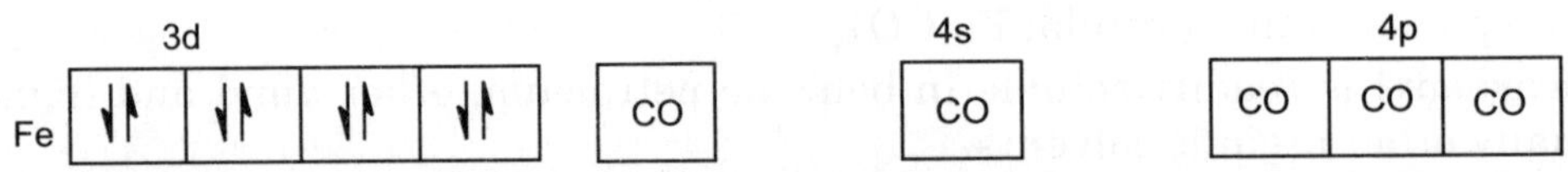

Detection and Estimation of Iron: (1) When a solution of ferric chloride is treated with potassium thiocyanate it produces a deep red colour as blood red ferric tetrathionate is produced (2) when a solution of ferric salt is treated with potassium ferricyanide it produces deep blue coloured complex ion.

Estimation: The Gravimetric estimation of iron consists in oxidizing in the ferrous state by means of nitric acid and treating the resulting solution with ammonium hydroxide and hydroxide and ammonium chloride. The precipitate being Fe(OH)$_3$ is filtered off, washed, dried ignited and weighted as Fe_2O_3.

The volumetric method, much commonly used for the estimation of iron is as follows ferrous salt is triturated against a known volume of potassium permanganate solution.

For the estimation of an iron salt it the presence of chloride, stannous chloride is added slightly in excess. The excess of the stannous salt is then destroyed by the means of a few drops of mercuric chloride this process is done in the presence of potassium ferricyanide as an indicator.

COBALT

Melting Point : 1495°C

Boiling Point : 3000°C

Atomic Number : 27

Atomic Weight : 58.9332

History and Occurrence

Blue glasses called faience belonging to the 14th century has been found among the products of ancient Egypt and Babylon. Some specimens of blue glass consisting of cobalt were prepared before the middle of 18th century.

Cobalt is more rare element than iron almost associated with nickel and mainly combinations with arsenic. The proposal ores of cobalt are *smallite, speiss cobalt* ($CoAs_2$) and *Cobalt glance or Cobaltite* ($CoAsS$). Many cobalt ores contain an appreciable amount of silver, small amounts in Congo and Northern Rhodesia. In India it is available in Raka Mines with sulphite ore, it is not freely available the percentage is less than one.

Small quantities of metallic cobalt in the free state are found in metallic iron. It is also present in soil in traces and as a constituent of Vitamin B-complex.

Extraction

Cobalt is extracted by the process, the ore is at first crushed and then roasted to remove the arsenic and sulphur. The roasted ore is at first crushed and then roasted to remove arsenic and sulphur. The ore is then sent to blast furnace with the addition of sand and limestone to form a speiss or a mass of impure cobalt and nickel arsenide and antimonide, with sulphide ores a matte or number of sulphide is formed. The speiss or the matte is roasted in air to yield a mixture of oxides and free metals. The roasted product is leached with dilute sulphuric acid when iron, nickel and cobalt, present as oxides, dissolve and the copper (present as the metal) is left behind.

From the filtered solution iron is precipitated by the addition of calcium carbonate and sodium chlorate. The filtrate is then cautiously heated with bleaching powder, which precipitates nearly all the cobalt as sesquioxide, leaving nearly all the nickel in the solution. The metallic or by reduction with hydrogen, or by Goldschmidt's Thermit process, or by electrolysis of its solution of dilute acids.

Reaction with Acids

Cobalt is slightly electropositive, the standard oxidation potential reaction.

$$Co \rightarrow Co^{2+} + 2e^-$$

Being 0.28volt. It is therefore, slowly attacked by dilute hydrochloric acid and sulphuric acid evolving hydrogen and cobaltous salts.

$$\text{Cobalt} + \text{Sulphuric Acid} \longrightarrow \text{Cobalt Sulphate} + \text{Hydrogen}$$

$$Co + H_2SO_4 \longrightarrow CoSO_4 + H_2$$

It is attacked directly by nitric acid.

Cobalt + Nitric Acid ⟶ Cobultous Nitrate + Water

$$3Co + 2HNO_3 \longrightarrow 3Co(NO_3)_2 + H_2O$$

It becomes passive on treatment with concentrated nitric acid.

Reaction with Non-Metals

It does not react with hydrogen or nitrogen directly. It combines with phosphorous and sulphur on heating.

Cobalt + Sulphur ⟶ Cobaltous Sulphide

$$Co \quad S \longrightarrow CoS$$

Cobalt Phosphorous ⟶ Cobaltous Diphosphide

$$Co + P_2 \longrightarrow CoP_2$$

It reacts with atmospheric oxygen only when heated giving the oxide CoO. This oxide is used in ceramic industry as a blue pigment.

ALLOYS

The bulk consumption of cobalt is now in the form of alloys while many of these alloys carry only a minor constituent, the number in which cobalt is a major constituent has increased greatly in recent years and they are now account for a major portion of the consumption. Data on the percentage consumption of more important cobalt alloys are shown in the table below. The table is designed to show the extent of which cobalt has been included in alloys for various uses. It is by no means a complete cobalt alloys. Although the table consists of resent alloys of cobalt but some of the older alloys also been included.

Trade Name	Cobalt	Chromium	Nickel	Tungsten	Molybdenum	Iron	Others
Stellite	75%	25%	—	—	—	—	—
Original Stellite 23	65%	27%	1%	5%	—	1%	—
Barrio Soft	60%	20%	—	20%	6%	1%	—
Multimet	20%	20%	20%	2%	3%	33%	—
Cochrome	60%	12%	—	—	—	24%	—
Permedur	50%	—	—	—	—	50%	—
Stellite 30	50%	27%	15%	—	6%	1%	—
S-816	43%	19%	20%	4%	4%	4%	Cobalt 4
J-Metal	40%	35%	—	20%	—	—	—
Cobaflux	37%	5%	4%	—	—	53%	—

Also in most cases the alloys listed are not unique, but more often than are not representative examples selected from a group of similar composition.

The alloys listed in the above table include not only these in which cobalt content. Their inclusion is justified by the fact that the cobalt content, even though less that that of some other metal is still a major factor in determining the characteristics of the resulting alloys. In addition to especially in the field of high-speed steels, tool-steel, magnet-steel and magnet alloys. Besides these there are many more alloys, which are not so important, and not used frequently.

The properties of cobalt alloys may be classified according to the type of use for which they are designed, even though there is a considerable degree of overlapping between some categories.

Grouping of Cobalt Alloys

Separated according to magnitude of the cobalt content, the group may be subdivided as follows:

Magnet Alloys: Although the magnetic properties of cobalt are markedly less than those of iron, there have been numerous cobalt alloys developed in which magnetic properties for exceed those of iron. Alloys of this type carry amounts of cobalt varying from a fraction of a per cent up to as iron in the minority and some have no iron except such small amounts as are present as an impurity. When the iron increases to the point where it is the major consistent, the material is then classed as magnet steel.

Magnet Steel: Cobalt is extremely used as an alloying agent in both permanent and soft magnet steels, in amounts varying from a fraction of a percent up to amounts to that make cobalt the major constituent and sift of the material over into the category of magnet alloys. The simplest of the magnet alloys and magnet steels, is permendur, which contains 50% cobalt and 50% iron. The chief alloying agents used with cobalt in magnet steels include chromium, nickel, tungsten, molybdenum, vanadium, titanium and aluminium.

A latest development has occurred in the manufacture of permanent magnet; it is called Vectolite, with a composition of 30%Fe_2O_3, 44% Fe_3O_4 and 20% Ca_2O_3.

High-Temperature Alloys: Probably the important group of cobalt alloys at the present time is that characterized by desirable high-temperature properties including high strength, low values at high temperature and high resistance to oxidation, corrosion, erosion and abrasion. This phase of cobalt alloy utilization was not particularly more important than others until the aircraft program of World War II placed special emphasis on the development of gas turbines and jet engines, a major factor of which is the operating characteristics of the group of cobalt alloys. These alloys use cobalt in proportions from 20 to 65% along with varying amounts of chromium, nickel, tungsten, molybdenum, titanium and columbium.

Tool Alloys: The original commercial development of a cobalt alloy was for use as a high speed, heavy duty cutting tool and this use still accounts for a large percentage of the demand for the metal. Alloys of this type carry from 40% to 55% of cobalt; about two thirds as much chromium and smaller amount of tungsten and for molybdenum.

Tool Steels: The high degree of reel hardness that cobalt imparts to steel accounts for its inclusion in a wide variety steels for different uses. Cobalt is used in high speed tool steels to the extent of 3 to 15% but rarely extends 1 to 5% in other types, such as tools steels, machine steels, machine steels.

Hard Facing Alloys: As satisfactory welding produces become available cobalt alloys of the same general type as the tool alloys same into use as hard facing materials. Where their high hardness and resistance to abrasion can be advantageously utilized in a tin layer applied by welding; there it would be impractical, or uneconomical to make the entire part of the alloy in question.

Low Expansion Alloy: There are so many applications where a minimum or controlled coefficient or expansion is desirable; while nickel is the chief alloying agent used with iron in the production of controlled expansion. Cobalt has been used along with nickel in a number of cases, with cobalt percentages around 18% another type uses 53% to 55% cobalt in combination chromium instead of nickel, while a third type uses 25% cobalt with both nickel and chromium.

Cemented Carbides: The developments of the methods of powder metallurgy has made it possible to use finely divided cobalt metal as a binder to hold tungsten carbide and produce cemented carbide, for use as a cutting tool. The proportion of cobalt in this combination usually falls within the limit of 3 to 13 % and while there is no actual alloying between the cobalt and the carbide. The mixture is very intimate, that it is regarded as a cemented alloy or carbide.

Stellite and Stellite Type Alloys: The earliest important cobalt alloy to be developed was stellite on which the original the original patient was issued in 1907. The original alloy contained 75% cobalt and 25% chromium but later increasing amounts of tungsten and or molybdenum were added to increase the hardness. The alloy retains its forgeability* with up to 20% of tungsten, but beyond this must be cast and ground to desired shape. The stellite alloys do not tarnish and is resistant to many corrosive chemical agents, while this hardness at room temperature is some what less than hat of ordinary steels. It is resistant up to a visible red heat and alloys are highly resistant to corrosion and abrasion. Other modifications of composition have been made and that the greatest pressure for future development will be concentrated for future progress in the use of turbo super-charger, gas turbines and jet engines depends largely upon the success attending the efforts to improve alloys of this type.

The important member of this group have been included in the table below; identified according to their identification nomenclature, the table below shows the different designation for the same type of alloy:

IDENTIFICATION OF COBALT HIGH-TEMPERATURE ALLOYS

[a]Stellite Alloy	[b]N.D.R.S.	[c]A.M.S.	Development Designation
21	NR-10	AMS-5385	Vitallium[d]
23	NR-63	AMS-5375	61 Alloy
27	NR-60	AMS-5378	6059 Alloy
30	NR-12	AMS-5380	422-19 Alloy
31	NR-71	AMS-5372	X-40 Alloy
Multimet	NR-21	——	N-155 Alloy

(*a*) Hayens Stellite Co
(*b*) National Defence Research Council
(*c*) Aeronautical Material Specification
(*d*) Trade Mark Name.

---

* Forgeability: Shape by heating

Compounds of Cobalt

Under this subhead we are going to discuss only the inorganic compounds of cobalt.

The best known compounds of cobalt are cobaltous compounds, which consists of bivalent cobalt, which has tendency to form so called 'basic salts' for *e.g.* carbonates and sulphates, in addition to the normal salts. Trivalent cobalts are commonly called cobaltic, but the simple cobaltic compounds are usually unstable. Of the very numerous complex co-ordination number of six) have been studied more than the complexes of any metal.

Although the use of cobalt compounds in the blue glazes, glasses and porcelain enamels was continued and although small quantities of cobalt compounds were used as early as 1705 in sympathetic ink (in which the blue colour develops on heating), no important news uses develop until the late 19th century. The use of cobalt compounds began in the preparation of dryers, which are catalysts used in paints and varnishes, instantly dries the paints and polished surfaces. Usually these cobalt catalysts, as finally developed, where metallic soaps prepared from other cobalt compounds such as the hydroxide and sulphate. Latter, cobalt compounds were used as electrolytes for cobalt plating although no large volume consumption developed. Cobalt compounds also have come to be used in the correction of cobalt nutritional deficiencies in cattle and sheep raised on cobalt deficient soils. A more recent development in the catalytic application of cobalt compounds arose when the Fischer- Tropsch synthesis was developed for the reaction of carbon monoxide and hydrogen to form new useful organic compounds.

Cobaltous Oxide (CoO): This is best obtained by reducing cobaltic oxide in a current or hydrogen, it is a powder when heated in air, it is conversion into cobalto-cobaltic oxide.

$$\text{Cobalt} + \text{Oxygen} \longrightarrow \text{CobalticOxide}$$

$$6Co + 4O_2 \longrightarrow 2Co_3O_4$$

Cobaltous oxide is a basic oxide in character and gives rise to cobaltous salts.

When an alkali is added to a solution of a cobaltous salt a blue precipitate of cobaltous hydroxide is produced.

$$\text{Cobaltous Chloride} + \text{Sodium Hydroxide} \longrightarrow \text{Cobaltic Oxide} + \text{Sodium Chloride}$$

$$CoCl_2 + 2NaOH \longrightarrow Co(OH)_2 + 2NaCl$$

On boiling the liquid the solution mixture turns rose pink like ferrous hydroxide; it also absorbs oxygen from air to give brown cobaltic hydroxide.

Cobaltic Oxide (CO_2O_3): This is obtained as a black powder by gently heating cobaltous nitrate.

$$\text{Cobaltous Nitrate} \longrightarrow \text{Cobaltic Oxide} + \text{Nitrogen Dioxide} + \text{Oxygen}$$

$$4Co(NO_3)_2 \longrightarrow 2Co_2O_3 + 8NO_2 + O_2$$

When strongly heated (dry) cobalto-cobaltic oxide.

$$\text{Cobaltic Oxide} \longrightarrow \text{Cobalto-Cobaltic Oxide} + \text{Oxygen}$$

$$6Co_2O_3 \longrightarrow 4Co_3O_4 + O_2$$

When an alkaline oxidizing agent, such as sodium hypochlorite is added cobaltic hydroxide is obtained. This is soluble in cold acids giving brown solutions, probably containing cobaltic salts. On warming the liquid decomposition of the cobaltic to form the cobaltous salt occurs.

Cobaltic Hydroxide + Hydrochloric Acid ⟶ Cobaltous Chloride + Water + Chlorine

$$2Co(OH)_3 + 6HCl \longrightarrow 2CoCl_2 + H_2O + Cl_2$$

Cobalto-cobaltic Oxide (Co_3O_4): This oxide of cobalt is obtained by strongly heating any oxide of the metal in air or by strong ignition of cobaltous nitrate.

Cobaltous Nitrate ⟶ Cobalto-Cobaltic Oxide + Nitrogen Dioxide + Oxygen

$$3Co(NO_3)_2 \longrightarrow Co_3O_4 + 6NO_2 + O_2$$

Cobalto-cobaltic oxide is a black powder; towards acids it acts as a super oxide and gives cobaltous compounds.

Cobalto-cobaltous Oxide + Hydrochloric Acid ⟶ Cobaltic Chloride + Water

$$Co_3O_4 + 8HCl \longrightarrow 3CoCl_2 + 4H_2O$$

Cobaltic Acetate [$Co(C_2H_3O_2)_3$]: This compound is obtained by the process of electrolysis of cobaltous acetate and acetic acid using platinum electrodes.

By evaporation in a vacuum over phosphorous pentoxide, well defined green crystals of cobaltic acetate are found. An excess of acetic acid will cause an aqueous solution of the cobaltic acetate to loose oxygen until it is converted to cobalto-cobaltic oxide (Co_3O_4). The solution is also reduced by light and ferrous sulphate.

Cobaltous Carbonate ($CoCO_3$): This compound is produced by the reaction of cobaltous oxide and sodium bicarbonate.

Cobaltous Oxide + Sodium Bacarbonate ⟶ Sodium Hydroxide + Conbaltous Carbonate

$$CoO + NaHCO_3 \longrightarrow NaOH + CaCO_3$$

Cobaltous Sulphide (CoS): When hydrogen sulphide passes through a solution of cobaltous oxide and sodium bicarbonate.

Cobaltous Hydroxide + Hydrogen Sulphide ⟶ Cobaltous Sulphide + Water

$$Co(OH)_2 + H_2S \longrightarrow CoS + H_2O$$

It is also prepared by the action cobaltous salt and sodium sulphide.

Cobaltous Hydroxide + Sodium Sulphide ⟶ Sodium Hydroxide + Cobaltous Sulphide

$$Co(OH)_2 + Na_2S \longrightarrow 2NaOH + CoS$$

Cobaltous sulphide is insoluble in dilute acids but soluble in concentrated acids and Aqua Regia.

Cobaltous Sulphate ($CoSO_4$, $7H_2O$): This compound is obtained when cobalt metal is allowed to react with dilute sulphuric acid.

Cobalt + Sulphuric Acid ⟶ Cobaltous Sulphate + Hydrogen

$$Co + H_2SO_4 \longrightarrow CoSO_4 + H_2$$

These crystals are similar to ferrous sulphate, when heated about 250°C they loose water and become anhydrous.

Cobaltous Chloride ($CoCl_2$): The anhydrous salt can be obtained by heating metallic cobalt in chlorine or by heating the hexahydrate $CoCl_26H_2O$ in the stream of hydrogen chloride. This compound as pale blue mass. The common hydrate $CoCl_2H_2O$ is obtained in the form of dark red deliquescent crystals by evaporating a solution of the metal oxide or carbonate in hydrochloric acid. The hexahydrate looses water on heating to form lower hydrates and at about 120°C forms the blue anhydrous salt with partial decomposition. Anhydrous cobalt chloride dissolves in alcohol either and other organic solvents yields blue solutions.

Cobalt Bromide ($CoBr_2$): This is obtained when cobalt is heated in bromine or oxide in hydrogen bromide.

$$\text{Cobaltous Oxide} + \text{Hydrogen Bromide} \longrightarrow \text{Cobaltous Bromide} + \text{Water}$$

$$CoO + HBr \longrightarrow CoBr_2 + H_2O$$

The anhydrous compound is dark green in colour and very hydroscopic. The hydrate $CoBr_26H_2O$ is reddish purple and $CoBr_22H_2O$ is bluish-purple.

Cobaltous Iodide (CoI_2): It is obtained in the form of black crystals when cobalt is heated in iodine in iodine vapour.

$$\text{Cobalt} + \text{Iodine} \longrightarrow \text{Cobaltous Iodide}$$

$$Co + I_2 \longrightarrow CoI_2$$

It is also obtained when cobalt reacts with a saturated solution of iodine water. From the solution the hydrate separates out as dark red crystals on evaporation.

Cobaltous Fluoride (CoF_2): This compound is obtained as anhydrous salt by heating the anhydrous salt of cobaltous chloride, $CoCl_2$ in a current of hydrogen fluoride. It forms a red powder, moderately soluble in water. The hydrate, CoF_24H_2O can be obtained from a solution of cobaltous carbonate in aqueous hydrofluoric acid.

$$\text{Cobaltous Chloride} + \text{Hydrogen Fluoride} \longrightarrow \text{Cobaltous Fluoride} + \text{Hydrogen Chloride}$$

$$CoCl_2 + 2HF \longrightarrow CoF_2 + 2HCl$$

Cobaltic Fluoride (CoF_3): When fluoride reacts with cobaltous chloride, cobaltic fluoride is produced.

$$\text{Cobaltous Chloride} + \text{Florine} \longrightarrow \text{Cobaltic Floride} + \text{Chlorine}$$

$$2CoCl_2 + F_2 \longrightarrow CoF_3 + Cl_2$$

A green hydrate CoF_35H_2O is also obtained when electrolysis is done on cobaltous fluoride with concentrated hydrochloric acid in cold condition.

On heating cobaltic fluoride it is converted by water with the separation of cobaltic hydroxide.

$$\text{Coblaltic Fluoride} + \text{Water} \longrightarrow \text{Cobaltic Hydroxide} + \text{Hydrogen Fluoride}$$

$$CoF_3 + H_2O \longrightarrow Co(OH)_3 + 3HF$$

Cobalt Sulphate [$Co(SO_4)_2 18H_2O$]: This compound is deposited in the form of blue needle shaped crystals at the anode by electrolysis of cobaltous sulphate in 40% sulphuric acid in a separated cell. It is readily decomposed by water with the evolution of oxygen.

Cobaltous Sulphate + Water $\longrightarrow$ Cobaltic Sulphate + Sulphuric Acid

$$2Co_2(SO)_3 + H_2O \longrightarrow 4CoSO_4 + H_2SO_4$$

But in dilute sulphuric acid it dissolves without decomposition with sodium or potassium sulphates it forms blue cobaltic atoms.

$$M'Co(SO_4)_3 12H_2O$$

Where M' is Na, K, Rb, or NH_3.

These alums are similar to other common alums. Cobaltic alums are also unstable and decomposed by water with evolution of oxygen.

Cobalt Complexes

The commonest cobalt compounds are complex; the most important among them are the cobaltamines, cobaltinitrates and cobalticyanides.

Precipitated cobaltous hydroxide dissolves in excess of ammonia to a yellowish brown solution deposits cobaltous hydroxide on dilution. On exposure to air the solution absorbs oxygen, more rapidly on boiling or heating with lead dioxide and forms a pink solution of a cobaltamine. More than 2000 are known but here some of them the typical.

1. A solution of ammonium carbonate and ammonia is added to a solution of cobalt nitrate solution and air is drawn through the violet liquid; oxidation occurs and blood red solution of the compound *tetrammine-carbonate-cobaltic nitrate* $[CO(NH_3)_4CO_3]NO_3$ is formed, which gives purple crystals on evaporation.
2. On acidifying a solution of this with hydrochloric, heating with excess of ammonia and adding concentrated hydrochloric acid *pentammine-chloro.cobaltic dichloride* $Co(NH_3)_5ClCl_2$ is formed.
3. On heating pentammine-chloro.cobaltic dichloride with ammonium chloride and ammonia in a pressure bottle, and adding hydrochloric acid and cooling *hexammine cobaltic trichloride* $Co(NH_3)_6Cl_2$ is formed.
4. From a solution of $Co(NH_3)_6C1_2$ in ammonia cooled in ice, slow addition of concentrated hydrochloric acid precipitates *aquo-pentammine cobaltic trichloride,* $CO(NH_3)_5H_2OCl_3$.

Cobalticyanide: Consists of a stable complex anion $CO(CN)_6''$ containing 3-valent cobalt, potassium, Cyanide solution gives with a cobalt salt solution a reddish brown precipitate of cobaltous cyanide. $Co(CN)_2 3H_2O$ soluble in excess to a yellow solution of potassium cobaltocyanide contains 2-valent cobalt. This is precipitated as a deep purple coloured powder in the alcohol. On adding a little acetic acid on hydrochloric acid to the cobaltiyanide is precipitated.

Potassium cobaltocyanide forms stable yellow crystals similar to potassium ferrocyanide. With silver nitrate it gives a white precipitate of silver cobaltiyanide from which by precipitation of copper by hydrogen sulphide gas a colourless crystalline *hydrocobaltocyanic acid* $H_2Co(CN)_6$ is formed. Cobalticyanides offer no reaction of cobalt or cyanide and are not decomposed by concentrated nitric acid or hypochlorite.

Cobaltinitrites: Contain the complex anion $CO(NO_2)_6$ containing 3-valent cobalt, potassium nitrate gives with acidulated cobaltous salt solution a yellow precipitate of potassium cobaltinitrate. $K_3Co(NO_2)_3$ Fischer's Salt which is scarcely soluble in water.

This complex develops only in the acid solutions otherwise the *coblatinitrite* containing 2-valent cobalt, is produced. The cobaltinitrite is less stable than the cobalticyanide and is decomposed by ammonium sulphide, it is used as a yellow pigment called cobalt yellow and is used for painting.

Reagent for Detection of Potassium dissolve 30 g of cobalt nitrate and 50g of sodium nitrite in 150ml of water and add 10ml of glacial acetic acid. The addition of silver nitrate makes it more sensitive, as the salt $K_2AgCo(NO_2)_6$ and $KAg_2Co(NO_2)_6$ are less soluble than $K_3Co(NO_2)_6$ 1 part of potassium in 10,000 of water may be detected.

If cobaltous salt is converted to cobaltic salt by precipitating cobaltic hydroxide and dissolving this in acetic acid, precipitation of $Co(C_{10}H_{60}N)_3 2H_2O$ is the result.

COBALT CARBONYLS

We are now going to discuss the cobalt carbonyls under this subhead.

Cobalt Octacarbonyl (Co_2CO_8): It can be made by four different methods:

1. By the direct action of carbon monoxide on metallic Cobalt: Reaction between carbon monoxide and reduced metallic cobalt occurs most rapidly at 220°C and no reactions takes place below this and the pressure being 253.5MPa.

2. By the action of carbon monoxide on certain compounds of Cobalt in the presence of copper: Dry carbon monoxide at 200°C and 202.8MPa in the presence of metallic copper, reacts with certain binary compounds of cobalt to yield carbonyl.

Compound	CoS	CoI_2	$CoBr_2$	$CoCl_2$	CoF_2	CoO
Heat of Formation in Kcal	20	24.5	55	78	148.95	60
*Lattice of Compound	NiAs	CdI_2	CdI_2	CdI_2	Rutile	Rock salt
Relative amount of carbonyl formed Under same condition	100	100	9	3.5	0	0

*Lattice: The structural arrangements of the atoms in a crystal.

The above table shows that the compounds of cobalt having low thermal stability, forming nonionic lattices are the most efficient, after the reaction the sulphur or halogen is present in combination with cuprous copper, for example.

Cobaltous Sulphide + Carbon Monoxide $\longrightarrow$ Cobalt Octacarbonyl + Cuprous Sulphide

$$2CoS + 8CO \longrightarrow Co_2(Co)_8 + 2Cu_2S$$

The reaction continues even if a screen is placed between the cobalt compounds and the copper, thus showing that the transference of the sulphur or halogen from the cobalt to the copper must be due to the formation of available intermediate such as $Co(CO)I_2$. The variation in the efficiency above table may be due to the variation in the stability of the intermediates, which may be formed from them.

3. By the Action of an Acid on a solution of Cobalt Carbonyl Hydrate: Hydrogen is evolved and $Co_2(CO)_8$ is formed.

4. By the thermal decomposition of Cobalt Carbonyl Hydride: Cobalt carbonyl hydride is cooled at –79°C in a closed vessel. It is allowed to warm up to –30°C. Some of it decomposes. The vessel is recooled and the hydrogen, which has been liberated, is pumped off. This process is repeated until all the cobalt carbonyl hydride is decomposed.

Cobalt carbonyl is unstable in water but it is stable in alcohol to some extent, also in ether, carbon disulphide and naphtha. Either pure or in solution it decomposes on standing or at temperature above 51°C to yield black crystals of the tetracarbonyl, $Co_4(CO)_{12}$. At 60°C these crystals themselves decompose to give carbon monoxide and metallic cobalt. The chemical reactions of cobalt octahedral is summarized in the chart I.

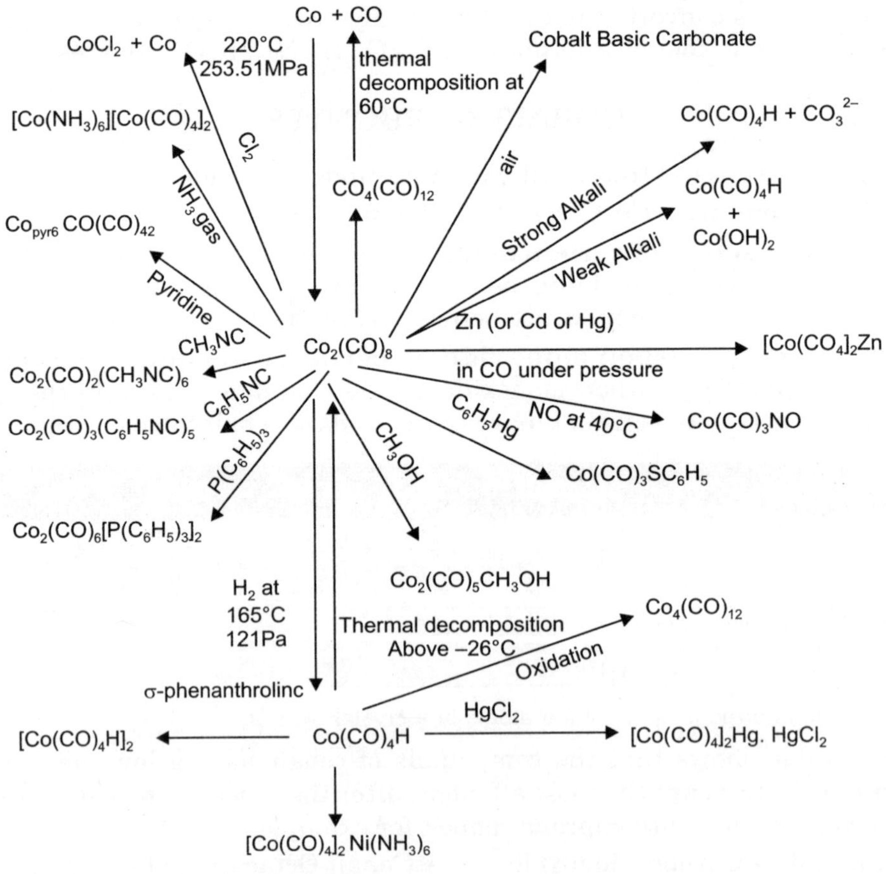

Chart I : Summary of Cobalt Complex (Octacarbonyl)

Hydrolysis and Reactions with Acids and Bases

By exposure to air $Co_2(CO)_8$ is conversion to a deep violet basic carbonate of cobalt. It is only very slowly attached by non-oxidizing acids, such as hydrochloric acid sulphuric acid. It is hydrolyzed by bases in two ways.

Strong bases like hydroxide convert it into a mixture of cobalt carbonyl hydride and cobalttricarbonyl.

Cobalt Octacarbonyl	+	Barium Hydroxide	$\longrightarrow$	Cobalt Carbonyl Carbonate	+	Barium Hydride	+	Cobaticarbonyl
$3CO_2(CO)_8$	+	$Ba(OH)_2$	$\longrightarrow$	$4Co(CO)_4H$	+	$BaCO_3$	+	$\frac{1}{2}CO_4(CO)_{12}$

Weak bases, such as ammonia, convert it to a mixture, of cobalt carbonyl hydride and cobaltous hydroxide.

Cobalt Octacarbonyl	+	Water	$\longrightarrow$	Cobalt Carbonyl Hydride	+	Cobaltous Hydroxide	+	Carbon Dioxide
$Co_2(CO)_8$	+	H_2O	$\longrightarrow$	$4Co(CO)_4H$	+	$2Co(OH)_2$	+	8CO

Reaction with Halogens

No Cobalt carbonyl is formed during the reaction with Octacarbonyl and bromine

The halogens decomposes cobalt octacarbonyl thus:

Cobalt Octacarbonyl	+	Bromine	$\longrightarrow$	Cobaltous Bromide	+	Carbon Monoxide
$Co_2(CO)_8$	+	2Br	$\longrightarrow$	$2CoBr_2$	+	8CO

Reaction with Hydrogen

At 165 and 121.68MPa hydrogen (in the presence of carbon monoxide at 152.1MPa to prevent decomposition) partly converts cobalt octacarbonyl to convert carbonyl hydroxide.

Cobalt Octacarbonyl	+	Hydrogen	$\longrightarrow$	Cobalt Carbonyl Hydroxide
$Co_2(CO)_8$		H_2		$2Co(CO)_4H$

Reaction with Ammonia and the Amines

Gaseous ammonia reacts directly with cobalt octacarbonyl to yield cobalt hexammine salt of cobalt carbonyl hydride.

Pheantroline yields cobalt hexammine which, is insoluble in water and can be used for gravimetric determinations. Pyridine yields $CO_2(CO)_{5pry}$ with the evolution of carbon monoxide.

Reaction with Nitric Oxide

Nitric oxide reacts with slowly at room temperature, brownish-yellow prisms of Cobalt Dicarbonyl Hexammine crystals. Phenol isonytrile similarly yields Cobalt Tricarbonyl Pentaammine which crystalline as brownish-yellow rhmbohedra. These compounds are non-ionic and compared with carbonyl they are more stable and deeper in colour.

The summery of the reactions of cobalt octacarbonyl is shown in chart I.

Detection and Estimation of Cobalt

Cobalt is detected by adding any cobalt salt to a solution of ammonium thiocyanate. A blue coloured cobalt complex is formed. It can also be detected by producing it in a borax head test where it produces a blue flame both at reducing.

NICKEL

Melting Point : 1455°C
Boiling Point : 2900°C
Atomic Number : 28
Atomic Weight : 58.69

Isolation and Occurrence

Nickel was isolated by Cronstelt in 1751. Richter prepared the first pure metal in 1804 and described its properties with considerable accuracy by 1840; production from Norwegian deposits was adequate for European needs. The copper-nickel zinc alloy in England as early as 1835. Faraday developed nickel plating in 1843. Copper nickel alloys were used in Belgium for coinage. About 1860 malleable nickel was produced by Flietmann in 1870 by the addition of magnesium and in 1839 Rilley described the effects of nickel on steel. In 1876 production of nickel from the New Caledonian deposits exceeds that from Norway.

The largest of the known ore bodies of nickel was discovered in the district of Sudbury Ontario Canada in 1833.

Nickel occurs after iron cobalt in the periodic table just before copper. The stable isotopes of nickel, in order of abundance, have atomic weights.

Some important and readily available ores of nickel and their physical state of these are shown in the table below:

Name	Chemical Composition	Colour	Crystalline From
Pentlandite	$(Fe, Ni)_9S_8$	Bronze-Yellow	Isometric
Millerite	NiS	Brass-Yellow	Hexagonal
Niccolite	$NiAs$	Copper-Red	"
Brethauptite	$NiSb$	"	"
Maucherite	$Ni_{11}As_8$	Platinum-Grey	Tetragonal
Gersdorffite	$NiAsS$	Steel-Grey	Isometric

Other well-known nickel minerals are Polymite, Ni_3S_4, Violarite $(Ni, Fe)_3S_4$ Browite $(Ni, Fe)S_2$. Ultimanite, NiSbS, etc.

Nickel is also found in a number of less common secondary silicate minerals and is known to substitute of magnesium in certain primary minerals (olivine, hypersthene and biotite) Native nickel approaching purity is unknown. Small quantity of nickel combined with iron have been found in India at laduguda mines also at Mosabani and Cuttack district of Orissa. It is also found in Newzealand, Canada and some parts of U.S.A.

ORES

There are three principal classes of nickel ores: (1) sulphide ores, found in India, Norway Japan, Sweden and Finland, (2) silicate(oxide) found in Cuba, Brazil Spain and Greece and (3) Arsenide ores found in Continental Europe, Canada and U.S.A.

1. Arsenide Ores: These ores are found principally in Saxony and Canada having a nickel content of 3.5 to 14% and are of little commercial importance. The chief nickel bearing arsenides are Niccolite (NiAs) and Chloanthite ($NiAs_2$).

2. Sulphide Ore: Sulphide ores occur associated with basic, intensive rocks of the Gabbro or periodotite groups. The largest known deposits are in the Sudbury area of Ontario Canada. The nickel carrying material is pentlandite [$(Ni,Fe)2S_8$] which occurs along with the pyrrhotite (FeS to FesS6) chalcopyrite ($CuFeS_2$) and considerable rocky gangue (norite). The average nickel plus copper content is about 2.5%. The ore content is small amount of cobalt, selenium, tellurium, silver gold and five of the platinum metals ruthenium, rhodium, palladium, iridium and platinum.

3. Silicate(oxides) Ores: Oxide nickel ores result from the breakdown of nickel in typical climates bearing basic rocks in tropical climates, which produce laterite weathering. They are of two general types, known as silicate, limonite, geolite, matite and silica. The nickel occurs usually in the silicate crystal structure and may be oxides. The best known deposits are in New Calandina and supplies about 7% of worlds demand. The nickel content ranges upto about 7.5% and the ores contain as much as 20% of total moisture and approximately 10% of which is combined chemically with the nickel silicate. The principal nickel mineral is gemietite a complex hydrated nickel-magnesium silicate in which nickel and magnesium contents vary approximately. The latarite ores in Cuba are extensive contains about 1% of nickel and have been exploited.

4. Other Sources: Small percentages of nickel are present in many copper ores and in some manganese ores.

Smelting and Refining

Sulphur Ores: Production of nickel, consists of crushing; wet grinding; floatation to reject rock and float a sulphide concentrate containing the bulk of nickel, copper, and differential floatation to produce a nickel concentrate and a copper concentrate.

The concentrate is roasted for partial oxidation of iron and partial removal of sulphur and smelted with a flux to remove all of the rock content and part of the iron in the form of slag. It also forms a copper bearing nickel iron matte, consisting largely of sulphide of these metals. The matte is bessmerised for further removal of iron and sulphur. Final separation of nickel, copper and precious metal is achieved by subjecting the Bessemer matte to controlled cooling, fine grinding, magnetic separation and differential floatation. The resulting nickel sulphide is sintered* to nickel oxide and either marketed as such are reduced to metallic nickel that is cast to anodes and refined electrically or by Mond Process.

---

* Sintered: Heating mineral under pressure.

This process is based on the production of sintering nickel sulphide and reducing oxide, combines readily with carbon monoxide at 60°C to form nickel carbonyl gas, $Ni(CO)_4$ leaving the impurities including cobalt as a solid residue. Subsequent heating of the nickel and copper which are separated by roasting to remove most of the sulphur, leading with dilute sulphuric acid to remove most of the copper reduction melting the insoluble nickel residue, casting into anodes and refining electrolytically.

Silicate Ores: The New Caledonian silicate ores are dried and converted into a brick shaped mass, by mixing the ore with calcium sulphate and limestone fluxes to take care of the siliceous gangue. Smelting with coal or coke produces a nickel sulphite matte (39 - 40%) which is processed further by Bessemerizing to oxidize some of the sulphur, remove some of the iron and produce a low iron nickel sulphide containing about 77% of nickel. The nickel sulphide is grounded and roasted to remove sulphur, mixed with a carbon material to reduce it to metallic nickel in the form of rods by heating in retorts. A ferronickel averaging about 37% nickel is also produced.

Cuban Ores: The lataritic chromium-nickel-iron ores of Cuba were refined as an emergency operation during World War II by an ammonia leaching process. Which includes grinding; selective reduction of the ore by producer gas at a temperature well below fusion; leaching with aerated; ammonical, ammonium carbonate; removal of iron by precipitation of nickel oxide, in which form it was marketed.

Alloys of Nickel

Nickel has many alloys with iron and other different metals. They are all listed below in the table:

Principal Elements	Nickel Content (%)	Alloy	Use
Nickel	99.4%	Commerically pure Wrought Nickel	Corrosion resistance, radio and electronic tube parts, coinage
Nickel and Iron	0.5 – 9%	Wrought Alloy Steel	Transportation, Low temperature application.
	30 – 90%	Magnetic Alloys	Communication and Electrical equipment
	5 – 40%	Clad Steel	Petroleum, chemical and food processing instruments.
Nickel and Copper	2 – 13% 10 – 30%	High-Copper alloys, Cupronickels	Tubes and plates in capacitors and electronic parts.
	25%	Coinage	Coinage
Nickel and Chromium	80%	Electrical resistance	Heating coils and resistance.
Nickel copper and Zinc	5 – 30%	Nickel Silver	All types of hardware, which requires strength.

Physical Properties

Melting Point	1455°C
Boiling Point	2900°C
Density	8.908 g/cc
Lattice Constant, Face-centred	3.5168
Specific Heat at 100°C	0.1123
Latent Heat of Fusion	73 cal/g
Thermal Expansion Co-efficient at 0-100°C Thermal Conductivity	0.0000133
At 100°C	0.198
At 300°C	0.152
At 500°C	0.148
Electrical Resistivity at 20°C Temperature Co-efficient of electrical	6.844 μΩ/cm
Resistivity (0 – 100°C) per degree centigrade Modulus Elasticity	0.0068
In tension	2.06^2 Pa
In Torsion	7584 Pa
Poisson's Ratio	0.31

Chemical Properties

1. Action of Air: Nickel is less reactive among iron, cobalt and nickel. In bulk it is quite resistant to attack by air or by water at ordinary temperature. It is due to formation of a thin layer of oxide, which protects the metal from further reaction. Nickel is particular resistant to attack in the alkaline media. In powder, state it is fairly reactive towards air. It may ignite in certain conditions; it reacts with steam at high temperature giving hydrogen.

$$\text{Nickel + Steam} \longrightarrow \text{Nickel Oxide + Hydrogen}$$

$$Ni + H_2O \longrightarrow NiO + H_2$$

2. Oxidation: Nickel is also more inert to chemical oxidation than iron and cobalt. The standard oxidation potential is 0.25v, which is lowest of all the three metal.

3. Action of acids and Alkalis: When added to concentrated nitric acid, the metal becomes passive like iron and cobalt. It is readily soluble in aqua regia. It resists the attack of alkali.

4. Action of Hydrogen: It is readily attracted by hydrogen forming metal halides, for *e.g.*

$$\text{Nickel + Hydrogen} \longrightarrow \text{Nickel Hydride}$$

$$2Ni + H_2 \longrightarrow 2NiH$$

Electronic Configuration and Oxidation State

The outer electronic configuration of Ni is $3d^8 4s^2$. It shows oxidation states of 0, +1, +2, +3, and +4. The +1 state ($3d^2 4s^1$) arises from the removal of one of the two 4s-electrons. This state is very rare. The +2 state ($3d^8$) arises from the removal of 4s electrons. It is the only stable state. A few compounds of trivalent nickel ($3d^7$), which are known as very strong oxidizing agents.

Since nickelous compounds has a strong tendency to change into nickelic compounds.

Compounds of Nickel

Nickel Monoxide (NiO): It is prepared as a green precipitate by strongly heating the hydroxide carbonate or nitrate of nickel.

Nickel Hydroxide ⟶ Nickel Monoxide + Water

$$Ni(OH)_2 \longrightarrow NiO + H_2O$$

Nickel Hydroxide [Ni(OH)$_3$]: This is obtained as a green precipitate by the addition of caustic alkalis to a nickel salt.

Nickel Sulphate + Sodium Hydroxide ⟶ Nickel Hydroxide + Sodium Sulphate

$$NiSO_4 + 2NaOH \longrightarrow Ni(OH)_2 + Na_2SO_4$$

It dissolves in ammonia to form complex nickel ammines.

Nickel Hydroxide + Ammonia ⟶ Nickel Ammine

$$Ni(OH)_2 + NH_3 \longrightarrow [Ni(NH_3)_4]\,(OH)_2$$

Nickel Sesquioxide, Nickelic Oxide (Ni$_2$O$_3$): It is a black powder which is produced, when nickel carbonate or nitrate is heated in air at a temperature of 300°C.

Nickel Nitrate ⟶ Nickel Sesquioxide + Nitric Oxide + Nitrogen Dioxide + Oxygen

$$2Ni(O_3)_2 \longrightarrow Ni_2O_3 + 3NO + NO_2 + 2O_2$$

It is probably a mixture of oxides, NiO and Ni_2O_3. It dissolves in hydrochloric acid with the evolution of chlorine and in sulphuric acid with the precipitation of nickel sulphate.

Nickel Sesquioxide + Hydrochloric Acid ⟶ Nickel Chloride Water + Chlorine

$$Ni_2O_3 + 6HCl \longrightarrow 2NiCl_2 + 3H_2O + Cl_2$$

Nickel Monoxide + Sulphuric Acid ⟶ Nickel Sulphate + Water

$$2NiO + 2H_2SO_4 \longrightarrow 2NiSO_4 + 2H_2O$$

The composition of the product is however, somewhat variable and does not correspond exactly to the formula Ni_2O_3 containing trivalent nickel.

A black precipitate with the approximate composition of hydrated nickel oxide N_2O_3 is obtained by the action of chlorine or bromine on a suspension of nickel hydroxide ($Ni(OH)_2$] in water. The precipitate is however found to contain more active oxygen than what corresponds to nickelic oxide (Ni_2O_3) and is in fact, a mixture of Hydrated (Ni_2O_3 and NiO). A similar product is obtained by the electrolytic oxidation of nickelous hydroxide in potassium hydroxide.

The formation of hydrated nickel sesquioxide, containing some nickel dioxide by anodic oxidation, forms the basic of the Edison accumulator cell. It consists of nickelic hydroxide anode and an iron cathode in electrolysis of potassium hydroxide solution. The reactions that occur when cell discharged or charged are being stated below:

$$\text{Iron} + \text{Nickelic Hydroxide} \longrightarrow \text{Ferrous Hydroxide} + \text{Nickelous Hydroxide}$$

$$Fe + 2Ni(OH)_3 \longrightarrow Fe(OH)_2 + 2Ni(OH)_2$$

Nickel Carbonate ($NiCO_3 6H_2O$): This compound is obtained in the hydrated from as apple green precipitate by the addition of sodium bicarbonate to a nickel salt. With a large excess of alkali bicarbonate, a double salt like $KHNI(CO_3)_2 4H_2O$ is precipitated. A basic carbonate is precipitated if a solution of normal alkali carbonate is used.

Nickel Nitrate [$Ni(NO_3)_2$]: A solution of nickel in dilute nitric acid when evaporated yields nickel in dilute nitric acid when evaporated yields nickel nitrate which is found obviously in hydrated from $Ni(NO_3)_2 6H_2O$. At the ordinary temperature, which is changed, to lower hydrate when the temperature rises.

Nickel Sulphide (NiS): This compound of nickel is formed as a black precipitate on the addition of ammonia and ammonium sulphate to a nickel salt solution. It is also not soluble in dilute acids, but it gets slowly dissolved in concentrated hydrochloric acid. It is readily soluble in nitric acid and aqua regia.

Nickel Sulphate ($NiSO_4$, $7H_2O$): It is obtained as green prismatic crystals similar to ferrous sulphate. It separates on evaporation of a solution of the oxide or basic carbonate of nickel in dilute sulphuric acid at room temperature. The crystals loose water gradually, when heated, yielding hex, tetra and dihydrate and finally the yellow anhydrous salt ($NiSO_4$).

With alkali sulphates it forms double sulphates like $Ni(SO)_2 6H_2O$ similar to the double salt of cobalt and iron nickel ammonium sulphate is used in the nickel plating.

Nickel Fluoride (NiF_2): This is produced in the anhydrous state, like the cobalt fluoride by sublimating a mixture on nickel chloride and ammonium fluoride. It forms yellow crystals soluble in water. The hydrate (NiF_2 $3H_2O$) is obtained in the form of bluish green crystals by evaporating a solution of nickel hydroxide or carbonate in hydrofluoric acid.

Nickel Chloride ($NiCl_2 6H_2O$): The hydrated salt ($NiCl_2 6H_2O$) is obtained in the form of green deliquescent crystals by evaporating a solution of nickel hydroxide and hydrochloric acid. The hexahydrate looses water on heating, to form the lower hydrate with $4H_2O$ at 70°C. At a higher temperature it is completely dehydrated into a basic chloride. When this hydrated product is heated in a current of hydrogen chloride the anhydrous $NiCl_2$ sublimes in the form of yellow scales.

Nickel Bromide ($NiBr_2$): It is yellow in the anhydrous state, from an aqueous solution the green hydrate ($NiBr_2$ $6H_2O$) as usual.

Nickel Iodide (NiI_2): It is a black substance when in a hydrated state. The green hexahydrate ($NiI_2 6H_2O$) is obtained by evaporating its aqueous solution.

Nickel Cyanide [$Ni(CN)_2$]: When a solution of potassium cyanide is added to a nickel salt solution a green precipitate of the nickel cyanide is produced. This dissolves in an excess of the alkali cyanide to form a red solution. On evaporation, the solution deposits $K_2Ni(CN).JI_2O$. Potassium Nickelocyanide, yellow crystals unlike potassium cobaltocyanide show no tendency to oxidation. It is however decomposed easily by acids. When its aqueous solution is treated with chlorine or bromine in presence of alkali, the black hydrated (NhO_3) is precipitated. In this it differs from cobalt and iron.

Nickel Carbonyl [$Ni(CO)_4$]: It is obtained in the form of a colourless volatile strongly refracting liquid by passing carbon monoxide over finely powdered nickel at a temperature of 40°C, usually a higher temperature is used under a high pressure. Having a melting point of 25°C and boiling point of 43°C it is insoluble in water but dissolves in many organic solvents, dilute acids and alkalis have got no action on it. It bums in a luminous flame and forms an explosive mixture with oxygen. When heated at a temperature of 150°C it decomposes into metallic nickel and carbon monoxide. It plays an important role in the production of metallic nickel. Nickel carbonyl is obviously covalent through nickel appears to have valency 0 in the compound (that is the outermost orbit of nickel atom will be complete in this case).

Nickel Ammine: The green solution of nickel salts in water changes to blue on treatment with an excess of ammonia. From these solutions, complex nickelammine salts in beautiful blue colour crystallizes out. Salts having composition $NiCl_2$ $6NH_3$ or Ni $(NH_3)_6Cl_2$ $NiSO_4$ $4NH_32H_2O$ or $Ni(NH_3)_4(H_2O)_2SO_4$ etc have prepared. In these the co-ordination number of nickel is apparently 6. But in many of its other complexes, particularly with organic molecules, nickel shows a co-ordination number of 4 only.

Detection and Estimation

A very highly sensitive reagent for the detection of and also estimation of nickel has been found dimethylglyoxime. It gives with a solution of nickel has in the presence of ammonia or a solution of sodium acetate a bright red, silky voluminous precipitate of nickel dimethylglyoxime is produced. This permits the detection of nickel, even in traces and its estimation in the presence of many metals, including cobalt. Nickel can also detected and estimated in the presence of cobalt by heating their solution with excess of potassium cyanide to form complex cyanides of the two metals, now if the mixture is treated with caustic alkali and bromine or chlorine, black hydrated nickel is produced.

COPPER

Melting Point : 1083°C	Atomic Number : 29
Atomic Weight : 63.54	Boiling Point : 2595°C

History

Copper was one of the first metals known to man, occurring in the glacial drift as dark brown heavy nuggets*. The early Egyptians made copper knives and weapons eight

*Nuggests: A solid Lamp.

thousands years ago and pipes or tubes as early as 2750 BC .These ancient metallurgists soon found that alloys copper possessed more attractive properties than the pure metal. (A bronze mirror has been found which was probably made as early as 1800 BC). The Roman supply was known from Cyprus, it became known as *aos cyprum* and from this the Latin word cuprum has derived and hence the English word copper and the symbol Cu.

In 1809, the Calunett and Hecla Company (Michigan) became the largest individual copper producer of the world, although its production was less than 62,99,200kg of copper. By 1877, the mines of Rio Tinto (Spain) touched the leading position with an output of 38,10,00,00kg. The first mine in the V.S. to produce more than 50,8,00,000kg a year was Anaconda Copper Company (Montona) in 1896 and with the exception of the year 1907. When Calumet and Hecla again led and in 1908 and 1909 when the Copper Queen Anaconda led it maintained its position as the worlds greatest copper mine.

Occurrence

Copper is widely distributed in nature and occurs both in the free and the combined state. The more important ores both in the free and the combined state. The more important ores of copper may be divided into two classes:

(a) Sulphate Ores: The most important of these is *Copper Pyrites,* (Cu_2S, Fe_2S_3)

(b) Oxidized ores: These include *(Cuprite,* Cu_2O*) Malachite* [Cu_2O_3, $Cu(OH)_2$] *and Azurite* [$2CuCO_3$ $Cu(OH)_2$]

The sulphide ores are far more important and yield about 75% of the worlds production of copper.

Extraction

Copper is extracted from its ores by roasting, smelting, converting and refining.

Roasting: In roasting, the ore is heated in the presence of air to an elevated temperature without fusion or at minimum fusion. Sulphur is oxidized and particularly eliminated and the object of roasting is to control the amount of sulphur in the ore so that the charge when smelted will yield a matte that can be treated efficiently and economically in a converter. In addition the removal of arsenic and anatomy is often selected because it is easier and less expensive to remove them later after this point.

In sulphurisation of the roast the object is not to eliminate sulphur, but to convert the sulphide minerals to copper sulphate, which can be done by leaching*. In general the formation of sulphate, is forward by low roasting temperature and high connections of sulphur trioxide. In the roasting process, the temperature must maintained as high as 950°C to ignite the sulphides or other oxidizable constituents and to keep them above ignition point some massive sulphide ores and concentrates, however, liberates enough heat, a process known as **autogenous roasting**. The lower unit for autogenous roasting of copper ores is about 24% of sulphur. In most cases, however, this heat is not sufficient and some fuel must be used in roasting process.

---

* Leaching: Percolating.

Multiple Hearth Roasters: The good example of this type of roasting is the seven-hearth Wedge furnace is shown in Fig. 16.18. The furnace consists of a vertical cylinder of boilerplate lined with refractory bricks. The hearth floor, walls and furnace roof are lined with firebricks, in order to protect the metal parts as well as to conserve heat. There is a central vertical saft of boiler plate (about 2 m in diameter) lined on the outside with brick; to which are attached are rabble arms, two for each hearth (in the fig the rabble arms are shown only on every other hearth). The central shaft also contains pipes, which conduct cooling air or water to the arms. To each arm are fixed seven to nine racks or rubbles. The rotation this shaft, which is driven by motor and train of gears beneath the furnace, is of the order of 3rd r.p.m. One distinct advantage of this type of furnace lies in the fact that the operator may find the work easy in the order to remove and repair the rabble arms, which tend to get wrap and corrode, if high temperature is applied.

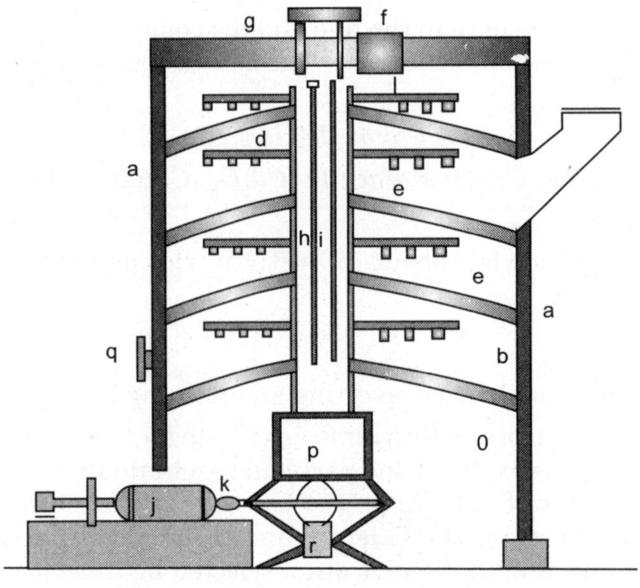

(*a*) Furnace Shell, (*b*) Refractory Lining (*c*) Rabble Ann, (*d*) Rabble Blades (*e*) Central Shaft (*f*) Air Outlet (*g*) Air Inlet (*h*) Supply Air Duct (*i*) Discharge air Duct (*j*) Motor (*k*) Bevel Gears (*l*) Drying Hearth (*m*) Gas Outlet (*n*) Air holder (*o*) Calcine Discharge (*p*) Manhole (*q*) Inspection door (*r*) Main bearing.

Fig. 16.18. The Wedge Roasting Furnace

The ore is fed upon the upper hearth, which serves as the dryer of the crude material. The rubbles so adjusted that the ore is gradually moved from the outer edge of the upper hearth toward the centre and falls through a drop hole upon hearth I. The rakes there move it across this hearth to a slot hear the periphery, through which it drops onto hearth II. This in a zigzag fashion the ore progresses through the furnace the ore until finally it drops into a car or conveyer beneath the last hearth. Until finally it drops on each hearth for visual observation, repair and admission of air. In an effort to conserve heat, flux for the subsequent smelting operation is sometimes preheated by adding one of the lower hearths.

There are seven factors which govern the roasting, they are

The seven factors of roasting

(*i*) Rate of Agitation
(*ii*) The Depth of the Bed of Ore
(*iii*) The kind of Fed
(*iv*) Number of Hearths
(*v*) The number and type of Drop holes
(*vi*) The final Temperature, attained and
(*vii*) The amount of air admitted to the furnace.

The sizes of the particles are also the factor because the larger particles take longer time and the smaller particles take lesser time. In the roasting of copper ore the temperature usually ranges from 200°C in the upper hearth and 800°C to 850°C upto 950°C as stated earlier. Once the furnace starts working the temperature and ret of admission of air offer the readiest means of control of the operation. When large amount of heat is required it is usually added on the lower hearths by means of oil burners, although coal may be burnt in an outside firebox if necessary. The capacity of a Wedge, roasting a high sulphur charge varies from 10,16,00 to 31,15,00kg, per day.

Reactions of Roasting (Oxidizing Roast)

$$2Cu_2S + 3O_2 \longrightarrow 2Cu_2O + SO_2 \quad ...(1)$$

$$2CuO + SO_2 \longrightarrow Cu_2O + SO_3 \quad ...(2)$$

$$CuO + SO_3 \longrightarrow CuSO_4 + SO_3 \quad ...(3)$$

$$Cu_2S + 3SO_3 \longrightarrow Cu_2O + 4SO_2 \quad ...(4)$$

$$\text{Arsenic Trisulphide} + \text{Oxygen} \longrightarrow \text{Arsenious Oxide} + \text{Sulphur Dioxide} \quad ...(5)$$

$$2As_2S_3 + 9O_2 \longrightarrow 2As_2O_3 + 6SO_2$$

$$\text{Antimony Trisulphide} + \text{Oxygen} \longrightarrow \text{Antimony Trioxide} + \text{Sulphur Dioxide}$$

$$2Sb_2S_3 + 9O_2 \longrightarrow As_2O_3 + 6SO_2$$

$$\text{Antimony Trisulphide} + \text{Oxygen} \longrightarrow \text{Antimony Trioxide} + \text{Sulphur Dioxide} \quad ...(6)$$

$$2Sb_2S_3 + 9O_2 \longrightarrow As_2O_3 + 6SO_2$$

$$2FeS_2 \longrightarrow 2FeS + 2S \quad ...(7)$$

$$2S + 2O_2 \longrightarrow 2SO_2 \quad ...(8)$$

$$2FeS + 3O_2 \longrightarrow 2FeO + 2SO_2 \quad ...(9)$$

$$2SO_2 + O_2 \longrightarrow 2SO_3 \quad ...(10)$$

$$FeO + SO_3 \longrightarrow FeSO_4 \quad ...(11)$$

Reactions (1) (8) and (9) are strongly exothermic the hearts of reaction being respectively 188, 900, 70, 920, 224 and 200 calories per mole.

The principal reaction of sulphurizing roast are:

$$CuS + 2O_2 \longrightarrow CuSO_4 \quad ...(12)$$

$$CuSO_4 + CuS + O_2 \longrightarrow 2CuO + 2SO_2 + CO_2 \quad ...(14)$$

$$2CuFeS_2 + 4CuO + 17O_2 \longrightarrow 8CuSO_4 + 2FeO_3 \quad ...(15)$$

$$4FeS + 4CuO + 9O_2 \longrightarrow 4CuSO_4 + 2Fe_2O_3 \quad ...(17)$$

$$CuSO_4 + 2C \longrightarrow Cu + SO_2 + 2CO \quad ...(17)$$

With these reaction equations as a basis the following conclusions can be drawn about the roasting process:

1. Sulphate and not oxides are formed primarily; both oxides and sulphur dioxide are secondary products. Iron oxide probably acts as a catalyst.
2. Iron is probably the only constituent of common base metal ores that will readily pass wholly to the oxide from, Fe_2O_3 in a sufficiently oxidizing atmosphere.
3. Iron sulphate is not decomposed normally at a temperature sufficiently far below the decomposition temperature of copper sulphate iron oxide. As a matter of fact, none of these sulphates alone is decomposed at 700°C to 800°C or even higher.
4. A 550°C and above, the formation of cupric ferrite (CuO,Fe_2O_3) begins this compound is insoluble in conventional acid leaching solutions.
5. At 400°C reaction 13 begins and thus prevents the formation of "all water soluble" copper. The furnace gases provide carbon monoxide and carbon for reactions with copper sulphate according to equations 14 and 17. This cannot be prevented in an ordinary rabble furnace.
6. It follows that sulphatizing will be difficult or impossible under usual furnace conditions. One must (*a*) properly proportion the change so that neither to much not too little will be present, (*b*) not allow the products of combustion to mix with the ore and (*c*) maintain relatively low temperature in the furnace. In oxidizing roasting, this one is less important because the ultimate aim is to eliminate sulphur and convert the copper to an oxide.

Smelting

Before conversion, the ore is after roasting is smelted to matte* in reverberatory furnace, which uses pulverized+ coal, fuel oil or natural gas as fuel.

A reverberatory furnace for the smelting of sulphide copper ores, ore concentration is shown in the Fig. 16.19 and 16.20.

The furnace is 27 to 35m high and 5.5 to 8m in width. The shorter furnaces provide adequate setting area and are cheaper to build and maximum modem furnaces smelt upto 21,33,600kg of charge daily. The foundation of the furnace is .5m to .7m for it must withstand the heavy load imposed by the furnace as well as the .6m of the molten matte and slag

* Matte: Metal consisting of sulphide ore as one of its composition.

+ Pulverized: Powdered.

that collect on the hearth during the smelting operation. The hearth proper must be built for some refractory practice. However, certain smelters notably one operating on a copper-nickel ore, have found a basic hearth of magnesite brick be very successful and chrome brick is often used as the slag line to keep down the corrosion due to the formation of ferrous oxide at the slag gas interface.

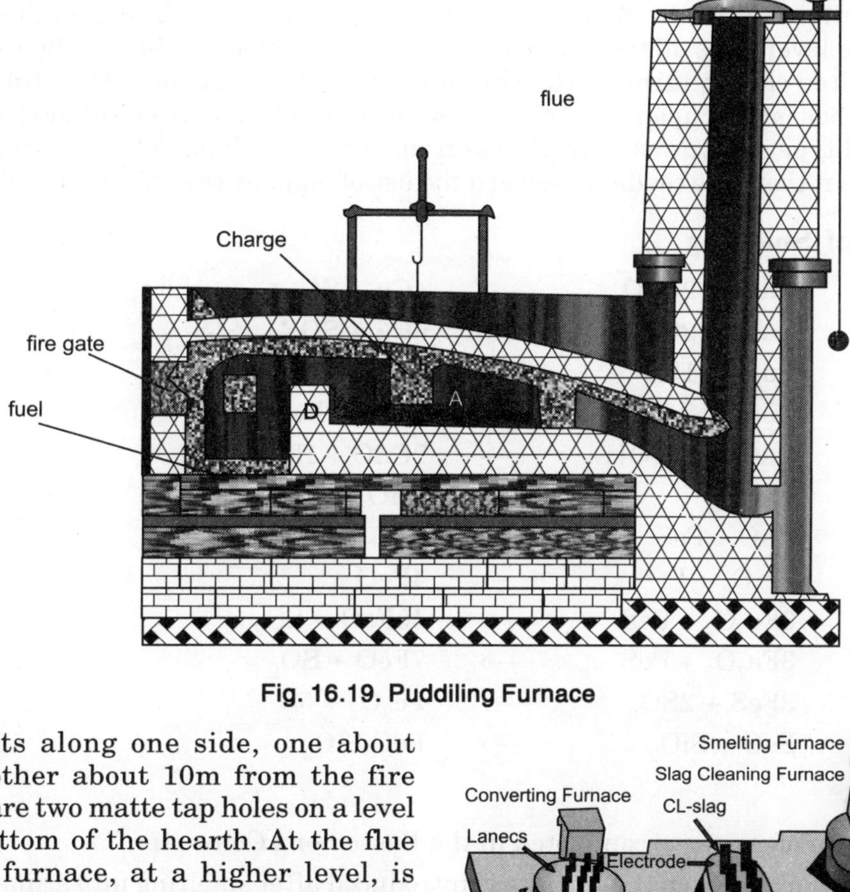

Fig. 16.19. Puddiling Furnace

At points along one side, one about 7.6m and other about 10m from the fire bridge and are two matte tap holes on a level with the bottom of the hearth. At the flue and of the furnace, at a higher level, is provided a skimming hole for the slag is tapped at a point along the side. About 3.8 m from the flue end water jackets are used at some plants for cooling the hearth, ports and other special parts. On a level with top of the hearth and along both the sides of the furnace, are small openings with tightly fitting sliding doors. Through these, the operators may watch the changes in the content of the furnace, rabbling the charge if found necessary and making temporary repairs to the bottom and sidewalls.

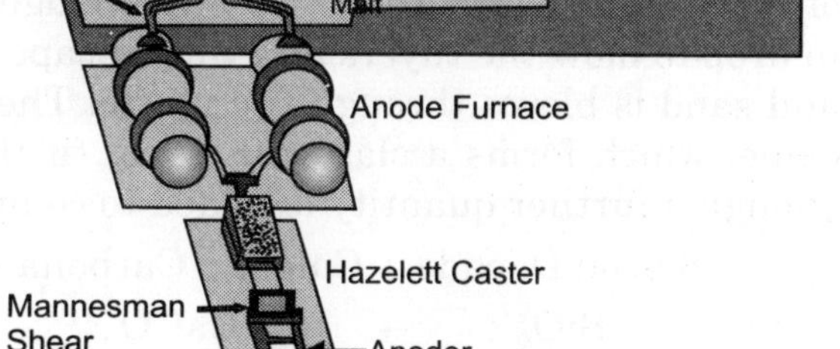

Fig. 16.20. The Smelting of Copper

The roof reflects the flame down onto the charge on the hearth hence, the name reverberatory. It is the rule to construct the furnace section dome shaped at the combustion and specially to provide sufficient combustion space raising the roof about 6m higher than the normal arch. This cannot be done because of the nature of the fine hot calcines. These have such a flat angle of response that any great increase in the elevation of the drop holes is left unchanged and increase in cross sectional area is obtained by using a raised flat section of work in the centre 6m of the furnace, with an abrupt drop at each side down to the drop hole level. This offers a wide channel in the centre for the free flow of converter slag and melted charge through the skimming end of the furnace. The efficiency of the furnace depends largely upon the structure of this roof, which may be either the spring arch or supported type and varies in thickness from .3 to .5m. This roof was formerly build of silica brick, but there is a tendency toward the use of magnesite roofs of the supported type.

Reactions of Smelting

$$Cu_2S + 2CuO \longrightarrow 4Cu + SO_2$$
$$Cu_2S + 2Cu_2O \longrightarrow 6Cu + SO_2$$
$$Cu_2O + FeS \longrightarrow Cu_2S + Fe$$
$$CuSiO_3 + FeS \longrightarrow Cu_2S + FeSiO_3$$
$$CuCO_3 \cdot Cu(OH)_2 \longrightarrow 2CuO + H_2O + CO_2$$
$$CaCO_3 \longrightarrow CuO + CO_2$$
$$CuO + FeS \longrightarrow CuS + FeO$$
$$Fe_2O_4 + Fe \longrightarrow 4FeO$$
$$3Fe_3O_4 + FeS \longrightarrow 10FeO + SO_2$$
$$3Fe_2O_3 + FeS \longrightarrow 7FeO + SO_2$$
$$3FeS + 2SO_2 \longrightarrow Fe_3O_4 + 5S$$
$$FeO + SiO_2 \longrightarrow FeO.SiO_2$$

Conversion

The conversion of copper is conducted in the Bessemer's Converter.

The molten matte from the furnace is introduced after smelting into a small Bessemer converter exactly used in the preparation of steel but having the tuyers in the side (instead of at the base), which are placed high enough. This arrangement enables the metal produced to drop to blow the tuyers and thus escape the oxidizing action of the blast. A blast of air and sand is blown through the matte. The iron sulphide, that escapes oxidation to iron oxide, which forms a slag with silica (if the quantity of silica present in the ore is not enough a further quantity is added to completely slag out the iron.

Silicon Dioxide + Calcium Carbonate ⟶ Calcium Silicate + Carbon Dioxde

$$SiO_2 + CaCO_3 \longrightarrow CaSlO_3 + CO_2$$

After this the cuprous sulphide gets oxidized to cuprous oxide, which then reacts with remaining cuprous sulphide for forming copper metal.

Cuprous Sulphide + Oxygen $\longrightarrow$ Cuprous Oxide + Sulphur Dioxide

$$2Cu_2S + O_2 \longrightarrow 2Cu_2O + 2SO_2$$

Cuprous Sulphide + Curpous Oxide $\longrightarrow$ Copper + Oxygen

$$Cu_2S + 2Cu_2O \longrightarrow 6Cu + SO_2$$

The reaction between the oxide and sulphide is very vigorous and audible and the heat set free is sufficient to keep the charge in the fused state. When the reaction is over, the molten copper is run off into sand moulds. The products, when solid, is known as blister copper, because of the blistered appearance that is assumes during solidification.

Blister copper consists of 98% copper, the rest being iron together with small amounts of silver and gold originally present in the ore. It is refined as described.

Refining of Copper (electrolytic Process): The modern method of refining crude copper containing 98% of the metal is by electrolysis. For this purpose the crude metal is cast into plates, each .2 sq m in area and 10 mm thick. These are hung as regular intervals in huge, lead lined tanks containing consists of thin plates of pure copper which are coated with graphite to permit the metal deposited on them during electrolysis to be easily stripped off. As the electrolysis proceeds, the copper forms the anodes together with traces of more active metals *e.g.* zinc, iron, tungsten etc. present as impurities, goes into the solution, while the less active metals; gold and silver are left undissolved and settle to the bottom of the tank as anode mud. At the cathodes, copper alone is deposited, because the voltage applied (1.3V) is not sufficient to deposit the more active metals present in the solution. The copper obtained is almost 100% pure.

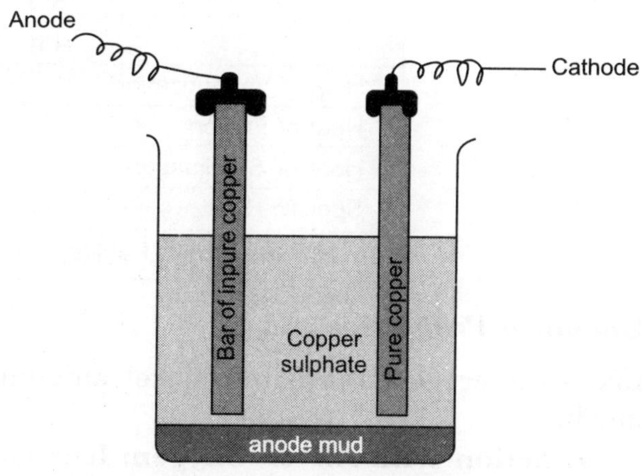

Fig. 16.21. The refining Process of Copper by Electrolysis is shown in the Diagram

Alloys of Copper

Alloy	Composition %		Temper	Rockwell Hardness	Tensile (MPa) Strength
	Cu	Zn			
Yellow Brass	66	34	Light Anneal	78(F)	.365
Cartidge Brass	70	30	Light Anneal	78(F)	.365
Commercial Brass	90	10	Light Anneal	71(F)	.310
Gliding Metal	95	5	Light Anneal	55(F)	.262

Physical Properties

Melting Point		1059°C
Boiling Point		2055°C
Denisty		8.96 g/cc
Atomic Radius		1.28 Å
Ionic Radius		0.96(+1) Å
First Ionization Energy		178
Oxidation Potential		0.52
$M - M^{++}$		
Important Oxidation States		
Ionization Potential(eV)	1st	7.72
	2nd	26.29
	3rd	36.83
	4th	58.9
Heat of Vapourization		81.0 Kcal/mol
Heat of Fusion		3.11 Kcal/mol
Heat of Sublimation		81.1
Specific Heat		0.092
Electronegativity (Pauling's Scale)		1.8

Chemical Properties

Like all other metals such iron, nickel, aluminium copper is also chemically active, but not much.

1. Action with Air or Oxygen: It is not attacked by air in dry state at ordinary temperature, but when heated to redness in air or oxygen it first forms.

$$\text{Copper + Oxygen} \longrightarrow \text{Cupric Oxide}$$
$$Cu + O_2 \longrightarrow 2CuO$$

In the presence of atmosphere carbon dioxide and moisture it is covered with a green layer of basic carbonate of copper [$CuCO_3.Cu(OH)_2$]. This layer protects the metals from further corrosion.

2. Action with Water: Copper does not decompose water or steam at ordinary temperatures, however, at white heat, steam acts upon it to give the oxide,

$$\text{Copper + Water} \longrightarrow \text{Cupric Oxide + Hydrogen}$$
$$Cu + H_2O \longrightarrow CuO + H_2$$

3. Action with Acids: Non-oxidizing acids (*e.g.* hydrochloric acid and concentrated sulphuric acid) have no effect in the absence of air or oxygen. In the presence of air, such acids dissolve the metal forming the corresponding salt and water.

$$\text{Copper + Hydrochloric Acid + Oxygen} \longrightarrow \text{Cupric Chloride + Water}$$
$$Cu + 2HCl + 1/2\ O_2 \longrightarrow CuCl_2 + H_2O$$
$$\text{Copper + Sulphuric Acid + Oxygen} \longrightarrow \text{Copper Sulphate + Water}$$
$$Cu + H_2SO_4 + 1/2\ O_2 \longrightarrow CuSO_4 + H_2O$$

4. Action with Non-metals: Copper combines with chlorine and other halogen to form halides.

$$\text{Copper + Chlorine} \longrightarrow \text{Cupric Chloride}$$
$$Cu + Cl_2 \longrightarrow CuCl_2$$

Moist chlorine rapidly attacks copper even at ordinary temperature. It combines with sulphur; when heated gives a compound $Cu_{1.87}S$.

5. Action of Ammonia: Copper does not combine with nitrogen directly, but when ammonia gas is passed over red-hot copper, the metals having hydrogen free absorb the nitrogen.

Copper gets dissolved in ammonia solution in the presence of air or oxygen, giving a complex, $Cu(NH_3)_4$.

$$\text{Copper + Water + Oxygen + Ammonia} \longrightarrow \text{Cuprammonium Hydroxide}$$
$$2Cu + 2H_2O + O_2 + 8NH_3 \longrightarrow 2Cu(NH_3)_4\ (OH)_4$$

6. Action of Cyanide: It also dissolves in aqueous solution of potassium cyanide yields the complex $Cu(CN)_4^{2-}$

7. Displacement of Silver and other less positive Metals: Copper displaces silver and mercury from solutions of their salts.

$$\text{Copper + Silver Nitrate} \longrightarrow \text{Cupric Nitrate + Silver}$$
$$Cu + 2AgNO_3 \longrightarrow Cu(NO_3)_3 + 2Ag$$

Electronic Configuration Oxidation States

The outer electronic configuration of copper is $3d^{10}4s^1$. It shows oxidation states of + 1 ($3d^{10}$) and 2 + ($3d^9$) in its compounds. The removal of the solitary 4s-electron gives Cu^+ is rather small, another electron can be taken out easily yielding Cu^{2+} complete disproportionation as

$$2Cu^+ \rightleftharpoons Cu^{2+} + Cu$$

The +2 state, therefore, is much more stable; the + 1 state is more stable only in insoluble and complex compounds.

Uses of Copper

(*a*) The chief uses of copper lies in the high electrical conductivity. Enormous quantities of copper are used for making electric cables and other electrical goods.

(*b*) As it is not attracted by water or air it is used in making utensils containers, kettles, evaporating pans, coins.

(*c*) It is also used in making electrodes.

Compounds of Copper

Copper forms two series of compounds cuprous and cupric, in which the metal behaves as a univalent and a bivalent element respectively.

Cuprous Compounds

All the cuprous compounds, with the exception of the oxide, which is red, and the sulphide, which is black, are white (the colour of the compounds is mentioned are all of anhydrous). All the compounds except the acetate and the sulphate are all soluble in water. The latter two, however are immediately cupric salt (disproportionation phenomenon). The insoluble cuprous compounds are readily oxidized in air to the state.

Cuprous Oxide (Cu_2O): In nature it occurs as cuprite or red copper ore when a solution of cuprous salt is treated with a reducing agent like sugar salt is deposited as precipitate. In the laboratory cuprous oxide is usually prepared by taking Fhelling's Solution and glucose when they are heated cuprous oxide is deposited.

Cuprous oxide is also prepared by the direct action of copper and cupric oxide.

$$\text{Copper Oxide} + \text{Copper} \longrightarrow \text{Cupric Oxide}$$

$$CuO + Cu \longrightarrow Cu_2O$$

It is red powder insoluble in water but soluble in nitric acid. It is also soluble in connected ammonia to form a complex colourless in absence of air and oxygen. The solution consists of $Cu(NH_3)_2OH$, which turns blue in air by oxidation ($[Cu(NH_3)_4](OH)_2$). With dilute hydrochloric acid cuprous oxide forms insoluble cuprous chloride, which dissolves in the concentrated acid forming a complex anion ($[CuCl_4]^{-3}$ of H_3CuCl_4). With dilute sulphuric acid it reacts to form metallic copper and cupric sulphate.

$$\text{Cuprous Oxide} + \text{Sulphuric Acid} \longrightarrow \text{Copper} + \text{Copper Sulphate} + \text{Water}$$

$$Cu_2O + H_2SO_4 \longrightarrow Cu + CuSO_4 + H_2O$$

With nitric acid it reacts to form cupric nitrite and nitric oxide evolved.

Cuprous Oxide is used for making ruby red glass and enamel paint.

$$\text{Cuprous Oxide} + \text{Nitric Acid} \longrightarrow \text{Curpic Nitritite} + \text{Nitric Oxide} + \text{Water}$$

$$3Cu_2O + 14HNO_3 \longrightarrow 6Cu(NO_3)_2 + 2NO + 7H_2O$$

Cuprous Chloride ($CuCl_2$): It is prepared by heating copper turnings with concentrated hydrochloric acid, with addition of potassium chlorate. The dark coloured solution : first formed finally becomes colourless. The solution, which contains the cuprous chloride in the forms of complex ion. It is mixed with large volume of water, when the cuprous chloride separates out as a snow-white precipitate. But it rapidly turns greenish, due to the oxidation by air.

The same compound can be obtained by heating cupric oxide and copper mixture with strong hydrochloric acid.

Another method of preparing the compound consists in the reductions of cupric chloride in solution by sulphur dioxide.

Cupric Chloride	+ Sulphur Dioxide	+ Water	$\longrightarrow$	Cuprous Chloride	+ Sulphuric Acid	+ Hydrogen Chloride
$2CuCl_2$	+ SO_2	+ $2H_2O$	$\longrightarrow$	$2CuCl$	+ H_2SO_4	+ $2HCl$

In place of cupric chloride an equivalent mixture of copper sulphate and sodium chloride can be also used.

Cuprous chloride forms a snow-white solid m.p. 422°C. It volatilizes above 1366°C. But by lowering of the freezing point solution of the substance in some organic solvents shows that it exists in single molecules, as CuCl. The determination of the valency of the ion by E.M.F measurements gives a value of one. Hence it is better represented by the unimolecular formula CuCl. In moist air it is oxidized to the basic cupric chloride, having the same composition as actmite [$CuCl_2.3Cu(OH)_2$]. Cuprous chloride is insoluble in water, but dissolves in concentrated hydrochloric acid or sodium chloride solution forming complexes $CuCl_2^-$ $CuCl^-$. It is also dissolves in aqueous ammonia to form a colourless carbonyl compound which consists of complex diamino cuprous chloride [$Cu (NH)_3Cl$]. The solution of cuprous chloride in hydrochloric acid or ammonia can absorb carbon monoxide. Ammonical cuprous chloride solution can also react with acetylene gas to give a precipitate of cuprous acetalide (Cu_2C_2) that is explosive in dry state.

Cuprous Iodide (CuI): When a solution of potassium iodide is added to a solution of copper sulphate, cuprous iodide is formed, with the liberation of iodine. If some sulphur dioxide is added previously to the copper sulphate solution cuprous iodide is obtained as a white insoluble solid without the mixture being coloured by the liberation of iodine.

Copper Sulphate	+ Sulphur Dioxide	+ Potassium Iodide	+ Water	$\longrightarrow$	Sulphuric Acid	+ Potassium Sulphate	+ Cuprous Iodide
$2CuSO_4$	+ SO_2	+ $2KI$	+ $2H_2O$	$\longrightarrow$	H_2SO_4	+ K_2SO_4	+ 2CuI

The reaction is utilized for the estimation of copper by triturating the liberated iodine with a solution of sodium thiosulphate.

Cuprous Cyanide (CuCN): If a solution of potassium cyanide is added to a solution of copper sulphate are a yellow precipitate of cupric cyanide is produced which readily decomposes with evolution of cyanogen gas, however, is soluble in excess potassium cyanide forming a complex anion $Cu(CN)_4^{3-}$ potassium cuprocyanide.

Potassium Cyanide	+ Copper Sulphate	$\longrightarrow$	Cuprous Cyanide	+ Potassium Sulphate	+ Cyanogen
$4KCN$	+ $2CuSO_4$	$\longrightarrow$	$2CuCN$	+ K_2SO_4	+ $(CN)_2$

Potassium cuprocyanide is used for electroplating and other electrolysis.

Cuprous Thiocyanate [$Cu(SCN)_2$]: A solution of ammonium thiocyanate, when added to that of potassium sulphate, produces a black precipitate of cupric thiocyanate [Cu(CN)2], which changes slowly; but quickly on the addition of sulphuric acid, into the white cuprous thiocyanate (CuSCN).

Copper Sulphate	+ Sulphur Dioxide	+ Ammonium Thiocyanate	+ Water	$\longrightarrow$	Cuprous Thyocyanate	+ Ammonium Sulphate	+ Sulphuric Acid
$2CuSO_4$	+ SO_2	+ $2NH_4SCN$	+ $2H_2O$	$\longrightarrow$	$2CuSCN$	+ $(NH_4)_2SO_4$	+ $2H_2SO_4$

The precipitate of cuprous thiocyanate is soluble in water and dilute acids but dissolves in concentrated alkali thiocyanate solutions with the formation of alkali salts.

Copper is often estimated by precipitation as cuprous thiocyanate is insoluble in water and dilute acids but dissolves in concentrated alkali solutions with the formation of double salts. Copper is often estimated as the cuprous thiocyanate.

Cuprous Sulphide (CuS): This compound is obtained as blackish-grey crystalline powder by heating cupric sulphide mixed with some sulphur in a current of hydrogen. It can be obtained by passing H_2S through a solution of cuprous chloride in hydrochloric acid. In nature it occurs as copper glance. Cuprous sulphide is insoluble in water, as well as in solution of colourless ammonium sulphide.

Cuprous Sulphite (Cu_2SO_3): This is obtained as white powder by the action of barium sulphate on cuprous oxide at 160°C.

Cuprous Oxide + Barium Sulphate ⟶ Cuprous Sulphite + Barium Oxide

$$Cu_2O + BaSO_4 \longrightarrow Cu_2SO_3 + BaO$$

The salt is fairly soluble in the absence of moisture; it suffers immediately decomposition in contact with water, yielding metallic copper and cupric sulphate by disproportionation.

Cuprous Sulphite ⟶ Copper + Copper Sulphate

$$Cu_2SO_3 \longrightarrow Cu + CuSO_4$$

Cuprous Thiosulphate ($Cu_2S_2O_3$): This compound can be obtained in the form of a double salt with sodium thiosulphate [$Na_2S_2O_3$ or $Na_2Cu_2(S_2O_3)_2$]. When a solution of cuprous chloride is added to a solution of sodium thiosulphate. The solution on boiling or acidification produces cuprous sulphide.

Cuprous Thisulphate + Water ⟶ Cuprous Sulphide + Sulphuric Acid

$$Cu_2S_2O_3 + H_2O \longrightarrow Cu_2O + H_2SO_4$$

Cupric Compounds

Cupric Oxide (CuO): It is a compound of copper and oxygen, when they react directly with each other.

Copper + Oxygen ⟶ Cupric Oxide

$$2Cu + O_2 \longrightarrow 2CuO$$

It is also prepared by heating cupric nitrate.

Cupric Nitrate ⟶ Cupric Oxide + Nitrogen Dioxide + Oxygen

$$2Cu(NO_3)_2 \longrightarrow 2CuO + 4NO_2 + O_2$$

It can be prepared by the prolonged heating of copper turnings or wire in air or oxygen as said earlier by the product might contain cuprous oxide and metallic copper. A pure product may be obtained by the decomposition of the basic copper carbonate.

Copper Carbonate ⟶ Cupric Oxide + Carbon Dioxide

$$CuCO_3 \longrightarrow CuO + CO_2$$

It occurs in nature as black solid. Cupric oxide is black solid, which is having a melting point of 990°C, with decomposition at 1026°C, it is converted into cuprous oxide and oxygen. When hydrogen is passed over a hot cupric oxide, or carbon monoxide or it is heated with carbon, the compound is reduced to metallic copper. It dissolves in dilute acids to form cupric compounds.

Cupric oxide is used in the glass and blue colour. When used along with a reducing agent like stannous oxide, it produces the red colour in the glass.

Cupric Hydroxide [$Cu(OH)_2$]: This is obtained as light blue voluminous precipitate, when a solution of cupric salt is treated with an alkline solution in slight excess. The precipitate formed in the cold is gelatinous; but on prolonged boiling of the mixture, particularly in presence of an excess of alkline solution it turns black and granular with the partial formation of cupric oxide. Cupric hydroxide can however, be obtained in a crystalline form by adding ammonia to a boiling solution of copper sulphate till the green precipitate turns blue and then digesting and washing the product with strong sodium hydroxide solution at 25°C - 30°C. Cupric hydroxide dissolves in dilute acids forming cupric hydroxide [$Cu(NH_3)_4(OH)_2$].

Cupric Carbonate ($CuCO_3$): Only the basic carbonates of copper have been obtained. Many of these occur as minerals. The normal carbonate is unknown. A basic carbonate is obtained as a green precipitate by adding a solution of sodium carbonate to that of a cupric salt. A blue crystalline double carbonate ($Na2CO_3.CuCO_32H_2O$) is formed by warming the precipitate of basic carbonate with sodium carbonate and bicarbonate solutions.

Cupric Chloride ($CuCl_2$): The anhydrous chloride is obtained as a dark brown deliquescent mass by heating the hydrated chloride ($CuCl_2H_2O$) in a current of hydrogen chloride gas at 150°C. When heated strongly cupric chloride looses chlorine and changes to cuprous chloride. The blue crystalline dihydrate can by prepared by concentrating a solution of cupric oxide in strong hydrochloric acid. Cupric chloride is freely soluble in water. The dilute solution of the salt is blue; concentrated aqueous solutions are green and in presence of concentrated hydrochloric acid the colour becomes yellow. This change in colour is responsible for the formation of different complex ions. The blue solution contains the complex cation ($Cu.4H_2O^{++}$) and the brownish-yellow solution the complex anion (CuC_4^-) while the green solution contains both in the equilibrium.

Cupric Nitrate [$Cu(NO_3)_23H_2O$]: This is obtained in the form of blue deliquescent prismatic crystals by evaporating a solution of metallic copper or cupric oxide, or the basic carbonate, in dilute nitric acid. Hydrates with more water has also been prepared, when heated, hydrated salts of cupric nitrate decomposes at first with the formation of a basic salt and finally leaves a residue of copper oxide.

Copper Nitrate $\longrightarrow$ Cupric Oxide + Nitrogen Dioxide + Oxygen + Water

$$2Cu(NO_3)_2,\ 3H_2O \longrightarrow 2CuO + 4HO_2 + O_2 + 6H_2O$$

Cupric Sulphite (CuS): This compound is obtained as a black mass by heating copper powder with excess of sulphur at about 400°C. In impure state, mixed with some cuprous sulphide and sulphur, it is precipitated from an acid solution of cupric salt by the passage of hydrogen sulphide gas. In moist state it rapidly oxidizes in air to form the sulphate when heated, cupric sulphide breaks upon into cuprous sulphide and sulphur. It is insoluble in dilute hydrochloric acid or dilute sulphuric acid as in alkalis; it is only dissolved in the hot and dilute nitric acid.

Cupric Sulphate, Copper Sulphate, Vitrium Lividum ($CuSO_4$, $5H_2O$): It is usually prepared by the action of sulphuric acid and basic copper carbonate. It consists of 5 molecules of water of crystallization ($CuSO_4.5H_2O$) and it is known as *Vitrium Lividum (blue crystals).*

$$\text{Cupric Sulphide} + \text{Oxygen} \longrightarrow \text{Copper Sulphate}$$

$$CuS + 2O_2 \longrightarrow CuSO_4$$

It is also prepared by the weather of moist pyrites in large scale, after preliminary roasting.

$$\text{Copper Sulphide} \longrightarrow \text{Copper Oxide} + \text{Sulphur Trioxide}$$

$$CuSO_4 \longrightarrow CuO + SO_3$$

The commercial product from the pyrites usually contains ferrous sulphate as an impurity, with which it can form mix crystals. Hydrated cupric sulphate is similar with hydrated ferrous sulphate. Mixed crystals having the composition (Cu, Fe)SO_48H_2O are deposited from a solution of cupric sulphate and ferrous sulphate, containing an excess of the former, white crystals of the composition (Cu, Fe)SO_47H_2O are detained from a mixture of two sulphates containing an excess of ferrous sulphate.

Copper sulphate forms large blue crystals which efflorescence to a pale blue powder $CuSO_4.3H_2O$ exposure to air and forms bluish white monohydrate. $CuSO_4$, H_2O when heated to 100°C. The anhydrous salt $CuSO_4$ is obtained as a white powder by heating the hydrated salt to 250°C. When heated beyond 640°C, the salt decomposes into cupric oxide and sulphur trioxide.

The anhydrous salt rapidly dissolves water to regenerate the blue pentahydrate. An aqueous solution of copper sulphate gives a pale blue precipitate of basic copper sulphate, $CuSO_4$ $3Cu(OH)_2$ when treated with sodium hydroxide.

Copper sulphate forms double sulphates with alkali sulphates *e.g.* $K_2SO_4CuSO_4.6H_2O$, $(NH_4)_2SO_4CuSO_46H_2O$ etc similar double sulphates with many bivalent elements e.g. Zinc, Manganese, Cobalt, Nickel and Magnesium with 5 and $7H_2O$ are also known.

A solution of copper sulphate turns deep blue with ammonia from which deep blue crystals of complex tetrammino-cupric sulphate $Cu(NH_3)_4SO_4H_2O$ may be obtained.

Copper sulphate is used as a fungicide in agriculture as it is toxic to lower organisms under the name of *Bordeaux mixture* (a mixture of copper sulphate and lime).

Relative Stability of Cuprous and Cupric Compounds

A study of the compounds shows that cupric salts with certain anions are very unstable and change, as soon as they are formed in solutions into those of univalent copper. This can be explained from a consideration of the electronic structures of the cuprous and the cupric ions. The atomic number of copper is 29. In the cuprous ion there are therefore 28 electrons representing the closed shell, in which 1, 2 and 3 are completely filled up. In the cupric ion there will be 27 electrons, one less that in cuprous state. This electron has to be removed obviously from the completed third quantum level during the formation of cupric compound. This generally effected only by radicals or atoms, of strongly electropositive character, where affinity for electron is very high. Hence, weakly electronegative radicals or atoms when combining with copper will tended to form preferably the cuprous compounds. This is, indeed, actually the use. For it has been observed that cupric compounds with weak electronegative radicals, I, CN, SCN and SO_3 are very unstable and readily change to the corresponding cuprous compounds. On the other hand, strongly electropositive atoms, O, S, F and Cl can form quite stable cupric compounds though these cupric compounds are also fairly stable. The solution however seems to be reverse in the case of the complex compounds. In general, the higher valency of an element is found to be stabilized by complex formation. This may be attributed to the formation of a shell of dipole molecules around the central ion (here the cupric ion) as a result of attraction on them, which prevents the central ion from capturing an electron in order to be reduced to a lower valency state (here the cuprous state). Thus, though cupric iodide cannot exist in the free state (here the cuprous state), tetrammine cupric iodide $Cu(NH_3)_4 I_2$ or diethyleneammine cupric iodide Cu_2enI_2 can be prepared without difficulty (en = one molecule of diethylene diamine). Similarly cupric sulphite $Cu_2en (SCN)_2$ have been prepared as stable compounds.

Detection and Estimation

All the compounds produce a green colour to the non-luminous flame. Copper liberates iodine from potassium iodide. The formation of deep blue colour when ammonia is added to a copper salt.

Copper is generally estimated as its oxide or sulphide by gravimetric method, or electrolytically by deposition as metallic copper it is also estimated volumetrically by triturating the iodine liberated from the potassium iodide solution by cupric salts.

ZINC

Melting Point : 419.5°C — Atomic Weight : 66.39

Boiling Point : 906°C — Atomic Number : 30

Occurrence

Zinc is widely distributed in nature and occurs in small amounts in nearly all igneous rocks, mainly as a substitute for iron; its abundance is usually estimated to be 0.013% of the earths crust. The principal ore material of zinc is *Sphalarite*, cubic zinc commonly known as *zinc Blende*. Zinc Blende is found in the Zawar and Udaypur and parts of Kashmir. Ideally Zinc Blende consists of 68% zinc. A hexagonal form of the sulphide,

Wurtzite, is much less available. Individual specimens of both forms have been known to fluorescence under ultraviolet light, but not all specimens do so.

Sphalarite has an resin like appearance, the colour varying from light yellowish-brown to black in natural specimens (pure artificial zinc sulphide is a white pigment of high hiding power) depending upon the content of iron crystallized substitutionally in the lattice. Above a ratio of Fe : Zn of 1 : 5 the mineral is known as *marmelite*. Above a ratio of5:6 equivalent to a total iron content of 265 the spalarite structure no longer exists. There are some oxidized zinc minerals present they are:

$ZnO \longrightarrow$ Zincite

$ZnSO_4 \cdot 7H_2O \longrightarrow$ Gaslerite

$ZnCO_3 \longrightarrow$ Sithosmite (Calamine in Europe)

$Zn_4Si_2O_7(OH)_2H_2O \longrightarrow$ Hemimophite (or Calamine in America)

$Zn_2SiO_4 \longrightarrow$ Willmite

$(Zn, Mn)OFe_2O_3 \longrightarrow$ Franklinite

$2ZnCO_3 \cdot 3Zn(OH)_2 \longrightarrow$ Hydrozincite

Most of these oxidized minerals are minor sources of zinc although Franklinite and Zincite are important in the ores.

Zinc sulphate is more commonly found in limestone, white carbonate is more regularly found in the dolomite rock. This is perhaps explained on the basis that the sum of heats of the formation of manganese sulphate and zinc carbonate is greater than that of zinc sulphate and calcium carbonate. Here the magnesium is present, as in dolomite, a double decomposition reaction appears to occur between zinc sulphide and magnesium carbonate.

Extraction of Zinc

In India zinc is extracted from Zinc Blende (ZnS) by Carbon Reduction Process, also known as 'Dry Process'. It has been described below.

Removal of impurities from Zinc by Oil Filtration: Zinc Blende contains lead sulphide, cadmium sulphide, etc. as impurities, to remove them ore is concentrated some pine or eucalyptus oil, sodium xanthanate and a little acid. It is stirred up thoroughly by a stream of air in a tank. At first lead sulphide is collected in the nrst chamber formed on the surface of the liquid and removed. Zinc Blende collects in the second chamber, produced by stirring with a little more acid. The impurities such as sand, earthy materials sink to the bottom, from the second chamber the zinc blende is separated and dried. (see Fig. 16.21).

Roasting: Roasting of concentrated zinc blende is done in a multiple hearth furnace at a temprature between 815°C – 895°C, through which also a current of hat air circulates. It is carefully watched that no zinc blende remains as partly sulphide or sulphate, but all of it is converted to oxide, however any zinc oxide at this temperature (See Fig. 16.22).

> If the ore is calamine this above process is cancelled because it does not contain lead sulphide.

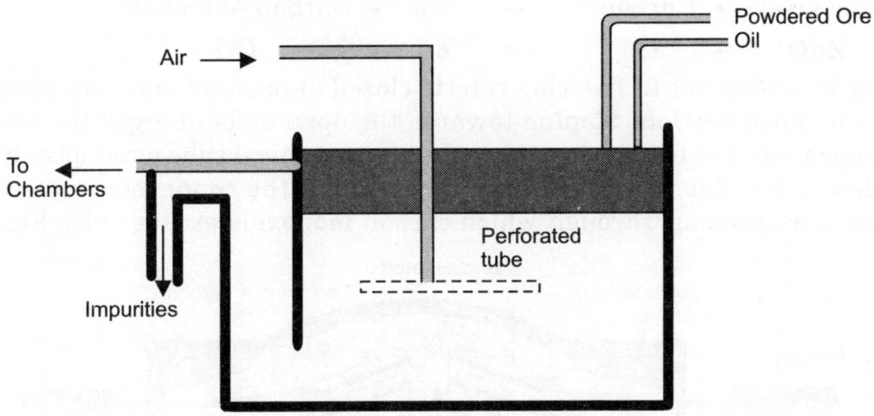

Fig. 16.22. The Oil Floatation Method

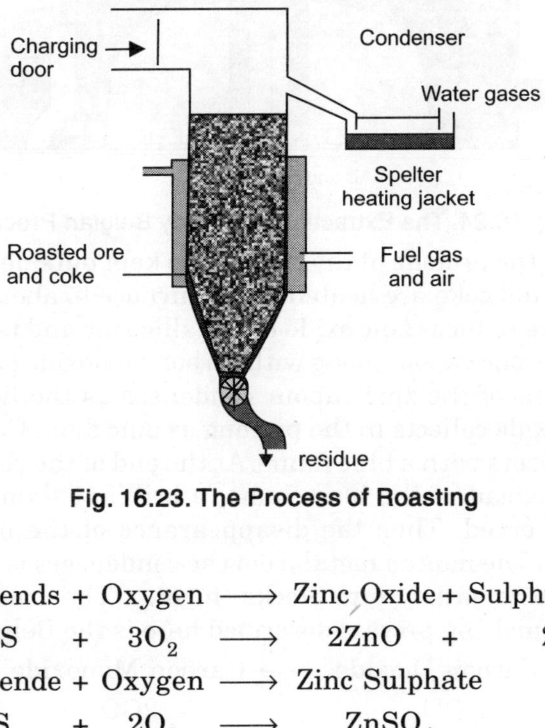

Fig. 16.23. The Process of Roasting

Equations

$$\text{Zinc Blends} + \text{Oxygen} \longrightarrow \text{Zinc Oxide} + \text{Sulphur Dioxide}$$
$$2ZnS + 3O_2 \longrightarrow 2ZnO + 2SO_2$$
$$\text{Zinc Blende} + \text{Oxygen} \longrightarrow \text{Zinc Sulphate}$$
$$ZnS + 2O_2 \longrightarrow ZnSO_4$$
$$\text{Zinc Sulphate} \longrightarrow \text{Zinc Oxide} + \text{Sulphur Dioxide} + \text{Oxygen}$$
$$2ZnSO_4 \longrightarrow 2ZnO + 2SO_2 + O_2$$

Smelting: Zinc oxide obtained after roasting is then mixed with coke about half of the weight of the ore and the mixture is heated at a temperature of about 1350°C. During heating zinc oxide is reduced to metallic zinc by carbon. The other product is gaseous carbon monoxide.

$$\text{Zinc Oxide} + \text{Carbon} \longrightarrow \text{Zinc} + \text{Carbon Monoxide}$$
$$ZnO + C \longrightarrow Zn + CO$$

Smelting is carried out in five clay retorts closed at one end and arranged in three or four rows in inclined portions sloping towards the open ends in a gas-fired furnace. The open end of each retort is fitted with an earthen ware conical tube or adopter which serves as the condenser for zinc vapours. At the other end of the condenser is attached a tube (nozzle) known as 'prolong' through which carbon monoxide escapes (see Fig. 16.24).

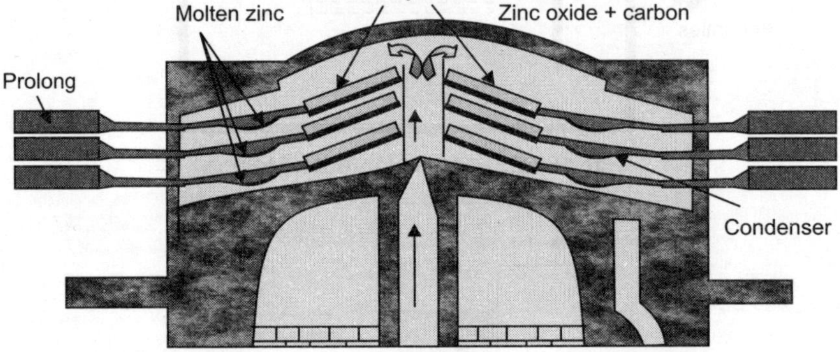

Fig. 16.24. The Extraction of Zinc by Belgian Process

The condensers and the prolong of the retorts are kept outside the furnace. The retorts packed with zinc oxide and coke are heated in the furnace to about 1340°C by combustion of producer gas. The coke reduces zinc oxide to metallic zinc and is itself oxidized to carbon monoxide as vapour. The zinc vapour along with carbon monoxide passes into the condensers where the major portions of the zinc vapour condenses as the liquid metal and the rest mixed with some zinc oxide collects in the prolong as zinc dust. Carbon monoxide bums at the mouth of the condensers with a blue flame. At the end of the reaction, carbon monoxide ceases to evolve and instead of the blue flame, a brilliant flame due to the burning of metallic vapours is observed. Thus the disappearance of the blue flame indicates the completion of reduction. The molten metal from the condensers is transferred form time to time moulds and cooled to satisfy into blocks (ingots). The crude zinc thus obtained is called the spelter. The smelting process described here is the Belgian Process.

$$\text{Carbon} + \text{Carbon Dioxide} \longrightarrow \text{Carbon Monoxide}$$
$$C + CO_2 \longrightarrow 2CO$$
$$\text{Zinc} + \text{Carbon Dioxide} \longrightarrow \text{Zinc Oxide} + \text{Carbon Monoxide}$$
$$Zn + CO_2 \longrightarrow ZnO + CO$$

Refining: The crude zinc of spelter containing about 2% lead and traces of iron, cadmium, arsenic as impurities is refined by fractional distillation. The spelter is at first heated in a retort at a temperature slightly above the boiling point of zinc (910°C) when zinc together with cadmium distils over leaving behind lead and iron. The distillate that is the molten mixture of zinc and cadmium is taken redistilled in a second similar retort at a temperature slightly above 767°C when more volatile cadmium is removed as a by product. Chemically pure zinc is obtained by electrolysis of pure zinc sulphate solution acidified with dilute sulphuric acid using impure zinc (spelter) rod as the anode and a pure thin sheet of aluminium as the cathode. During the electrolysis, zinc from the anode dissolves and the metal in the very pure state is deposited on the aluminium cathode.

Physical Properties

Melting Point	419.5°C
Boiling Point	906°C
Heat of Fusion	24.09 cal/g
Heat of Capacity	5.375 cal deg^{-1} mol^{-1}
Thermal Conductivity	0.265 cal/sqcm/cm/°C/sec
Co-efficient of Linear Expansion	3.97×10^{-6}
Electrical Resistivity	5.916 micro ohm/cm
Magnetic Scusceptibility	-0.15×10^{-6} C.G.S.
Surface Tension (liquid) at 450°C	755 dynes/cm
Tensile strength of Commercial Zinc	16.50×10^{3}
Crystalline structure	Close packed Hexagonal

Chemical Properties

1. **Action of Air:** When in air to about 500°C, it burns to form a white cloud of zinc oxide, which settles to form a wooly flock commonly called 'Philosophers Wool'.

$$\text{Zinc} \quad \text{Oxygen} \longrightarrow \text{Zinc Oxide}$$

$$2Zn + O_2 \longrightarrow 2ZnO$$

2. **Action of Water:** Pure zinc does not react with water but impure zinc decomposes steam quite readily with the evolution of steam.

$$\text{Zinc} + \text{Water (steam)} \longrightarrow \text{Zinc Oxide} + \text{Water}$$

$$Zn + H_2O \longrightarrow ZnO + H_2O$$

It forms basic compounds in moist, which dulls its lusture and forms a blackish-grey film on its surface.

$$\text{Zinc} + \text{Water} + \text{Oxygen} + \text{Carbon Dioxide} \longrightarrow \text{Basic Zinc Carbonate}$$

$$4Zn + 3H_2O + 2O_2 + CO_2 \longrightarrow ZnCO_{3.3}Zn(OH)_2$$

3. **Action of Acid:** Pure zinc does not react with acids but in impure state it reacts with many acids. The non-reactivity of pure zinc is due to the very large concentration of the hydrogen on its surface. But when mixed with copper, silver, and platinum, on the surface of which the hydrogen gas has low concentration zinc reacts readily with the evolution of hydrogen on the surface of the metal with the distillation of zinc.

The amalgamation of the impure zinc stops such action as it gives uniformity to the surface.

If pure zinc is coated with copper, over which hydrogen has very low concentration, it reacts readily with dilute acids evolving hydrogen. Zinc-copper couple is therefore, used as reducing agent in number of organic reactions.

The reaction with sulphuric acid depends upon the concentration of the acid and the temperature of the reaction. If the acid dilute and cold, zinc sulphate and hydrogen are formed.

Zinc + Sulphuric Acid ⟶ Zinc Sulphate + Hydrogen

$$Zn + H_2SO_4 \longrightarrow ZnSO_4 + H_2$$

If the acid is strong and hot, hydrogen sulphide is evolved with or without separation of sulphur.

Zinc + Sulphuric Acid ⟶ Zinc Sulphate + Hydrogen Sulphide + Water

$$Zn + 5H_2SO_4 \longrightarrow 4ZnSO_4 + H_2S \quad H_2O$$

Zinc + Sulphuric Acid ⟶ Zinc Sulphate + Sulphur + Water

$$3Zn + 4H_2SO_4 \longrightarrow 3ZnSO_4 + S + 4H_2O$$

The reaction with nitric acid is also completed. If the acid is very dilute, ammonium nitrate results.

Zinc Nitric Acid ⟶ Zinc Nitrate + Ammonium Nitrate + Water

$$Zn + 5H_2SO_4 \longrightarrow 4Zn(NO_3)_2 + NH_4NO_3 + 3H_2O$$

If the acid is concentrated and hot, nitrogen dioxide is mainly liberated.

Zinc + Nitric Acid ⟶ Zinc Nitrate + Water + Nitrogen Dioxide

$$Zn + 4HNO_3 \longrightarrow Zn(NO_3)_2 + 2H_2O + 2NO_2$$

4. Action of Alkali: It reacts with moderately strong alkali solutions producing zincates.

Zinc + Sodium Hydroxide ⟶ Sodium Zincate + Hydrogen

$$Zn + 2NaOH \longrightarrow Na_2ZnO_2 + H_2$$

Uses of Zinc

1. Zinc is used in the preparation of alloys such as brass and silver.
2. On account of its resistance to rusting and corrosion, zinc is used to coat iron. This is done by dipping the clean metals in molten zinc or by spraying molten zinc on the iron sheet. This process is called galvanizing. An alternate method is to heat the iron articles with zinc dust. This process is known as Sherardising. The galvanized iron sheets are used for roofs gutters etc.
3. Amalgamated zinc rods are used as electrodes in Leclanche and Daniel cells.
4. Zinc dust is used as reducing agent in the laboratory as well as the manufacture of drops, dyestuffs, and chemicals.
5. Zinc dust is also used as a precipitating agent in the cyanide process in the precipitating gold silver.

6. Zinc oxide (white zinc) is used in the paints. Its superior to white lead, as it does not blacken lithopone a mixture of zinc sulphide and barium sulphate is supporter to both on account of greater covering power. Zinc chromate is used as a yellow pigment.

Electronic Configuration and Oxidation States

The outer electronic configuration of zinc is $3d^{10} 4s^2$. It exhibits and oxidation state of +2 involving the loss of both the 4s electrons. This Zn^{+2} ion, having the loss of both the 4s electrons. This Zn^{+2} ion, having completely filled d-orbital (3d10) is highly stable. Zinc, therefore, shows neither variable valency nor forms coloured salts.

Compounds of Zinc

Zinc produces many compounds like other elements.

Zinc Oxide (ZnO): This compound is formed as white powder by burning zinc in air, by heating carbonate of zinc. When heated it becomes yellow, but turns again white when cooled. The oxide is reduced to metallic zinc when heated to redness with carbon. Zinc oxide dissolves in acid to form zinc salts and in alkalis to form zincates. It is therefore, an amphoteric oxide.

When zinc oxide is heated with cobalt nitrate as in the blowpipe test for zinc, a green coloured product is obtained. This is a solid solution of cobalt zincate $CoZnO_2$, in zinc oxide and is used as a pigment Rinman's Green. Zinc oxide itself is used as a white pigment, known as a zinc white. It is also used in surgical dressings.

Zinc Hydroxide ($Zn(OH)_2$): When a solution of alkali hydroxide e.g. sodium hydroxide is added to a zinc salt in a calculated quantity is added to that of zinc salt, zinc hydroxide is produced, as a white precipitate.

$$\text{Zinc Chloride} + \text{Sodium Hydroxide} \longrightarrow \text{Zinc Hydroxide} + \text{Sodium Chloride}$$

$$ZnCl_2 + 2NaOH \longrightarrow Zn(OH)_2 + 2NaCl$$

The precipitate is soluble in of alkali hydroxide to form a zincate also in ammonia to form complex ion of zinc-ammine.

$$\text{Zinc Hydroxide} + \text{Potassium Hydroxide} \longrightarrow \text{Potassium Zincate} + \text{Water}$$

$$Zn(OH)_2 + 2KOH \longrightarrow K_2ZnO_2 + 2H_2O$$

$$\text{Zinc Hydroxide} + \text{Ammonia} \longrightarrow \text{Zinc Ammine}$$

$$Zn(OH)_2 + 4NH_3 \longrightarrow Zn(NH_3)_4(OH)_2$$

Zinc hydroxide dissolves in acids to form zinc salts. On heating, it looses water to form zinc oxide.

Zinc Carbonate ($ZnCO_3$): It occurs in nature as elements, or smithsenite. It is precipitated as an insoluble white powder by adding a solution of sodium bicarbonate to that of a zinc salt.

$$\text{Zinc Chloride} + \text{Sodium Carbonate} \longrightarrow \text{Zinc Carbonate} + \text{Sodium Chloride}$$

$$ZnCl_2 + Na_2CO_3 \longrightarrow ZnCO_3 + 2NaCl$$

By the addition of the sodium carbonate solution, a precipitate of basic carbonate of variable composition is produced. Both the carbonate and basic carbonate are converted by heating into zinc oxide with loss of carbon dioxide.

Zinc Chloride ($ZnCl_2$): When metallic oxide or zinc carbonate is dissolved in hydrochloric acid and the solution when evaporated a syrupy mass is obtained from which crystals of the hydrated chloride ($ZnCl_2$ H_2O) separated on standing. The anhydrous chloride can be prepared by the action of chlorine on zinc oxide or zinc at 200°C.

$$\text{Zinc Oxide} + \text{Chlorine} \longrightarrow \text{Zinc Chloride} + \text{Oxygen}$$

$$2ZnO + 2Cl_2 \longrightarrow 2ZnCl_2 + O_2$$

Zinc Carbonate, when heated in a stream of hydrogen chloride gas.

If a solution of zinc chloride is evaporated to dryness, the oxychloride [Zn(OH)Cl] and (Zn_2OCl_2) are partially formed.

Anhydrous zinc chloride is a white deliquescent substance and readily absorbs water. It is highly soluble in water and also in alcohol, acetone and in ether. When fused, it may be cast into sticks. Zinc chloride is used as a dehydrating agent an as condensing agent in the organic chemistry. A solution of zinc chloride prepared by dissolving zinc in hydrochloric acid is used as flux in soldering. Zinc chloride is used for filling cotton goods. An oxychloride formed by mixing syrupy zinc chloride solution with zinc oxide, which sets as a hard mass, is used as a dental filling. Zinc chloride is a good disinfecting agent; it is thus used for protecting timber.

The manufacturing process of zinc chloride is discussed below:

The Mass process

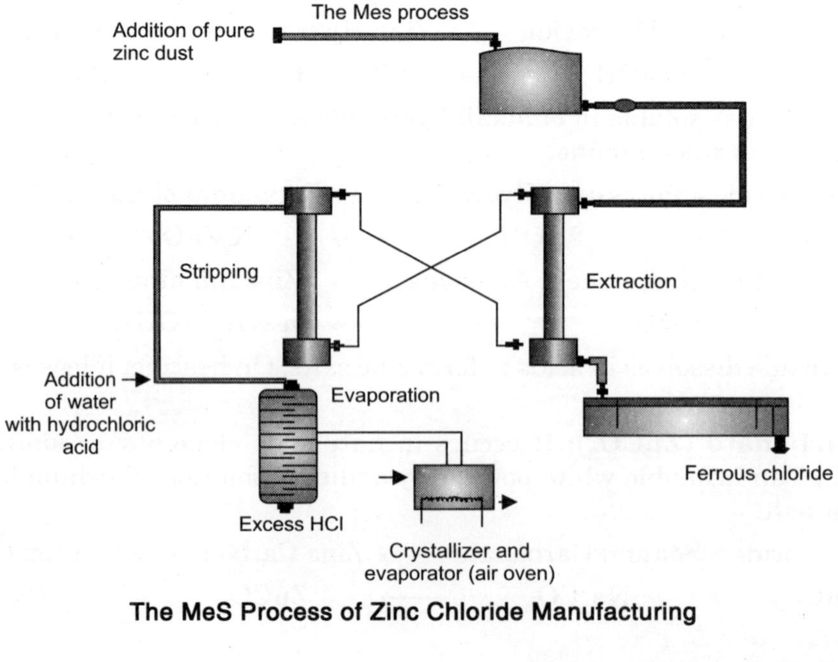

The MeS Process of Zinc Chloride Manufacturing

Recovery of Zinc from Spent Hydrochloric Acid Pickling Liquors in Recovery of Metals from Liquid Effluents, Handbook of Solvent Extraction 1982.

Spent hydrochloric acid pickling liquors, emanating from the galvanizing industry, contain 100 - 130 g/l iron (mainly $FeCl_2$) and 20 - 70 g/l zinc. The iron and zinc concentrations balance, *e.g.* high zinc concentration will be accompanied by low iron concentration.

From these spent pickling liquors, zinc is separated from iron and recovered. Using solvent extraction, zinc is first extracted as a zinc chloride complex to an organic solution containing TBP. To avoid co-extraction of iron, the iron is reduced to its II-valent state. Then, zinc is stripped from the organic solution with water or diluted hydrochloric acid. The resulting strip solution is evaporated, either (1) after addition of sulphuric acid, giving a diluted hydrochloric acid condensate and a zinc sulphate precipitate or (2) directly without any addition, giving a diluted hydrochloric acid condensate and a concentrated zinc chloride mother liquor.

Three product streams are the possible process result:

- Iron (II) chloride solution - possible to treat in a pyrolysis plant or, more probable, for the production of flocculation chemicals, used in sewage water treatment
- Diluted hydrochloric acid condensate - mainly used as strip solution
- Zinc sulphate (solid precipitate) or a concentrated (about 250 g/l) zinc chloride solution

A pilot plant, containing two pulsed columns and a batch-vice evaporation was in operation at MX-Processer AB in Sweden 1981. Pilot plant operations has also been performed in Holland and Germany. The experience from these tests are that the process concept is technically practicable and well proven, however, the economy is strongly dependent on the cost for deposit of spent pickling liquors.

$$\text{Zinc Oxide} + \text{Sulphur} \longrightarrow \text{Zinc Sulphide} + \text{Oxygen}$$

$$2ZnO + 2S \longrightarrow 2ZnS + O_2$$

Zinc Sulphide (ZnS): It occurs in native as zinc blende and wurtzite. It is obtained as white precipitate when ammonium sulphide is added to a zinc salt. The precipitate is soluble in water, in alkali solution and in acetic acid. This compound is also prepared by passing hydrogen sulphide over a zinc salt. Zinc sulphide dissolves in dilute acids, with the evolution of hydrogen sulphide. It is also easily obtained by heating zinc oxide and sulphur.

When zinc sulphide contains traces of heavy metals, like bismuth, lead, copper, manganese etc. as impurities it shows phosphorescent properties. Hence the crystalline zinc blende which is far form pure is phosphorescent. The pure sulphide, on the other hand, is not phosphorescent. One part of copper present in one million part of sulphide is quite sufficient to execute green phosphorescence in the material. The visible light does not only produce this phosphorescence, but also by the ultraviolet rays and X-Rays. The light energy absorbed by the substance raises some of the outer most electrons in the molecule to a higher level. There, on returning to the energy level, they re-emit the energy in the form of light. Alpha-rays and Beta-rays emitted from the radioactive sources can produce vivid spot of light or scintillation*, when they strike a zinc sulphide screen.

Zinc sulphide is therefore used for making phosphorescent screen for X-Rays and radioactive screen. It is largely used for the manufacture of white paint the lithopone, which is the mixture of barium sulphate and zinc sulphide. This is prepared by precipitating zinc sulphate solution with barium sulphate and heating the mixture.

$$\text{Zinc Sulphate} + \text{Barium Sulphide} \longrightarrow \text{Zinc Sulphide} + \text{Barium Sulphate}$$

$$ZnSO_4 + BaSO_4 \longrightarrow 2ZnS + BaSO_4$$

Zinc Sulphate, Vitrium Album ($ZnSO_4$): When metallic zinc, zinc oxide or zinc carbonate is dissolved in dilute sulphuric acid and the solution is evaporated and allowed to crystallize below 39°C, the heptahydrate crystals ($ZnSO_4.7H_2O$) are deposited, this is similar to heptahydrate of Cadmium, magnesium, nickel and cobalt. Zinc sulphate heptahydrate is also known as white vitriol. Zinc also resembles these metals in forming double sulphates with potassium and ammonium sulphates, having the composition $R_2SO_4ZnSO_46H_2O$ where R = NH_4 or K. When heated above 39°C but below 70°C the heptahydrate of zinc sulphate looses a molecule of water to form $ZnSO_46H_2O$; this changes to the monohydrate above 70°C. The anhydrous salt $ZnSO_4$ formed when heated at 300°C is a current of dry air. The hexahydrate when heated to 800°C the zinc sulphate decomposes to zinc and sulphur dioxide.

$$\text{Zinc Sulphate} \longrightarrow \text{Zinc Oxide} + \text{Sulphur Dioxide} + \text{Oxygen}$$

$$2ZnSO_4 \longrightarrow 2ZnO + 2SO_2 + O_2$$

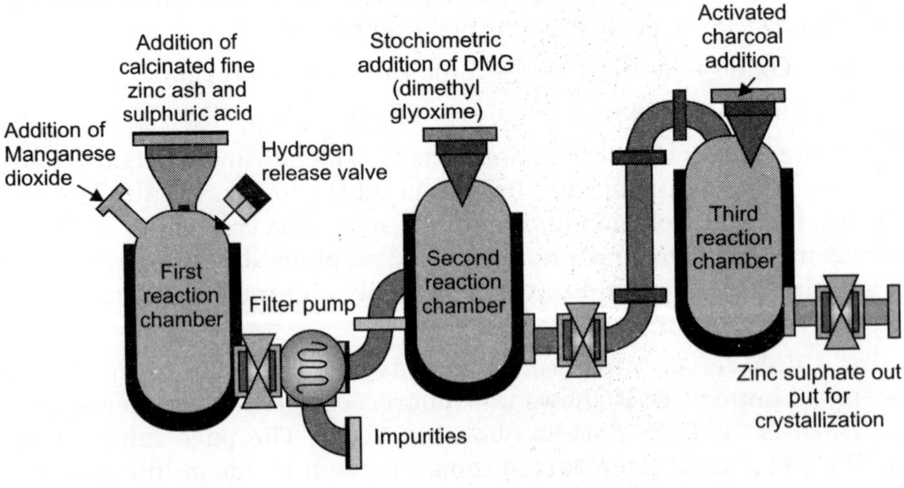

Schematic Diagram for the Zinc Sulphate Manufacture

* Scintillation: Sparkling.

RECYCLING OF ZINC

Secondary Zinc

15-20% of zinc demand in India is met through secondary production. There are 20 secondary zinc units (with a production capacity of about 3000 tonnes of zinc ingot per month) and more than 200 zinc chemical units (zinc oxide, zinc sulphate, zinc chloride etc.). These play an important zinc supply-demand-balancing role in the country. Most of the secondary zinc units use both mechanical and electrolytic methods while some units recover metallic zinc from zinc ash by mechanical method and sell-out fines (mainly oxide with 50-60% zinc) to zinc chemical manufacturer. One pilot scale unit in Gujarat is also reported to be using a plasma process for zinc. The small size of secondary zinc units has restricted the use of advanced technology.

Electrolytic Process for Zinc Ash

The process of extraction of zinc, using zinc ash as raw material, constitutes mainly following stages:

Material Preparation

Zinc ash is generally available in the form of lumps and chips. It is, therefore, first crushed and then pulverized to separate out zinc metallic from fine ash. On melting and casting this metallic, zinc is obtained.

Fine ash is first subjected to calcination in an oil fired rotary kiln at 900oC. Particle retention time in the kiln is decided by the percentage of chloride present in the ash. The rotary kiln is associated with cyclones, settling chamber, and water and lime scrubbers. All the evolved chlorine is absorbed in lime or NaOH scrubber. Calcined ash lumps are pulverized again to get particle size of (–) 100 mesh.

Leaching

The calcined fine ash is then treated with sulphuric acid and / or spent electrolyte, generated during electrolysis for leaching operation. Zinc is rendered soluble according to the following reaction, using agitated reactors:

$$ZnO + H_2SO_4 \longrightarrow ZnSO_4 + H_2O$$

During this operation, compressed air and pyrolusite (MnO_2) are added to oxidize ferrous iron to ferric state. Small quantity of steam, if required, is also added to leach zinc at about 60°C to get better extraction. Leaching is continued till pH of the slurry reaches 4.5 to 5 where the oxidized impurity of iron is hydrolyzed to ferric hydroxide precipitate and is thus removed from the solution:

$$3ZnO + Fe_2(SO_4)_3 + 3H_2O \longrightarrow 3ZnSO_4 + 2Fe(OH)_3$$

The slurry is then discharged to a thickener for solid-liquid separation. For quicker settling, suitable flocculent is added to slurry at the time of discharge. The thickener, overflow, which is a clear solution of $ZnSO_4$, is sent for purification.

The thickener underflow is pumped to another reactor, where it is subjected to water washing. The pulp is filtered in a drum filter to get soluble zinc.

The residual cake is discarded to secured landfill facility and the filtrate goes to effluent treatment plant for making zinc hydroxide.

Purification

The primary object of purification is to prepare the solution for electrolysis by removing all those impurities, which are harmful to the process of electrolysis. This operation is done in three stages.

First Stage: In first stage, copper is cemented out with the addition of zinc dust/ powder. The resultant pulp is filtered in filter press (FP) to recover copper as copper cement and the filtrate is taken to second stage of purification.

Second Stage: Solution is treated with stoichiometric requirement of DMG to remove the impurity of nickel. Addition of excess quantity of DMG is avoided as it increased chemical oxygen demand (COD). Once the nickel removal is achieved the solution is filtered through a FP and is sent to third stage of purification.

Third Stage: To absorb organic impurities, activated charcoal is added here. This purification stage is also considered as the polishing stage. Any unwanted impurity, escaping through 1st and 2nd stage will be removed here. It is thus ensured that $ZnSO_4$ solution is crystallized properly.

Electrolysis and Melting

The purified solution, mixed with spent electrolyte, and cooled in atmospheric cooler to a temp. of 35°C is electrolyzed in electrolytic cells, using lead anodes and aluminum cathodes. Zinc metal deposits on cathodes, whereas oxygen is given off at anodes. During electrowinning of zinc, sulphuric acid is generated as per the reaction given below:

$$ZnSO_4 + H_2O \longrightarrow H_2SO_4 + Zn + \frac{1}{2} O_2$$

Zinc so deposited on aluminum cathodes is stripped-off manually after every 24 hours and is melted in oil fired crucible furnace. Molten Zinc metal is cast as zinc ingots. Spent acid is pumped to leaching section for reuse.

Bleed-Off

Even with the removal of maximum amount of chlorides and other impurities from zinc ash, there will always be build-up of impurities in the system. To lower down the level of impurities in the system some solution need to be removed from it. Bleed-off solution is used to manufacture zinc sulphate crystals.

Effluent Treatment

The only waste in the system would be solid cake containing mainly silica, gypsum etc. which is disposed-off in secured landfill facility. Wash solution is converted into zinc hydroxide, which is reused in the system.

Process for Zinc Sulphate

Code of practice for environmentally sound management for zinc ash/skimmings.

1. For obtaining ESM registration, the industrial units shall have "consent" and "authorisation" for disposal of hazardous wastes from respective State Pollution Control Board/Committee.
2. During transportation, processing for recovery of zinc and production of zinc oxide/zinc sulphate, the following measures will be taken for environmental safeguards:
   (*a*) To avoid dust emissions during handling and transport, properly covered containers should be used.
   (*b*) In the rotary furnaces, the industry should install properly designed bag filter followed by a scrubber. The emission of lead and total particulate matter should not be more than 10 $\mu g/Nm^3$ and 50 $\mu g/Nm^3$ respectively.
   (*c*) The effluent from scrubber should be properly treated and disposal of the same, as may be required, shall meet the standards stipulated by the concerned State Board/Committee.
   (*d*) In the refining section, the solid waste generated from drum filter cake which may contain lead, cadmium etc., shall be properly neutralised and disposed in a secure landfill to be provided as per the guidelines issued by MoEF for implementation of Hazardous Waste (Management & Handling) Rules, 1989 and amendments thereof. The design of secure landfill shall be approved by SPCB/PCC. The leachate collection system shall be provided and leachate shall be treated so as to conform the standards stipulated by the SPCB/PCC.
   (*e*) In the electrolysis section, the fugitive emission generated by electrolysis process, *i.e.* acid fumes, should be properly controlled. The industry should install canopy hood for canalization of emission and it should be connected to a properly designed scrubber before discharging through a stack, as may be necessary.
   (*f*) The industry should utilize the solid waste as far as possible under intimation to the SPCB/PCC.
   (*g*) The container of zinc ash/skimming (after taking the material) should be properly cleaned and the washing, if any, should be treated in the effluent treatment plant as directed by the SPCB/PCC.
   (*h*) Analysis report (from the State Pollution Control Board or any recognized testing laboratory) regarding the compliance of emission control system and effluent treatment equipment provided by the user to meet the standards.
   (*i*) Treated wastewater from the industry should be fully recycled.

Zinc Phosphate, Zinc Orthophosphate [$Zn_3(PO_4)_2$]: It is a white powder insoluble in water, it is prepared by the reaction of zinc sulphate and disodium phosphate.

Zinc Sulphate + Sodium Diphosphate ⟶ Zinc Orthophosphate + Sodium Sulphate

$$2ZnSO_4 + 2Na_2PO_4 \longrightarrow Zn_3(PO_4)_2 + 3Na_2SO_4$$

It also forms as zinc ammonium phosphate when the solution of a zinc salt is treated with a solution of ammonium phosphate.

Zinc Chloride + Ammonium Phosphate ⟶ Zinc Ammonium Phosphate + Ammonium Chloride + Hydrochloride

$$ZnCl_2 + (NH_4)_2HPO_4 \longrightarrow Zn(NH_4)_2PO_4 + NH_4Cl + HCl$$

When dried at 102°C it becomes a solid, which is used for the gravimetric detection of zinc. But if heated above this temperature (200°C) in an open crucible, zinc pyrophosphate ($Zn_2P_2O_7$) is produced.

Zinc Ammonium Phosphate ⟶ Zinc Pyrophosphate + Ammonia + Water

$$Zn(NH_4)_2PO_4 \longrightarrow Zn_2P_2O_7 + 2NH_3 + H_2O$$

A mixture of zinc oxide and phosphoric acid, forming basic zinc phosphate is used as dental cement.

Zinc Phosphide (Zn_3P_2): It is grey solid, formed when zinc oxide reacts with phosphorous pentoxide.

Zinc Oxide + Phosphorous ⟶ Zinc Phosphide + Oxygen

$$3ZnO + P_2O_5 \longrightarrow Zn_3P_2 + 4O_2$$

It evolves phosphene when reacts with acids.

Zinc Phosphide + Hydrochloric Acid ⟶ Zinc Chloride + Phosphene

$$ZnP_3 + 6HCl \longrightarrow 3ZnCl_2 + 2PH_3$$

Zinc Nitrate [$Zn(NO_3)_2 4H_2O$]: This compound is obtained by dissolving the metal, oxide or carbonate in dilute nitric acid and crystallizing. It is deliquescent and is soluble in alcohol. The anhydrous nitrate is formed by heating the hydrate at 130 to 135°C in a current of carbon dioxide and nitrogen pentoxide vapour. It decomposes at 140°C.

The action of zinc on nitric is little but complicated; with concentrated nitric acid nitrogen dioxide is evolved and with dilute acid, nitric oxide is produced. But with very dilute acid nitrous oxide may be evolved and the acid is also reduced to ammonia.

Zinc + Nitric Acid ⟶ Zinc Nitrate + Nitrous Oxide + Water

$$4Zn + HNO_3 \longrightarrow 4Zn(NO_3)_2 + N_2O + 3H_2O$$

Zinc + Nitric Acid ⟶ Zinc Nitrate + Ammonium Nitrate + Water

$$4Zn + 10HNO_3 \longrightarrow 4Zn(NO_3)_2 + NH_4NO_3 + 3H_2O$$

Ammonia and a solution of zinc ethyl and dry ether gives amide, $Zn(NH_2)_2$

Zinc Ethyl + Ammonia ⟶ Zinc Amide + Ether

$$Zn(C_2H_5)_2 + 2NH_3 \longrightarrow Zn(NH_2)_2 + 2C_6H_6$$

On heating to dull redness this forms zinc nitride Zn_3N_2

Zinc Amide ⟶ Zinc Nitride + Ammonia

$$Zn_3N_2 \longrightarrow Zn_3N_2 + 4NH_3$$

Zinc nitride is grayish-green in colour, vigorously decomposed in water.

$$\text{Zinc Nitride} \quad \text{Water} \longrightarrow \text{Zinc Oxide} + \text{Ammonia}$$

$$Zn_3N_2 + 3H_2O \longrightarrow 3ZnO + 2NH_3$$

Detection and Estimation

Zinc is estimated by the gravimetric analysis where zinc ammonium phosphate is prepared and the zinc is estimated form that.

Zinc is detected by the Rinman's Reagent where the sample consisting of zinc turns green in the presence of the reagent.

YTTRIUM

Melting Point : 1500°C

Boiling Point : 2927°C

Atomic Weight : 88.9059

Atomic Number : 39

Introduction and Occurrence

Yttrium is one of the rare elements and is very similar to Lanthanides. This element was invented by Ghldl~Jlin in the year 1794 in the form of rare earth called Yttria. The metal yttrium was isolated by Wi)neJr in the year 1828.

Physical Properties

Melting Point	1500°C
Boiling Point	2927°C
Other Electronic Configuration	$4d^1 5s^2$
Density	4.46 g/cc
Atomic Volume	19.8cc
Atomic Radius	1.80Å
Ionic Radius	.93Å
First Ionization Energy	152 Kcal/mole
Stable Oxidation States	+3
Electronegativity (Pauling's Scale)	93 kcal/g atom
Heat of Fusion	2.7 Kcal/g atom

Chemcial Properties

Yttrium is soluble in dilute acids and alkalis forming different compounds. It is decomposed by water.

Compounds of Yttrium

Yttrium Oxide, Yttria (Y_2O_3): This compound of yttrium is obtained by the ignition of yttrium nitrate.

$$\text{Yttrium Nitrate} \longrightarrow \text{Yttrium Oxide} + \text{Nitric Oxide} + \text{Oxygen}$$

$$Y_2NO_3 \longrightarrow Y_2O_3 + NO + O_2$$

It is yellowish-white powder insoluble in water, but soluble in dilute acids, with a melting point of 2400°C. It is used as phosphor in colour picture tubes (alloy with Europium). Yttrium-Iron alloy is used for microwave filters.

Yttrium Phosphide (YP): This compound is prepared when yttrium oxide reacts with phosphorous pentoxide.

Phosphorous Pentoxide + Yttrium Oxide ⟶ Yttrium Phosphide + Water

$$P_2O_5 + Y_2O_3 \longrightarrow 2YP + 3H_2O$$

This reaction produces lots of heat. The compound is used for semi-conductor.

Yttrium Chloride (YCl_2): This compound is prepared by the action of hydrochloric acid on yttrium oxide.

Yttrium Oxide + Hydrogen Chloride ⟶ Yttrium Chloride + Water

$$Y_2O_3 + HCl \longrightarrow 2YCl_3 + 3H_2O$$

This gets dissolved in alcohol and water. It is deliquescent, red and prismatic crystals. It decomposes at 100°C. This compound is used mainly in analytical chemistry.

Yttrium Bromide (YBr_39H_2O): When hydrogen bromide reacts with yttrium oxide this compound is produced.

Yttrium Oxide + Hydrogen Bromide ⟶ Yttrium Bromide + Water

$$Y_2O_3 + 6HBr \longrightarrow 2YBr_3 + 3H_2O$$

It is a colourless crystals hydroscopic; the melting point of the anhydrous salt is 905°C it is soluble in water.

Yttrium Acetate [$Y(C_2H_3O_2)_3.9H_2O$]: This hydrated salt is produced by the action of glacial acetic acid on yttrium oxide.

Yttrium Oxide + Acetic Acid ⟶ Yttrium Acetate + Water

$$Y_2O_3 + 6C_2H_4O_2 \longrightarrow 2Y(C_2H_3O_2)_3 + 3H_2O$$

Yttrium Antimonide (YSb): When yttrium oxide is allowed to react with antimony trioxide, yttrium antimonide is produced.

Yttrium Oxide + Antimony Trioxide ⟶ Yttrium Antimonide + Oxygen

$$Y_2O_3 + Sb_2O_3 \longrightarrow 2YSb + 3O_2$$

It is also used as semi-conductor.

Yttrium Arsenide (YAs): This compound is prepared by the action of arsenious trioxide and yttrium oxide.

Yttrium Oxide + Arsenic Trioxide ⟶ Yttrium Arsenide + Oxygen

$$Y_2O_3 + As_2O_3 \longrightarrow 2YAs + 3O_2$$

Yttrium Sulphate [$Y_2(SO_4)_38H_2O$]: This compound of yttrium is obtained by the action of sulphuric acid on yttrium Phosphate.

Yttrium Phosphate + Sulphuric Acid ⟶ Yttrium Sulphate + Phosphene + Oxygen

$$2YPO_4 + 3H_2SO_4 \longrightarrow Y_2(SO_4)_3 + 2PH_3 + 4O_2$$

PALLADIUM

Melting Point : 1554°C Atomic Weight : 106.42
Boiling Point : 2870°C Atomic Number : 46

History and Occurrence

Palladium was discovered by Wollaston in the year 1803 and named after planetoid Pallas.

Palladium is rare than noble like other members of the platinum group. There are a very few sources from which palladium can be obtained. It occurs with gold as Propesite in the Brazilian sand and also accompanies native platinum. But the most important source is the residue from the extracted nickel ore and also from the anode slime from the refining of the nickel by electrolytic process.

Extraction

Palladium is easily separated from the other members of the platinum group by the preparation as cyanide [$Pd(CN)_2$] by the addition of mercuric cyanide, as PdI_2 by potassium iodide or as $PdCl_2.2NH_3$ by adding hydrochloric acid to an ammonical solution of its chloride. The solution of the crude platinum metals obtained by treatment with aqua regia is first formed from platinum by precipitation with ammonium chloride. Palladium is then prepared from the filtrate as cyanide iodide or dichlrodiammine. The metal is obtained by simply igniting any of the precipitation.

Physical Properties

Boiling Point	1554°C
Melting Point	2870°C
Outer electronic Configuration	$4d^{10}\, 5s^0$
Stable Oxidation States	+2
First Ionization Energy	192 Kcal/mol
Oxidation Potential ($M - M^{++} + 2e^-$)	–1.2 volt
Atomic Radius	0.50 (+)Å
Atomic Volume	8.6 cc
Density	12.0 g/cc
Heat of Vapourization	91 kcal/mol
Heat of Fusion	40 kcal/mol

Chemical Properties

It is not affected by air but when heated in air a blue film of palladium oxide (PdO) is formed.

$$\text{Palladium} + \text{Oxygen} \longrightarrow \text{Palladium Oxide}$$

$$2Pd + O_2 \longrightarrow 2PdO$$

It resists the attack of chlorine unless the temperature is raised to red hot state. It combines directly with sulphur at a high temperature forming the sulphide (PdS).

It dissolves in dilute nitric acid forming palladous nitrate [$Pd(NO_3)$] and in dilute sulphuric acid containing a small amount of nitric acid to form palladous sulphate ($PdSO_4H_2O$).

It dissolves in aqua regia to give H_2PdCl_4 which on evaporation and subsequent heating decomposes to give $PdCl_2$.

The most remarkable property of the palladium is that of absorption of hydrogen, when it is heated to redness it absorbs about 1000 times its own volume of hydrogen. If the temperature is raised further the hydrogen is expelled. It is believed that Occlusion in the absorption process the hydrogen molecules break down into hydrogen atom to fit into the palladium lattice.

Uses of Palladium

Palladium is used in jewellery and in the formation of a white alloy with gold, which serves as the protective coating of the gold. This is because it is inert towards air and can hold several hydrogenous reactions.

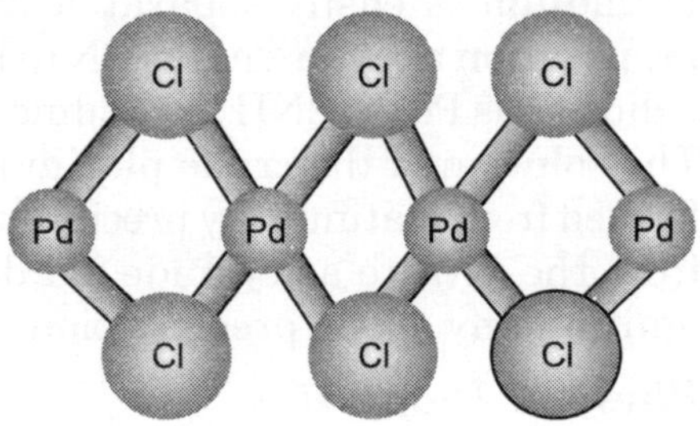

It appears that the catalytic performance is due to the dissociation of the molecular hydrogen into atomic state.

The atomic hydrogen being far more active, the reaction is brought out very efficiently.

Palladium exists in the +2 state in most of its compounds such as the oxide, PdO, the chloride ($PdCl_2$) the fluoride (PdF_2) the sulphate ($PdSO_4$) and the nitrate [$Pd(NO_3)_2$ PdF_2] is an ionic solid ($PdCl_2$) however, exists as a linear polymer in the solid state.

$$\text{Hydrogen Molecule} \longrightarrow \text{Atomic Hydrogen}$$
$$H_2 \longrightarrow 2H$$

$PdCl_2$ is soluble in hydrochloric acid forming the complex $[PdCl_4]_2^-$.

Palladium ion Pd^{2+} in aqueous solution is hydrated and as the complex $Pd(H_2O_4)^{2+}$. It also forms complex with ammonia $[Pd(NH_3)_4]^{2+}$.

A few +3 compounds of palladium are also known as PdF_3, which is an important example. It is formed on heating the metal in a current of fluorine, at 500°C.

Compounds of Palladium

Palladium in its compound is mostly bivalent. Except in some complex forming, palladium rarely occurs in the quadravalent state. It its fluoride palladium shows a valency of three. A univalent chloride has also been reported. But the bivalent palladium compounds both simple and complex are most common and stable.

Palladous Oxide (PdO): It is obtained as a black powder in oxygen or by igniting palladous nitrate. The oxide decomposes, when strongly heated, producing the metal and liberating oxygen. In a hydrated form it is obtained by precipitating solutions of palladous salts with solution on boiling. The hydrated oxide forms a brown precipitate.

Palladous Dioxide (PdO_2): This has been prepared in a hydrated form as a dark-brown precipitate by the addition of caustic alkalis to the solution of a chloropalladate (Na_2PdCl_6). It is unstable and possesses powerful oxidizing properties. The same product is obtained by the anodic oxidation of palladous nitrate solution.

Palladous Sulphide (PdS): It is formed as a dark-brown precipitate by passing hydrogen sulphide through a solution of palladous salt. The precipitate is soluble in dilute acids and ammonium sulphide solution prepared in the dry way by heating ammonium chloropalladate, with sulphur, it forms hard shinning bluish crystals, insoluble in nitric acid and even in aquaregia. A disulphide (PdS_2) has also been described as a dark brown crystalline product easily soluble in aqua regia.

Palladium Trifluoride (PdF_3): It is formed as a black product by the action of fluorine of palladous chloride.

$$\text{Palladium Chloride} + \text{Fluorine} \longrightarrow \text{Palladium Fluoride} + \text{Chlorine}$$

$$2PbCl_2 + 3F_2 \longrightarrow 2PbF_3 + 2Cl_2$$

Palladous Chloride ($PdCl_2$): It is obtained in the form of dark-red crystals deliquescent in nature, by the action chlorine on palladium, but can be distilled in a current of chlorine. Palladous chloride is readily soluble in water and form the solution of reddish-brown crystals of the hydrate ($PdCl_22H_2O$) separate on slow evaporation in the cold. A solution palladous chloride is used as a very sensitive test for carbon monoxide. A neutral solution is turned intense black in the presence of carbon monoxide, because colloidal palladium is separated.

Palladium is used for detection of carbon monoxide.

$$\text{Palladium Chloride} + \text{Carbon Monoxide} \longrightarrow \text{Palladium} + \text{Hydrogen Chloride} + \text{Carbon Dioxide}$$

$$PdCl_2 + CO \longrightarrow Pd + 2HCl + CO_2$$

When palladium is dissolved in hydrochloric acid in presence of an oxidizing agent and then the solution is evaporated, chloropalladous acid (H_2PdCl_4) is produced.

$$\text{Palladium} + \text{Hydrochloric Acid} + \text{Potassium Chloride} \longrightarrow \text{Chloropalladous Acid} + \text{Potassium}$$

$$Pd + 2HCl + 2KCl \longrightarrow H_2PdCl_4 + 2K$$

With alkali chlorides it forms red or brown complex chloropalladites R_2PdCl_4(R = Na, K, NH_4 Rb, Cs). The chloropalladites are oxidized to chloropalladates (R_2PdCl_6) similar to chloroplatinates. The chloropalladates forms red or brown crystals. Potassium or ammonium chloropalladates are sparingly soluble in water like the corresponding chloroplatinates.

Palladous chloride combines with ammonia to form the sparingly soluble complex dichlorodiammine palladium [$Pd(NH_3)_2Cl_2$]. This occurs in three forms.

$Pd(NH_3)PdCl_4$ (Pink) $Pd(NH_3)_2Cl_2$ (Yellow) $[Pd(NH_3)Cl]H_3N$ (Yellow)

Palladous Iodide (PdI_2): It is obtained as black precipitate by the addition of potassium iodide to a solution of palladous chloride. The precipitate is soluble in an excess of potassium iodide.

Palladous Cyanide [$Pd(CN)_2$]: It is produced as yellowish-white gelatinous precipitate by the addition of mercuric cyanide to a solution of palladous chloride.

Palladous Chloride + Mercuric Cyanide ⟶ Palladous Cyanide + Mercuric Chloride

$$PdCl_2 + Hg(CN)_2 \longrightarrow Pd(CN)_2 + HgCl_2$$

The precipitate dissolves readily in potassium cyanide solution, when evaporated colourless crystals of $K_2Pd(CN)_4\ 3H_2O$ can be obtained.

Palladous Nitrate [$Pd(NO_3)_2$]: This compound is obtained when metallic palladium reacts with concentrated nitric acid.

Nitric Aicd + Palladium ⟶ Palladium Nitrate + Hydrogen

$$2HNO_3 \quad Pd \longrightarrow Pd(NO_3)_2 \quad H_2$$

There crystals are hydroscopic in nature and yellowish brown in colour.

Palladous Sulphate ($PdSO_4.H_2O$): Palladium dissolves in dilute sulphuric acid containing nitric acid to form palladous sulphate in solution. Form the solution the hydrated salt $PdSO_4.2H_2O$ can be obtained in olive green deliquescent crystals.

SILVER

Melting Point : 961°C
Boiling Point : 2212°C
Atomic Number : 47
Atomic Weight : 107.868

History and Occurrence

Silver was known in Predynamic Egypt (4000BC) but was very rare. A fire caladian vase of 2850 B.C is in the Louvre. The oldest silver probably came from North Syria and Asia Minor, which were centers of Hittite Civilization. Silver mine in Spain were worked at an early date.

Silver occurs mainly in the form of ores, some of them are *Argentite (silver glance)* Ag_2S this is the commonest ore, *Chlorargyrite (horn-silver) AgCl, Pyrargyrite (or ruby silver)* Ag_3SbS_3, *Stromeyrite (or silver copper glance)* $(Cu, Ag)_2S$ and *Stephanite,* $AgjSbS_4$.

In India, mainly in Rajasthan, silver is obtained as a by-product during the concentration of Galena (ore of Zinc) in Hindusthan Zinc Smelter.

It is now reported that lead ores from Cuddapah, Guntur and Kumool districts in the State of Andhra Pradesh (India) containing silver in fair proportions.

Extraction

1. Cyanide Process: Silver is extracted from its ores generally by the cyanide process. In this process the finely powdered ore is leached for several hours with a dilute (0.5%) solution of sodium cyanide; the mixture being all the while agitated by a current of air, metallic silver, silver chloride and silver sulphide as pass into solution with formation of the complex sodium dicyanoargentate, $Na[Ag(CN)_2]$.

Silver + Sodium Cyanide + Water + Oxygen (air) $\longrightarrow$ Sodium Dicyanoargentate + Sodium Hydroxide

$$4Ag + 8NaCN + 2H_2O + O_2 \longrightarrow 4Na[Ag(CN)_2] + 4NaOH$$

Silver Chloride + Sodium Cyanide $\longrightarrow$ Sodium Dicyanoargentate + Sodium Chloride

$$AgCl + 4NaCN \longrightarrow Na[Ag(CN)_2] + NaCl$$

Silver Sulphide + Sodium Cyanide $\longrightarrow$ Sodium Dicyanoargentate + Sodium Sulphide

$$Ag_2S + 4NaCN \longrightarrow Na[Ag(CN)_2] + Na_2S$$

Sodium sulphide is at largely oxidized to sodium sulphate by the air current.

Sodium Sulphide + Water + Oxygen $\longrightarrow$ Sodium Sulphate + Sodium Hydroxide + Sulphur

$$4Na_2S + 2H_2O + 5O_2 \longrightarrow 2Na_2SO_4 + 4NaOH + 2S$$

The solution containing the silver complex is removed and treated with zinc dust when silver precipitates out.

Sodium Dicyanoargentate + Zinc $\longrightarrow$ Sodium Tetracyanozincate + Silver

$$Na[Ag(CN)_2] + Zn \longrightarrow Na_2[Zn(CN)_4] + 2Ag$$

The solution containing the silver complex is removed and treated with zinc dust when silver is finally with borax or potassium nitrate to purify it.

2. Amalgamation Process: This method though suppressed by the cyanide process but still in use in some of the countries. The sulphide ore is crushed and converted into slime with a solution of cupric chloride.

Silver Sulphide + Cupric Chloride $\longrightarrow$ Silver Chloride + Cuprous Sulphide

$$Ag_2S + CuCl_2 \longrightarrow 2AgCl + CuS$$

Some mercury is then also added to the product when silver is precipitated out.

Silver Chloride + Mercury $\longrightarrow$ Silver + Mercuric Chloride

$$2AgCl + 2Hg \longrightarrow 2Ag + Ag_2Cl_2$$

The silver set free dissolves in excess of mercury to form amalgam. The amalgam is washed and then distilled to recover the mercury while the silver remains in the retort.

3. From Argentiferrous Lead: Lead prepared from its sulphide or galena, contains traces of silver and is called argentiferrous lead. It is economical to recover silver from thin material. The process involves enrichment of silver in the first instance. The enrichment is carried out by Pattinson's Process. In this process, advantage is taken of the fact that when a molten lead-silver alloy is allowed to cool. Lead separates out until concentration of silver rises to 2.6% when the entire mass solidifies.

Lead consisting of 2.6% of silver is then method and a small amount of molten zinc is added to it, since silver is about 300 times more soluble in zinc than lead, most of the silver passes into the zinc layer which rises to the top. On cooling, the zinc-silver alloys solidifies and is skimmed* off. White the lead still remains in the molten state. This process of recovering silver is known as **Parke's Process.**

Some of the lead, however, may be related to zinc-silver alloy. It is eliminated by heating the alloy in a sloping furnace, when lead melts and drains always from the solid mass (liquidation process). The purified alloy is finally heated strongly in a fine-clay retort. Zinc being more volatile distils off. The small amount of lead which may still remains related to silver is removed by Cupillation.

In this process, the impure silver is heated in a small bone-ash crucible (cupola) in a blast of air in a small reverberatory furnace. Lead gets slowly oxidized to litherage (PdO) which is blown away. After some time when the shinning surface of silver because visible. The silver thus obtained may be further purified by electrolytic process.

Physical Properties

Boiling Point	1554°C
Melting Point	2870°C
Outer electronic Configuration	$4d^{10}\ 5s^0$
Stable Oxidation States	+2
First Ionization Energy	192 Kcal/mol
Oxidation Potential ($M - M^{++} + 2e^-$)	–1.2 volt
Atomic Radius	0.50 (+2)Å
Atomic Volume	8.6 cc
Density	12.0 gm/cc
Heat of Vapourization	91kcal/mol
Heat of Fusion	40 kcal/mol

Chemical Properties

Silver has a poor chemical activity. Therefore, it is known as noble metal. Following are some of its chief chemical properties

1. Action of Air: Silver is not attacked by air or moisture, but when exposed to air with traces of hydrogen sulphide it turns black forming a thin layer of silver sulphide.

$$\text{Silver + Hydrogen Sulphide} \longrightarrow \text{Silver Sulphide}$$

$$2Ag + H_2S \longrightarrow Ag_2S$$

2. Action of Halogen: When heated strongly in a current of a gas or vapour it forms tile corresponding halides, *e.g.*

* Skim: Layer.

Silver Chlorine ⟶ Silver Chloride

$$2Ag + Cl_2 \longrightarrow 2AgCl$$

Silver + Bromine ⟶ Silver Bromide

$$2Ag + Br_2 \longrightarrow 2AgBr$$

3. Action of Sulphur: Silver has a tendency to react with sulphur when heated with sulphur powder, it forms silver sulphide.

Silver + Sulphur ⟶ Silver Sulphide

$$2Ag + S \longrightarrow Ag_2S$$

4. Action of Acids: The standard oxidation potential for the reaction below is -.80volt which is even lower than that of copper.

$$Ag(s) \longrightarrow Ag + e^-$$

Hence, it cannot displace hydrogen from dilute acids. Silver is thus, not attacked by non-oxidizing acids, like dilute hydrochloric acid and sulphuric acid.

> This is the process why silver does not react with any non-oxidizing acids like hydrochloric acid

However, it reacts with nitric acid and concentrated sulphuric acid forming nitrate and silver sulphate respectively.

It is not attached by aquaregia. Advantage is taken from this fact in separating it form gold (which gets dissolved in aquaregia).

> Silver does not get dissolved in aquaregia

5. Action of Alkali Cyanides: Silver dissolves in sodium or potassium cyanide in the presence of air to form the complex ion.

Silver + Water + Oxygen + Potassium Cyanide ⟶ Silver Cyanide + Potassium Hydroxide

$$2Ag + 2H_2O + O_2 + 4KCN \longrightarrow 2Ag(CN)_2 + 4KOH$$

Reduction of Silver Nitrate to Metallic Silver

This phenomenon is also known as 'silverization of glass'. A mixture of silver nitrate, glucose and potassium hydroxide is boiled in a test tube over a water bath. After five or six minutes when the mixture is poured out the glass walls are left away with a shinning silver coating.

Uses of Silver

Silver is mainly used for making coins, jewellery and ornamental utensils for electroplating on cheaper metals or alloys and for mirroring.

Silver is one of the best conductors of electricity, so it is used for making electrical connectors and switches.

Silver is also used for making ornaments (used with copper).

Silver is also used in photography and in making Dental fillings.

Compounds of Silver

With the exception of a very few instances, silver is dedicatedly univalent in its simple compounds. In this respect silver differs from copper, which exists a stable state of bi-valency in a large majority of its simple compounds.

Silver Oxide (Ag_2O): It is also known as argentenous oxide, it is obtained as a brown precipitate when silver nitrate is added to a solution of sodium hydroxide.

Silver Nitrate + Sodium Hydroxide ⟶ Silver Oxide + Sodium Nitrate + Water

$$2AgNO_3 + 2NaOH \longrightarrow Ag_2O + 2NaNO_3 + H_2O$$

No silver hydroxide is formed, the brown precipitate turns black on drying at 60°C to 75°C.

Silver oxide is very sparingly soluble in water and gives an alkaline solution. It can precipitate many heavy metals as oxide or hydroxides from solution of their salts. It absorbs carbon dioxide to form silver carbonate ($AgCO_3$) and reacts with acids to the formation of complex diammino-silver hydroxide (Ag_2NH_3) of fulminating silver, containing silver nitride, a very explosive body when dry.

Silver Carbonate ($AgCO_3$): It is formed as light yellow precipitate when sodium carbonate is added to a solution of silver nitrate.

Silver Nitrate + Sodium Carbonate ⟶ Silver Carbonate + Sodium Nitrate

$$2AgNO_3 + NaCO_3 \longrightarrow 2AgCO_3 + 2NaCO_3$$

When heated it looses carbon dioxide and decomposes completely leaving behind metallic silver.

Silver Carbonate ⟶ Silver + Carbon Dioxide + Oxygen

$$2Ag_2CO_3 \longrightarrow 2Ag + 2CO_2 + O_2$$

Silver Fluoride (AgF): It is only silver halide. It is prepared by dissolving silver oxide hydrofluoric acid.

Silver Oxide + Hydrogen Fluoride ⟶ Silver Fluoride + Water

$$Ag_2O + 2HF \longrightarrow 2AgF + H_2O$$

The solution on evaporating in vacuum deposits yellow crystals of $AgF.H_2O$. Silver fluoride melts at 436°C

Silver Chloride (AgCl): When a solution of sodium chloride is added to a solution of silver nitrate, a white precipitate of silver chloride is found.

Silver Nitrate + Sodium Chloride ⟶ Sodium Nitrate + Silver Chloride

$$AgNO_3 + NaCl \longrightarrow NaNO_3 + AgCl$$

It is also available as the ore of *Cerargyrite.* Silver chloride fuses at 435°C to a dark yellow liquid, which solidifies to a colourless mass. When heated to a very high temperature it volatilizes away and the vapour density corresponds to that of Agel. The solubility of silver chloride in water is very small; but it dissolves in hydrochloric acid and also in the strong solution of sodium chloride due to the formation of complex $AgCl^-$ ions. Silver chloride also gets dissolved in the boiling concentrated sulphuric acid.

Silver Chloride + Sulphuric Acid ⟶ Silver Sulphide + Hydrogen Chloride

$$2AgCl + H_2SO_4 \longrightarrow Ag_2SO_4 + 2HCl$$

It is readily soluble in aqueous solution of ammonia forming as complex $Ag(NH_3)^{2+}$ ion. In alkali cyanide it gets readily dissolved and forms concentrated anions (ions) $[Ag(CN)^{2-}$ and $[Ag(S_2O_3)^{2-}$ respectively. Silver Chloride is reduced to silver when heated in a current of hydrogen.

Silver Chloride + Hydrogen $\longrightarrow$ Silver + Hydrogen Chloride

$$2AgCl + H_2 \longrightarrow 2Ag + 2HCl$$

Silver Bromide (AgBr): This is obtained as a pale yellow precipitate by the addition of a solution of a potassium bromide to that of a silver nitrate. The precipitate is insoluble in dilute nitric acid and only sparingly soluble in dilute ammonia, but readily soluble in concentrated ammonia. It melts at 435°C. Silver bromide is largely used in photography.

Silver Iodide, AgI: It is a light yellow precipitate produced by the addition of potassium iodide to that of silver nitrate, it is insoluble in nitric acid and very less soluble in concentrated ammonia. It however, soluble in sodium thiosulphate and melts at 560°C. It is not completely reduced by hydrogen even at high temperature.

Silver Thiocyanate (AgSCN): It is obtained as a cardy white insoluble precipitate by the addition of potassium thiocyanate to a solution of silver nitrate. It is insoluble in nitric acid but dissolves in an excess of potassium thiocyanate to form complex potassium silver thiocyanate.

Silver Cyanide (AgCN): It is formed as; white precipitate form a soluble silver salt and potassium cyanide solution. It is also dissolved in the excess cyanide solution, forming a complex anion of $Ag(CN)^{2-}$. The complex alkali salts are also known in the solid state.

Silver Sulphide (Ag_2S): This compound occurs as argentite, it can be prepared directly from the two elements (silver and sulphur) or by passing hydrogen sulphide through silver nitrate. The colour of this compound is black, it is insoluble in ammonia and in the hot dilute acid. When heated in hydrogen it is reduced to metallic silver.

Silver Nitrate ($AgNO_3$): It is prepared by dissolving metallic silver in concentrated nitric acid.

Silver + Nitric Acid $\longrightarrow$ Silver Nitrate + Water + Nitric Oxide

$$3Ag + 4NHO_3 \longrightarrow 3AgNO_3 + 2H_2O + NO$$

The metals at 218°C, when fused it can be cast into sticks called *lunar caustic.* Organic matter readily reduces silver nitrate. Silver nitrate is freely soluble in water. With ammonia it forms complex ammine ($Ag[2NH_3]NO_3$). Silver nitrate serves as the starting point for the preparation of other silver compounds; particularly the halides used in photography. It is also used for making standard solutions for volumetric analysis and as a reagent in organic detections. As already stated ammonical solution of silver nitrate is used for making glass mirrors, it is also used for making linen.

Silver Sulphate ($AgSO_4$): It can be obtained by the reaction of sulphuric acid. It is however, soluble in both the sulphuric acid and nitric acid. It decomposes at a high temperature.

Silver Nitrate + Potassium Chromate $\longrightarrow$ Silver Chromate + Potassium Nitrate

$$2AgNO_3 + K_2CrO_4 \longrightarrow Ag_2CrO_4 + 2KNO_3$$

Action of Light upon Silver Salts

Silver salts are very much sensitive to light. That is why these compounds are always kept in coated bottles covered with a black paper in the laboratory. The photochemical reactions are stated below:

$$AgNO_3 + h\nu \longrightarrow AgO + NO_2 + e^-$$

$$AgO + H\nu \longrightarrow Ag^+ + e^- \text{ [where } h\nu \text{ is Photon]}$$

PHOTOSENSITIVE

Detection and Estimation

Silver can be detected and estimated by its precipitation of silver chloride. Volumetrically, it can be estimated by titration against chloride or thiocyanate using internal indicators like potassium chromate for chloride and ferric alum for thiocyanate.

Silver salts are detected by the silver mirror test as stated earlier. Also by adding potassium chromate to a solution of silver compound which produces silver chromate, brown precipitate.

Silver Nitrate + Potassium Chromate ⟶ Silver Chromate + Potassium Nitrate

$$2AgNO_3 + K_2CrO_4 \longrightarrow Ag_2CrO_4 + 2KNO_3$$

Now-a-days silver in a solution can also be tested by using silver estimation papers, (DYE & STAIN or INDICATOR company), these are available along with a colour chart from which the quantitative results can be obtained.

CADMIUM

Melting Point : 320.9°C
Boiling Point : 767°C

Atomic Number : 48
Atomic Weight : 112.411

History and Occurrence

Zinc oxide which had a yellow colour although free from iron was found by Someyer in 1817 to contain the oxide of a few metal to which he gave the name 'cadmium' from the word *cadmia*, the zinc ore (kadmea).

Most of the zinc ores contain small amount of cadmium. Blende may contain 2-3% of cadmium and calamine up to 5% but the average is less than 0.5%. The only mineral, the sulphide *greencite* CdS is rare.

Extraction

Cadmium is more volatile than zinc and its oxide are more easily reducible than oxide of zinc it is found concentrated in the first fraction of the distillate of zinc dust, collected in the receiver of zinc furnace. Most of the cadmium is present in the dust as cadmium oxide; CdO remained mixed with zinc oxide. It is mixed with coal and heated in retorts at 800°C – 900°C. Cadmium distills over into the adopter cones fitted to the retorts. At this temperature zinc is neither reducible from zinc oxide, nor volatile.

Physical Properties

Melting Point	320.9°C
Boiling Point	767°C
Outer Electronic Configuration	$4d^2\ 5s^2$
Density	8.65 g/cc
Atomic Volume	13.1 cc
Atomic Radius	1.54 Å
Ionic Radius	.97 (+2) Å
First Ionization Energy	207 Kcal/mol
Oxidation Potential ($Cd - Cd^{2+}\ 2e^-$)	0.40 volt
Stable Oxidation State	+2
Heat of Vapourization	23.9 Kcal/mol
Heat of Fusion	14.6 Kcal/mol
Electronegativity (Pauling's scale)	1.7
Specific Heat	0.055

Chemical Properties

Chemically cadmium follows zinc quite closely in its properties. Cadmium is slowly oxidized by moist air and its vapour reacts with steam to form cadmium oxide and hydrogen. At red-hot condition it burns in air to form cadmium oxide just as zinc oxide looks like in the smelter plants. It reacts readily with mineral acids and more slowly with organic acids. Cadmium may be precipitated from solution by metallic zinc. It is insoluble in alkali hydroxides, which distinguishes cadmium from zinc.

Compounds of Cadmium

Cadmium Oxide (CdO): Cadmium oxide can be prepared by heating it to a temperature of decomposition.

$$\text{Cadmium Nitrate} \longrightarrow \text{Cadmium Oxide} + \text{Nitrogen Dioxide}$$

$$CdNO_3 \longrightarrow CdO + NO_2$$

Cadmium oxide is yellowish-brown powder; it volatilizes at 650°C. Cadmium oxide is used for cadmium plating. It is also used for sensitizing the photochemical reaction for the preparation of water from hydrogen peroxide and as a catalyst for the preparation of highly unsaturated alcohol. It is also used in the manufacture of paints pigments etc.

Cadmium Hydroxide [$Cd(OH)_2$]: Addition of ammonium hydroxide to a solution of cadmium oxide result in the formation of cadmium hydroxide.

$$\underset{\text{Chloride}}{\text{Cadmium}} + \underset{\text{Hydroxide}}{\text{Amminium}} \longrightarrow \underset{\text{Chloride}}{\text{Ammonium}} + \underset{\text{Hydroxide}}{\text{Cadmium}}$$

$$CdCl_2 + 2NH_4OH \longrightarrow 2NH_4Cl + Cd(OH)_2$$

Cadmium hydroxide is a white powder, soluble in ammonium salt solutions, but insoluble in water. It readily absorbs carbon dioxide to form carbonate.

Cadmium Hydroxide + Carbon Dioxide ⟶ Cadmium Carbonate + Water

$CdCl_2 + CO_2 \longrightarrow CdCO_2 + H_2O$

Cadmium hydroxide at 300°C. It is used for the preparation of the negative electrode of the nickel cadmium battery.

Cadmium Sulphate ($CdSO_4$): It forms colourless efflorescent crystals that are soluble in water. Cadmium sulphate is prepared by the reaction of cadmium metal and sulphuric acid.

Cadmium + Sulphuric Acid ⟶ Cadmium Sulphate + Hydrogen

$Cd + H_2SO_4 \longrightarrow CdSO_4 + H_2$

It is also prepared in the anhydrous state by melting cadmium and potassium persulphate.

Cadmium sulphate is used for the fluorescent screen, which is used for the study of the spectral distribution of light emitted by different radiant sources. It is also used in the medicine as an antiseptic and astringent. Cadmium sulphate is also used as an electrolyte in the nickel-cadmium battery.

Cadmium Sulphide (CdS): It is a yellow solid occurs naturally as the mineral greenockite. It is not found in free state to become the direct source of this compound. Cadmium sulphide appears in the crystalline and amorphous form. The amorphous from is yellow, orange or brown in colour. Cadmium sulphide can be prepared by fusing the cadmium oxide with sulphur.

Cadimum Oxide + Sulphur ⟶ Cadmium Sulphate + Oxygen

$2CdO + 2S \quad 2CdS \quad O_2$

Cadmium sulphide is used as a pigment in yellow paints. It is used for colouring vulcanized rubber it is also used in the preparation of artists colour. With Ultramarine it is used for making a fluorescent pigment a green colour.

Cadmium Carbonate ($CdCO_3$): It can be prepared by the action of sodium carbonate and cadmium oxide under carefully controlled conditions.

Sodium Carbonate + Cadmium Oxide ⟶ Cadmium Carbonate + Sodium Oxide

$Na_2CO_3 + CdO \longrightarrow CdCO_3 + N_2O$

Cadmium carbonate is insoluble in water, but soluble in ammonia, potassium cyanide and dilute acids. It decomposes at 470°C.

Cadmium Chloride ($CdCl_2$): When metallic cadmium dissolves in hydrochloric acid cadmium chloride is produced.

Cadmium + Hydrochloric Acid ⟶ Cadmium Chloride + Hydrogen

$Cd + 2HCl \longrightarrow CdCl_2 + H_2$

It is colourless crystalline solid with a specific gravity of 4.05 and melts at 569°C and boils at 962°C. It is soluble in water, ethyl alcohol. It is used in photography dyeing and calico printing.

Cadmium Nitrate [$Cd(NO_3)_24H_2O$]: Cadmium nitrate is made by dissolving the oxide of cadmium in nitric acid and evaporated to the beginning of the crystallization.

$$\text{Cadmium Oxide} + \text{Nitric Acid} \longrightarrow \text{Cadmium Nitrate} + \text{Water}$$
$$CdO + 2HNO_3 \longrightarrow CdN(NO_3)_2 + 2H_2O$$

It is a white hydroscopic crystal, soluble in water, alcohol and ammonia.

Cadmium Iodide (CdI_2): It is obtained as colourless shinning crystals, when cadmium oxide reacts with hydroiodic acid and the solution is evaporated. Cadmium iodide is soluble in alcohol and acetone.

Cadmium Acetate [$Cd(CH_3COO)_2$]: When cadmium oxide reacts with acetic acid the compound is produced.

$$\text{Cadmium Oxide} + \text{Acetic Acid} \longrightarrow \text{Cadmium Acetate} + \text{Hydrogen}$$
$$CdO + 2CH_3COOH \longrightarrow Cd(CH_3COO)_2 + H_2$$

Colourless crystals with a melting point of 250°C.

It is used in ceramics, also used in dyeing and printing textiles and as a laboratory reagent.

Cadmium Bromate [$Cd(BrO_3)_2.H_2O$]: When barium bromate is allowed to react with cadmium sulphate cadmium bromate is produced.

$$\text{Barium Bromate} + \text{Cadmium Sulphate} \longrightarrow \text{Cadmium Bromate} + \text{Barium Sulphate}$$
$$Ba(BrO)_3 + CdSO_4 \longrightarrow Cd(BrO_3)_2 + BaSO_4$$

It is a white crystalline powder soluble in water and insoluble in alcohol. It is used is photography. It is dangerous when comes is contact with organic compound. It is highly toxic and irritant oxidizer.

Cadmium Chlorate, $Cd(ClO_3)_2$: When sodium chlorate reacts with cadmium sulphate cadmium chlorate is produced.

$$\text{Cadmium Sulphate} + \text{Sodium Chlorate} \longrightarrow \text{Cadmium Chlorate} + \text{Sodium Sulphate}$$
$$CdSO_4 + 2NaClO_3 \longrightarrow Cd(ClO)_3 + Na_2SO_4$$

It is a colourless hydroscopic crystal, soluble in alcohol and water and also in acetone. It is also dangerous when comes in contact with organic matter.

Detection and Estimation

To a solution of cadmium salt a solution of potassium cyanide is added to the sample and then some hydrogen sulphide is allowed to pass, an yellow precipitate of cadmium sulphide will be produced.

TIN

Melting Point : 232°C
Boiling Point : 2260°C
Atomic Number : 50
Atomic Weight : 118.710

History and Occurrence

It is supposed that some Egyptian Hierographics and the word bechil refer in the Old Testament is thin and tin articles occur at least as early as the Eighteen Dinestry (1400B.C)

Egypt. The metal is mentioned by Homer as Hassiteres and Pliny speaks about Plumbum nigrum (Black lead) and plumbum candicum (Shining lead or tin) observing that the later was brought from Islands and the Island Iktes in the coast of Britain which (according to Diodorous Siculus) was separated from the mainland only at high water, is St Michels Mount, Cornwell. The metal was afterwards called Stannum.

Traces of the element does exist in Siberia, Guinea, Bolivia and New South Wales. But the only important source is the ore *tinstone or Cassiterite* the crystalline SnO_2 is found in alluvial deposits. The chief ore is tinstone, which is also, exists at small proportions in Topaz, Zircon and Monazite ores.

Extraction

This is extracted from tinstone, which contains 10% of tin. The following steps are taken for extraction:

1. Crushing and Concentration: The ore is crushed and finely powdered in ball mills. This then concentrated by washing is current of water and after that the major portion of the liquid paste is removed by electromagnetic separation.

2. Roasting: The concentrated ore is roasted in an inclined revolving furnace when sulphur arsenic is removed as sulphur dioxide and arsenic trioxide. While arsenic trioxide As_4O_6 volatilizes and its vapour are cooled and collected in special chambers iron and copper pyrite are oxidized to the corresponding oxides and sulphates.

3. Washing: The roasted ore is washed with water to remove copper sulphate and iron sulphate, which dissolve. The light ferric oxide and other light matter are washed away. The washed ore contains 60-75% of tin as oxide and is called Black Tin.

4. Smelting: Black tin is mixed with 1/5th its weight of powdered anthracite coal and smelted in a reverberatory furnace between 1250°C to 1350°C. A little lime or fluspar is used as flux material if silica is present. Tin oxide is reduced while impurities if any are removed off as slag. Excess of lime must be avoided; otherwise it would react with tin to form calcium stannate.

$$\text{Stannous Oxide} + \text{Carbon} \longrightarrow \text{Tin} + \text{Carbon Monoxide}$$

$$SnO_2 + 2C \longrightarrow Sn + 2CO$$

The molten metal which collects on the bottom of the furnace is dropped off and cast into blocks or ingots. It is called block tin and contains about 99% of metallic tin. The slag obtained also contains about 35-45% of the metal and arsenic as impurities. These can be removed by purification.

5. Purification: Black tin contains iron, lead, sulphur and arsenic as impurities which are removed by the following process:

(a) ***Liqufication:*** Black tin is heated at 300°C on the sloping hearth furnace (reverbatory). When tin melts and flows down leaving the impurities behind.

(b) ***Poling:*** The metal purified above is melted and stirred with green poles. The gases evolved bring the impurities to the surface where they are oxidized by air. The impurities, which collect on the surface, are skimmed off. The refinery impurities are re-smelted with ore to recover the large percentage of tin they contain.

(c) ***Electrolytic Purification:*** Tin may be purified still further by electrolysis. In a cell containing hydroflurosilicic acid (H_2F_6Si) tin sulphate and sulphuric acid as electrolyte, block tin is made the anode white cathode consists of pure tin sheets. On passing electric current tin is dissolved from anode and gets deposited on the cathode.

Physical Properties

Melting Point	232°C
Boiling Point	2260°C
Density	7.29 g/cc
Atomic Volume	16.3 cc
Standard Electrode Potential	–0.136 V
Specific gravity (grey tin)	5.77
(white tin)	7.29
Latent Heat of Fusion	14.2 cal/g
Latent Heat of Vapourization	532 cal/g
Heat of Transformation	4.2 cal/g
Thermal Conductivity (white Tin) at 0°C	0.150
Co-efficient of Linear Expansion	19.9×10^{-6}
Shrinkage on Solidification	2.8%
Resitivity of white tin at 0°C	$11.0\ \mu\Omega/cm^2$
Brinell hardness 10 kg/5mm (180 sec)	3.9
Tensile Strength at 15°C	2100
Standard Free Energy	$1.1\ kcal/mol^{-1}$
Standard Heat of Formation	$0.6\ kcal/mol^{-1}$
Heat of Capacity	$6.16\ cal\ deg^{-1}\ mol^{-1}$
Standard Entropy	$10.7\ cal\ deg^{-1}\ mol^{-1}$

Chemical Properties of Tin

1. Action of Air: Air has action on tin at 1600°C when white hot, but not at ordinary temperature, in presence of oxygen at that high temperature it burns with a bright flame giving stannic oxide.

2. Action of Halogens and Sulphur: If the metal is heated in the atmosphere of chlorine or sulphur vapour, it readily combines to form stannic chloride and stannic sulphide, respectively.

$$\underset{\text{Tin}}{Sn} + \underset{\text{Chlorine}}{2Cl_2} \longrightarrow \underset{\text{Stannic Chloride}}{SnCl_4}$$

$$\underset{\text{Tin}}{Sn} + \underset{\text{Sulphur}}{2S} \longrightarrow \underset{\text{Stannous Sulphide}}{SnS_2}$$

3. Action of Acids: Tin is slowly attacked by dilute hydrochloric acid but the reaction rapid on heating with concentrated hydrochloric acid forming stannous chloride and evolving hydrogen.

$$\text{Tin} + \text{Hydrochloric Acid} \longrightarrow \text{Stannous Chloride} + \text{Hydrogen}$$
$$Sn + 2HCl \longrightarrow SnCl_2 + H_2$$

Dilute sulphuric acid has no action but hot concentrated acid dissolves the metal forming stannous sulphate and evolving sulphur dioxide.

$$\text{Tin} + \text{Nitric Acid} \longrightarrow \text{Stannous Sulphate} + \text{Sulphur Dioxide} + \text{Water}$$
$$4Sn + 10HNO_3 \longrightarrow SnSO_4 + SO_2 + 3H_2O$$

Dilute Nitric acid attacks the metal forming stannous nitrate and ammonium nitrate.

$$\text{Tin} + \text{Nitric Acid} \longrightarrow \text{Stannous Nitrate} + \text{Ammonium Nitrate} + \text{Water}$$
$$4Sn + 10HNO_3 \longrightarrow 4Sn(NO_3)_2 + NH_4NO_2 + 3H_2O$$

Hot concentrated nitric acid produces heavy fumes of nitrogen dioxide and meta-stannic acid.

$$\text{Stannous Oxide} + \text{Nitric Acid} \longrightarrow \text{Metastannic Acid} + \text{Nitrogen Dioxide}$$
$$2SnO + 4HNO_3 \longrightarrow 2H_2SnO_3 + 4NO_2$$

Organic acid has no action on tin. That is why tin cans are used for food packing.

4. Action of Alkalis: When the metal is heated with alkali solution it reacts to evolve hydrogen.

$$\text{Tin} + \text{Sodium Hydroxide} + \text{Water} \longrightarrow \text{Sodium Stannate} + \text{Hydrogen}$$
$$Sn + 2NaOH + H_2O \longrightarrow Na_2SnO_3 + 2H_2$$

Name	Percentge Composition	Uses
Solder	Tin 67% Lead 33%	In Soldering
Pewter	Tin 75% Lead 25%	In making Tin utensils
Babbit Metal	Tin 90% Antimony 75 Copper 3%	Machinery Parts
Britannia Metal	Tin 90% Antimony 8% Copper 2%	For Making Crockery
White Metal	Tin 82% Antimony 12% Copper 6%	" " "
Bell Metal	Tin 25% Copper 75%	For making Bell and Gongs
Rose Metal	Tin 28% Lead 22% Bismuth 50%	For making Fuse wires

It is used in the preparation of collapsible tubes for toothpaste and other ointments.

The metal is extensively used in tinning brass utensils. The vessel to be tinned is perfectly cleared and heated. A pinch of ammonium chloride is sprinkled to remove the oxide film. A small amount of the metal is then fused on the clean surface and rubbed over it with the help of a cotton rag until a uniform layer of tin is formed on the surface. To avoid the formation of any oxide, which dulls the lusture, the vessel is immediately dipped in cold water.

It is largely used in forming a protective coating over iron. The process is known as *tin plating*. It consists in dipping the clean sheets of iron or steel in a bath of molten tin, when the sheet is removed the tin is deposited as a layer on the surface of the sheet. Such sheets are used for making tin cans, kerosene oil containers and other similar items.

Tin amalgam is used for making mirrors.

Tin foils are used for rapping cigarettes and other articles.

Compounds of Tin

Stannous Oxide (SnO): This compound is obtained as a dark grey powder by heating stannous hydroxide or stannous oxalate in absence of air.

Stannous Oxalate ⟶ Stannous Oxide + Carbon + Carbon Dioxide

$$SnC_2O_4 \longrightarrow SnO + C + CO_2$$

Stannous hydroxide is formed as a white precipitate by the action of sodium hydroxide on solution of stannous chloride. It has a formula of $3SnO.2H_2O$.Stannous oxide and hydrated stannous oxide dissolve in acid to form stannous salts and in alkalis to form stannates, e.g. Sodium stannates, Na_2SnO_3; these are powerful reducing agents. Highly concentrated alkali converts stannous oxide to a stannate and spongy tin.

Stannous Oxide + Sodium Hydroxide ⟶ Sodium Stannate + Tin + Water

$$SnO + 2NaOH \longrightarrow Na_2SnO_3 + Sn + H_2O$$

Dry stannous is readily ionized and when exposed to air it changes to stannic oxide, SnO_2.

Stannous Oxide + Oxygen ⟶ Stannic Oxide

$$2SnO + O_2 \longrightarrow 2SnO_2$$

Stannous Chloride ($SnCl_2$): Tin foil or granulated tin readily dissolves in hydrochloric acid with formation of stannous chloride. The solution on cooling deposit crystals of the salt, hydrated stannous chloride ($SnCl_22H_2O$). This is also known as *tin salt.* On heating the hydrated salt decomposes with loss of hydrogen chloride and the formation of a basic salt. The anhydrous salt can be prepared by heating tin in a current of hydrogen chloride.

Tin + Hydrochloric Acid ⟶ Stannous Chloride

$$Sn + HCl \longrightarrow SnCl_2$$

Anhydrous stannous chloride forms a transparent glossy mass having a melting point of 247°C. It is soluble in alcohol and ether. It dissolves in a small quantity of water, but is readily hydrolyzed with more water to form a white precipitate of the basic stannous oxychloride ($2Sn(OH)Cl.\ H_2O$). A clear solution of the salt is obtained in the presence of hydrochloric acid. The acid solution becomes turbid due to oxidation, until metallic tin was previously added; stannic chloride is precipitated and stannic chloride remains in solution.

For this reason stannous chloride is used with hydrochloric acid to reduce oxidation.

Stannous Chloride + Water + Oxygen ⟶ Stannic Chloride + Stannous Oxychloride

$$6SnCl_2 + H_2O + 2H_2O \longrightarrow 2SnCl_2 + 4Sn(OH)Cl$$

It is already stated that stannous chloride is a powerful reducing agent. In strong hydrochloric acid it forms stannic acid (H_2SnO_4) in solution, from which crystals of the hydrated chlorostannous $HSnCl_3H_2O$ may be obtained. Salts of chlorostannous acid have been prepared for example ammonium chlorostannate [$(NH_4)_2SnC_42H_2O$]. Ammonium chlorostannate is also known as *pink salt* and is used as mordent.

Stannous Sulphide (SnS): It is obtained as a grey crystalline mass on heating tin with sulphur, or as brown precipitate by passing hydrogen sulphide into an acid solution of stannous chloride. The sulphide is soluble in concentrated hydrochloric acid, forming stannous chloride and yellow ammonium sulphide. It dissolves to form ammonium thiostannate [$(NH_4)_2SNS_3$].

Stannous Sulphide + Ammonium Sulphide ⟶ Ammonium Thiostannate + Sulphur

$$SnS + (NH_4)_2S_2 \longrightarrow (NH_4)_2SnS_3 + S$$

From which the hydrochloric acid precipitates stannic sulphide.

Ammonium Thiostannate + Hydrochloric Acid ⟶ Stannic Sulphide + Ammonium Chloric + Hydrogen Sulphide

$$(NH_4)_2SnS_3 + 2HCl \longrightarrow SnS_2 + NH_4Cl + H_2S$$

Stannous Sulphide is first oxidized to stannic sulphide ammonium sulphide or polysulphide by sulphur present in yellow ammonium sulphide or polysulphide

Stannic Hydride (SnH_4): It is formed in very small quantities mixed with hydrogen by the action of hydrochloric acid on alloy of tin and magnesium or by electrodes. Tin hydride is a very unstable compound and decomposes rapidly into tin and hydrogen.

Stannic Hydroxide [$Sn(OH)_4$]: Hydrated stannic oxide as it is also called, appears m two forms, both of which are distinguished as α-stannic acid and β-stannic acids.

Alpha Stannic Acid: It is obtained as white gelatinous precipitate by a stannate solutions. It is also known as orthostannic acid and has a formula of $Sn(OH)_4$ or H_2SnO_3. The precipitate is soluble m dilute acids. The solution does not gelatinize on boiling but slowly deposits beta stannic acid on standing.

Beta Stannic Acid: This obtained as cardy white precipitate when tm reacts with concentrated nitric acid, it is also known as *meta-stannic acid.* This compound is soluble m nitric acid or concentrated sulphuric acid. Concentrated hydrochloric acid slowly changes it into a substance, which dissolves in water to form a colloidal solution. Beta-stannic acid is soluble in alkali hydroxide solutions to form the corresponding alkali, alpha-stannate is produced. The composition of the meta-stannic acid or beta stannic acid represented by the formula $H_2Sn_5O_{11}.4H_2O$ and the corresponding salt by $Na_2Sn_5O_{11}. 4H_2O$.

Stannic Chloride ($SnCl_4$): It is prepared by passing a current of chlorine over metallic tin. It forms a colourless mobile strongly fuming fluid, which is purified by distillation. It boils at 114°C, with a small quantity of water it forms a clear solution, from which hydrated crystals which 3, 5 and $8H_2O$ can be obtained. It is used as a mordent with excess of water it is hydrolyzed to stannic hydroxide.

Stannic Chloride + Water ⟶ Stannic Hydroxide + Hydrochloric Acid

$$SnCl_4 + H_2O \longrightarrow Sn(OH)_4 + 3HCl$$

The anhydrous chloride is soluble in organic solvents like alcohol and ether. Hydrated stannic chlorine combines with hydrogen chloride in the cold state to form crystals of hydrochlorostannic acid ($H_2SnCl_6H_2O$) it produces chlorostannates when reacts with ammonia and phosphorous pentachloride to from respectively to form respectively the compounds $SnCl_44NH_3$ and $SnCl_4PCl_5$.

Stannic Sulphide (SnS): When hydrogen sulphide is passed through a solution of stannic salt, the yellow precipitate of stannic sulphide (SnS_2) is formed. This becomes black on dying. The precipitate formed is pure and consists of a mixture of stannic oxide and sulphide. Pure crystalline golden-yellow, shinning scales of stannic sulphide, known as *mosaic gold* is obtained, by heating a mixture of tin fillings, sulphur and ammonium chloride.

Tin + Ammonium Chloride ⟶ Ammonium Chlorostannate + Ammonia + Hydrogen

$$Sn + 4NH_4Cl \longrightarrow (NH_4)_2SnCl_4 + 2NH_3 + H_2$$

Ammonium Chlorostannate + Sulphur ⟶ Stannic Sulphide + Ammonium Hexachlorostannate + Ammonium Chloride

$$2(NH_4)_2SnCl_4 + 2S \longrightarrow SnS + (NH_4)_2SnCl_6 + 2NH_4Cl$$

Mosaic gold is used for gliding. It is insoluble m acids, but dissolves m alkalis to forms a mixture of stannate and thiostannate and m alkali sulphide to form a thiostannate.

Stannic Chloride + Water ⟶ Stannic Hydroxide + Hydroichloric Acid

$$SnCl_4 + H_2O \longrightarrow Sn(OH)_4 + 3HCl$$

Stannic Sulphide + Sodium Sulphide ⟶ Sodium Thiostannate

$$3SnS_2 + Na_2S \longrightarrow Na_2SnS_3$$

Detection and Estimation

1. When the sample containing any tin compound is placed in crucible covered with a few ml of hydrochloric acid and a test tube containing cold water is dropped in the solution mixture, and the mixture is heated it shows fluorescence, this is the most sensitive test of the tin.
2. To a solution of a tin compound a portion of mercuric chloride is added which turns white due to the formation of mercurous chloride.

Tin is estimated by the deposition of Meta stannic acid.

TUNGSTEN

Melting Point : 3410°C
Boiling Point : 5927°C
Atomic Number : 74
Atomic Weight : 183.85

History and Occurrence

Scheele in 1871 discovered a new metallic oxide combined with lime in a mineral, was known as tungstein (heavy stone), now known as Scheelite after the name of Scheele. In 1883, the two Spanish chemist brothers d'Elhuyer found the same oxide in the mineral wolframite, where it was associated with iron and manganese oxides. They succeeded in extracting metallic tungsten from the mineral.

The two important ores of tungsten are wolframite (Fe, Mn)WO_4 and Scheelite, $CaWO_4$. They are chiefly found in Myamar, Siam, Malay, States of Korea, Australia and U.S.A. But more than 50% of world's supply come from China.

Extraction

For the extraction of the metal the minerals are crushed and roasted with sodium carbonate. The mass is dissolves while iron, manganese or calcium remain behind in the form of thin oxide or carbonate. The solution is treated with acids to precipitate the hydrated trioxide (WO_3), which is washed and dehydrated by heating. The oxide is then reduced to the metal by heating with carbon or in a current of hydrogen or by Thermit Process. The metal can also be obtained by the electrolysis of a solution of tungstic acid in a fused mixture of alkali metal chlorides. On account of its very high melting point the metal is generally obtained in a power form.

Wolframite + Sodium Carbonate + Oxygen ⟶ Sodium Woeframite + Ferric Oxide + Carbon Dioxide

$$2FeWO_4 + Na_2CO_3 + \tfrac{1}{2} O_2 \longrightarrow 2Na_2WO_4 + Fe_2O_3 + 2CO_2$$

Sodium Wolframite + Hydrogen Chloride ⟶ Sodium Chloride + Tungstic Acid

$$Na_2WO_4 + 2HCl \longrightarrow 2NaCl + H_2WO_4$$

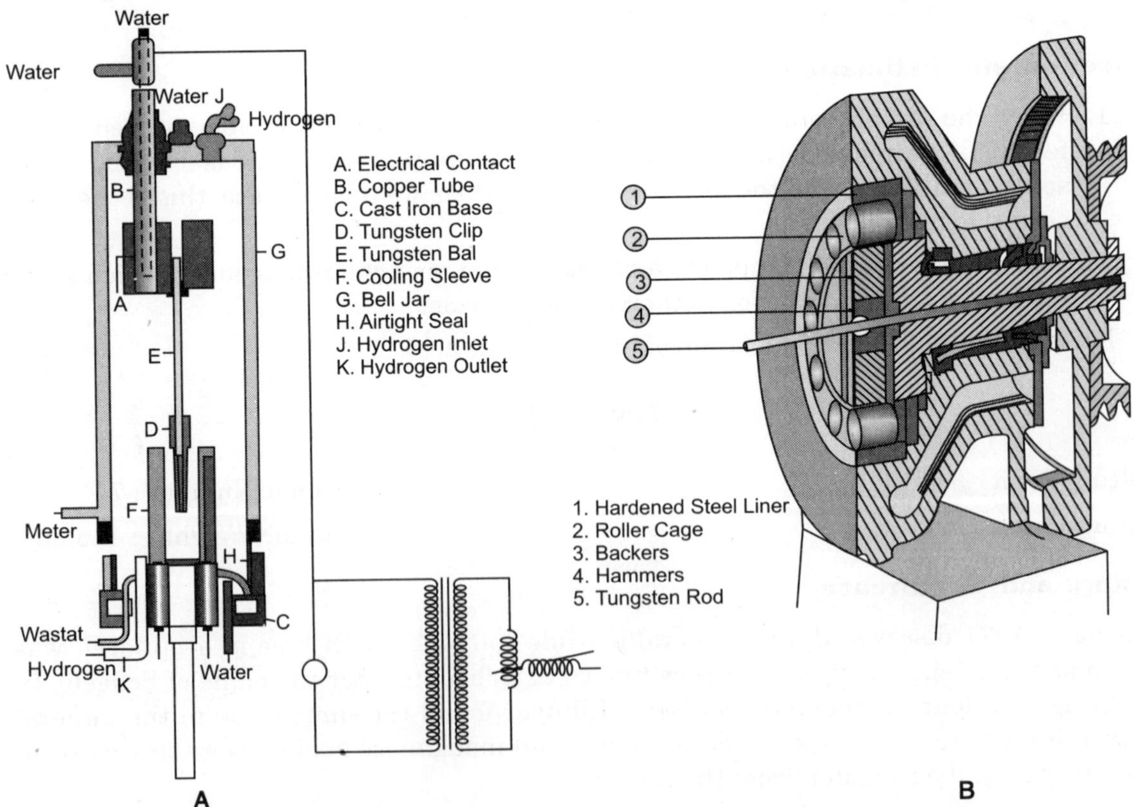

The Tungsten Preparation in Industry and Tungsten Wire Preparation

Tungstic Acid ⟶ Tungsten Trioxide + Water

$$H_2WO_4 \longrightarrow WO_3 + H_2O$$

Tungsten Trioxide + Hydrogen ⟶ Tungsten + Carbon Monoxide

$$WO_3 + 3H_2 \longrightarrow W + 3CO$$

Tungsten Trioxide + Carbon ⟶ Tungsten + Carbon Monoxide

$$WO_3 + 3C \longrightarrow W + 3Cl_2$$

Very pure tungsten can be obtained by the decomposition hexachloride (WCl_6) vapour or glowing tungsten wire.

Tunsten Hexachloride ⟶ Tungsten + Chlorine

$$WCl_6 \longrightarrow W + 3Cl_2$$

For making filaments of electric lamp the tungsten powder, obtained by the reduction of the oxide in hydrogen, is pressed into rods; these are sintered by heating to 2500°C in hydrogen. Then rolled and hammered at a very high temperature and finally drawn through fine bores in a piece of diamond or of tungsten carbide at about 600°C.

Physical Properties

Melting Point	3410°C
Boiling Point	5927°C
Density	19.3 g/cc
Outer Electronic Configuration	$5d^46s^2$
Specific Heat	0.032
Vapour Pressure	6.5×10^{-7} Pa
Electrical Resistivity	5.65 μΩ/cm
Thermal Conductivity	0.43 cal/sec/sqcm/°C/cm
Atomic Volume	9.5 cc
Atomic Radius	1.39Å
First Ionisation Energy	184 Kcal/mol
Stable Oxidation State	+6
Heat of Vapourization	185 cal/mol
Heat of Fusion	8.0 Kcal/mol

Chemical Properties

Tungsten is not affected by air at room temperature, but when in powder form, it may by pyrophoric. Tungsten begins to oxidize at 400°C.

Tungsten resists water and rapidly oxidizes at red heat. It is attacked quite readily by molten sodium nitrate and this may be used for etching pr pointing rod or wire for drawing.

Dry chlorine free from air, attacks tungsten at about 250 – 300°C forming the hexachloride, but if air or moisture is present the oxychloride are formed. Fluorine reacts with tungsten at ordinary temperature. The pentabromide is formed is formed at red-hot condition of the tungsten in the absence of moisture. The di-iodide is formed when tungsten at red-hot condition and treated with iodine.

Tungsten is not attracted by hydrofluoric acid. Warm nitric acid easily oxidizes tungsten to the yellow trioxide WO_3. Hot concentrated sulphuric acid reacts only slightly with the metal. A mixture of nitric acid and hydrochloric acid react with tungsten to form trioxide. Aquaregia forms a superficial* oxide on the tungsten at room temperature.

Tungsten is resistant to molten sodium hydroxide, but the presence of hydracids will cause rapid corrosion. Alkali solutions and ammonia do not corrode tungsten in the absence of oxygen.

The gases like carbon dioxide, carbon disulphide, sulphur vapour, nitric oxide and nitrogen dioxide reacts with tungsten at high temperature. Molten sulphur vapour reacts with tungsten at high temperatures. Nitrogen does not attack tungsten upto 1500°C but the dinitride is formed when nitrogen is present in an incandescent lamp (tungsten filament). At 2300°C hydrogen does not react with tungsten. Molten nitrates and peroxide reacts violently with tungsten.

Compounds of Tungsten

Tungsten Oxide, Tungsten Trioxide (WO_3): It is the most important oxide of tungsten which is extracted from the tungsten ore (as described before). In pure state it can be prepared by strongly heated ammonium tungstate in air. It forms a dense yellow powder insoluble in water and acids, but dissolves in alkalis to form tungstates.

When a tungstate solution is treated with acids in the dilute cold state, a white precipitate of the hydrated trioxide or tungstic acid of the composition $WO_3.2H_2O$ is first formed this becomes yellow, when the solution is heated to form WO_3H_2O or H_2WO_4. These substances give the trioxide on gentle ignition. Tungstic acid resembles molybdic acid in the series of salts; polytungstens derived from these in the presence of hydrogen ion; Na_2WO_4 and $Na_2W_4O_{13}$ are known as *common* tungstates.

Tungsten Pentoxide (W_2O_5): It is obtained as a blue substance by heating the trioxide gently in a current of hydrogen or by heating it with metallic tungsten at about 1000°C. It is difficult to obtained when a tungstate is reduced. A similar intense blue coloured product is produced when a tungstate is reduced by stannous chloride in hydrochloric acid solution. This usually known as tungsten blue and serves as a delicate test of tungsten

This is the most sensitive test of tungsten

Tungsten Dioxide (WO_2): It is formed as a brown powder by reduction of the trioxide in a current of hydrogen at about 900°C. A product of the same composition is also obtained in red leaflets by reducing a solution of any tungstate compound, with zinc and hydrochloric acid. The dioxide is insoluble in water, but contains quadravalent tungstate salts with reducing properties.

* Superficial: Surface.

Tungsten Hexachloride (WCl_6): It is formed in black or dark violet crystals melting at 275°C when tungsten is heated in dry air free chlorine. It is particularly insoluble in cold water, but is decomposed by warm water. It tends to form oxychloride (WOC_4) and WO_2Cl_2.

Tungsten Pentachloride (WCl_5): It is formed by the thermal dissociation of the hexachloride. It is obtained in dark-green shinning needles by repeated distillation of the hexachloride strongly in a current of hydrogen; it is less volatile than the higher chlorides and is hydrolyzed water to the dioxide. The trichloride is not known but the double chloride like $K_3W_2Cl_6$; are obtained as greenish-yellow crystals by reducing alkali tungstate solutions in hydrochloric acid with tin. They are strong reducing agents and contain tetravalent tungsten.

Tungsten Carbide (WC): It is prepared by the reduction of the WO_3 with carbon or carbon dioxide at high temperature.

$$\text{Tungsten Trioxide + Carbon Monoxide} \xrightarrow{1500°C} \text{Tungsten Carbide + Oxygen}$$

$$WO_3 + CO \longrightarrow WC + 2O_2$$

Tungsten Dichloride (WCl_2): It is formed as an amorphous, grey powder by the reaction of tungsten hexachloride and dry hydrogen. It is a powerful reducing agent and is extremely hydroscopic. In air and water it suffers decomposition even at room temperature.

$$\text{Tungsten Dichloride + Water} \longrightarrow \text{Tungsten Dioxide + Hydrogen Chloride + Hydrogen}$$

$$WCl_2 + H_2O \longrightarrow WO_3 + HCl + H_2$$

Complex tungstacyanic acids and their salts in which tungsten shows a valency of four and five are known, similar to those of molybdenum. Also pertungstates like Na_2WO_5 have been prepared in the crystalline state by the action of hydrogen peroxide on alkali tungstate.

Tungsten Bromide (WBr_2): It is prepared by heating tungsten and bromine in an electric arc furnace.

$$\text{Tunsten + Bromine} \longrightarrow \text{Tungsten Bromide}$$

$$W + Br_2 \longrightarrow WBr_2$$

It is silver white solid, melting point of 2900°C and insoluble in water but soluble in aquaregia. It is decomposed by chlorine at 100°C with the formation of tungsten dichloride.

It is used for refractory works in furnace

$$\text{Tungsten Bromide + Chlorine} \longrightarrow \text{Tungsten Dichloride + Bromine}$$

$$WBr_2 + Cl_2 \longrightarrow WCl_2 + Br_2$$

Tungsten Diselinide (WSe_2): It is prepared by the direct combination of tungsten and selenium. It has high and low temperature stability. It is used in different industrial activities.

Tungsten Disulphide (WS_2): Tungsten and sulphur when heated at high temperature a grayish-black solid called tungsten disulphide is produced. It is having a melting point of 1840°C. It is attacked by fluorine and hot sulphuric acid.

Tungten Disulphide + Fluorine ⟶ Tungsten Fluoride + Sulphur

$$WS_2 + F_2 \longrightarrow 2WF + S$$

Tunsgen Disulphide + Sulphuric Acid ⟶ Tunsten Sulphate + Hydrogen Sulphide + Sulphur

$$WS_2 + H_2SO_4 \longrightarrow WSO_4 + 4S + S$$

It is used as a lubricant and aerosol.

Tungsten Hexafluoride (WF_6): It is prepared by distillation under pressure.

Tungsten + Fluorine ⟶ Tungsten Hexafluride

$$W + 3F_2 \longrightarrow WF_6$$

It is a colourless gas or light yellow liquid with a melting point of 2.5°C and a boiling point of 19.5°C, it is decomposes in water. It is used as fluorinating agent.

Tungsten Oxychloride (WOCl): It is prepared by the action of chlorine on tungsten trioxide.

Tungsten Trioxide + Chlorine ⟶ Tungsten Oxychloride + Oxygen

$$WO_3 + 2Cl_2 \longrightarrow WOCl_2 + O_2$$

It is a dark red acicular* crystals, gets decomposed in water or moisture, that is why it is kept in tightly stoppered bottle. It melts at 210°C and boils at 228°C. It is soluble in carbon disulphide.

Tungstic Acid, Wolframic Acid, Orthotungstic Acid (H_2WO_4): It is prepared by the reaction of sodium tungstate and sulphuric acid.

Sodium Tungstate + Sulphuric Acid ⟶ Tungstic Acid + Sodium Sulphate

$$Na_2WO_4 + H_2SO_4 \longrightarrow H_2WO_4 + Na_2SO_4$$

It is a yellow powder soluble in alkalis and hydrogen fluoride. It is used as colourgard in plastics and used for the preparation of tungsten wires.

Sodium Tungstate (Na_2WO_4): It is prepared by dissolving tungsten trioxide in ceramic soda.

Tungsten Trioxide + Sodium Hydroxide ⟶ Sodium Tungstate + Water

$$WO_3 + 2NaOH \longrightarrow Na_2WO_4 + H_2O$$

It is obtained as colourless crystals soluble in water; insoluble in acids and alcohol, the melting point of the compound is 695°C. It is non-combustible. It is used for making tungsten compounds and fireproof fabrics and cellulose.

Barium Tungstate ($BaWO_4$): It is prepared by the action of barium sulphate and tungstic acid at high temperature.

Tungstic Acid + Barium Sulphate ⟶ Barium Tunsgtate + Sulphuric Acid

$$H_2WO_4 + BaSO_4 \longrightarrow BaWO_4 + H_2SO_4$$

---

* Acicular: Needle like sharp.

It is white powder insoluble in water. It is used as a pigment in X-Ray and for manufacturing fluorescent screens.

Estimation and Detection

When a tungsten salt is heated with hydrochloric acid, a white precipitate is produced, when the precipitate mixture is boiled it is converted to yellow anhydrous tungstic acid. It an alkali tungstate is treated with hydrochloric acid and aluminium it turns blue.

OSMIUM

Melting Point : 3000°C | Atomic Number : 76
Boiling Point : 5500°C | Atomic Weight : 190.2

Occurrence

Osmium occurs in crude platinum and in *osmoiridium* and alloy of osmium and iridium. Osmium is a bluish-grey metal. The physical properties are being stated below.

Physical Properties

Melting Point	3000°C
Boiling Point	5500°C
Density	22.5 g/cc
Linear Coefficient of Thermal Expansion	5.8×10^{-6}
Specific Heat at 0°C	0.031 cal/g
Vapour Pressure at melting point	13.5 microns
Electrical Resistivity at 0°C	8.8 microohm
Temperature Coefficient of Electrical Resitivity	0.0042/°C
Mass Susceptibility	0.04
Young's Modulus of Elasticity	558 MPa
Vickers Hardness	400

Chemical Properties

Osmium oxidizes very readily, producing the poisonous and volatile tetraoxide. It is resistant to non-oxidizing agents specially acids at room temperature but is generally inferior to ruthenium in corrosion resistance. Nitric acid oxidizes it to the volatile tetraoxide and this behavior is unique. Oxidizing alkaline solutions such as hypochlorites attack osmium. Fused sodium peroxide and potassium nitrate plus potassium hydroxide also effectively attack osmium. The resulting osmate can be treated with chlorine or nitric acid to yield osmium tetraoxide, which is readily distilled and recovered by adsorption. In general osmium compounds are reduced to the metal by heating in hydrogen.

Uses

Osmium alloys are noted principally for their hardness and natural and synthetic osmium alloys have been used for tipping fountain-pen nibs and for long life phonograph pins and pivots. At present osmium is replaced by ruthenium as because it is harder than osmium.

Compounds of Osmium

Osmium Halides

Osmium Tetrafluoride (OsF_4): This compound is prepared by the direct reaction of osmium and limited amount of fluorine.

$$\text{Osmium + Fluorine} \longrightarrow \text{Osmium Fluoride}$$
$$Os + 2F_2 \longrightarrow OsF_4$$

It is black solid and decomposes in water.

Osmium Hexafluoride (OsF_6): This compound is also prepared by the direct action of fluorine and osmium.

$$\text{Osmium + Fluorine} \longrightarrow \text{Osmium Fluoride}$$
$$Os + 3F_2 \longrightarrow OsF_6$$

It has a boiling point of 200°C. It is intermediately volatile in between the tetra and octafluoride and thus can be separated from the mixture of fluorides.

Osmium Octafluorides (OsF_8): It is also prepared from the direct reaction between fluorine and osmium.

$$\text{Osmium + Fluorine} \longrightarrow \text{Osmium Hexafluoride}$$
$$Os + 2F \longrightarrow OsF_8$$

This compound is extremely reactive and also volatile having a boiling point at 47°C. It is also extremely irritating. It forms a relatively stable combination with alkalines and alkali fluorides.

Osmium Dichloride ($OsCl_2$): This compound is obtained from the osmium trichloride by heating it in vacuum at a temperature of 500°C.

$$\text{Osmium Trichloride} \longrightarrow \text{Osmium Dichloride + Chlorine}$$
$$2OsCl_3 \longrightarrow OsCl_2 + Cl_2$$

It can also be produced by heating ammonium chloroosmate and chlorine at a temperature of 350°C.

$$\text{Ammonium Chlorosmate} \longrightarrow \text{Ammonium Chloride + Osmium Dichloride + Chlorine}$$
$$2(NH_4)_2OsCl_6 \longrightarrow 4NH_4Cl + 2OsCl_2 + 2Cl_2$$

The solution in water or in alcohol is very much stable.

Osmium Tetrachloride ($OsCl_4$): It is produced from the elements at a temperature of 650°C. It is insoluble but slowly gets decomposed by water.

Hexachlorosmic Acid (H_2OsCl_6): It is prepared by reaction of hydrochloric acid and osmium tetrachloride.

$$\text{Osmium Tetrachloride + Hydrochloric Acid} \longrightarrow \text{Hexachloroosmic Acid}$$
$$3OsCl_4 + 6HCl \longrightarrow 3H_2OsCl_6$$

Osmium Nitrate [$Os(NO_2)_3$]: It is brown soluble compound, prepared by the reaction of barium osmonitrate and dilute sulphuric acid.

Barium Osmonitrate + Sulphuric Acid ⟶ Osmium Nitrate + Barium Sulphate + Water

$$Ba_2Os(NO_2)_3 + 2H_2SO_4 \longrightarrow Os(NO_2)_3 + 2BaSO_4 + 2H_2O$$

Sodium Pentanitroosmate [$Na_2Os(NO_2)_5$]: This compound is prepared by the action of osmium trichloride and sodium nitrite; other alkali earth metal osmonitrates are also prepared in the same way.

Sodium Nitrite + Osmium Trichloride ⟶ Sodium Pentanitroosmate + Sodium Chloride

$$5NaNO_2 + OsCl_3 \longrightarrow Na_2Os(NO)_5 + 3NaCl$$

When potassium pentaosmonitrate is added to hydrochloric acid a stable complex is produced.

Potassium Pentaosmonitrate + Hydrochloric Acid + Oxygen ⟶ Potassium Nitroosmiumpentachloride + Nitric Acid

$$K_2Os(NO_2)_5 + 5HCl + 3O_2 \longrightarrow K_2[Os(NO)Cl_5] + 5HNO_3$$

Osmonitrates

Potassium Osmiamates [$K(OsO_3N)$]: This compound is produced by the reaction of potassium hydroxide and osmium tetraoxide and ammonia.

Potassium Hydroxide + Ammonia + Osmium Tetraoxide ⟶ Potassium Osmiamates + Water

$$KOH + NH_3 + OSO_4 \longrightarrow K(OSC_3N) + 2H_2O$$

Many of these osmiamates explodes, therefore any kind of provocation or manipulations like heating etc. should be avoided.

Osmium Oxides

Osmium Dioxide (OsO_2): This compound is produced by heating osmium in a limited quantity of oxygen or in osmium tetraoxide. By treating a solution of sodium osmate with ethyl alcohol as a reducing agent and then with sufficient sulphuric acid just to neutralize the black hydrated oxide, OsO_2.2HzO is precipitated. This should be dehydrated in an oxygen-free atmosphere as it may be pyrophoric, oxidizing to the volatile and poisonous tetraoxide.

Osmium Tetraoxide, Osmic Acid OsO_4: This is most important compound of osmium, prepared by the direct reaction of osmium and oxygen.

Osmium + Oxygen ⟶ Osmium Tetraoxide

$$Os + 2O_2 \longrightarrow OsO_4$$

It is very poisonous and irritating to eyes and mucous membranes also. In preparation of the compound, at first the osmium is fused with sodium peroxide and sodium perosmate,

$NaOsO_4$ is produced; it is dissolved in water and the washing residue along with sodium hypochlorite extracts osmium. The osmium solution also consists of ruthenium mixed with sodium hypochlorite can be treated with chlorine by slowly bubbling through. The evolved vapours pass through an all glass system consists of a series of absorption flasks, which is progressively warmed. The first four flasks contain progressively diluted hydrochloric acid, which serves to reduce the ruthenium. This osmium tetraoxide is carried through these flasks as they are progressively heated and the stream of chlorine is eventually replaced by an air current. The osmium is finally absorbed in an alcoholic sodium hydroxide solution, forming the osmate. The osmate may be converted to the hydrated dioxide and reduced to metal by heating in hydrogen and cooling in carbon dioxide.

Potassium Osmate ($K_2OsO_4.2H_2O$): It is obtained by the reaction of potassium hydroxide and sodium osmate. The potassium osmate separates, as its solubility in potassium hydroxide is very low.

Sodium Osmate + Potassium Hydroxide ⟶ Potassium Osmate + Sodium Hydroxide

$$Na_2OsO_4 + 2KOH \longrightarrow K_2OsO_4 + 2NaOH$$

Test and Estimation

Treatment of an osmate salt with ammonium chloride yield orange yellow complex, $OsO_2(NH_3)_4Cl_2$ which is easily reduced to osmium by hydrogen. This serves as the analytical test for osmium.

Also if Osmic acid is treated with a little coconut inside scales, it produces a black strain due to presence of oil in the coconut after washing.

IRIDIUM

Melting Point : 2443°C — Atomic Number : 77

Boiling Point : 4500°C — Atomic Weight : 192.2

Occurrence

Iridium occurs in crude platinum and stays with osmium as an alloy called *osmoiridium* or *iridosmine*.

It occurs in Ural, Bringeria (New South Wales) Brazil and Canada.

Physical Properties

Melting Point	2443°C
Boiling Point	4500°C
Density	22.5 g/cc
Linear Coefficient of Thermal Expansion at 20°C per °C	6.6×10^{-6}
Specific Heat at 0°C	0.031

Thermal Conductivity, at 20°C	cal/(sec) (sqcm) (°C/cm)
Vapour Pressure at Melting Point	3.5 microns
Electrical Resistivity	4.9 μΩ/cm
Temperature Coefficient for Electrical Resistivity	0.0042
Tensile Strength	2.48 MPa
Young's Modulus of Elasticity	5.24 MPa
Vickers's hardness	220
Isotopes	191, 193

Chemical Properties

Iridium oxidizes visibly on heating in air in the range 600–1000°C, but at higher temperatures it remains bright and looses weight rapidly. The presence of oxygen increase the rate of loss even in rather dilute alloys with platinum, so that the iridium-platinum alloys are not generally used at high temperatures in air or other oxidizing atmospheres. Iridium is not attracted by any of the acids and not also by aquaregia by it is attacked by molten salts like sodium chloride and sodium cyanide and is oxidized and made soluble in aquaregia by fusion with sodium peroxide or a mixture of caustic potash and potassium nitrate. Mercury does not attack iridium, at higher temperature, at a very high temperature iridium is soluble in lead but gradually precipitates when the temperature is lowered.

The resistance of pure iridium, high-iridium synthetic alloys and some natural minerals to attack by aqueous corrosives is so high that it has afforded a problem in dissolving everything but the iridium. After dissolving lead, the residue may be attracted by oxidizing alkline fusion, or heated with sodium chloride and chlorine.

An ahemative procedure for laboratory use is now available form the National Bureau of Standards, which involves a treatment with hydrochloric acid and chlorine in a sealed tube at 300°C and 2.79MPa pressure. Under these circumstances even pure iridium dissolves at a considerable high rate (160, mg/sqdm per day) and the resulting solution is free from complicating diluents.

Metallic iridium is obtained by heating appropriate complexes such as ammonium chloroiridate; final reduction with hydrogen may be required to produce an oxygen free product. Aqueous reducing agents do not reduce the iridium compounds to the metal but only to the trivalent state. It can be separated out of the solution by zinc or magnesium but this is not complete.

As pure iridium is quite hard, it can be only worked hot and even then with great difficulty. It can be melted in a thorium crucible, but there is no question about melting the pure metal, as because most of it is employed in the platinum alloys in which iridium powder readily dissolves in molten platinum. These alloys behave well on melting and processing and are even more resistant to corrosion than platinum at moderate temperatures.

Compounds of Iridium

Iridium Halides

Iridium Monochloride (IrCl): This compound is produced when iridium trichloride reacts with chlorine at 790-795°C, it is obtained as red crystals.

$$\text{Iridium Trichloride + Chlorine} \longrightarrow \text{Iridium Monochloride + Chlorine}$$

$$IrCl_3 + Cl_2 \longrightarrow IrCl + 2Cl_2$$

Iridium Trichloride ($IrCl_3$): When iridium in metallic state reacts with chlorine at 550°C this compound is produced as green crystals.

$$\text{Iridium + Chlorine} \longrightarrow \text{Iridium Trichloride}$$

$$2Ir + 3Cl_2 \longrightarrow 2IrCl_3$$

It decomposes in chlorine at 700°C. The hexachloroiridates such as $Na_3IrC_6\ 6H_2O$ are made by reducing the corresponding chloroiridate with alcohol or sulphur dioxide.

Iridium tetrachloride ($IrCl_4$): It is difficult to prepare and isolate and therefore unimportant. The related hexachloroiridates such as the very soluble brown $Na_2IrCl\ 6H_2O$ and the quite insoluble $(NI_4)2IrCl_6$ are important. The sodium salt is prepared by passing chloride over a mixture of the fine powder of iridium and sodium chloride, at red heat.

Iridium Tribromide ($IrBr_3$, $4H_2O$): It is obtained from a solution of iridium trihydroxide in hydrogen bromide.

$$\text{Iridium Trihydroxide + Hydrogen Bromide} \longrightarrow \text{Iridium Tribromide + Water}$$

$$Ir(OH)_3 + 3HBr \longrightarrow IBr_3 + 3H_2O$$

The solution also can be converted to green crystals, which are soluble in water.

Iridium Triiodide (Ir_3): This compound is also prepared like tribromide by dissolving iridium trihydroxide in hydrogen iodide and the compound is obtained as green solid. It is slightly soluble in water. The anhydrous compound is obtained by heating the hydrated salt in vacuum at a temperature of 200°C.

$$\text{Iridium Trihydroxide + Hydrogen Iodide} \longrightarrow \text{Iridium Triiodide + Water}$$

$$Ir(OH)_3 + HI \longrightarrow Irl_3 + H_2O$$

Iridium Tetrafluoride (IrF_4): It is a yellow oily liquid with a pungent odour, which decomposes in water. It is prepared by heating finely divided iridium powder with iridium hexafluoride. In a closed tube to about 200°C.

Iridium Hexafluoride (IrF_6): It is a volatile yellow solid with a melting point of 44°C and a boiling point of 53°C which reacts with water. It is formed when iridium powder directly reacts with fluorine at 260°C.

Iridium Nitro Compounds

Trisodium Hexanitroiridate, Sodium Iridonitrate [$Na_3Ir(NO_2)6H_2O$]: It is very much soluble in water. It results in boiling a solution of sodium chloroiridate with sodium nitrite. The ammonium and potassium salts are slightly soluble in cold water, but the solubility increases with the temperature.

Iridium Oxides and Hydroxides

Iridium Trihydroxide, $Ir(OH)_32H_2O$: It is a green or black solid which is insoluble in water. It can be prepared by adding potassium hydroxide to iridium trichloride.

Iridium Trichloride + PotassiumHydroxide ⟶ Iridium Hydroxide + Potassium Chloride

$$IrCl_3 + 3KOH \longrightarrow Ir(OH)_3 + 3KCl$$

It is the key compound for many iridium compounds. It also forms iridium sesquioxide, when potassium chloroiridate reacts with an alkali.

Potassium Chloroiridate + Potassium Hydroxide ⟶ Iridium Hydroxide + Potassium Chloride + Iridium Sesquioxide + Water

$$3K_2IrCl_6 + 12KOH \longrightarrow Ir(OH)_3 + 18KCl + I_2O_3 + 3H_2O$$

Iridium trihydroxide is also produced in the above reaction, but both of them gets decomposed into oxygen and the metal when made yellow hot.

Iridium Hydroxide ⟶ Iridium + Hydrogen + Oxygen

$$2Ir(OH)_3 \longrightarrow Ir_2 + 3H_2 + 3O_2$$

Iridium Sesquioxide ⟶ Iridium + Oxygen

$$2Ir_2O_3 \longrightarrow 2Ir_2 + 3O_2$$

Iridium Tetrahydroxide [$Ir(OH)_4$ or IrO_22H_2O]: It is produced when sodium bromate reacts with sodium chloroiridate in the presence of water.

Sodium Bromate + Sodium Chloro iridate + Water ⟶ Iridium Tetrahydroxide + Sodium Chloride + Hydrogen Chloride + Oxygen + Sodium Bromite

$$2NaBrO_3 + Na_2IrCl_6 + Ir(OH)_4 \longrightarrow Ir(OH)_4 + 2NaCl + 4HCl + O_2 + 2NaBrO_2$$

When iridium tetrahydroxide is heated in an atmosphere of carbon dioxide a black iridium dioxide is produced.

Iridium Tetrahydroxide + Carbon Dioxide ⟶ Iridium Oxide + Carbon Acid

$$Ir(OH)_4 + 2CO_2 \longrightarrow IrO_2 + 2H_2CO_3$$

At a considerably high temperature, all the oxygen is evolved, leaving the pure metal.

Iridium Sesquisulphate [$Ir_2(SO_4)_3$]: It is yellow solid, prepared by dissolving iridium trihydroxide in sulphuric acid.

Iridium Trihydroxide + Sulphuric Acid ⟶ Iriidium Sesquisulphate + Water

$$2Ir(OH)_3 + 3H_2SO_4 \longrightarrow Ir(SO_4)_3 + 6H_2O$$

Iridium sesquioxide is soluble in water and forms many alums with alkalis.

Iridium Sesquisulphide (Ir_2S_3): It is produced by the reaction of iridium trihydroxide and hydrogen sulphide.

$$\text{Iridium Trihydroxide} + \text{Hydrogen Sulphide} \longrightarrow \text{Iridium Sesquisulphide} + \text{Water}$$
$$2Ir(OH)_3 + H_2S \longrightarrow Ir_2S_3 + H_2O$$

Iridium Monosulphide (IrS): It is a black lustrous compound, which is insoluble in water acids and aquaregia. It is formed by heating the iridium disulphide and nitrogen (air) at a temperature of 700°C.

$$\text{Iridium Disulphide} \xrightarrow{\text{Nitrogen}} \text{Iridium Monosulphide} + \text{Sulphur}$$
$$IrS_2 \longrightarrow IrS + S$$

Iridium Disulphide (IrS_2): It is brown solid insoluble in water, this compound is formed when iridium is heated in a quartz pressure tube above 800°C. It is probable that one or two higher sulphates also exists under special conditions.

Iridium Sulphite [$Ir_2(SO_3)_3 6H_2O$]: It is yellow solid which is very slightly soluble in water. Treating a suspension of trihydroxide of iridium and sulphur dioxide produces it.

$$\text{Iridium Trihydroxide} + \text{Sulphurous Acid} \longrightarrow \text{Iridium Sulphide} + \text{Water}$$
$$2Ir(OH)_3 + 3H_2SO_4 \longrightarrow Ir_2(SO_3)_3 + 6H_2O$$

Uses

Iridium is used as an alloy with platinum, for use as a catalyst in ammonia fuel cell. It is also used in electrical contacts as thermocouples. Commercial electrodes and resistance wires. Laboratory ware, extrusion* dyes for glass fibres, jewellery. Primary standard weight and length.

Detection and Estimation

When the iridium compounds are treated with sodium hydroxide solution, different colourful complex compounds of iridium are produced (for this it got the name the iridium as iris means rainbow).

PLATINUM

Melting Point : 1769°C	Atomic Number : 78
Boiling Point : 3827°C	Atomic Weight : 195.08

History and Occurrence

The name platinum has been derived from the Spanish word **plata,** which means silver. Platinum occurs in the Ural Mountains in Siberia in the form of an alloy containing about 80% of platinum, the rest consisting of iron, copper a little gold and other metals of the

* Extrusion: Process of crystallization at high temperature.

platinum group. Columbia in South America is also a good source of platinum. Nowaday's Canada has become the very important source of platinum.

Extraction

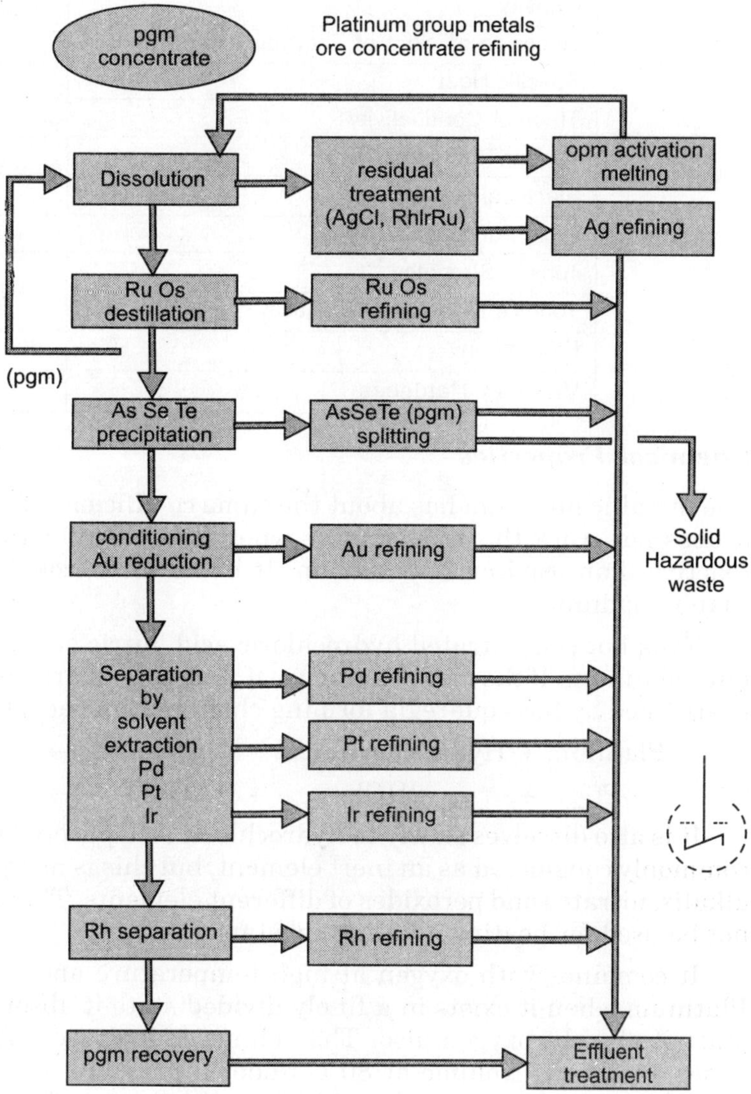

Crude platinum from the mentioned sources in the digested with aquaregia. The solution is diluted and filtered from the insoluble residue, which includes osmoiridium. The filtrate is evaporated repeatedly with hydrochloric acid to expel all the nitric acid and finally the solution of the chlorides is evaporated to dryness. The dried mass is dissolved in a little water and treated with a strong solution of ammonium chloride. This precipitates the sparingly soluble ammonium chloroplatinate, $(NH_4)_2PtCl_6$ with some ammonium chloroiridate $(NH_4)_2Ir_2Cl_6$ The precipitate is suspended in water and treated with sulphur dioxide, which reduces the chloroiridate to soluble chloroiridite $(NH_4)_2IrCl_6$ leaving behind the insoluble ammonium chloroplatinate unchanged. The latter is not easily reduced. The undissolved ammonium chloroplatinate is filtered and heated strongly, when spongy platinum is formed. This metal is then fused in the oxy-hydrogen flame or in an electric arc furnace.

Physical Properties

Melting Point	1769°C
Boiling Point	3827°C
Density	21.4 g/cc
Linear co-efficient of Thermal Expansion	90×10^{-6}
Specific Heat	0.032 cal/g
Thermal Conductivity	0.17 cal/(sec) (sqcm) °C/cm
Vapour Pressure	0.16 microns
Electrical Resistivity	9.8 μΩ/cm
Work Function	5.32 eV
Tensile Strength	.1378 MPa
Young's Modulus of Elasticity	1516 MPa
Elongation	25 – 40%
Vickers's Hardness	39

Chemical Properties

The metallic platinum has about the same co-efficient of expansion of glass and therefore it does not crack the glass handle when it is used for the flame test. Platinum is more reactive than osmium and iridium. It is, however, considerably more resistant to acids than palladium.

Even hot concentrated hydrochloric acid, nitric acid and sulphuric acid do not attack pure platinum. If the metal is impure it is slightly attracted by sulphuric acid. However, it is attacked by hot aquaregia forming chloroplatinic acid (H_2PtCl_6).

$$\text{Platinum} + \text{Hydrochloric Acid} + \text{Chlorine} \longrightarrow \text{Chloroplatinic Acid}$$

$$Pt + 2HCl + 2Cl_2 \longrightarrow H_2PtCl_6$$

It is also dissolves slowly in hydrochloric acid purely in the presence of air. Platinum is commonly considered as an inert element, but this is not quite true. It is attacked by fused alkalis, nitrates and peroxides of different elements. Therefore, platinum crucibles should not be used for heating alkaline substances.

It combines with oxygen at high temperature and pressure to give the oxide (PtO). Platinum when it exists in a finely divided state it absorbs hydrogen and when in fused state it absorbs oxygen also. The volume of hydrogen absorbed may be as high as 900 times as it's own volume at 80°C under a pressure of 1 atmosphere. Platinum is used, therefore, as a catalyst like palladium in the number of hydrogenation process. The mechanism of absorption of hydrogen by platinum is the same as that by palladium.

It may be pointed out incidentally that the chemistry of platinum metal has not yet been successfully systemized and as a matter of fact in some respects it has been not yet properly studied.

Platinum forms many alloys with different metals.

Uses

1. Since air, oxygen or water does not attack platinum it is used in jewellery and on account of its bright lusture, it is used for ornamental purposes. It is also use in dentistry.
2. As it is not attacked by strong acids even at high temperature, it is used in chemical apparatus, such as crucible, dishes etc. For the same reason it is used for making electrodes in various electrolytic process.
3. Platinum can be drawn into very thin wires and foils used in making a number of instruments required for various experiments.
4. It is used as an important catalyst in the form of platinized absorbs, which is prepared by soaking adsorbs in a solution of platinic chloride containing little hydrochloric acid and ammonium chloride acts as a catalyst in some of the chemical reactions they are as follows:
   (*a*) Combination of sulphur dioxide and oxygen for the preparation of sulphur trioxide for the manufacture of sulphuric acid.
   (*b*) Oxidation of ammonia to nitric oxide in the Ostwalds Process for the manufacture of nitric acid.
   (*c*) Decomposition of hydrazine to ammonia.

Compounds of Platinum

Though platinum is not much reactive chemically but it produces many chemical compounds they are as follows:

Platinum Monoxide (PtO): It is obtained as black powder by gently heating the hydrated oxide or platinous hydroxide separates $Pt(OH)_2$; the hydroxide separates out as a black precipitate, by heating a solution of platinic chloride and a solution of dilute alkali. It is soluble in acids, but suffers disproportionation on heating into platinum dioxide and platinum.

$$\text{Platinum Monoxide} \longrightarrow \text{Platinum Dioxide}$$
$$PtO \longrightarrow PtO_2$$

Platinum Dioxide (PtO_2): It is a black powder produced by heating platinic acid about 1000°C, platinic acid or platinic hydroxide $[H_2Pt(OH)_6]$ or $[Pt(OH)_4 2H_2O]$ is formed as a pale yellow precipitate when a solution of chloroplatinic acid is boiled with sodium carbonate and then acidified with acetic acid to remove the alkali.

$$\underset{\text{Acid}}{\text{Platinic}} + \underset{\text{Carbonate}}{\text{Sodium}} + \text{Water} \longrightarrow \underset{\text{Hydroxide}}{\text{Platinic}} + \underset{\text{Chloride}}{\text{Hydrogen}} + \underset{\text{Dioxide}}{\text{Carbon}}$$
$$H_2PtCl_6 + 2Na_2CO_3 + 3H_2O \longrightarrow H_2Pt(OH)_6 + 6HCl + 3CO_2$$
$$\text{Platinic Hydroxide} \longrightarrow \text{Platinum Dioxide} + \text{Water}$$
$$H_2Pt(OH)_6 \longrightarrow PtO_2 + 4H_2O$$

Platinic hydroxide dissolves in dilute hydrochloric acid to form the complex acid called hydrochloroplatinic acid, $H_2[PtC_4(OH)_2]$. It dissolves in platinum hydroxide forming potassium platinate $[K_2Pt(OH)_6]$ similar with potassium stannate $[K_2Sn(OH)_6]$.

Potassium platinate separates from solution in golden yellow crystals on concentration.

Platinic Sulphide (PtS_2): It is produced as a black precipitate when hydrogen sulphide gas is passed through a solution of chloroplatinic acid.

Platinic Acid + Hydrogen Sulphide ⟶ Platinic Sulphide + Hydrogen Chloride

$$H_2PtCl_6 + 2H_2S \longrightarrow PtS_2 + 6HCl$$

The precipitate gets dissolved in yellow ammonium sulphate probably producing a thioplatinate.

Platinum Sulphide + Ammonium Sulphate ⟶ Ammonium Thioplatinate

$$2PtS_2 + 2(NH_4)_2S \longrightarrow (NH_4)_4Pt_2S_6$$

Platinous Chloride ($PtCl_2$): It is produced as a brown insoluble product by heating platinic chloride or chloroplatinic acid to a temperature of 500°C in a current of chlorine.

Platinic Chloride + Chlorine ⟶ Platinous Chloride + Chlorine

$$3PtCl_4 + Cl_2 \longrightarrow 3PtCl_2 + 4Cl_2$$

It dissolves in hydrochloric acid and produces chloroplatinous acid (H_2PtCl_4).

Platinous Chloride + Hydrochloric Acid ⟶ Chloroplatinous Acid

$$PtCl_2 + HCl \longrightarrow H_2PtCl_4$$

The same acid is obtained also by heating chloroplatinic acid and sulphur in aqueous medium. On evaporation of the solution of chloroplatinic acid gives an amorphous dark brown mass, which cannot be crystallized. But a number of its salts has been obtained in the from of crystals by reducing the corresponding chloroplatinates. The red potassium chloroplatinate (K_2PtCl_4). is obtained by boiling a solution of chloroplatinites with potassium oxalate.

Platinous chloride combines with carbon monoxide at elevated temperatures (150°C to 250°C) to form carbonyls such as $PtCl_2CO$, $PtCl_22CO$ and $2PtCl_23CO$.

Platinum Iodide (PtI_2): This compound or platinum is produced as black powder when platinous chloride is heated with potassium iodide.

Potassium Iodide + Platinous Chloride ⟶ Platinium Iodide + Potassium Chloride

$$2KI + PtCl_2 \longrightarrow PtI_2 + 2KCl$$

Platinic Chloride ($PtCl_4$): It is obtained as reddish-brown hydroscopic solid by heating chloroplatinic acid at 260°C in a current of chlorine.

Chloroplatinic Acid ⟶ Platinic Chloride + Hydrogen Chrlode

$$H_2PtCl_6 \longrightarrow PtCl_4 + 2HCl$$

It dissolves in water to form a strongly acidic solution which consists of $H_2PtCl(OH)_2$. In hydrochloric acid it dissolves to form chloroplatinic acid (H_2PtCl_6) which is also produced when platinum is dissolved in aquaregia. The solution of chloroplatinic acid gives yellow crystals of the hexahydrate (H_2PtC_4 $6H_2O$) on concentration. Potassium chloroplatinate (K_2PtCl_6) like the ammonium, rubidium and cesium salts of the acid forms slightly yellow crystals. These are precipitated when this chloride salts are added. Potassium is sometimes estimated by the precipitation as potassium chloroplatinate, which is particularly insoluble in aqueous alcohol.

Complex Comounds of Platinum

As already stated platinum shows a great tendency to form complex compounds both in its bivalent and quadrivalent state, with a co-ordination number of four and six co-ordinated complexes are octahedral in structure. Platinum complexes show a very high degree of stability like the cobaltic complexes. Complexes of chlorides and chloroplatinates and chloroplatinites have been described above.

Complex Amines

By the action of ammonia on platinum chloroplatinate a series of ammonical platinous complex has been prepared with the gradual replacement of chlorine atoms by ammonia molecule. With the progressive replacement of the chlorine atoms in the complex by the neutral ammonia molecule, there occurs progressive variations in the charge on the complex ion for *e.g.*

$$K_2[PtCl_4]^{2-} \rightarrow K[PtCl_3NH_3]^{1-} \rightarrow [PtCl_2(NH_3)_2]^0 \rightarrow [PtCl(NH_3)_3]^{+1} \rightarrow [PtCl(NH_3)_4]^{2+}\ Cl_2$$

In this way mono-ammine, diammine and triammine and tetraammine complex arises. Of these dichloroammine platinum, which is a non-electrolytic and has been obtained in three different modifications can be represented as follows:

$Pt(NH_3)_4PtCl_4$

Magnus Salt
(green)

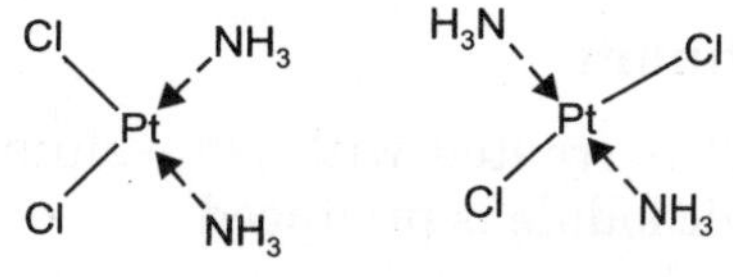

A similar series of complex is also known in the case of quadrivalent platinum as shown in the complexes below:

$$K_2[PtCl_6]^{2-} \rightarrow K[PtCl_5NH_3]^{1-} \rightarrow [PtCl_4(NH_3)_2]^0 \rightarrow [PtCl_3(NH)_3]^0 \rightarrow [PtCl_3(NH_3)_3]^{1+}\ Cl$$

$$[PtCl_2(NH_3)_4]^{2+}\ Cl_2 \rightarrow [PtCl(NH_3)_5]^{3+}\ Cl_2 \rightarrow [PtCl(NH_3)_5]^{3+}\ Cl_2 \rightarrow [Pt(NH_3)_6]^{4+}\ Cl_4$$

All the complexes from monoamine to hexammine are known as Cis-trans isomers also here in the case of tetrachloroammine platinum $PtCl_4(NH_3)_2$.

Complex Cyanides

Platinum in its trivalent states forms in combinations with cyanogen the remarkably stable platinocyanides, which contain the complex anion Pt(CN)'. These are not even decomposed by aquaregia.

Potassium Platinocyanide [$K_2Pt(CN)_4$]: This compound is prepared by the reaction of platinous chloride and potassium cyanide.

$$\underset{\text{Chloride}}{\text{Platinous}} + \underset{\text{Cyanide}}{\text{Potassium}} \longrightarrow \underset{\text{Platinocyanide}}{\text{Potassium}} + \underset{\text{Chloride}}{\text{Potassium}} + \text{Cyanogen}$$

$$H_2PtCl_6 + 6KCN \longrightarrow K_2Pt(CN)_4 + 4KCl + C_2N_2$$

This resembles the action of potassium cyanide on copper sulphate. Ammonium chloroplatinate and potassium cyanide is used for the reaction. From the clear solution potassium platinocyanide crystallizes out in the form of long rhombic yellow coloured prisms, exhibiting a blue metallic lusture by reflecting light. The crystals are very readily soluble in water.

Barium Platinocyanide [$BaPt(CN)_4 4H_2O$]: It is prepared by treating a warm solution of chloroplatinic acid with barium hydroxide and hydrocyanic acid and then passing sulphur dioxide into the mixture till it becomes colourless.

$$\underset{\text{Acid}}{\text{Chloroplatinic}} + \underset{\text{Hydroxide}}{\text{Barium}} + \underset{\text{Cyanide}}{\text{Potassium}} + \text{Sulphur} \longrightarrow \underset{\text{Platino-cyanide}}{\text{Barium}} + \underset{\text{Sulphate}}{\text{Barium}} + \underset{\text{Chloride}}{\text{Barium}} + \text{Water}$$

$$H_2PtCl_6 + 5Ba(OH)_2 + 4KCN + SO_2 \longrightarrow BaPt(CN)_4 + BaSO_4 + 3BaCl_2 + 8H_2O$$

Barium sulphate is removed by filtration and the filtrate is allowed to crystallize. The barium salt forms a lemon yellow powder, soluble in water.

The platinocyanides show a strong fluorescence in the ultra-violet rays or when exposed to X-Rays or radiation of any radioactive substances. The barium salt is used for making fluorescent screens of X-Ray works.

The tree platinocyanic acid can be obtained by decomposition of the barium salt by sulphuric acid and then evaporating the filtrate from barium sulphate. The forms are prismatic.

Detection of Platinum

When a platinic salt is treated with potassium iodide and the mixture is heated a black precipitate of platinic iodide is produced.

GOLD

Melting Point : 1063°C
Boiling Point : 2800°C

Atomic Number : 79
Atomic Weight : 196.97

History of Occurrence

Gold has been known from the prehistoric age and valued as a precious metal for use in jewellery and ornament. It is indeed interesting to note that the development of chemistry is closely associated with the search of gold and its artificial production of gold; this was the favorite choice and one of the fundamental objectives of the alchemists. The whole alchemical literature is full of recipes for gold making. In all most all countries in early ages, the desire for its possession supplied the incentive for attempts at the transmission of baser metals into gold, which gave rise to alchemy.

Gold occurs in native state and also in ores and amalgams. The main ores of gold are *Sylvanaite, (Au, Ag)* Te_2, *Petzite, (Au, Ag)*$_2$*Te.*

The amalgams of gold are found with silver and mercury. The important sources of gold are Karnataka (Kolar Area) Andhra Pradesh (Ramgiri Fields in India). The other sources are South Africa, Russia, Mexico, Australia and U.S.A.

Extraction

Different methods are adopted for the extraction of gold according to nature of sources.

Extraction of Alluvial or Placer Gold: When gold is mixed with sand, salt and gravel, its separation is affected by taking advantage of its sources higher specific gravity (19.3). The ore sand mixture is thoroughly washed with powerful water current in long sluices. A sluice is an inclined channel with a number of small troughs, provides with cleats along the bottom (see Fig. 16.25).

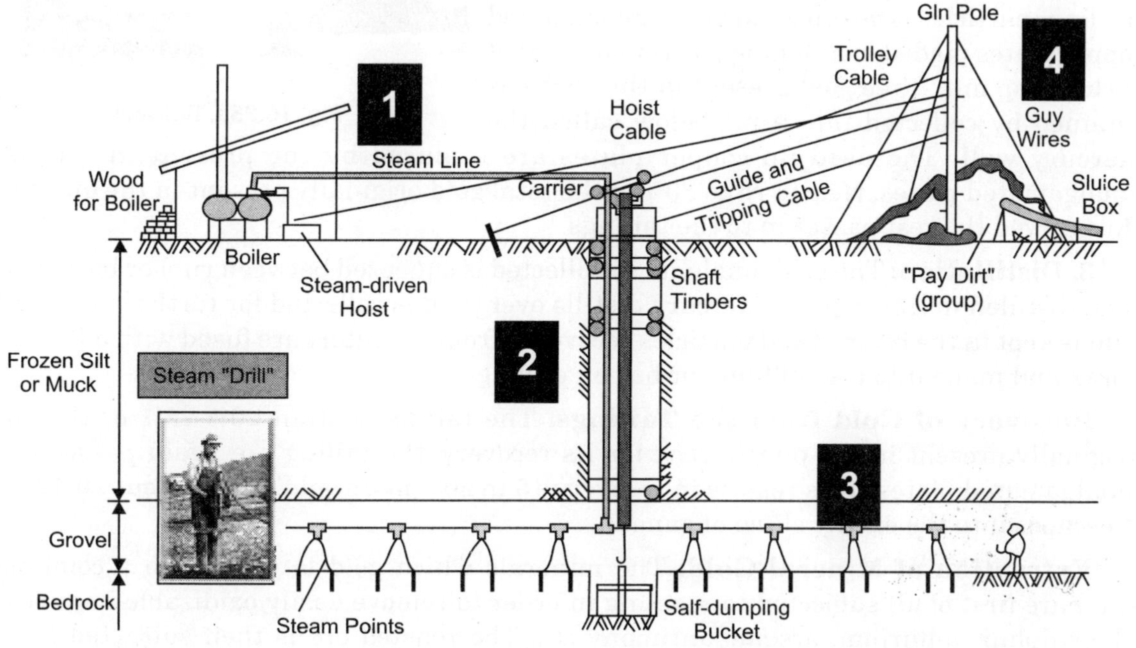

Fig. 16.25. The Placer Mining in a Diagram

The heater particles of gold settle down and collect in the troughs while the higher particles of slit sand any gravel are carried away by the current of water.

This process is called *placer mining.* Sometimes it is necessary to throw water current on the deposits to wash away the impurities. The process is known as *hydraulic mining.*

Extraction of Vein Gold: It is popularly known as vein mining and employed for the extraction of gold from hard auriferous quartz rocks. The rock is blasted at the quarry and the bigger lumps are broken to a smaller size by means of rock crusher. It is then carried to the factory and subjected to the following treatment for the extraction of gold from it.

1. Purification: The smaller pieces of auriferous quartz are finely powdered by means of a group of stamp mills. Each of these mills consists of mortar box enclosing a die and a removing stamp. The stamp weighs about 453.5924kg and they fall from a height of 8" about 100 times per minute. The auriferous quartz crystals pieces are fed from one side into the mill and pounded by the stamp. A current of water carries off the powdered matter through a screen provided at the other end of the mill. The coarser particles, if any are retained back and powdered again.

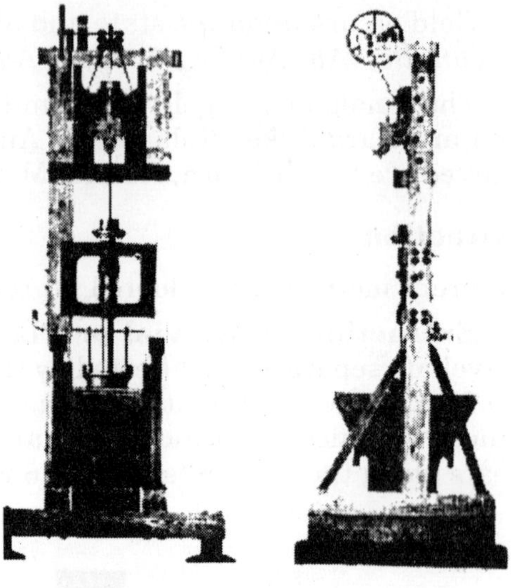

Fig. 16.25a. Ball Mill

2. Amalgamation: The water slurry carrying the finer particles is next passed over amalgamated copper plates pladed in a sloping position in front of each stamp mill. Free gold present in the slurry is retained by collected into a pit below called the mercury well. The used up copper plates are replaced by the present in the new amalgamated plates. Nearly 75 to 80% of the total gold originally present in the quartz is thus served the rest passes in to the tailings.

3. Distillation: The gold amalgam so collected is squeezed between rubber blocks and then distilled in iron retorts. Mercury distills over (and is collected for further use) while gold is kept in the retort. Gold particles recovered from the retort are fused with a flux like borax and made into a solid lump or bar on cooling.

Recovery of Gold from the Tailings: The tailing contain 20%-25% of the gold originally present in the quartz rock for its recovery, the tailing are again passed over amalgamated plates. This results in recovery 15 to 20% more gold, while about 10-12% of it escapes into the earthy slime of gangue.

Extraction of Mineral Gold: The minerals which gold is present in a combined state are first of all subjected to roasting in order to remove easily oxidizable impurities like sulphur, tellurium, arsenic antimony etc. The roasted ore is then subjected to the *MacAurther's Forest Cyanide* process or the *Plattner's Chlorine Process;* which are as follows:

(a) MacAurtber's Cyanide Process: This process is based upon the fact that in the presence of atmospheric oxygen gold dissolves in a fact that in the presence of atmospheric oxygen, gold dissolves in a solution of potassium cyanide the formation of the cyanide complex.

$$\text{Gold} + \underset{\text{Cyanide}}{\text{Potassium}} + \text{Water} \longrightarrow \text{Oxygen} + \underset{\text{Aurocyanide}}{\text{Potassium}} + \underset{\text{Hydroxide}}{\text{Potassium}}$$

$$4Au + 8KCN + 2H_2O \longrightarrow O_2 + 4[KAu(CN)_2] + 4KOH$$

The finely powdered and roasted ore is filtered in large vats made of cement in iron consisting of also wooden bottoms with holes. A solution of potassium cyanide (0.5%) percolates all through for 12-24 hours. From the complex cyanide solution is run through wooden through contain zinc shavings.

$$\text{Potassium Aurocyanide} + \text{Zinc} \longrightarrow \text{Potassium Zinc Cyanide} + \text{Gold}$$

$$2K[Au(CN)_2] + Zn \longrightarrow K_2[Zn(CN)_4] + 2Au$$

(b) ***Plattner's Chloride Process:*** The roasted ore is placed in vats with false bottoms moistened with chlorine and saturated from underneath and left as such for a day. The gold chloride ($AuCl_3$) thus formed is leached with water. The solution is transferred to separate tanks where it is treated with ferrous sulphate to precipitate gold.

$$\text{Gold Chloride} + \text{Ferrous Sulphate} \longrightarrow \text{Gold} + \text{Ferric Chloride} + \text{Ferric Sulphate}$$

$$AuCl_3 + 3FeSO_4 \longrightarrow Au + FeCl_3 + Fe_2(SO_4)_3$$

Refining of Gold: Gold obtained by any of the above process is impure and generally contaminated with other metals like silver copper lead and some other baser metal lead and the other baser metals are removed by Cupillation while silver and possibly a small amount of copper remains. The removal of silver and copper from gold is known as parting and can be achieved by following methods:

1. Chlorine Parting: The impure gold is melted in fire clay crucible and dry chlorine is bubbled through the melt by means of a clay pipe. At first fumes of base metal chlorides are evolved then silver chloride is converted to silver and gold remains unaffected. Silver forms a scum on the surface of the metal and therefore can be easily skimmed off. The gold is now in pure state and is tapped off and solidified in small ingots.

2. Parting with Sulphuric Acid: It is found, however, that if the percentage of gold in the impure gold is more than 30%, the acid does not attack the impure gold. More silver is added till the mixture contains 25% of gold is obtained. This is treated with boiling sulphuric acjd when silver and copper dissolve and gold is left behind.

$$\text{Silver} + \text{Sulphuric Acid} \longrightarrow \text{Silver Sulphate} + \text{Water} + \text{Sulphur Dioxide}$$

$$2Ag + 2H_2SO_4 \longrightarrow AgSO_4 + 2H_2O + SO_2$$

In this process, since the gold content in the sample has to be 25% or one quarter before the acid was used for this purpose; but it has been replaced now by much cheaper sulphuric acid.

3. Electro-refining: Blocks of impure gold are made the anodes in an electrolytic cell containing a solution of gold chloride ($AuCl_3$) in hydrochloric acid. Thin plates of pure gold consists of the cathode. On electrolysis pure gold gets deposited on the surface of the cathode. Silver forms the silver chloride, which remains near the anode. Gold thus obtained to a second electrolytic refining.

Physical Properties

Melting Point	1063°C
Boiling Point	2800°C
Density	19.3 gm/cc
Atomic Radius	1.44 Å
Ionic Radius	1.37 Å
First Ionization Energy Oxidation Potential	213 Kcal/mol
Au .... Au	–1.70
Au .... Au	–1.50
Important Oxidation States	+1, +3
Heat of Vapourization	87.3 Kcal/mol
Heat of Fusion	3.93 Kcal/mol
Specific Heat	0.0312
Electronegativity	2.4
Linear Co-efficient of Thermal Expansion	14.2×10^{-6}
Temperature Coefficient of Resistance	0.004
Thermal Conductivity	0.71 cal/(sec) (sqcm) (°C/cm)
Other Properties	
Resistivity at 0°C	2.10×10^{-6} ohmem
Brinell Hardness	25
Rockwell Hardness	22
Tensile Strength	1.31 MPa
Elongation Pecentage	45%

Chemical Properties

Chemically gold is inert. It has been valued for nobility that is for its resistance to chemical attack of air, moisture, carbon dioxide, hydrogen sulphide and even most of the acids.

1. Action of Acids: The oxidation potential of gold is extremely low being –1.42 volts. Hence, it is not attacked by hydrofluoric acid, hydrochloric acid and sulphuric acid. It is dissolved in aquaregia and produces chloroauric acid.

Nitric Acid + Hydrochloric Acid + Gold ⟶ Chloroauric Acid + Water + Nitrogen Oxychloride

$$3HNO_3 + 11HCl + 2Au \longrightarrow 2HAuCl_4 + 6H_2O + 3NOCl$$

2. Action of Halogen: Gold is attacked readily by halogen forming the respective halides, AuX3. When heated with iodine, gold forms aurous iodide, AuI (auric iodide is unstable).

3. Action of Potassium Cyanide: - Gold dissolves in alkali cyanide ion, [Au(CN)z], or potassium aurocyanide.

Gold	+ Potassium Cyanide	+ Water	+ Oxygen	$\longrightarrow$	Potassium Aurocyanide	+ Potassium Hydroxide
4Au	8KCN	+ H_2O	O_2	$\longrightarrow$	$4K[Au(CN)_2]$	4KOH

Colloidal Gold

Gold is available in the form of colloidal solutions by the process of Bredig's Method or by reducing against such as formaldehyde, hydrazine hydrate, yellow phosphorous, carbon monoxide etc. Solution of different colours red, blue, purple or reddish purple. The difference in colour is due to the different size in colloidal gold particles are obtained.

When gold is prepared along with another colloid, a solid colloidal solution of purple of Cassius; this is used for making ruby glass and high-class pottery.

Use of Gold

Gold is a precious metal, which is used for making ornaments and jewellery. It is generally mixed with small quantity of copper for making it hard, it is also used for making medals and coins.

Purity of gold is generally measured in carats, all the gold used for making ornaments are of 22 carats. Gold is also used for plating baser metals like silver, copper etc.

Oxidation States

Gold exhibits oxidation states of +1 and +3. The +3 State is more stable. Gold (I) ion is unstable as it goes disproportionation in solution.

$$3Au^+ \rightleftharpoons Au^{3+} + 2Au$$

The equilibrium constant of the reaction.

$$K = \frac{[Au^{3+}]}{[Au^+]^3}$$

Compounds of Gold

Gold forms two type of compounds the *aurous* and the *auric.*

Aurous Compounds

Aurous Hydroxide (AuOH): This compound is obtained as a violet powder by the action of cold dilute alkali solution on aurous chloride.

Auric Chloride + Sodium Hydroxide $\longrightarrow$ Aurous Hydroxide + Sodium Chloride

$$AuCl_3 + NaOH \longrightarrow AuOH + NaCl$$

When heated at 200°C aurous hydroxide loses water and forms aurous oxide Au_2O

Aurous Hydroxide $\longrightarrow$ Aurous Oxide + Water

$$2AuOH \longrightarrow Au_2O + H_2O$$

Aurous Chloride (AuCl): It is obtained as a yellow powder by heating, auric chloride at 175°C when heated aurous chloride is decomposed to gold and chlorine, with water it suffers disproportionation into metallic gold and auric chloride.

Aurous Iodide (AuI): This compound is produced as a greenish yellow powder by the action of chlorine on gold at about 110°C, or by adding a solution of potassium iodide to that of auric chloride.

Auric Chloride + Potassium Iodide $\longrightarrow$ Aurous Iodide + Potassium Iodide + Iodine

$$AuCl_3 + 3Kl \longrightarrow Aul + 3KCl + I_2$$

It, however, decomposes when warm with water or heated alone, forming metallic gold. It is, therefore, more stable than chloride and aurous bromide.

Aurous Sulphide (Au_2S): It is produced as a steel grey precipitate by passing hydrogen sulphide through a solution of potassium aurocyanide and acidifying the solution with sulphuric acid. All aurous salts are soluble in alkali cyanide, alkali cyanides or alkali thiosulphate and alkali polysulphide solutions forming complex anions.

Auric Compounds

Auric Hydroxide $Au(OH)_3$: It is found as a reddish brown precipitate by treating a solution of auric chloride with alkali. When dried, it has the composition [Au(OH)O]. It is soluble in both as a weak base as well as a weak acid and as an amphoteric body. From its solution in alkali, crystal of an alkali aurite (e.g. $KAuO_23H_2O$) may be obtained by evaporation in vacuum.

When heated at 145°C -150°C auric hydroxide loses water and is changed into auric oxide (Au_2O_3) which on further heating looses oxygen with the formation of metallic gold.

Ammonia reacts with auric hydroxide from an oleic green powder of the explosive substance, known as fulmination of gold, which has a viable composition approximately $NH = Au\text{-}NH_2\ 1.5H_2O$.

Auric Chloride, $AuCl_3$: It is obtained as red needle shaped crystals by passing over gold leaves or finely powdered gold at about 200°C. It sublimes at this temperature in the current of chlorine.

Auric chloride decomposes when heated to aurous chloride, which at higher temperature is reduced to gold. It dissolves on water forming a complex anion. With hydrochloric acid in aqueous solutions it forms chloroauric acid.

Auric Chloride + Hydrochloric Acid $\longrightarrow$ Chloroauric Acid

$$AuCl_3 + HCl \longrightarrow H[AuCl_4]$$

Auric Iodide (AuI_3): As stated it is very unstable and decomposes into aurous iodide and iodine. But in an excess of hydroiodic acid of potassium iodide, it forms complex ions of $[AuI_4]$ from which black crystals of $HAuCl_4$ or $KAuCl_4$ can be obtained.

Chloroauric Acid [H[AuCl]4H_2O]: Yellow delequescent crystals of chloroauric acid are obtained by evaporating a solution of gold in aquaregia. The crystals loose hydrogen chloride at 120°C to form auric chloride. The potassium salt $[2K[AuC_4]H_2O]$ is precipitated as light yellow crystals by adding potassium chloride and concentrated hydrochloric acid to a solution of chloroauric acid.

Potassium Aurocyanide ($K[Au(CN)_4]$): When a hot concentrated solution of potassium cyanide is added to that of auric chloride, no precipitation occurs, but the solution becomes colourless and form the solution colourless crystals of potassium aurocyanide ($K[Au(CN)_4]$) can be obtained on cooling. No gold is precipitated from a solution of the aurocyanide by the addition of ferrous sulphate.

Detection of Gold

The gold is detected usually by the formation of Purple of Cassius. This is the most sensitive test of gold, where the oxide of gold is heated with ammonia when sesquiauramine (NAu_3,NH_3) is formed, which on boiling with water gives aurous nitride (Au_3N). The whole solution is boiled with stannous chloride, when stannous hydroxide is produced. This is purple in colour and that is why, it is termed as Purple Cassius

MERCURY

Freezing Point : –38.85°C
Boiling Point : –357°C
Atomic Number : –80
Atomic Weight : –200.59

History and Occurrence

Mercury has been known in the preliminary state form the early age. Theophastrus has described the methods of the extraction of mercury in his book (about 300B.C). The metal received a great attraction and importance from the alchemists in India, Arabia and Europe. Crystalline sulphide, it finds a prominent place in the ancient Indian medicine where the red crystalline sulphide was used as *swamasindur* or *makaradhwaj.* It was used as master medicine for all the diseases. The alchemists regarded Mercury as the seed of all metals, which were believed by them to be compounds of mercury and sulphur, in India the mercury and sulphur together was regarded as *Kadjjali* (900 A.D) by Nagarjuna.

Mercury occurs only at a very few places in the form of its sulphide are, chinnabar, HgS, which is usually remains mixed with a considerable amount of worthless materials. Small quantities also occur in the native state or as amalgams. The most important deposits of cinnabar are found in Almaden (Spain), in California and in Mexico. The richest ore contain only 7% of mercury but usually the ore consists of less than 1% of hydrogen sulphide.

Extraction

In the older process used in Almaden the ore is roasted in a current of air in a shaft furnace, the vapours pas through a series of skin ware receivers called aludels from which the condensed mercury runs out into a channel.

In modern process the ore is mixed with charcoal and then roasted in a shaft furnace, and the mercury vapour condensed in a series of earthen ware pipes, cooled in water. (see Fig. 16.26).

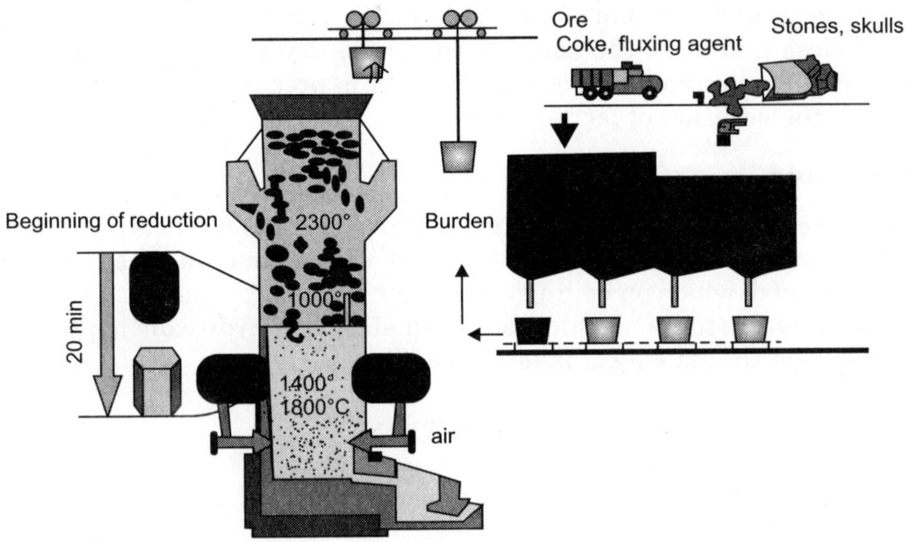

Fig. 16.26. The Extraction of Mercury by Saft Furnace

Reaction Equation

$$\text{Mercuric Sulphide} \quad \text{Oxygen} \longrightarrow \text{Mercuric Oxide} + \text{Sulphur Dioxide}$$

$$2HgS + 3O_2 \longrightarrow 2HgO + 2SO_2$$

$$\text{Mercuric Oxide} \longrightarrow \text{Mercury} + \text{Oxygen}$$

$$2HgO \longrightarrow 2Hg + O_2$$

$$\text{Mercuric Oxide} + \text{Carbon} \longrightarrow \text{Mercury} + \text{Carbon Monoxide}$$

$$HgO + C \longrightarrow Hg + CO$$

Rich ores and flue dust from the furnace are distilled with quick lime or iron fillings.

$$\text{Mercuric Sulphide} + \text{Calcium Oxide} \longrightarrow \text{Mercury} + \text{Calcium Sulphide} + \text{Calcium Sulphate}$$

$$4HgS + 4CaO \longrightarrow 4Hg + 3CaS + CaSO_4$$

$$\text{Mercuric Sulphide} + \text{Iron} \longrightarrow \text{Mercury} + \text{Ferrous Sulphide}$$

$$4HgS + Fe \longrightarrow Hg + FeS$$

Commercial mercury usually contains impurities of other metals like lead copper and leaves tail when allowed to run over a glass surface. The impurities get oxidized in air and produce a black coating over the surface of mercury. So the impure mercury is therefore filtered by passing it through a 1.5m long glass tube (see 16.27A and B). The metal is dried and filtered through chamois leather and finally distilled in vacuum.

Fig. 16.27A. The Filtration of Mercury

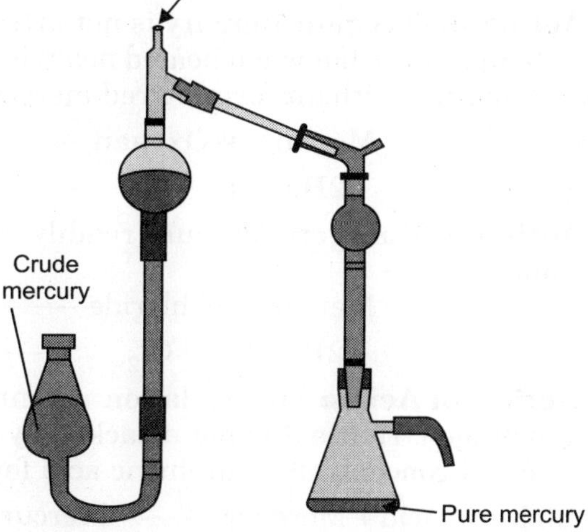

Fig. 16.27B. The Vacuum Distillation of Mercury

Physical Properties

Freezing Point	–38.85°C
Boiling Point	356.6°C
Density	13.6 g/cc
Outer Electronic Configuration	$5d^{10}6s^2$
Atomic Radius	1.57Å
Ionic Radius	1.10(+2) Å
First Ionization Energy Oxidation Potential	24 Kcal/mol
(Hg ..... $2Hg^{2+}$ + 2e)	–8.5 volt
Stable Oxidation States	+2
Heat of Vapourization	13.9 Kcal/mole
Heat of Fusion	0.56 kcal/mole
Specific Heat	0.038
Electronegativity (Pauling's Scale)	1.9

Chemical Properties

1. It is a silvery white metal (from this the name Hydragyrum means silver in liquid form). It is the liquid metal at room temperature. When it is cooled below –39°C it is transferred to solid crystal, which is malleable and ductile. The metal boils at 356°C, giving a colourless vapour. The vapour is not at all a conductor of electricity, but when a arc is send it produces a green light and with it U-V radiation. Mercury forms amalgams with other metals except with platinum and iron.

> Mercury vapour lamps are used for the source of U-V rays

2. Action of Oxygen: Mercury is not attracted by air at ordinary temperature but when heated nears its melting point it slowly combines with air forming red mercuric oxide.

$$\text{Mercury} + \text{Oxygen} \longrightarrow \text{Mercuric Oxide}$$

$$2Hg + O_2 \longrightarrow 2HgO$$

3. Action of Halogen: Mercury readily combines with halogens forming the halides for example

$$\text{Mercury} + \text{Chloride} \longrightarrow \text{Mercuric Chloride}$$

$$2Hg + Cl_2 \longrightarrow HgCl_2$$

4. Action of Acids: The oxidation potential of mercury -0.85 volt that is much less than of hydrogen. Hence it is not attacked by dilute hydrochloric acid and sulphuric acid in cold but hot concentrated sulphuric acid forms mercurous sulphate (Hg_2SO_4).

$$\text{Sulphuric Acid} + \text{Mercury} \longrightarrow \text{Mercurous Sulphate} + \text{Water} + \text{Sulphur Dioxide}$$

$$H_2SO_4 + Hg \longrightarrow Hg_2SO_4 + 2H_2O + SO_2$$

With excess of sulphuric acid mercurous sulphate is dissolved and mercurous sulphate is produced.

$$\text{Mercurous Sulphate} + \text{Sulphuric Acid} \longrightarrow \text{Mercurous Sulphate} + \text{Sulphur Dioxide} + \text{Water}$$

$$Hg_2SO_4 + H_2SO_4 \longrightarrow 2HgSO_4 + SO_2 + 2H_2O$$

The reaction with acid is quite complicated. The products obtained depend upon the temperature and concentration of the acid. Warm dilute nitric acid from mercurous nitrate and nitric oxide.

$$\text{Mercury} + \text{Nitric Acid} \longrightarrow \text{Mercuric Nitrate} + \text{Water} + \text{Nitric Oxide}$$

$$6Hg + 8HNO_3 \longrightarrow 3Hg(NO_3)_2 + 4H_2O + NO$$

With concentrated nitric acid, mercuric nitrate is produced and nitrogen dioxide is evolved.

$$\text{Mercury} + \text{Nitric Acid} \longrightarrow \text{Mercuric Nitrate} + \text{Water} + \text{Nitric Oxide}$$

$$Hg + 8HNO_3 \longrightarrow Hg(NO_3)_2 + 2H_2O + 2NO_2$$

Uses of Mercury

It is used in the medicine (calomel is used as purgative). Mercury is also used for U-V lamps. It is also used in making mercury cells. It is also used for making mercury switches and making mercury cells and mercury-platinum electrodes. Mercury is mainly used as the thermometer.

Compounds of Mercury

Mercuric Oxide (HgO): It is also known as red oxide of mercury or red precipitate of mercury. It is formed by heating mercury at a temperature of 300°C; the reaction is reversible and at a higher temperature the oxide decomposes. It is prepared on a large scale as red crystalline powder by heating a moderate temperature mercuric nitric or a mixture of mercuric nitrite and mercury.

$$\text{Mercuric Nitrate} + \text{Mercury} \longrightarrow \text{Mercuric Oxide} + \text{Nitrogen Dioxide}$$
$$Hg(NO_3)_2 + Hg \longrightarrow 2HgO + 2NO_2$$
$$\text{Mercuric Nitrate} \longrightarrow \text{Mercuric Oxide} + \text{Nitrogen Dioxide} + \text{Oxygen}$$
$$2Hg(NO_3)_2 \longrightarrow 2HgO + 4NO_2 + O_2$$
$$\text{Mercury} + \text{Oxygen} \longrightarrow \text{Mercuric Oxide}$$
$$2Hg + O_2 \longrightarrow 2HgO$$

It is precipitated as a yellow powder by adding caustic soda to a solution of mercuric chloride.

$$\text{Mercuric Chloride} + \text{Sodium Hydroxide} \longrightarrow \text{Mercuric Oxide} + \text{Water} + \text{Sodium Chloride}$$
$$HgCl_2 + 2NaOH \longrightarrow HgO + H_2O + 2NaCl$$

It is a basic oxide and is soluble in acids. It is readily dissolved in potassium iodide giving an alkaline solution.

$$\text{Mercuric Oxide} + \text{Water} + \text{Potassium Iodide} \longrightarrow \text{Potassium Mercurous Tetraiodide}$$
$$HgO + H_2O + 4Kl \longrightarrow K[Hgl_4]$$

Mercuric oxide is used as a rust protector in the metals and also used as paint.

Mercurous Carbonate ($HgCO_3$): It is prepared by adding potassium carbonate in mercuric nitrate solution ($KHCO_3$) gives a brown precipitate of $HgCO_33H_2O$.

Mercurous Halides

The halides of mercury are all crystalline solids heir melting points are between 250°C to 500°C. Mercurous fluoride is only soluble but other halides are insoluble.

Mercurous Fluoride (HgF_2): It is a yellow compound prepared by the reaction of mercurous carbonate and hydrofluoric acid.

$$\text{Mercurous Carbonate} + \text{Hydrofluoric Acid} \longrightarrow \text{Mercurous Fluoride} + \text{Carbonic Acid}$$
$$HgCO_3 + 2HF \longrightarrow Hg_2F_2 + H_2CO_3$$

It blackens when exposed to light. It is soluble in water in which it is rapidly hydrolyzed.

$$\text{Mercurous Hydroxide} \longrightarrow \text{Mercury} + \text{Mercuric Oxide} + \text{Water}$$
$$Hg_2(NO_3)_2 \longrightarrow Hg + HgO + H_2O$$

Mercurous Bromide ($HgBr_2$): It is obtained as a yellowish white precipitate on adding potassium bromide to a solution of mercurous nitrate. It sublimes at 350°C.

Mercurous Nitrate + Potassium Bromide ⟶ Mercurous Bromide + Potassium Nitrate

$$Hg_2(NO_3)_2 + 2KBr \longrightarrow Hg_2Br_2 + 2KNO_3$$

Mercurous Chloride, Calomel (Hg_2Cl_2): It is prepared by heating an intimate mixture of (*a*) Mercuric Chloride and mercury.

Mercuric Chloride + Mercury ⟶ Mercurous Chloride

$$Hg_2Cl_2 + Hg \longrightarrow Hg_2Cl_2$$

(*b*) Mercuric sulphate, mercury and common salt in an iron pot.

Mercuric Sulphate + Mercury + Sodium Chloride ⟶ Mercurous Chloride + Sodium Sulphate

$$HgSO_4 + Hg + 2NaCl \longrightarrow Hg_2Cl_2 + Na_2SO_4$$

Mercurous chloride sublimes on the lid, the crust of the sublimate is collected ground to powder and boiled with water to completely remove the mercuric chloride. It is obtained as a white precipitate when dilute hydrochloric acid is added to mercurous nitrate solution.

Mercurous Nitrate + Hydrochloric Acid ⟶ Mercurous Chloride + Nitric Acid

$$Hg_2(NO_3)_2 + 2HCl \longrightarrow Hg_2Cl_2 + 2HNO_3$$

Or, by reducing mercuric chloride with sulphur dioxide.

Mercuric Chloride + Sulphur Dioxide + Water ⟶ Mercurous Chloride + Sulphuric Acid + Hydrogen Chloride

$$2HgCl_2 + SO_2 + 2H_2O \longrightarrow Hg_2Cl_2 + H_2SO_4 + 2HCl$$

It is an amorphous powder, almost insoluble in water (about OAmg/litre at 20°C) and in dilute acids, for example nitric acid.

Mercurous Chloride + Nitric Acid ⟶ Mercuric Chloride + Mercurous Nitrate + Nitric Oxide + Water

$$3Hg_2Cl_2 + 8HNO_3 \longrightarrow 3HgCl_2 + 3Hg(NO_3)_2 + 2NO + 3H_2O$$

It blackens with ammonia the black mass is a mixture of metallic mercury and infusible white precipitate of ammonomercuric chloride.

Mercurous Chloride + Ammonia ⟶ Ammonomercuric Chloride + Mercury + Ammonium Chloride

$$Hg_2Cl_2 + 2NH_3 \longrightarrow Hg(NH_2Ck) + Hg + NH_4Cl$$

Mercurous chloride received its name the calomel in Greek it means beautiful black as it turns deep black when treated with ammonia.

Calomel turns yellowish when gently heated (or rubbed) and darkens when exposed to light as a partial decomposition into mercury and mercuric chloride.

It sublimes at 385°C. It's vapour amalgamates with gold leaf, which indicates the presence of free mercury vapour formed by thermal dissociation.

Mercurous Chloride ⟶ Mercury + Mercuric Chloride

$$Hg_2Cl_2 \longrightarrow Hg + HgCl_2$$

From the calomel vapour in a porous earthenware tube mercury separates by diffusion and deposits on a cold surface, the residue in the tube containing excess of mercuric chloride. The vapour density corresponds with complete dissociation into mercury and mercuric chloride. But since the vapour is diamagnetic, the formula HgCl (which has an odd electron) is discarded.

According to Baker the vapour density of perfectly dry calomel corresponds to the formula H_2Cl_2. The freezing point depressing in fused mercuric chloride is given the linear formula Cl-Hg-Hg-Cl.

Calomel is used in medicine as purgative and making calomel electrode (in the pH meter).

Mercurous Iodide (H_2I_2): It is formed as green precipitate on adding little potassium iodide to mercurous chloride solution acidified with hydrochloric acid.

Mercurous Chloride + Potassium Iodide ⟶ Mercurous Iodide + Potassium Chloride

$$Hg_2Cl + 2KI \longrightarrow Hg_2I_2 + 2KCl$$

The pure compound is formed as a bright yellow crystal darkens on exposure to light due to partial decomposition.

It sublimes at 140°C. It is insoluble in water and ethanol. It reacts with potassium iodide forming soluble potassium mercuric iodide with the separation of finely divided mercury.

Mercurous Iodide + Potassium Iodide ⟶ Potassium Mercurous Iodide

$$Hg_2I_2 + 2KI \longrightarrow K_2[HgI_4]$$

Mercuric Halide

The properties of mercuric halides include the charge from mercuric fluoride, which possesses a high degree of ionic character, to mercuric chloride, which is purified covalent. Their melting points are between 260°C – 600°C.

Mercuric Fluoride (HgF_2): It is prepared by passing chlorine in mercurous fluoride at 275°C.

Mercurous Fluoride + Chlorine ⟶ Mercuric Fluoride + Mercuric Chloride

$$Hg_2F_2 + Cl_2 \longrightarrow HgF_2 + HgCl_2$$

Or by heating mercurous fluoride alone in vacuum at a temperature of 450°C

Mercurous Fluoride ⟶ Mercury + Mercuric Fluoride

$$Hg_2F_2 \longrightarrow Hg + HgF_2$$

It can be prepared by passing fluorine over mercuric chloride. It forms colourless crystals with ionic lattice. It is aqueous solution, though strong electrolyte it is extensively hydrolyzed, giving a yellow basic fluoride [Hg(OH)F] which on further farther hydrolysis gives mercuric oxide.

Mercuric Fluoride + Water ⟶ Mercuric Oxyflurodie + Hydrogen Fluoride

$$2HgF_2 + H_2O \longrightarrow 2Hg(OH)F + 2HF$$

On further hydrolysis

Mercuric Oxyfluoride + Water ⟶ Mercuric Oxide + Hydrogen Fluoride + Water

$$2Hg(OH)F + 2H_2O \longrightarrow 3HgO + 3HF + H_2O$$

The solubility of HgF_2 in water is decreased by the addition of KF, potassium fluoride including the fluoride ion produced from both salts. Either in the case of other mercuric salts, or in the case of other mercuric halides, the solubility is. increased due to the formation of complex when potassium fluoride is added. The hydrate ($HgF_2 .2H_2O$) crystallizes from a solution of mercuric oxide in excess of hydrofluoric acid. The ability of the fluoride to form a hydrate. Mercuric fluoride does not have any complex.

Mercuric Chloride ($HgCl_2$): It is also known as corrosive sublimate, it is prepared by the reaction of equal amounts of sodium chloride and mercuric sulphate in the presence of a little manganese dioxide to prevent the formation of mercuric chloride.

Mercuric Sulphate + Sodium Chloride ⟶ Mercuric Chloride + Sodium Sulphate

$$HgSO_4 + 2NaCl \longrightarrow HgCl_2 + Na_2SO_4$$

It forms colourless needle like crystal having a melting point of 275°C. It is sparingly soluble in cold water but highly soluble in hot water.

The aqueous solution has a distinct acid reaction due to hydrolysis. It has a very poor electrical conductivity.

It is negligibly small and for the largest part of the solute remains in the solution as undissociated state (as mercuric chloride is strongly covalent). The solution contains the complex ions $HgCl^+$, $HgCl_3$and $HgCl_4$.

It is remarkable that the mercuric chloride is having the property covalent compound rather than that of an electrovalent one. Thus it has a low melting point 320°C, is more soluble in ether and in alcohol than in water and is very slightly ionized in aqueous solution. It is also distingly soluble in benzene.

It reacts with boiling solution of ammonia going the infusible white precipitate, which is hydrolyzed by deposition with water yielding chloride of Millon's Base.

Mercuric Chloride + Ammonia ⟶ Ammono Mercuric Chloride + Ammonium Chloride

$$HgCl_2 + 2NH_3 \longrightarrow Hg(NH_2)Cl + NH_4Cl$$

Mercuric chloride forms crystalline complex salts with chloride *e.g.* $K[HgCl_3]$ and $Na_2[HgCl_4]$ made from mixtures of equal weights of mercuric chloride and sodium chloride, is used as an antiseptic and does not coagulate protein. Mercuric chloride dissolves in concentrated Hydrochloric acid with the evolution of heat and the solution on cooling precipitates chloromercuric acid ($H[HgCl_3]$). The increase in solubility of mercuric chloride brought about by ammonium chloride other chlorides is evidence that complex ions such as $HgCl_3$ and $HgCl_4$ are formed in solutions of mercuric chloride consists of excess Cl^- ions.

$$HgCl_2 + Cl^- \longrightarrow HgCl_3^-$$

$$HgCl_2 + 2Cl^- \longrightarrow HgCl_4^{--}$$

Mercurous chloride reacts with potassium iodide solution giving a yellow precipitate (which rapidly turns red) of mercuric iodide, which dissolves in excess of potassium iodide, forming potassium mercuric iodide.

Mercuric Chloride + Potassium Iodide ⟶ Potassium Chloride + Mercuric Iodide

$$HgCl_2 + 2Kl \longrightarrow 2KCl + Hgl_2$$

Mercuric Iodide + Potassium Iodide ⟶ Potassium Mercuric Tetraiodide

$$Hgl_2 + 2Kl \longrightarrow K_2Hgl_4$$

Alkali reacts with mercuric chloride solution; giving an initial reddish-brown precipitate of basic chloride, converted into yellow mercuric oxide by excess of alkali.

Mercuric Chloride + Sodium Hydroxide ⟶ Mercuric Oxide + Sodium Chloride + Water

$$HgCl_2 + NaOH \longrightarrow HgO + 2NaCl + H_2O$$

Ammonomercuric Chloride + Water ⟶ Million's Base (chloride) + Ammonium Chloride

$$2Hg(NH_2Cl) + H_2O \longrightarrow Hg_2OClNH_2 + NH_4Cl$$

A grey film of mercury (which becomes silvery in appearance on rubbing) deposits on a bright copper foil when immersed in a solution of mercuric chloride solution.

Mercuric Chloride + Copper ⟶ Cupric Chloride + Mercury

$$HgCl_2 + Cu \longrightarrow CuCl_2 + Hg$$

Mercuric chloride is a deadly poison the lethal dose is between 0.2g to 0.5g for a human being. The antidote is egg white. The 1% of this solution is used for the sterilization of medical instruments. It is also used as a germicide and fungicide.

Mercuric Iodide (HgI_2): It is formed as a yellow precipitate, which rapidly turns red on adding potassium iodide in a small amount to a solution of mercuric chloride solution and as a red powder by grinding mercury with the correct amount of iodine. At 125°C the reverse change occurs, on cooling. It is also soluble in organic solvents e.g. benzene.

Mercuric iodide is ionized to a still lesser extent than other mercuric halides. Its solutions do not form mercuric oxide or basic salts on the addition of alkali; neither gives any precipitate of AgI on addition of silver nitrate. It readily forms complex salts like the chloride and bromide. Mercuric Iodide is used for making antiseptic ointments used for the skin diseases.

Mercuric Bromide ($HgBr_2$): It is very similar to chloride, but it is soluble in water less than chloride (about 0.6g in 100g at 25°C) and is weaklier ionized in water. It is prepared by the action of hydrogen bromide on mercuric oxide.

Mercuric Oxide + Hydrogen Bromide ⟶ Mercuric Bromide + Water

$$HgO + HBr \longrightarrow HgBr_2 + H_2O$$

Mercurous Nitrite [$Hg(NO_2)_2$]: It is a pale yellow coloured compound and is prepared by the action or dilute nitric acid on excess of mercury.

Sir P. C.Ray for which he was known as father of Nitrites developed this compound.

Mercury + Nitric Acid ⟶ Mercuric Nitrite + Nitric Oxide + Water

$$6Hg + 8HNO_3 \longrightarrow 3Hg_2(NO_2)_2 + 2NO + 4H_2O$$

Mercuric Nitrite [$Hg(NO_3)_2$]: It is prepared by the action of silver nitric acid on mercuric chloride, it forms yellow crystals, which are unstable.

Mercuric Chloride + Silver Nitrate ⟶ Silver Chloride + Mercuric Nitrite

$$HgCl_2 + 2AgNO_3 \longrightarrow 2AgCl + Hg(NO_2)_3$$

Mercurous Nitrate [$Hg_2(NO_3)_2$]: It is prepared in solution by the action of cold dilute nitric acid (sp.gr. 1.14) upon an excess of mercury. The solution deposits colourless crystals of the dihydrate $Hg_2(NO_3)_2 2H_2O$. It decomposes on heating this.

Mercurous Nitrate ⟶ MercuricOxide + Nitrogen Dioxide

$$Hg_2(NO_3)_2 \longrightarrow 2Hg + NO_2$$

Mercuric Nitrate [$Hg(NO_3)_2$]: This compound is prepared by dissolving mercury or mercuric oxide in hot concentrated nitric acid or by heating mercurous nitrite with nitric acid. It separates from solutions in colourless delequescent crystals of the monohydrate [$Hg(NO_3)2H_2O$]. It is decomposed by water forming an insoluble white basic nitrate, which on heating finally yields mercuric oxide. It is, therefore, dissolved on nitric acid.

Mercurous Sulphate (Hg_2SO_4): It is prepared by heating excess of mercury with concentrated sulphuric acid and deposited as white crystalline powder by diluting the acid solution with water, since it is sparingly soluble in water, but since it is concentrated sulphuric acid. It is hydrolyzed by water to a basic sulphate. It is used a depolarizer in standard western cell.

Mercuric Sulphate ($HgSO_4$): This is prepared by boiling mercury with one hand as halftimes it is weight of concentrated sulphuric acid, it crystallizes from its strong acid solution in silvery palates.

Mercury + Sulphuric Acid ⟶ Mercuric Sulphate + Water + Sulphur Dioxide

$$Hg + 2H_2SO_4 \longrightarrow HgSO_4 + 2H_2O + SO_2$$

It is hydrolyzed by water forming an insoluble lemon coloured basic sulphate.

Mercuric Sulphide, Vermilion (HgS): This occurs in nature as cinnabar and is the red pigment vermilion. It is made by grinding mercury and sulphur with a little caustic potash solution. The black mass or mercuric sulphide formed is heated at 50°C with the addition of a little more alkali when it is slowly becomes red and crystalline. The product is washed with water and dried.

Mecury + Sulphur ⟶ Mercuric Sulphide

$$Hg + S \longrightarrow 3HgS$$

If hydrogen sulphide is passed through a solution of a mercuric salt at first a white, then a yellow brown and finally a black precipitate of mercuric sulphide, HgS is formed the white precipitate of HgCl, 2HgS is decomposed by H_2S.

Mercuric Chloride + Hydrogen Sulphide ⟶ Mercuric Sulphide + Hydrogen Chloride

$$HgCl_2 + H_2S \longrightarrow 3HgS + 2HCl$$

The black precipitate becomes red on sublimation the insoluble black form being converted into the stable red form on heating.

Mercuric sulphide is insoluble in water, hot dilute nitric acid and in solution of ammonium sulphide and of caustic alkalis. It dissolves in aqua regia.

$$\underset{\text{Sulphide}}{\text{Mercuric}} + \underset{\text{Acid}}{\text{Nitric}} + \underset{\text{Acid}}{\text{Hydrochloric}} \longrightarrow \underset{\text{Chloride}}{\text{Mercuric}} + \underset{\text{Oxide}}{\text{Nitric}} + \text{Water} + \text{Sulphur}$$

$$3H_2S + 2HNO_3 + 6HCl \longrightarrow 3HgCl_2 + 2NO + 4H_2O + 3S$$

It dissolves in boiling concentrated sulphuric acid forming mercuric sulphate.

$$\underset{\text{Sulphide}}{\text{Mercuric}} + \underset{\text{Acid}}{\text{Sulphuric}} \longrightarrow \underset{\text{Sulphate}}{\text{Mercuric}} + \underset{\text{Dioxide}}{\text{Sulphur}} + \text{Sulphur} + \text{Water}$$

$$3HgS + 2H_2SO_4 \longrightarrow HgSO_4 + SO_2 + S + 2H_2O$$

It burns in air on heating.

$$\text{Mercuric Sulphide} + \text{Oxygen} \longrightarrow \text{Mercury} + \text{Sulphur Dioxide}$$

$$HgS + O_2 \longrightarrow Hg + SO_2$$

And forms mercury on distillation with quick lime or iron fillings. It dissolves in concentrated solution of alkali sulphide, with the formations of the salt K_2HgS_2, potassium mercurous thiosulphate, as a white delequescent substance these thiosulphate are only stable in the presence of an alkline medium

Mercurous Cyanide [$Hg(CN)_2$]: It is formed by the action of aqueous hydrogen cyanide on mercuric oxide or by the reaction between sodium cyanide and mercuric chloride

$$\text{Hydrogen Cyanide} + \text{Mercuric Oxide} \longrightarrow \text{Mercuric Cyanide} + \text{Water}$$

$$2HCN + HgO \longrightarrow Hg(CN)_2 + H_2O$$

It has strong tendency to form complex of the type M[Hg(CN)] and $M_2[Hg(CN)]_4$ where M is the external metal.

Mercuric Thiocyanate [$Hg(SCN)_2$]: It is similarly formed when mercuric salt solution reacts with sodium thiocyanate. It gives no precipitate with potassium hydroxide or potassium iodide since it is partially dissociated. On heating it yields mercury and cyanogen gas.

It forms colourless crystals fairly soluble in water. On heating it yields mercury and cyanogen gas.

$$\text{Mercurous Cyanide} \longrightarrow \text{Mercury} \quad \text{Cyanogen}$$

$$Hg(CN)_2 \longrightarrow Hg + (CN)_2$$

$$\text{Mercuric Chloride} + \text{Sodium Thyocyanate} \longrightarrow \text{Mercuric Thiocyanate} + \text{Sodium Chloride}$$

$$HgCl_2 + 2Na(SCN) \longrightarrow Hg(SCN)_2 + 2NaCl$$

This compound forms many complex like $M[Hg(SCN)_3]$ and $M_2[Hg(SCN)]_4$ (where M is the metal) are the best examples, thiocyanate of heavy metallic ions e.g. Fe^{++}, Co^{++} Ni^{++} and Zn^{++} forming insoluble precipitate with $Hg(SCN)_4$ ion.

Mercuric thiocyanate (made into small pills and when lit with a paper produces long snake like flake mass with cyanogen complex called (Pharaoh's Serpents).

Mercuric Fulminate [$Hg(CNO)_2$]: It is detonator and is used in the explosive industry, it is prepared by the reaction of nitric acid and mercuric nitrate in the presence of alcohol. (equation of this reaction is out of syllabus of this book being organic).

EXPLOSIVE

It can also be prepared by. heating mercuric oxycyanide.

Mercuric Oxychloride $\longrightarrow$ Mercuric Fulminate + Mercury

$$Hg_2(OCN)_2 \longrightarrow Hg(OCN)_2 + Hg$$

Mercurous-ammine Compounds

A variety of ammonomercuric compounds are known, they are classified into three groups.

(*i*) Ammines

(*ii*) Ammono-basic compounds and

(*iii*) Mixed ammonobasic compounds.

Amines are the additive compounds of mercury salts and ammonia fusible white precipitates [$Hg(NH_3)_2$]Cl (since it decomposes into mercurous chloride, ammonia and nitrogen, without melting).

Mercuric Chloride + Ammonia $\longrightarrow$ Mercuroammonium Chloride + Ammonia

$$HgCl_2 + NH_3 \longrightarrow Hg(NH_2)Cl + NH_4Cl$$

Aqueous ammonia do not precipitate mercuric oxide from mercuric chloride solution but gives the white precipitate. The formation of infusible white precipitate by the action of sodium amide on mercuric chloride shows that it contains an amino group and chloride atom directly attacked to mercury.

Mercuric Chloride + Sodium Amide $\longrightarrow$ Sodium Chloride + Mercuric Amine

$$HgCl_2 + NaNH_2 \longrightarrow NaCl + Hg(NH_2)Cl$$

Only mercuric chloride forms amines with ammonia. With aqueous ammonia mercurous chloride forms a black turns containing finely divided mercury and infusible white precipitate.

Mercurous Chloride + Ammonia $\longrightarrow$ Mercuroammonium Chloride + Ammonium Chloride + Mercury

$$Hg_2Cl_2 + NH_3 \longrightarrow Hg(NH_2)Cl + NH_4Cl + Hg$$

Mixed Aquo-ammonobasic Compounds: Both hydrolyzed and ammonolyzed, *e.g.* chloride of Millon's Base. NH_2= O – Hg – Cl.

When infusible white precipitate is digested with at 60°C – 70°C for 12h hydrolysis occurs forming the chloride of Millon's base in a similar manner.

Mercurous Ammonium Chloride Water $\longrightarrow$ Mercuroammonoxy Chloride + Ammonium Chloride

$$2[Hg(NH_3)_2Cl_2 + H_2O \longrightarrow NH_2(Hg_2O)Cl + 2NH_4Cl$$

With aqueous ammonia mercurous nitrate gives a black precipitate containing finely divides mercury and mercuric ammononitrate ($NH_2Hg\text{-}NO_3$) which is hydrolyzed, giving the nitrate of Millon's base.

Mercurous Nitrite + Ammonia ⟶ Mercuric Ammononitrate + Mercury + Ammonium Nitrate

$$Hg_2(NO_3)_2 + NH_3 \longrightarrow Hg(NH_2)NO_3 + Hg + NH_4NO_3$$

Mercuric Ammononitrate + Water ⟶ Millon's Base + Ammonium Nitrate

$$Hg(NH_2)NO_3 + H_2O \longrightarrow [OHg_2NH_2]NO_3 + NH_4NO_3$$

Nessler's Reagent (the alkaline solution of potassium mercuric iodide) reacts with ammonia forming the iodide of Millon's base.

Potassium Mercuric Iodide + Ammonia + Potassium Hydroxide ⟶ Millon's Base Iodide + Potassium Iodide + Water

$$2K_2[HgI_4] + NH_3 + 3KOH \longrightarrow OHg_2NH_2I + 7KI + 2H_2O$$

It is extremely necessary that the presence of the free alkali in the Nessler's reagent.

Millon's Base

This compound is formed when yellow mercuric iodide is wormed with ammonia. It has the composition of $NHg_2H_3O_2$ and then dark brown explosive NHg_2OH. Three structures have been proposed to explain its behaviours.

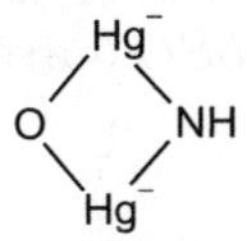

(a)

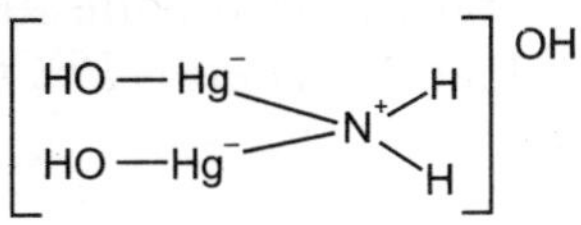

(b)

The above formula is the dimercurammonium Hydroxide (b) on drying in ammonia gas at 125°C gives the Compound NHg_2OH(a)

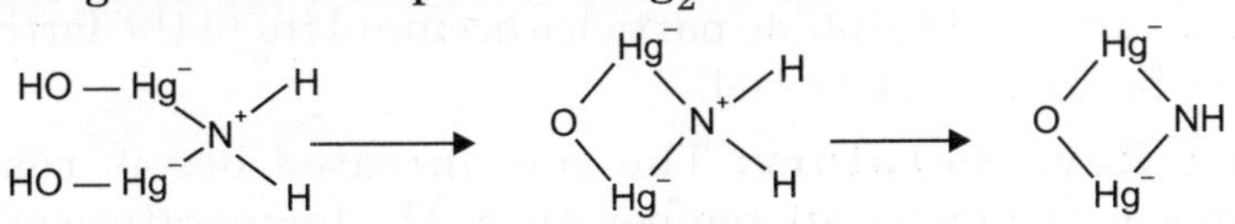

Dihydroammonomercury Ammonium Hydroxide — Loosing One H_2O molecule — Loosing another water molecule

It may be having the formula (b) Dihydroammonomercury ammonium hydroxide that on drying in ammonia first loses one molecule of water and then another molecule at 125°C, the steps in the dehydration is shown in the above reactions equations.

HO –Hg – Hg – NH_2H_2O a formula which is known and usually adopted since it looses H_2O when dried in ammonia and forms NH_2 – Hg – O – Hg which on further dehydration gives Ammonomercurous Oxide, which may have the formula (a).

Millon's base reacts with many acids forming salts of the composition Ohg_2NH_2X ($X = Cl, Br, NO_3$) having one molecule less water, the corresponding bases being NH_2 – Hg –Hg – OH.

Detection of Mercury

All the mercuric compounds produce a shinning mirror when heated with excess of soda lime or sodium carbonate in a bulb tube.

Mercuric Chloride + Sodium Carbonate ⟶ Mercury + Carbon Dioxide + Oxygen + Sodium

$$2HgCl_2 + 2Na_2CO_3 \longrightarrow 2Hg + 2CO_2 + O_2 + 4NaCl$$

LEAD

Melting Point : 327.4°C | Atomic Number : 82

Boiling Point : 1755°C | Atomic Weight : 207.2

History and Occurrence

Lead has been known from very early times. In ancient Egypt and Babilonia it was been as early as in 3000 B.C. In India it came in use during the Vedic period; we find its mention as ~ in Yajurveda and Atharvaveda of about 1000B.C. Lead mines were worked in Britain as early as the 1st century A.D.

The chief ore of lead is *Galena (PbS)*. It is found in the Zawar in Udaypur district of Rajasthan in India, it is also available in Australia, Mexico, Britain and America. It is generally associated with zinc blende and usually consists of 0.01 % to 0.1 % of silver. Among other ores of lead *Cerrasite* ($PbCO_3$) and *Anglesite* ($PbSO_4$) are important.

Extraction

Lead is extracted mostly from galena (PbS) which is the richest ore of lead. The ore is first crushed, sieved and concentrated by the forth floatation process. For this purpose the powdered material is mixed with water and pine oil in a big vessel and compressed air is pushed into the entire mass. The lead sulphide particles accumulate in the forth while the gangue settles down at the bottom of the vessel.

1. **By Melting in a Reverberatory:** The concentrated ore is roasted in a reverberatory furnace. At a moderate temperature. The temperature of furnace is controlled by regulating the air supply through the doors. During the roasting lead is partly oxidized to lead monoxide and partly to lead sulphate.

   Lead Sulphide + Oxygen ⟶ Lead Oxide + Sulphur Dioxide

   $$2Pbs + 3O_2 \longrightarrow 2PbO + 2SO_2$$

   Lead Sulphide + Oxygen ⟶ Lead Sulphate

   $$2PbS + 2O_2 \longrightarrow PbSO_4$$

   More of the sulphate (galena) is then added. The temperature is raised and simultaneously the air supply is reduced. Lead sulphide is allowed to react with two oxidizing products giving lead.

Lead Oxide + Lead Sulphide ⟶ Lead Sulphide + Sulphur Dioxide

$$2PbO + PbS \longrightarrow 3PbS + SO_2$$

Lead Sulphide + Lead Sulphate ⟶ Lead + Sulphur Dioxide

$$PbS + PbSO_4 \longrightarrow 2Pb + 2SO_2$$

Thus, in this process of roasting and smelting (reduction) are carried out separately. The following steps are also involved.

2. **Roasting and Smelting:** The concentrated ore is mixed with some lime, which acts as a flux. The mixture is roasted in a blast of air in a sinterer*. It is mounted on trunnions for charging and discharging. It is provided with a grating at the bottom through the lead sulphide is oxidized to lead oxide.

   Lime acts as a flux and prevents the formation of mass solidification. It also facilitates the forward reaction by combining with sulphur dioxide formed.

3. **Smelting** - The sintered ore is taken into small pieces and mixed with coke and flux (which consists of lime and ferric oxide) and smelted in a blast furnace. During smelting, lead oxide is reduced to metallic lead.

Lead Oxide + Carbon ⟶ Lead + Carbon Monoxide

$$PbO + C \longrightarrow Pb + CO$$

Lime serves to remove silica as calcium silicate

Calcium Oxide + Silicon Dioxide ⟶ Calcium Silicate

$$CaO + SiO_2 \longrightarrow CaSiO_3$$

It also prevents the formation of lead silicate, which would otherwise pass into a slag resulting in the loss of lead.

Ferric oxide reacts with any lead sulphide, which might have escaped oxidation during roasting.

Lead Sulphide + Ferric Oxide + Carbon ⟶ Ferric Sulphide + Lead + Carbon Monoxide

$$2PbS + Fe_2O_3 + 3C \longrightarrow 2FeS + 2Pb + 3CO$$

The molten lead is drawn off from the tapping hole. The slag containing calcium silicate and iron sulphide is removed from the slag hole.

Purification of Lead

Lead prepared by the above method always contains impurities such as silver, copper, tin and iron, which make metal hard and brittle. It is refined by following processes.

(a) ***Liqufication:*** Lead is heated on the sloping hearth of a furnace. It melts at a lower temperature thus many of the impurities do. Hence, when the temperature of the furnace apparatus reaches the melting point, the molten metal flows down the slope, while the impurities remain sticking the hearth.

---

* Sinterer: Roaster

(b) ***Oxidation:*** The base metal impurities (copper, tin and iron) are removed by softening. The impure lead is melted on the heat of the shallow flat-bottomed reverberatory furnace and air is passed over the molten mass. The impurities (copper, tin and iron are oxidized and form a scum on the surface which is removed.

(c) **Desilverization:** After softening silver is removed by well known Parker or Pattinson's process.

(d) ***Electrolytic Refining:*** The desilverized lead can be purified further by electrolyzing a solution of lead See Fig. 16.28. Fluorosilicate of lead (PbSiFe) and hydroflurosilicic acid (H_2SiF_6) containing a little gelatin. The desilverized lead is cast into anodes and while the cathodes are pure lead, Gelatin facilitates the formation of a simultaneous deposit of lead at the cathode. The metallic impurities, which are more electropositive, than lead, such as iron and tin go into solution while the rest of the impurities are thrown away in the anode mud.

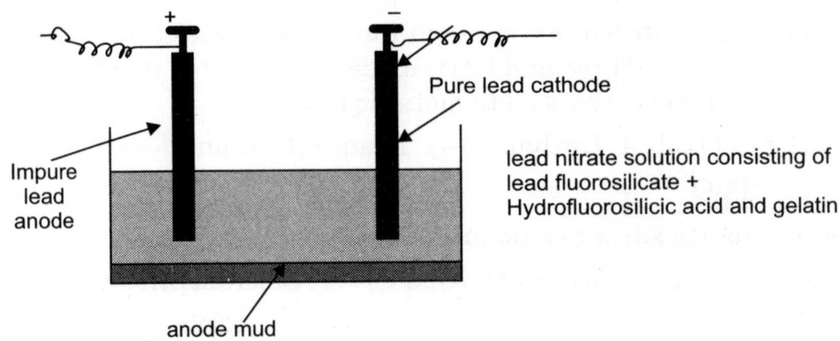

Fig. 16.28. The Electrolytic Refining of Lead

Physical Properties

Melting Point	327.4°
Boiling Point	1755°C
Density	11.35 g/cc
Heat of Capacity	6.4 cal deg^{-1} mol^{-1}
Standard Entropy	15.5 cal deg^{-1} mol^{-1}
Vapour Pressure	1.0 mm of Hg
Surface Tension	442 dynes/sqc
Viscosity	2.116 centippois
Specific Heat	0.0297 cal/g
Latent heat of Vapourization	207203 cal/g
Latent Heat of Fusion	5.86 cal/g
Thermal Conductivity	0.083 cal(sec) (sqcm) (°C/cm) at 20°C
Co-efficient of linear Expansion	27×10^{-6}
Resistivity	20.65 microohm/cm
Brinell Hardness	4.2
Tensile Strength	.1241 MPa

Chemical Properties

1. Action of Air: It is not attacked by dry air, but gets tarnished in moist air due to the formation of basic carbonate, which prevents it from further corrosion.

2. Action of Heat and Air: When it is heated with air lead is oxidized to lead oxide called litherage (PbO) and Red Lead ($Pb3O_4$).

$$\text{Lead} + \text{Oxygen} \longrightarrow \text{Litherage}$$
$$2Pb + O_2 \longrightarrow 2Pb_3O_4$$

3. Action of Water: Lead is not attracted by distil water (except at boiling point). Slowly decomposes steam at high temperature. It is readily corroded by water consisting of dissolved air forming lead hydroxide, which is approximately soluble in water. The solvent action of water, known a *Plumbo Solvency* is greater when it is not present with certain salts called phosphate, carbonates and sulphates since they form a coating of insoluble lead salts on the surface of the metal which protects it from further action. Thus hard water have no solvent action on lead. Lead pipes are often used for carrying water the formation of the soluble lead salts are minimized by passing water through the beds of limestone and chalk. The carbonates dissolved in water from the protective coating on the interior of the pipes, which protects the metal from the action.

4. Action of Acids: Lead is only slightly attacked by dilute hydrochloric acid or dilute sulphuric acid in the cold because a film of insoluble lead chloride or lead sulphate is formed on the surface, which prevents further action. Powdered lead is readily dissolved by boiling hydrochloric due to formation of the soluble complex H_2PbCl_4, Chloroplumbic acid.

$$\text{Lead} \quad \text{Hydrochrloric Acid} \longrightarrow \text{Lead Chloride} + \text{Hydrogen}$$
$$Pb + 2HCl \longrightarrow PbCl_2 + H_2$$

Concentrated sulphuric acid dissolves lead readily at 230°C- 240°C with the evolution of sulphur dioxide but the action is arrested by the formation of the insoluble layer (crust) of lead sulphate on the metal surface.

$$\text{Lead} + \text{Sulphuric Acid} \longrightarrow \text{Lead Sulphate} \quad \text{Water} + \text{Sulphur Dioxide}$$
$$Pb \quad 2H_2SO_4 \longrightarrow PbSO_4 + 2H_2O + SO_2$$

With cold and dilute nitric acid, it evolves brown fumes of nitrogen dioxide.

$$\text{Lead} + \text{Nitric Acid} \longrightarrow \text{Lead Nitrate} + \text{Nitrogen Dioxde} + \text{Water}$$
$$Pb + 4NHO_3 \longrightarrow Pb(NO_3)_2 + 2NO_2 + 2H_2O$$
$$\text{Lead Chloride} + \text{Hydrochloric Acid} \longrightarrow \text{Chloroplumbic Acid}$$
$$PbCl_2 + 2HCl \longrightarrow H_2PbCl_2$$

Lead dissolves in acetic acid in the presence of oxygen forming lead acetate.

$$\text{Lead} + \text{Acetic Acid} + \text{Oxygen} \longrightarrow \text{Lead Acetate} + \text{Water}$$
$$Pb + 2CH_3COOH + \tfrac{1}{2}O_2 \longrightarrow (CH_3COO)_2Pb + H_2O$$

5. Action of Alkalis: It is slowly attacked by caustic alkalis with the evolution of hydrogen.

$$\text{Lead} + \text{Sodium Hydroxide} \longrightarrow \text{Sodium Plumbate} + \text{Hydrogen}$$
$$Pb + 2NaOH \longrightarrow Na_2PbO_2 + H_2$$

6. Action of Chlorine and Sulphur: It also reacts with chlorine and sulphur to form lead tetrachloride and sulphide respectively.

$$\text{Lead} + \text{Chlorine} \longrightarrow \text{Lead Tetrachloride}$$
$$Pb + 2Cl_2 \longrightarrow PbCl_4$$
$$\text{Lead} + \text{Sulphur} \longrightarrow \text{Lead Sulphide}$$
$$Pb + S \longrightarrow PbS$$

Uses of Lead

- ➢ For making Telegraph and telephone wires which are to be buried under earth.
- ➢ For the construction of lead chambers in the industrial preparation of sulphuric acid.
- ➢ It is also used for making colour pigments and dies.
- ➢ It also used for making lead shots and bullets.
- ➢ Wood tanks with lead are frequently used for chemical operation in large scale.

Lead Poisoning

Acute poisoning is unusual but chronic poisoning due to absorption of small amounts over a period is less uncommon. This can occur after a regular contamination of lead articles. The effect of lead poisoning causes anæmia loss of appetite and the formation of blue line round the gums are characteristic symptom.

Lead Pigments

A few insoluble lead compounds are used as pigments in the preparation of oil paints, few of then are stated below:

(a) White Lead [Pb(OH)$_2$.2Pb(CO$_3$)]: It is prepared by the action of lead nitrate and sodium carbonate.

$$\text{Sodium Carbonate} + \text{Lead Nitrate} \longrightarrow \text{White Lead} + \text{Sodium Nitrate} + \text{Sodium Bicarbonate}$$
$$4Na_2CO_3 + 3Pb(NO_3)_2 \longrightarrow Ph(OH)_2\cdot 2Pb(CO_3) + 6NaNO_3 + 2NaHCO_3$$

(i) Read Lead (Pb3O$_4$): It is prepared by heating lead monoxide at 473°C – 450°C for a long time.

$$\text{Lead Oxide} + \text{Oxygen} \longrightarrow \text{Red Lead}$$
$$6PbO \quad O_2 \quad 2Pb_3O_4$$

Red lead is brilliant red powder. On heating it becomes violet and finally black. However, on cooling the original red colour is restored. It behaves as mixture of lead monoxide and dioxide.

$$\text{Red Lead} \longrightarrow \text{Lead Oxide} + \text{Lead Dioxide}$$
$$2Pb_3O_4 \longrightarrow 2PbO + PbO_2$$

On strong heating it decomposes giving lead monoxide and cooling oxygen.

$$\text{Red Lead} \longrightarrow \text{Lead Oxide} + \text{Oxygen}$$
$$2Pb_3O_4 \longrightarrow 6PbO + O_2$$

With nitric acid red lead reacts to form lead nitrate and lead dioxide.

$$\text{Read Lead} + \text{Nitric Acid} \longrightarrow \text{Lead Nitrate} + \text{Lead Dioxide} + \text{Water}$$
$$Pb_3O_4 + 4HNO_3 \longrightarrow 2Pb(NO_3)_2 + PbO_2 + 2H_2O$$

With concentrated hydrochloric acid, it forms lead chloride and chlorine.

$$\text{Lead Oxide} + \text{Hydrochloric Acid} \longrightarrow \text{Lead Chloride} + \text{Water} + \text{Chlorine}$$
$$Pb_3O_4 + 8HCl \longrightarrow 3PbCl_2 + 4H_2O + Cl_2$$

It is used in making flint glass, paints and glazes for pottery.

(ii) Chrome Yellow ($PbCrO_4$): It is obtained as a yellow precipitate by mixing chromate with lead acetate solution.

$$\text{Lead Acetate} + \text{Potassium Chromate} \longrightarrow \text{Lead Chromate} + \text{Potassium Acetate}$$
$$(CH_3COO)_2Pb + K_2Cr_2O_4 \longrightarrow PbCrO_4 + 2CH_3COOK$$

If the solutions are warmed before mixing the colour is much deeper. This is due to the granular particles. It is used as yellow pigment. If less deep is required white lead is added.

(iii) Chrome Red ($PbO, PbCrO_4$): It is obtained by heating a solution of sodium hydroxide and chrome yellow.

$$\text{Chrome Yellow} + \text{Sodium Hydroxide} \longrightarrow \text{Chrome Red} + \text{Sodium Chromate} + \text{Water}$$
$$2PbCrO_4 + 2NaOH \longrightarrow PbO.PbCrO_4 + Na_2CrO_4 + H_2O$$

It is used as a red pigment.

Oxide halides and some other compounds are to be discussed under this sub head which is following.

Compounds of Lead

Lead Sesquioxide (Pb_2O_3): It is reddish yellow powder, formed by the action of sodium hypochlorite upon a cold solution of lead monoxide in potassium hydroxide.

It reacts with dilute nitric acid forming lead dioxide and lead nitrate.

Suggesting the constitution of $PbO.PbO_2$. It may be reported as lead metaplumbate $[Pb(PbO_3)]$.

Lead Dioxide (PbO_2): It is a chocolate-brown powder, is made by the action of hot dilute nitric acid on red lead the lead dioxide is filtered off and it is washed and then dried.

$$\text{Read Lead} + \text{Nitric Acid} \longrightarrow \text{Lead Dioxide} + \text{Nitrogen Dioxide} + \text{Hydrogen}$$
$$Pb_3O_4 + 2HNO_3 \longrightarrow 3PbO_2 + 2NO_2 + H_2$$

It is insoluble in water also in nitric hydrochloric and dilute sulphuric acid. But concentrated sulphuric acid reacts with lead dioxide forming lead sulphate and sulphur dioxide.

Lead Dioxide + Sulphuric Acid ⟶ Lead Sulphate + Water + Oxygen

$$PbO_2 + H_2SO_4 \longrightarrow 2PbSO_4 + H_2O + O_2$$

It dissolves in hot sodium hydroxide solution forming metaplumbate.

Lead Dioxide + Sodium Hydroxide ⟶ Sodium Metaplumbate + Water

$$PbO_2 + 2NaOH \longrightarrow Na_2PbO_3 + H_2O$$

It is a powerful oxidizing agent, it oxidizes sulphur dioxide spontaneously forming lead sulphate.

Lead Dioxide + Sulphur Dioxide ⟶ Lead Sulphate

$$PbO_2 + SO_2 \longrightarrow PbSO_4$$

A mixture of lead dioxide and sulphur when inflamed combines to form lead sulphide in presence of nitric acid and oxalic acid and it also forms carbon dioxide.

Lead Dioxide + Nitric Acid + Oxalic Acid ⟶ Lead Nitrate + Carbon Dioxide + Water

$$PbO_2 + 2HNO_3 + H_2C_2O_4 \quad Pb(NO_3)_2 + CO_2 + H_2O$$

A pink solution of permanganic acid is produced when a solution of manganous sulphate reacts with lead dioxide in the presence of concentrated nitric acid.

Magnanous Sulphate + Lead Dioxide + Nitric Acid ⟶ Permanganic Acid + Lead Sulphate + Lead Nitrate + Water

$$2MnSO_4 + 5PbO_2 + 6HNO_3 \longrightarrow 2HMnO_4 + PbSO_4 + 3Pb(NO_3)_2 + 2H_2O$$

Lead dioxide is stable up to 300°C but at higher temperature it decomposes to lead oxide and oxygen.

Lead Hydroxide [$Pb(OH)_2$]: It is produced as a white precipitate by adding sodium hydroxide (not in excess) to a solution of lead salt. It is slightly soluble in water and the solution turns red litmus blue. It dissolves in acids to form lead salt and in alkalis to give potassium plumbates. The plumbate gets readily oxidized.

Lead Nitrate + Sodium Bicarbonate ⟶ Lead Carbonate + Sodium Nitrate + Water + Carbon

$$Pb(NO_3)_2 + 2NaHCO_3 \longrightarrow PbCO_3 + 2NaNO_3 + H_2O + CO_2$$

The precipitate of lead carbonate is slightly soluble in water (0.6/lit), but readily soluble in ammonium acetate or dilute acetic acid solution.

Lead Halides

	Lead Fluoride	Lead Chloride	Lead Bromide	Lead Iodide
Melting Point	820°C	498°C	373°C	402°C
Boiling Point	1284°C	950°C	916°C	954°C

Lead precipitate of lead carbonate is slightly soluble in water, but are more soluble in hot water than cold. Their melting point, boiling points are as follows:

Lead Fluoride (PbF_2): This compound is produced when sodium fluoride is added to a solution of lead nitrate.

Lead Nitrate + Sodium Fluoride ⟶ Lead Fluoride + Sodium Nitrate

$$Pb(NO_3)_2 + 2NaF \longrightarrow PbF_2 + 2NaNO_3$$

Lead Chloride ($PbCl_2$): It is obtained as a white precipitate by adding dilute hydrochloric acid to a solution of lead nitrate solution. It is nearly insoluble in cold water (nearly 1%) but more soluble in boiling water (4% nearly) from which it separates in the form of needle like structures on cooling. It dissolves in hot concentrated hydrochloric acid and in concentrated alkali chloride due to the formation of complex compounds these are decomposed on dilution with water with the separation of lead chloride.

Lead Chloride + Hydrochloric Acid ⟶ Chloroplumbic Acid

$$PbCl_2 + 2HCl \longrightarrow H_2PbCl_4$$

Complex crystalline salts such as K_2PbC_4 and KPb_2CIs are known as litherage reacts with sodium chloride solution on boiling, giving the yellow oxychloride (.$PbCl_2$.4PbO).

Lead Oxide + Sodium Chloride + Water ⟶ Sodium Hydroxide + Lead Oxychloride

$$PbO + 2NaCl + H_2O \longrightarrow 2NaOH \quad PbCl_2\cdot_4PbO$$

This is used as pigments under the name of Tumess-yellow, Cassel-yellow, $PbCl_27H_2O$, it made by heating lead oxide and ammonium chloride. Pallison's white is the white lead oxychloride [Pb(OH)Cl].

Lead Bromide ($PbBr_2$): It is obtained as white precipitate. It is approximately as soluble as chloride of lead.

It is prepared by the action of hydrogen bromide and lead oxide.

Hydrogen Bromide + Lead Oxide ⟶ Lead Bromide + Water

$$2HBr + PbO \longrightarrow PbBr_2 + H_2O$$

Calcium Orthoplumbate (Ca_2PbO_4): This compound is produced when a mixture of calcium oxide and lead oxide is heated in the presence of air.

Lead Oxide + Calcium Oxide + Oxygen ⟶ Calcium Orthoplumbate

$$2PbO + 4CaO + O_2 \longrightarrow 2Ca_2PbO_4$$

It can be easily decomposed by very easy means; it is dissolved in water containing carbon dioxide.

Calcium Orthoplumbate + Carbon Dioxide ⟶ Lead Dioxide + Calcium Carbonate

$$Ca_2PbO_4 + 2CO_2 \longrightarrow PbO_2 + 2CaCO_3$$

It is very slightly soluble in water. When its suspension is deposited with sodium peroxide it forms calcium metaplumbate (Ca_2PbO_3).

Calcium Orthoplumbate	+ Sodium + Peroxide	⟶	Calcium Metalplumbate	+ Sodium Oxide	+ Oxygen
$2Ca_2PbO_4$	+ Na_2O	⟶	$2Ca_2PbO_3$	+ 2NaO	+ O_2
Calcium Orthoplumbate	+ Sodium Peroxide	⟶	Calcium Metalplumbate	+ Sodium Oxide	+ Calcium Oxide
Ca_2PbO_4	+ Na_2O_2	⟶	$2Ca_2PbO_3$	+ 2NaO	+ CaO

Potassium Hexahydroplumbate [$K_2Pb(OH)_6$]: These compounds are obtained by adding lead dioxide PbO_2 with KOH yields potassium plumbate soluble in water.

Lead Dioxide	+ Potassium Hydroxide	⟶	Potassium Plumbate	+ Water
PbO_2	+ 2KOH	⟶	K_2PO_3	+ H_2O

Potassium Plumbate + Water ⟶ Potassium Hexahydrodroplumbate

PbO_2 + 2KOH ⟶ $K_2Pb(OH)_6$

Its solution in water containing excess of potassium hydroxide $K_2[Pb(OH_6]$ is separated by careful evaporation in vacuum. It is similar with $K_2Pt(OH)_6$, potassium hexahydroplatinate.

Lead Iodide (PbI_2): This compound is formed as a yellow precipitate. It differs from other lead halides as because it is bright yellow colour and being less soluble in water (0.06% at 15°C). It dissolves in boiling (4.34 g/lit) to give a colourless liquid from which it separates out as yellow powder. It dissolves in excess of potassium iodide solution forming of lead iodide.

Lead Nitrate [$Pb(NO_3)_2$]: It is made by dissolving metallic lead, the oxide or carbonate in hot dilute nitric acid. The solution when crystallized or carbonate in hot dilute nitric acid. The solution when crystallized it deposits in the form of milky white crystals. It readily soluble in water. On heating lead nitrate decomposes to nitrogen dioxide, oxygen

Lead Sulphate ($PbSO_4$): It is obtained as a heavy white precipitate by adding dilute sulphuric acid to lead oxide.

It is slightly soluble in water (1 in 12000) and particularly insoluble in a mixture of dilute sulphuric acid and alcohol (1 in 36000). It dissolves in hot ammonium acetate solution due to the formation of freely ionized lead acetate.

It also dissolves to some extent (6%) in hot concentrated sulphuric acid but deposits in rhombic crystals on cooling.

Sublimated white lead as a mixture of Lead Sulphate, Lead Oxide, and Zinc Oxide in the ratio of 15:4:1 is made by burning galena, containing zinc sulphide in an oxidizing atmosphere and collecting fumes.

Lead Sulphide (PbS): It occurs naturally as Galena. It is prepared as a black mass passing H_2S through a solution of lead salt. The precipitate often yellow or red in the presence of excess of hydrochloric acid due to the formation lead sulphochloride, which however is decomposed on dilution and by passing an excess of hydrogen sulphide giving the black sulphide.

Lead burns in sulphur vapour producing lead sulphide. It is insoluble in yellow ammonium sulphide separation of sulphur.

It also dissolves in hot concentrated hydrochloric acid; hydrogen peroxide converts it into lead sulphate.

Lead Tetrafluoride (PbF_2): If is the compound of lead, which is prepared by passing fluorine over lead dioxide or lead difluoride.

Lead Difluoride Fluorine $\longrightarrow$ Lead Tetrafluoride

$$PbF_2 + F_2 \longrightarrow PbF_4$$

Lead Oxide + Fluorine $\longrightarrow$ Lead Tetrafluoride + Oxygen

$$2PbO + F_2 \longrightarrow PbF_4 + O_2$$

Lead tetrafluoride decomposes at a high temperature, giving off fluorine and hence it is a good fluorinating agent.

Lead Tetrachloride ($PbCl_4$): It is present as the complex chloroplumbic acid, $H_2[PbCl_6]$ in the yellow solution of lead dichloride in ice cold concentrated hydrochloric acid saturated with chlorine.

Lead Dioxide + Hydrochloric Acid $\longrightarrow$ Lead Tetrachloride + Water

$$PbO_2 + HCl \longrightarrow PbCl_4 + 2H_2O$$

Lead Tetrachloride + Hydrochloric Acid $\longrightarrow$ Chloroplumbic Acid

$$PbCl_4 + 2HCl \longrightarrow H_2[PbCl_6]$$

Lead tetrachloride cannot be isolated from this solution as it readily breaks up into lead chloride and chlorine due to its insolubility. On addition of ammonium chloride to the complex of chloroplumbic acid, ammonium chloroplumbate is obtained.

Chloroplumbic Acid + Ammonium Chloride $\longrightarrow$ Ammonium Chloroplumbate + Hydrogen Chloride

$$H_2[PbCl_6] + 2NH_4Cl \longrightarrow (NH_2)_2PbCl_6 + 2HCl$$

The ammonium chloroplumbate is decomposed by adding it to cold and concentrated sulphuric acid, when lead tetrachloride (being insoluble in concentrated sulphuric acid) separates as heavy yellow oily liquid.

Ammonium Chloroplumbate + Sulphuric Acid $\longrightarrow$ Ammonium Sulphate + Hydrogen Chloride + Lead Tetrachloride

$$(NH_2)_2PbCl_6 + H_2SO_4 \longrightarrow (NH_4)_2SO_4 + 2HCl + PbCl_4$$

Ammonium Chloroplumbate is a liquid with a density of 3.2 g/cc at 0°C freezes at –15°C to a evolution of chlorine. At 150°C it explodes. It fumes strongly in air since it is hydrolyzed by water to form lead dioxide at ordinary temperature. The alkali hexachloroplumbate are similar with alkali hexachloroplatinates and hexachlorostannates.

Lead Acetate, Sugar of Lead [$Pb(CH_3COO)_2$]: This compound of Lead is prepared by the action of lead oxide and acetic acid.

POISON

Lead Oxide + Acetic Acid ⟶ Lead Acetate + Water

$$PbO + 2CH_3COOH \longrightarrow (CH_3COO)_2Pb + H_2O$$

It is sweet in taste but it is a poison. It is used for making a lot of lead compounds such as lead dichromate and lead chromate, it is also used for making the dyes and other chemical products. In laboratory it is used for the detection of hydrogen sulphide.

Lead Tetraacetate [$Pb(CH_3COO)_3$]: This compound is obtained by cooling a solution of lead oxide (red) in hot glacial acetic acid.

Lead Oxide + Galical Acetic Acid ⟶ Lead Tetraacetate + Lead Acetate + Water

$$Pb_3O_4 + 8CH_3COO1 \longrightarrow Pb(CH_3COO)_4 + 2Pb(CH_3COO)_2 + 4H_2O$$

Half of the lead acetate form simultaneously can be converted into lead tetraacetate by passing a stream of, chlorine the rest being precipitated at $PbCl_2$

Lead Acetate + Chlorine ⟶ Lead Tetraacetate + Lead Chloride

$$2(CH_3COO)_2Pb + Cl_2 \longrightarrow Pb(CH_3COO)_4 + PbCl_2$$

Lead tetraacetate melts at 175°C. It is more stable than the lead salts of stronger acids. It decomposes at 195°C giving the lead acetate. It is instantly hydrolyzed by the water to give lead dioxide. Its solvents include benzene and chloroform. Shows that it is a covalent compound. It is used as an oxidizing agent in organic chemistry.

Lead Sulphate [Pb(SO4)$_2$]: It is obtained by the electrolysis of 80% sulphuric acid in the presence of lead as anode positive electrode') keeping the temperature below 30°C. The compound separates out as yellow crystals. It is decomposed by water and lead dioxide is deposited.

Lead Sulphate + Water ⟶ Lead Dioxide + Sulphuric Acid

$$Pb(SO_4)_2 + 2H_2O \longrightarrow PbO_2 + 2H_2SO_4$$

Lead Hydride, Plumbane (PbH_4): It is an unstable gas and is obtained as small quantities with the action of dilute acids upon lead-magnesium alloy. It can be detected by the use of a radioactive indicator in the presence of thorium B (an isotope of lead) a radioactive gas is formed. The gas can also be produced by the action of lead and atomic hydrogen in the presence of carbon.

Lead + Atomic Hydrogen ⟶ Lead Hydride

$$Pb + 2H_2 \longrightarrow PbH_4$$

The gas may be condensed and re-evaporated. It behaves as a mirror of lead on passing through heated tube. Higher plumbanes do not exist.

Detection of Lead and Estimation

1. When lead compound is heated with sodium carbonate or charcoal a soft malleable bead of lead (which marks paper) surrounded by a yellow coloured hard coating is produced.
2. When potassium iodide is added lead iodide is produced.

   Lead is estimated volumetrically by dissolving the precipitated lead chromate in cold hydrochloric acid and estimating the liberated chromic acid eudiometrically.*

$$\text{Lead Chromate} + \text{Hydrochloric Acid} \longrightarrow \text{Lead Chloride} + \text{Chromic Acid}$$

$$PbCrO_4 + 2HCl \longrightarrow PbCl_2 + H_2CrO_4$$

* Eudiometry : This is discussed in tome 3.

CHAPTER

17

Elements of Semiconductor

Introduction

In this chapter we are going to discuss the elements from which the modern semiconductor is made. Now question is what is this semiconductor?

In electronics the electricity is carried out by the conductors, which offer minimum resistance to the flow of electricity, another type of medium is available which is called insulator, this offers maximum resistance. In between these there is some thing, which offers less resistance than insulator, but more resistance than conductor. These substances are called semiconductors (a conductor like material). Now these semiconductors are made from these elements which we are going to discuss.

SILICON

Melting Point : 14100C
Boiling Point : 2335°C

Atomic Number : 14
Atomic Weight : 28.05

History and Occurrence

In the year 1809 Gay Lussac and Themard prepared silicon by passing the vapour of silicon tetrafluoride over heated potassium.

$$\text{Silicon Tetrafluoride} + \text{Potassium} \longrightarrow \text{Silicon} + \text{Potassium Fluoride}$$

$$SiF_4 + K \longrightarrow Si + 4KF$$

After oxygen silicon is the most abundant element available on earth. But like oxygen and carbon it is not found in free state. Silicon occupies 25% of the earth's crust, where oxygen occupies 50%.

Silicon occurs mainly in the form of silicon dioxide or better known as silica, which is present in the sand.

Silicon Dioxide (SiO_2): This compound is found in the nature in different forms also in the form of quartz. This quartz is available in the form of alpha-quartz and beta-quartz. There is some other form of quartz also available they are called tridymite and crystobelite. The transformation is shown below :

$$\alpha\text{-quartz} \underset{}{\overset{575°C}{\rightleftharpoons}} \beta\text{-quartz} \underset{}{\overset{870°C}{\rightleftharpoons}} \text{tridumite} \underset{}{\overset{1470°C}{\rightleftharpoons}} \text{crystobetite} \underset{}{\overset{1710°C}{\rightleftharpoons}} \text{liquid}$$

Clear colourless crystal quartz is used for the preparation of lens in optical instruments. Many varieties of quartz possess delightful colours, owing to the presence of small amounts of impurities and are used as gems for example *amethyst, jasper opal.* Sandstone consists of sand grains cemented together with iron oxide. Flints are compounds of a mixture of quartz and amorphous silica.

Silica is also available in plants and animals. The stems and the resistant coating of straw and bamboo the quills of feather the claws of animal and the finger nails consist of a fair quantity of silica.

Preparation of Silicon

Silicon is prepared by the reaction of silicon dioxide and potassium in laboratory.

$$\text{Silicon Dioxide} + \text{Potassium} \longrightarrow \text{Silicon} + \text{Potassium Oxide}$$

$$SiO_2 + 4K \longrightarrow Si + 2K_2O$$

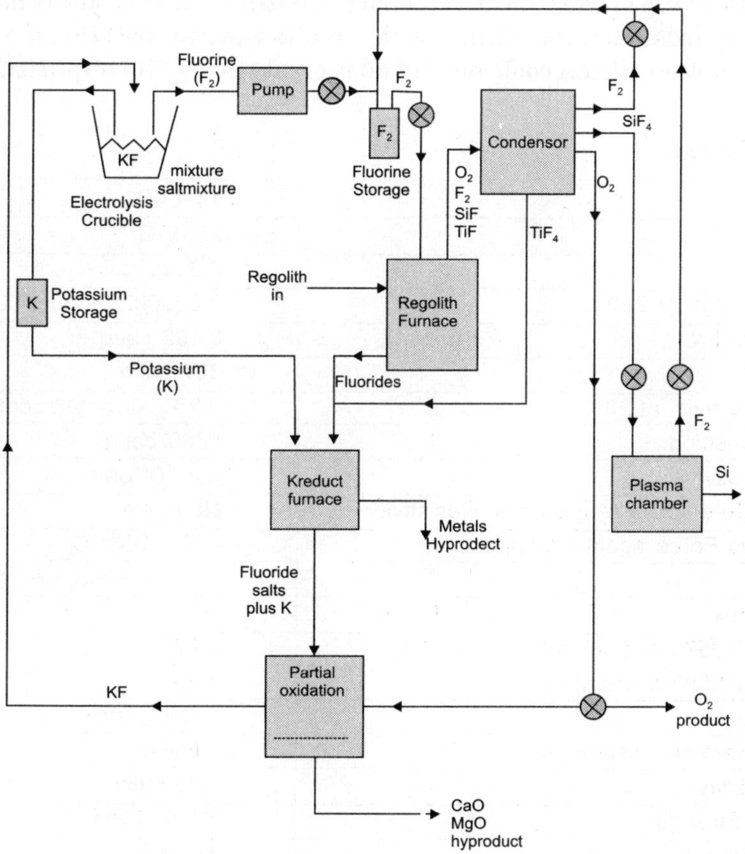

Fig. 17.1. Block Diagram for Silicon Extraction

Magnesium oxide and any unreacted silica are removed by treatment with hydrofluoric acid in polyethylene vessel. Silicon lest behind is a brown amorphous powder.

Production of Silicon

Silicon is obtained by reduction of silica by magnesium or carbon at high temperature. It has been observed that silicon obtained by this process consists of silicon carbide in high amount as impurities. The process here described in Fig. 17.2 is used for the preparation of silicon in industry at a pure state.

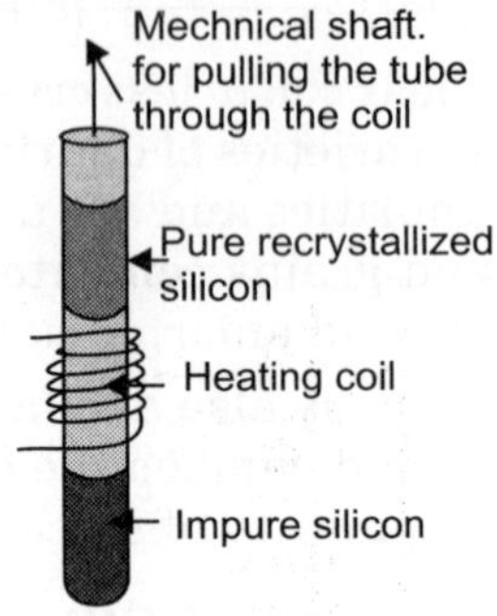

Fig. 17.2. The Refining of Silicon

Very pure silicon (99.99%) is required for the manufacture of semiconductors and integrated circuits. This type of pure silicon is available by zone filtering is the process in which the impurities are removed from the solid phase by melting and resolidifying the melt. Zone refining is the process of recrystallization.

The material to be purified is packed in a cylindrical tube and surrounded vertically (see Fig 17.2). At the top of the tube an electrical heating coil surrounds a portion of it. The tube is drawn upwards during refining, which melts the impure silicon and as the heated zone goes up the molten silicon cools down and recrystallizes, the impurities are left behind in the tube.

Physical Properties

Melting Point	14100C
Boiling Point	2335°C
Density	2.33 g/cc
Thermal Expansion at 20°C	0.0000028 – 0.0000073
Specific Heat at 0°C	0.162 cal/g
Latent Heat of Fusion	337 cal/g
Thermal Conductivity at 20°C	0.20 cal/(sec)(sqcm)(°C/cm)
Heat of Composition	6800 cal/g
Electrical Resistivity	6×10^4 ohm cm
Temperature Co-efficient of Electrical Resistivity	Negative
Thermo-electric Force against Copper	478×10^{-5}/°C
Hall Effect	41.0v-cm/ampere gausses
Refractive Index	
At Lambda = 0.0578 micrometre	3.87
At Lambda = 0.0598 micrometre	4.24
Crystal Structure	Cubic diamond a = 5.1417Å
Minimum Interatomic Distances	2.346Å
Modulus Elasticity	1.12 MPa
Compressive Strength	10767 MPa
Transverse Strength	159 Kg
Modulus Rupture	6.2 MPa

Chemical Properties

1. Amorphous Silicon is a light brown powder, while the crystalline silicon forms yellow to brown octahedral crystals with metallic lusture both the varieties are fusible at high temperature in an electric furnace.
2. It is a non-metal its oxide distinctly acid in character, but it is metallic in many of its physical properties. Crystalline silicon possesses metallic lusture and marked metallic conductivity that is why it is called as a semi-metal or metalloid.
3. Amorphous silicon burns vigorously when heated in oxygen and slowly burns when heated in air.

   $$\text{Silicon} + \text{Oxygen} \longrightarrow \text{Silicon Dioxide}$$
   $$Si + O_2 \longrightarrow SiO_2$$

   But the crystalline silicon does not burn in oxygen even at high temperature.
4. Silicon is insoluble in water, but it is slowly forming silicon tetrafluoride (SiF_4) these halides are decomposed by water.

   $$\text{Silicon Tetrafluoride} + \text{Water} \longrightarrow \text{Hydrogen Fluoride} + \text{Orthosilicic Acid}$$
   $$SiF_4 + 4H_2O \longrightarrow 4HF + Si(OH)_4$$
5. It is insoluble in any single acid but it is attached by a mixture of nitric acid and hydrofluoric acid.

   $$\text{Silicon} + \text{Hydrogen Fluoride} \xrightarrow{\text{Nitric Acid}} \text{Hydrogen} + \text{Hydrofluorosilicic Acid}$$
   $$Si + 6HF \xrightarrow{HNO_3} 2H_2 + H_2SiF_6$$
6. It dissolves in hot concentrated caustic alkalis or in the form of sodium carbonate forming sodium silicate.

   $$\text{Silicon} + \text{Sodium Hydroxide} + \text{Water} \longrightarrow \text{Sodium Silicate} + \text{Hydrogen}$$
   $$Si + 2NaOH + H_2O \longrightarrow Na_2SiO_3 + 2H_2$$
   $$\text{Silicon} + \text{Sodium Carbonate} \longrightarrow \text{Sodium Silicate} + \text{Carbon}$$
   $$Si + Na_2CO_3 \longrightarrow Na_2SiO_3 + C$$
7. A mixture of volatile hydrides of silicon, such as silico-methane (SiH_4) silico-ethane (Si_2H_6). These compounds can be prepared by the action of hydrochloric acid and magnesium silicide in the absence of air. The silicon hydrides are spontaneously inflammable in air.

   $$\text{Magnesium Silicide} + \text{Hydrochloric Acid} \longrightarrow \text{Magnesium Chloride} + \text{Silico-methane}$$
   $$MgSi + 4HCl \longrightarrow MgCl_2 + SiH_4$$
8. It combines with metals at high temperature, forming metallic silicides (MgSi).

Uses of Silicon

1. As a deoxidizer in making steel.
2. In the manufacture of alloys such as *silicon-bronze, manganese-silicon bronze, ferrosilicon* and acid resistant silicon steel, such as *ironac, tatiron,* and *durion.* Silicon is used for making semiconductors. Ferrosilicon is made by electromechanical reduction of silica and ferric oxide with carbon.

Compounds of Silicon

Silica, Silicon Dioxide (SiO_2) : Silica occurs in amorphous and crystalline form as stated earlier. Pure silica is a white solid but it looks yellowish and brown due to the presence of iron oxide as impurity, which can be removed by washing it with hydrochloric acid.

Pure amorphous silica is made from sand or silicate mineral by passing with excess sodium silicate on acidification with hydrochloric acid yields a gelatinous precipitate of silicic acid, which is washed with water and dried and ignited.

Magnesium Silicate + Sodium Carbonate ⟶ Magnesium Carbonate + Sodium Silicate

$$MgSiO_3 + Na_2CO_3 \longrightarrow MgCO_3 + Na_2SiO_3$$

Sodium Silicate + Hydrochloric Acid Sodium Chloride + Silicic Acid

$$Na_2SiO_3 + HCl \longrightarrow 2NaCl + 2H_2SiO_3$$

Silicic Acid ⟶ Water + Silicon Dioxide

$$H_2SiO_3 \longrightarrow 2H_2O + SiO_2$$

It may also be prepared by passing silicon tetrachloride through water the gelatenous mass of orthosilicic acid is formed is washed with water and dried and then ignited.

Silicon Tetrachloride + Water ⟶ Hydrogen Chloride + Orthosilicic Acid

$$SiCl_4 + H_2O \longrightarrow 4Cl + Si(OH)_4$$

Orthosilicic Acid ⟶ Water + Silicon Dioxide + Water

$$Si(OH)_4 \longrightarrow 2H_2O + SiF_4 + 2H_2O$$

Properties : 1. Silica is insoluble in water and in all acids but except hydrofluoric acid.

Silicon Dioxide + Hydrofluoric Acid ⟶ Silicon Tetrafluoride + Water

$$SiO_2 + 4HF \longrightarrow SiF_4 + 2H_2O$$

2. An acidic mixture of silica and coke heated to witness in a current of chloride yields silicon.

Silicon Dioxide + Carbon + Chlorine ⟶ Silicon Tetrachloride + Carbon Monoxide

$$SiO_2 + 2C + 2Cl_2 \longrightarrow SiCl_4 + 2CO$$

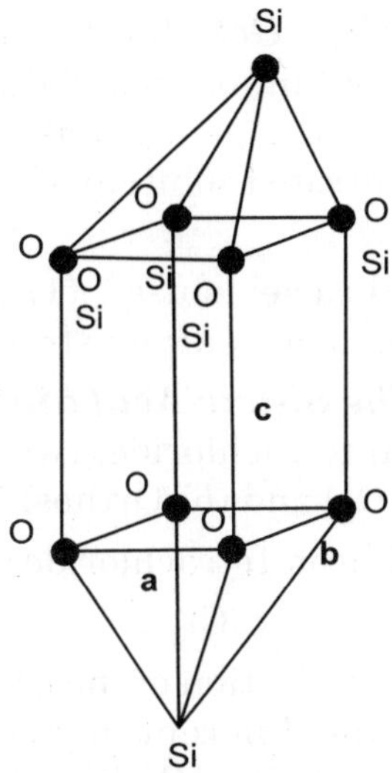

Fig. 17.3

3. All varieties of silica soften below 1600°C and fuse in the oxy-hydrogen flame at about 1700°C. They become plastic before fusing many working and blown like glass or drawn into threads.

Sodium Sulphate	+	Silica	$\longrightarrow$	Sodium Silicate	+	Sulphur Trioxide
Na_2SO_4	+	SiO_2	$\longrightarrow$	Na_2SiO_3	+	SO_3

4. Silica being non-volatile displaces volatile acid anhydrides from salts at high temperature.

5. Silica is *giant molecule* like diamond. It has a three-dimension structure, in which silicon is linked tetrahedral with four oxygen atoms by strong bonds and each oxygen atoms is common to two tetrahedra (See Fig. 17.3).

Silicon Monoxide (SiO): A very little is known about this compound of silicon. It is prepared by the reduction of silicon dioxide with silicon in a vacuum at a temperature of 1250°C

Silicon Dioxide	+	Silicon	$\longrightarrow$	Silicon Monoxide
SiO_2	+	Si	$\longrightarrow$	2 SiO

Use of Silica : Quartz or rock crystal occurs sometimes in transparent colourless crystals, but more frequently in opaque or coloured masses. Colourless quartz is used for making spectacles and for optical lens in different instruments for different purposes, because quartz has transparencies towards different wavelength of light. The coloured silica is used for making gems. Silica is used extensively used in the preparation of glass, glaze, porcelain, mortar, silica-bricks, etc and as building material. Sandstone is used in making milestones. Silicon carbide is an important abrasive*

Silica Glass: Also known as quartz glass may be prepared by fusing quartz or quartz in oxy-hydrogen flame. It possesses many remarkable properties. They are:

1. A very low co-efficient of expansion so that white hot silica may be quenched in water without cracking and therefore it is easy to make laboratory apparatus.
2. The softening point of the silica glass is 1500°C and that of ordinary glass is 500°C such a high softening point can withstand a wide range of experimental temperature.
3. Transparency to a wide range of wavelength allows us to make a variety of optical glasses.
4. It is free from alkalis and also very resistant to attack of acids and hence it is used in the construction of crucibles and evaporating dishes.

Vitrsol is a translucent variety of silica glasses another glass called *vycor* it is also a low expansion type.

---

* Abrasive : Smother or polisher used as knife grinder.

Silica Gel : Its chemical formula is also $SiO_2.H_2O$, it is completely dehydrated silicic acid containing about 6% mixture of hard granular transparent material. It made by heating the gelatenous mass of colloidal hydrated silica (formed by adding acid to a concentrated solution of sodium silicate) in vacuum at 300°C, it id broken up, washed and dried.

Silica gel is used: (1) as drying agent and demoisturizing agent and (2) it is also used for removing sulphur from petroleum.

The Silicic Acid and Silicates: The silicic acid is prepared by the action of water on silicon tetrachloride, two types of silicic acids are formed they are: (*a*) Meta-silicic Acid, (H_2SiO_3) and (*b*) Orthosilicic Acid [$Si(OH)_4$, H_2SiO_4]

$$\text{Silicon Tetrachloride} + \text{Water} \longrightarrow \text{Orthosilicic Acid} + \text{Hydrogen Chloride}$$

$$SiCl_4 + H_2O \longrightarrow Si(OH)_4 + 4HCl$$

Dehydration of the orthosilicic acid may give rise to the several silicic acids depending upon the different degrees of hydration of silica; *Orthosilicic acid* (H_2SiO_4), *Metasilicic acid* (H_2SiO_3), *Disilicic acid* ($H_2Si_2O_5$) *Trisilicic acid* ($H_2Si_3O_8$). The existence of the metasilicic acid disilicic acid has been confirmed from the vapour pressure dehydration curves.

Metasilicic Acid (H_2SiO_3): It is obtained as a white gelatenous mass containing varying proportions of water, by the action of acids upon a soluble silicate.

$$\underset{\text{Silicate}}{\text{Sodium}} + \underset{\text{Acid}}{\text{Hydrochloric}} \longrightarrow \underset{\text{Chloride}}{\text{Sodium}} + \underset{\text{Acid}}{\text{Metasilicic}}$$

$$Na_2SiO_4 + 2HCl \longrightarrow 2NaCl + H_2SiO_3$$

It is very weak acid, colloidal silicic acid is obtained by slightly adding sodium silicate to cold dilute hydrochloric acid solution. Many important silicates occur in rock minerals; examples are:

Tale(soapstone) [$H_2Mg_3(SiO_3)_4$]: It is used in face powder.

Asbestos [$Mg_3Ca(SiO_3)_4$]: It is used as a heat insulator in making fire-proof materials and the preparation of roofs and sheds it is also used in medicine. Common clay is impure aluminium silicate used in making earthen-ware and bricks etc.

Mica, Biotite [$MgCa(SiO_3)_4$] and Muscovite [$KH_2Al_3(SiO_4)_3$]: It is used as electrical insulator and also for making stove windows and lamp chimneys.

Garnet [$Ca_3Al_2(SiO4)_3$]: It is used for making sandpaper.

Beryl [($Be2Al_2Si_6O_{18}$)]: It is an important ore of beryllium. Many silicates e.g. zinc silicate, zinc beryllium silicate and cadmium silicate are used as fluorescent coating as the inside wall of fluorescent tubes containing a little inert gas colourless produced by the silicates being green, yellow white and pink respectively.

Silica Garden or Chemical Garden

This is a very beautiful experiment, which can be performed by any student, this experiment produces a beautiful garden of colloids. In an one litre beaker, some 1: 1 solution of sodium silicate is taken to it few crystals of cobaltous nitrate, zinc sulphate, manganese chloride, ferric sulphate, chromium nitrate are added. After an hour or so it will be seen that tree like structures are coming up to the mouth of the beaker.

Silicon Hydrides

The hydrides of silicon is called *silanesor silicanes* and has a general formula Si_nH_{2n+2}. Their compositions are as high as S_4H_{14} which have been proved. They are similar to paraffin hydrocarbons but they are less stable.

Monosilane and Disilane (Si_2H_6) are gases the higher are liquids with formula of Si_5H_{12} and Si_6H_4 having boiling points of above 100°C, Silico-methane (SiH_4) is slightly formed from its elements at a higher temperature.

Silicon	+	Hydrogen	$\longrightarrow$	Silico-Methane
Si	+	$2H_2$	$\longrightarrow$	SiH_4

But a mixture of volatile hydrides is usually prepared by the action of dilute hydrochloric acid upon magnesium silicate (made by heating 2 parts of magnesium powder and 1 part of silica in dry state). The mixture is taken in a flask from which the air has been driven out by hydrogen, the reaction is done in such a manor because the silanes are spontaneously inflammable in air.

FLAMABLE

Pure disilane is made by reaction between lithium aluminium tetrahydride and silicon hexachloride in the presence of ether.

Silicon Hexachloride	+	Lithium Aluminium Tetrahydride	$\longrightarrow$	Lithium Chloride	+	Aluminium Trichloride	+	Disilane
$2SiCl_6$	+	$2LiAlH_4$	$\longrightarrow$	$3LiCl$	+	$3AlCl_3$	+	$2Si_2H_6$

The silicon hydrides have strong reducing properties; copper sulphate is reduced to copper silicide and silver nitrate to metallic silver.

Copper Sulphate	+	Monosilane	$\longrightarrow$	Cuprous Sulphide	+	Sulphuric Acid
$2CuSO_4$	+	SiH_4	$\longrightarrow$	Cu_2S	+	$2H_2SO_4$

Silver Nitrate	+	Monosilane	$\longrightarrow$	Silver	+	Silicon	+	Nitric Acid
$4AgNO_3$	+	SiH_4	$\longrightarrow$	$4Ag$	+	Si	+	$4HNO_3$

Water containing a trace of alkali, but not pure water, decomposes the silanes with the evolution of hydrogen.

Silane	+	Sodium Hydroxide	$\longrightarrow$	Water	+	Sodium Silicate	+	Hydrogen
SiH_4	+	$2NaOH$	$\longrightarrow$	H_2O	+	Na_2SiO_3	+	$4H_2$
Disilane	+	Sodium Hydroxide	$\longrightarrow$	Water	+	Sodium Silicate	+	Hydrogen
Si_2H_6	+	$4NaOH$	$\longrightarrow$	$2H_2O$	+	$2Na_2SiO_3$	+	$7H_2$

Water Glass, Sodium Silicate (Na_2SiO_3): Chemically it is known as sodium silicate. It is made by the melting of powdered quartz (or pure white sand) with sodium carbonate and wood charcoal at a high temperature in a reverberatory furnace.

Sodium Carbonate	+	Silicon Dioxide	$\longrightarrow$	Sodium Silicate	+	Carbon Dioxide
Na_2CO_3	+	SiO_2	$\longrightarrow$	Na_2SiO_3	+	CO_2
Sodium Sulphate	+	Carbon	$\longrightarrow$	Sodium Sulphide	+	Carbon Monoxide
Na_2SO_4	+	4C	$\longrightarrow$	Na_2S	+	4CO

Sodium Sulphate	+	Sodium Sulphate	$\longrightarrow$	Silicon Dioxide	+	Sodium Silicate	+	Sulphur Dioxide
Na_2S	+	$3Na_2SO_4$	$\longrightarrow$	$4SiO_2$	+	$4Na_2SiO_3$	+	$4SO_2$

It forms lumps in colour due to the presence of traces of iron salts as impurities. It is broken up and heated with water under pressure in an autoclave, when it slowly dissolves into a thick solution known as water glass. It is used in paper industry, for fire proofing timber and textiles for preservation of eggs (4 - 10% solution) for joining pieces of china, as detergents in laundries and treating cement floors from dust and abrasion. Water glass gel is used as a filter in cheap soaps.

Sodium and potassium silicates are soluble in water and are known as soluble glass. They can be made by fusion of sand with sodium carbonate these solutions are strongly alkline due to hydrolysis.

Silicon Halides

Silicon forms a number of halogen compounds. These having more than one silicon atom and must consist of a chain of 4-valent silicon atoms. They are as follows:

Silicon Tetrafluoride (SiF_4): It is colourless compound available in gaseous state, which fumes in air. Ins prepared (1) by heating barium fluorosilicate the gas obtained is pure.

Barium Fluorosilicate	$\longrightarrow$	Barium Fluoride	+	Silicon Tetrafluoride
$BaSiF_4$	$\longrightarrow$	BaF_2	+	SiF_4

(2) It can also be prepared by spontaneously heating of silicon in fluorine

Silica	+	Hydrochloric Acid	$\longrightarrow$	Silicon Tetrafluoride	+	Water
SiO_2	+	4HF	$\longrightarrow$	SiF_4	+	$2H_2O$

(3) By the action of hydrofluoric acid on silica.

The gas is incombustible and it extinguishes a burning taper but heated sodium and potassium burn in the gas. It reacts with calcium and barium oxides producing incandescence.

Silicon Tetrafluoride	+	Potassium	$\longrightarrow$	Potassium Fluoride	+	Potassium Hexafluoride	+	Silicon
$2SiF_4$	+	4K	$\longrightarrow$	2KF	+	K_2SiF_6	+	Si

It is decomposed by water forming white gelatenous orthosilicic acid Si(OH)4 and hydroflurosilicic acid, a reaction utilized in testing silicates and fluorides.

Silicon Tetrafluoride + Water ⟶ Orthosilicic Acid + Hydrofluorosilicic Acid

$$3SiF_4 + 4H_2O \longrightarrow Si(OH)_4 + 2H_2SiF_6$$

Silicon Trifluoride (Si_2F_6, SiF_3): It is a colourless inflammable gas. It is produced by warming silicon trichloride and zinc fluoride.

Silicon Trichloride + Zinc Fluoride ⟶ Silicon Hexafluoride + Zinc Chloride

$$Si_2Cl_6 + 2ZnF_2 \longrightarrow Si_2F_6 + 2ZnCl_2$$

It is decomposed by water with liberation of hydrogen but some silicon oxalic acid is also formed.

Silicon Trifluoride + Water ⟶ Hydroflurosilicic Acid + Silicon Dioxide + Hydrogen

$$Si_2F_6 + H_2O \longrightarrow 2H_2SiF_6 + SiO_2 + H_2$$

Silicon Tetrafluoride + Water ⟶ Hydrogen Fluoride + Silicon Oxalic Acid

$$Si_2F_6 + H_2O \longrightarrow 6HF + H_2Si_2O_4$$

Silicon Tetrachloride ($SiCl_4$): It is colourless fuming liquid, it is made by the reaction of dry chlorine over heated amorphous silicon or a mixture of silica and magnesium powder.

Silicon + Chlorine ⟶ Silicon Tetrachloride

$$Si + 2Cl_2 \longrightarrow SiCl_4$$

Silicon Dioxide + Chlorine ⟶ Carbon + Silicon Tetrachloride + Carbon Monoxide

$$SiO_2 + 2Cl_2 \longrightarrow 2C + SiCl_4 + 2CO$$

Silicon Dioxide + Sulphur Monochloride ⟶ Chlorine + Silicon Tetrachloride + Sulphur Dioxide

$$2SiO_2 + S_2Cl_2 \longrightarrow 3Cl_2 + 2SiCl_4 + 2SO_2$$

The silicon tetrachloride vapour is condensed in a tube kept in a freezing mixture and then excess chlorine is irradiated from the liquid by distilling over mercury. It is kept in a sealed tube. It reacts with water violently forming silicon dioxide and hydrogen chloride.

Silicon Tetrachloride + Water ⟶ Hydrochloric Acid + Silicon Dioxide

$$SiCl_4 + 2H_2O \longrightarrow 4HCl + SiO_2$$

With ammonia it produces di-amide [$Si(NH)_2$]

Silicon Tetrachloride + Ammonium Hydroxide ⟶ Silicon Diamide + Ammonium Chloride + Water + Oxygen

$$SiCl_4 + 5NH_4OH \longrightarrow Si(NH)_2 + 4NH_4Cl + H_2O + 2O_2$$

Silicon Trichloride (Si_2Cl_6,$SiCl_3$): It is a colourless fuming liquid, is formed along with silicon tetrachloride and silicon oxychloride, by the action of chlorine on heated silicon and may be separated by fractionation. It is also formed by the action of chlorine on Si_2I_6, silicon tetraiodide or calcium silicide. Its vapour ignites if heated in air. The liquid fumes in moist air to produce siliconoxalic acid.

Silicon Trichloride	+	Water	⟶	Siliconoxalic Acid	+	Hydrogen Chloride
Si_2Cl_6	+	H_2O	⟶	$H_2Si_2O_4$	+	6HCl

Which is decomposed by alkali, liberating hydrogen.

Siliconoxalic Acid	+	Sodium Hyroxide	⟶	Sodium Silicate	+	Water
$H_2Si_2O_4$	+	4NaOH	⟶	$2Na_2SiO_3$	+	$2H_2$

The octachloride Si_3Cl_8 reacts with water forming silicon mesooxalic acid ($H_2Si_3O_5$) a white powder.

The halogen silanes are some of the important members of the silicon family which are similar to the chloroform they are called *silicofluoroform ($SiHF_3$)* and *silicochloroform, ($SiHCl_3$)* and *silicoiodoform ($SiHI_3$)* Silicofluoroform is a combustible gas with a boiling point of -80.2°C, is formed by the action of stannic fluoride upon silicochloroform.

Silicochloroform	+	Stannous Tetrafluoride	⟶	Silicofluoroform	+	Stannous Tetrachloride
$4SiHCl_3$	+	H_2O	⟶	$4SiHF_3$	+	$3SnCl_4$

It is decomposed by heat; it also reacts with water.

Silicochloroform ($SiHCl_3$): It is a colourless fuming liquid viscous in nature with a boiling temperature of 35°C, it is made by passing dry hydrogen chloride over red hot silicon.

Silicon	+	Hydrogen Chloride	⟶	Silicochloroform	+	Hydrogen
Si	+	2HCl	⟶	$SiHCl_3$	+	H_2

The vapour is condensed to a liquid by the freezing mixture and then fractionated to separate it from silicon tetrachloride. A highly inflammable substance burns with a green edged flame, giving white fumes of silica. It is more reactive than chloroform. Unlike chloroform, silicochloroform is hydrolyzed by water to form unstable silico-ortho-formic acid Si(OHh which readily dehydrates to silicon formic anhydride.

Silicochloroform	+	Water	⟶	Silico-ortho-formic Acid	+	Hydrogen Chloride
$4SiHCl_3$	+	$3H_2O$	⟶	$Si(OH)_3$	+	3HCl

Silico-Ortho-Formic Acid	⟶	Silicon-Formic Anhydride	+	Water
$Si(OH)_3$	⟶	$Si_2H_2O_3$	+	$3H_2O$

It possesses a strong reducing properties. Silicochloroform forms ethyl ortho-silicoformate $SiH(OC_2H_5)_3$.

Silicobromoform ($SiBBr_2$): It is a red liquid having a boiling point of 220°C, results on the action of hydrogen bromide on hot silicon.

Hydrogen Bromide	+	Silicon	⟶	Silicofluoroform	+	Hydrogen
3HBr	+	Si	⟶	$SiHBr_3$	+	H_2

Hydrofluorosilicic Acid : This compound of silicon halide is prepared by the action of silicon tetrafluoride and water, silicon tetrafluoride is prepared by the reaction of sulphuric acid sand and calcium fluoride at an elevated temperature (see Fig. 17.4).

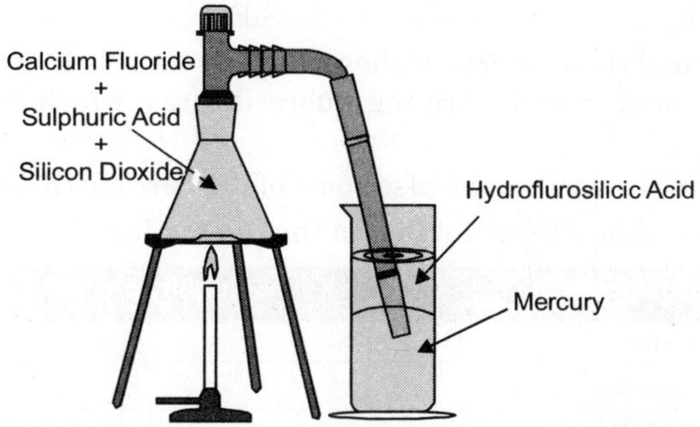

Fig. 17.4. The Laboratory Preparation of Hydrofluorosilicic Acid

Reaction Equation : Hydrofluorosilicic acid is also prepared by the action of silicon dioxide and hydrogen fluoride.

Calcium Fluoride + Sulphuric Acid + Silicon Dioxide ⟶ Calcium Sulphate + Hydrogen Fluoride + Silicon Tetrafluoride + Water

$$4CaF_2 + 4H_2SO_4 + SiO_2 \longrightarrow 4CaSO_4 + 4HF + SiF_4 + 2H_2O$$

Silicon Tetrafluoride + Water ⟶ Hydrofluorosilicic Acid + Silicon Dioxide

$$3SiF_4 + 2H_2O \longrightarrow 2H_2SiF_6 + SiO_2$$

Hydrofluorosilicic acid is also prepared by the action of silicon dioxide and hydrogen fluoride.

Silicon Dioxide + Hydrogen Fluoride ⟶ Hydroflurosilicic Acid + Water

$$SiO_2 + HF \longrightarrow H_2SiF_6 + 2H_2O$$

Precaution: *During the collection of the acid the mouth of the delivery tube should be kept dipped in the mercury to prevent the tube from blocking by the formation of silica gel. Also be careful the experimental set up must be of porcelain and plastic not glass (because fluorine etches glass).*

The acid is known only in the aqueous state. Silicon tetrafluoride (SiF_4) and hydrogen fluoride do not react and therefore the anhydrous acid is not possible to obtain. Its reaction with strong alkalis may be represented by the following equation:

Hydrofluorosilicic Acid + Sodium Hydroxide ⟶ Sodium Fluorosilicate + Water

$$H_2SiF_6 + 2NaOH \longrightarrow Na_2SiF_6 + 2H_2O$$

Sodium Fluorosilicate + Sodium Hydoxide ⟶ Sodium Fluoride + Silicon Dioxide + Water

$$Na_2SiF_6 + 4NaOH \longrightarrow 6NaF + SiO_2 + 2H_2O$$

Thus although the acid is dibasic it can neutralize six molecules of sodium hydroxide.

The pure acid has no reaction on glass and it decomposes at stated below:

Hydrofluorosilicic Acid ⟶ Sodium Tetrasfluoride + Hydrogen Fluoride

$$H_2SiF_6 \longrightarrow SiF_4 + 4KF$$

The hydrofluoric acid thus set free etches the glass vessels in which hydroflurosilicic acid is kept. It reacts with alkalis forming fluorosilicates, which are explosive, such as potassium,

sodium and barium fluorosilicates, also some of the rare earth fluorosilicates.

The table below shows the solubility of the fluorosilicates:

Name of the Fluorosilicate	Solubility per 100 g of Water at 20°C
Sodium Fluorosilicate	0.65
Potassium Fluorosilicate	0.12
Rubidium Fluorosilicate	0.0127
Barium Fluorosilicate	0.027

The fluorosilicate ion (SiF_6) is octahedral. The fluorosilicates are decomposed by concentrated sulphuric acid and also by heat.

Sodium Fluorosilicate + Sulphuric Acid ⟶ Sodium Sulphate + Silicon Tetrafluoride + Hydrogen Fluoride

$$Na_2SiF_6 + H_2SO_4 \longrightarrow Na_2SO_4 + SiF_4 + 2HF$$

Sodium Fluorosilicate ⟶ Silicon Tetrafluoride + Sodium Floride

$$Na_2SiF_6 \longrightarrow SiF_4 + 2NaF$$

Barium chloride gives crystalline precipitate of barium silicofluoride ($BaSiF_6$) with a solution of sodium fluorosilicate the precipitate is distinguished from barium sulphate by evolution of hydrogen fluoride and silicon tetrafluoride on heating with concentrated sulphuric acid. Magnesium fluorosilicate is painted over concrete to harden and polish the surface and sodium fluorosilicate is used as an insecticide in the poultry firms.

Silicon Nitride (SiN_4): It is made by heating silicon in nitrogen to 1350°C. In other the hydride or chlorides of silicon are allowed to react with ammonia and the amino and imino-silanes formed are heated. These give the polymers $Si(NH_2)_n$ which yields Ni_3N_4.

Silicon Disulphide (SiS_2): It is formed as a milky matter by heating silica in sulphur vapour. It is readily hydrolyzed to hydrogen sulphide and silica.

Silicon Disulphide + Water ⟶ Hydrogen Sulphide + Silica

$$SiS_2 + 2H_2O \longrightarrow 2H_2S + SiO_2$$

It reacts with alkalis to form thiosilicates.

Silicon disulphide forms fibrous crystalline macromolecules in which the sulphur atoms are arranged tetrahedrally round atoms.

Silicon Carbide (SiC): This is also known as carborundum, was invented by Achenson in 1891. The carborundum is prepared by the Achenson's Process. A charge consisting *of* coke 34% sawdust 10% salt 2% with silica. It is heated in an electrical arc furnace, which is 4.8m long and 1.5m wide and 1.4m deep and is made by furnace bricks (Fig. 17.6).

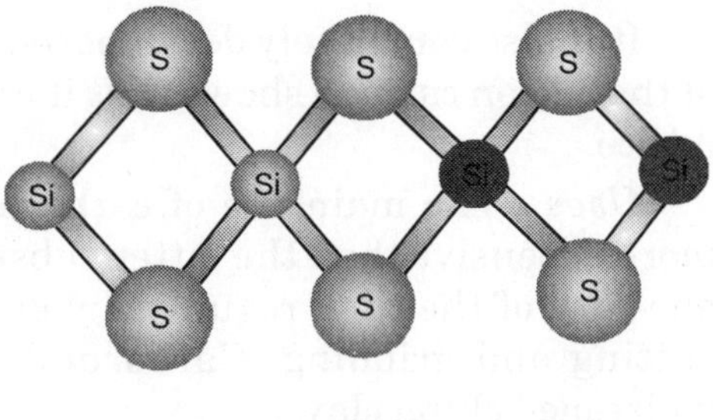

Fig. 17.5. The structure of Silicon Disulphide Molecule

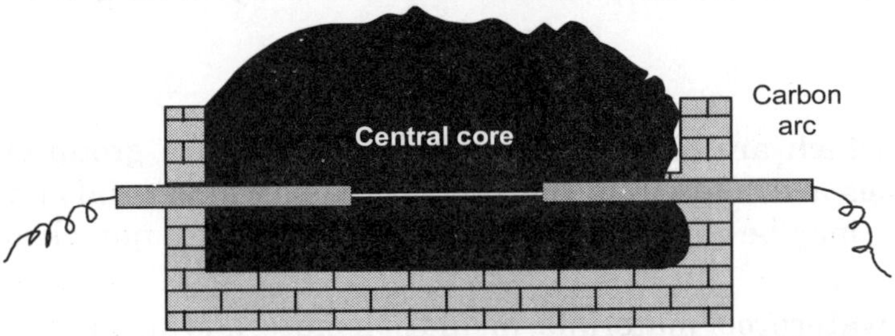

Fig. 17.6. The Acheson's Process of Carborundum preparation.

The end walls and the bed of the furnace are permanent; while the sides are built up with charge and are pulled down to take out the product. A central core made of carbon rods join the electrodes, a heavy current sets up arc between rod of coke and the temperature rapidly rises to over 2000°C. The equation for the reaction is stated below.

Silicon Dioxide + Carbon ⟶ Carborundum + Silica

$$SiO_2 + 3C \longrightarrow SiC + 2CO$$

The salt acts as a flux, while the saw dust increases the porosity of the charge, thus enabling steady escape of carbon monoxide, which bums at the top of the charge.

At the beginning, the e.m.f. is 165V but as the heat increases the resistance between the electrode decreases and the e.m.f. also decreases to 125V. During this period the current increases from 1,700 amps to 6,000 amps. The conditions then remain steady during the whole run of 36 hours.

The product consisting of a mass of black crystals showing beautiful iridescent* colours is crushed successfully in sulphuric acid and sodium hydroxide solution. Finally, it is dried in a kiln and graded through a system of success into powders of various degrees of fineness.

Properties of Silicon Carbide

When pure, silicon carbide is colourless though the commercial sample shows different colours. Chemically it is extremely inert and also inert to strong acids. Even a mixture of fuming nitric acid and hydrofluoric acid cannot dissolve silicon carbide, however silicon carbide is dissolved in the molten caustic soda, in the presence of air.

Silicon Carbide + Sodium Hydroxide + Oxygen ⟶ Sodium Silicate + Sodium Carbonate + Water $2H_2O$

$$SiC + 4NaOH \quad 2O_2 \longrightarrow Na_2SiO_3 + Na_2CO_3$$

*Iridescent: Producing a rainbow effect.

It is also completely decomposed by red-hot potassium dichromate. The X-Ray analysis of the silicon carbide shows that it consists of 24 SiC molecules arranged in a diamond like lattice.

Uses : The main use of carborundum as an abrasive in the place of emery. Though more expensive than the latter substance, it is far more effective taking about one-third to one-fifth of the time required by emery paper for the work. The powder is used in glass cutting and grinding. Carborundum is made into wheels horns etc. by mixing with moistened china clay.

Silicones

New families of synthetic resins, which are collectively called *silicones,* are group of silico-organic substances. All silicones are made from the same three basic materials for e.g. petroleum, brine and sand. They may be obtained in various physical forms namely a gas or mobile liquid also thick oil.

Silicones are used for preparing waterproof materials, insulation materials and also in the manufacture of vaselene and greeze, which do not freeze at –40°C.

Glass

It is the material, which is obtained in the form of a liquid and then super cooled to obtain an apparently solid.

Manufacture

The average composition of ordinary glass corresponds to the formula of Naz.CaO.6SiOz. A mixture of sodium carbonate, calcium carbonate and sand. After being powdered and intimately mixed it is fused in a tank furnace, which is heated by the producer gas and working on the regenerative system of the heat economy (see Fig. 17.7).

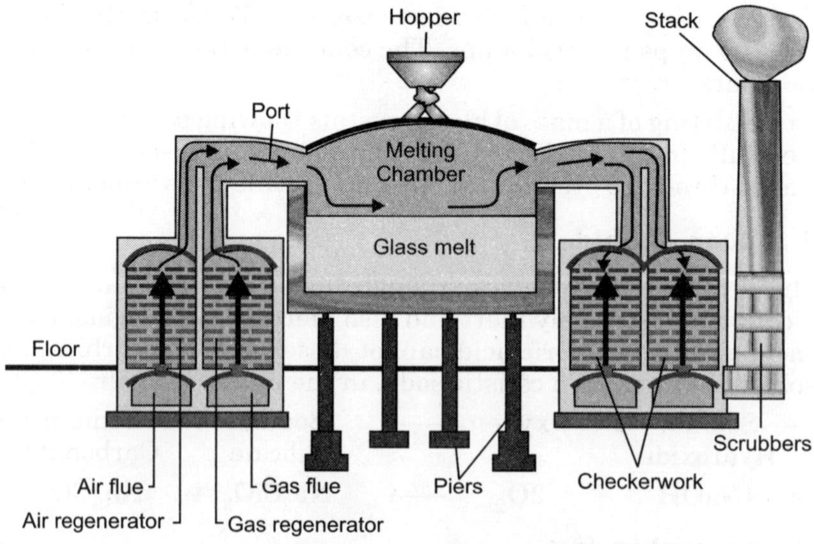

Fig. 17.7. The Glass Manufature

The silica decomposes the carbonates, setting free the carbon dioxide.

Sodium Carbonate	+	Silicon Dioxide	$\longrightarrow$	Sodium Silicate	+	Carbon Dioxide
Na_2CO_3	+	SiO_2	$\longrightarrow$	Na_2SiO_3	+	CO_2
Calcium Carbonate	+	Silicon Dioxide	$\longrightarrow$	Calcium Silicate	+	Carbon Dioxide
$CaCO_3$	+	SiO_2	$\longrightarrow$	$CaSiO_3$	+	CO_2

In few hours, when bubbles of carbon dioxide, no longer rise and the mass has melted down to a liquid it is allowed to cool and blown into or cast into moulds to produce different articles.

The glass articles obtained as above are allowed to cool extremely slowly through a tunnel like furnace, which is very hot at the entrance and much cooler at the exit. This process is called *annealing.*

Annealing is absolutely necessary because it makes the glass shock- proof without which the glass will be shattered due to a little shock.

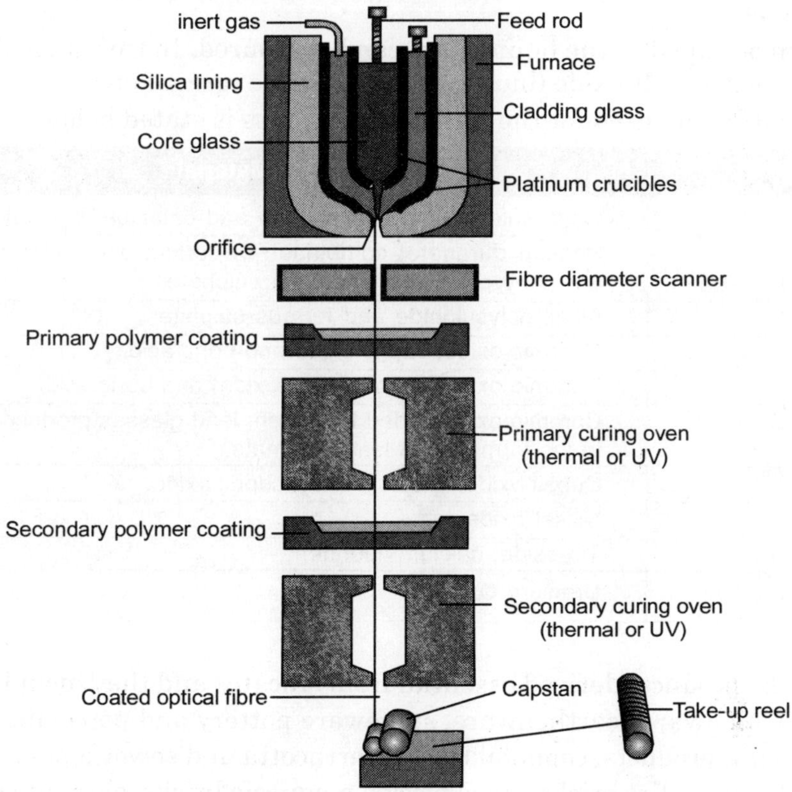

Fig. 17.8. The glass sheet production

The Chart showing various types of Glass with Composition Percentage of Different Compounds in it						
Lead Oxide	—	—	—	—	—	—
Zinc Oxide	10.90	—	—	—	—	80.5
Silicon Dioxide	64.70	77.90	77.90	70.64	69.65	3.8
Sodium Oxide	7.50	4.10	4.10	17.02	15.22	—
Magnesium Oxide	0.21	—	—	0.09	—	6.4
Potassium Oxide	0.37	2.30	2.30	—	—	—
Ferric	0.25	—	—	0.11	—	—
Oxide/Iron	—	—	—	—	1.82	—
Calcium Oxide	0.63	0.80	0.80	10.58	13.31	—
Boron Oxide	10.90	14.10	14.30	—	—	12.9
Aluminium Oxide	4.20	0.80	5.60	0.77	—	2.2
TYPE	Jena Duran	Pyrex Chemical Resistant	Vulcanite	Window Glasses	French Window Glasses	Borosilicate

Coloured Glass

The glass when obtained in the liquid state can be coloured. In this stage by adding some mineral oxides, e.g. cobalt oxide (blue) chromium oxide (green) etc.

Various components used for the colouring the glass is stated below in the table:

Colour	Colouring Material
Red	Gold, selenium, cuprous oxide and cadmium sulphate.
Yellow	Sodium duranate, combintion of cerium oxide with titatnium oxide, antimony oxide and cadmium sulphate.
Amber	Alkali polysulphide and ferrous sulphite.
Brown	Chromic oxide, cupric oxide and boric acid.
Green	Chromic oxide, (black cobalt oxide) and boric acid.
Orange Red	Chromic oxide added to potash; lead glasses produce this colour due to the formation of lead chromate.
Blue	Cobalt oxide, (black (Co_3O_4) cupric oxide.
Grey	Nickel oxide
White Opaque	Tin oxide, calcium fluoride.
Fluorescent Yellow	Uranium Oxide

Ceramics

Ceramics are the products derived essential from silicates and they include.

(*i*) White Chinawares earthenware, stoneware pottery and porcelain.

(*ii*) Heavy clay, products, common bricks, terracotta and sewer eipes.

(*iii*) Refractoriness; fire bricks, silica bricks, magnesia bricks, chromite bricks etc.

The chief raw materials for the common products are: (1) Clay, (2) Feldspar and (3) Sand also called Flint.

Porcelain: It is made from a mixture of Pure china clay finely powdered flint and feldspar.

Glaze: It is the thin coating of easily fusible glass melted on the surface of the more or less porous ceramic ware. The glaze composition includes feldspar, silica clay and fluxes in different proportions. The molar composition of typical glaze 0.3(Na, K)$_2$O, 0.3 CaO, 0.4 PbO, 0.28 Al_2O_3 3.0 SiO_2 and 0.5B_2O_3.

Refractoriness: Refractoriness are those materials, which are used as inside lining in furnace; they are required to withstand: (1) high temperature (2) mechanical contraction, wear and tear and (3) corrosive action of substances. Among the refractoriness, firebricks, and magnesite is produced.

Detection of Silicon

A silicate solution (1) gives a gelatinous precipitate of silicic acid with ammonium chloride otr ammonium carbonate solution. A silicate solution produces silver silicate with silver nitrate.

GALLIUM

Melting Point : 29.8°C — Atomic Number : 31
Boiling Point : 2403°C — Atomic Weight : 69.723

Occurrence

Froll Residue of Zinc Blende: Zinc blende contains lead as well as gallium. Lead is present in much larger quantity than gallium. After the recovery of zinc the residue is dissolved in nitric acid. Lead is removed by precipitation as lead sulphate in the addition of dilute sulphuric acid. If some copper is also present as an impurity this can be precipitated out by passing hydrogen sulphide gas as cuprous sulphide. The solution, after filtration, is mixed with aqueous ammonia when gallium and zinc present always in small quantities, get precipitated out as hydroxides. These are dissolved in sodium hydroxide to get sodium zincate and sodium gallate. Gallium is ultimately separated from zinc as its basic sulphate, which is very less soluble in water. The basic sulphate of gallium is fused with sodium gallate. The solution of sodium gallate is made slightly acidic by adding sulphuric acid and then electrolyzed by using platinum electrodes.

From Germanite Mineral: It is the germanium ore, the powdered from of which is distilled with concentrated hydrochloric acid and then a stream of chlorine is passed into it. In this way germanium gets volatilized as chloride. The distillate after filtration is treated with aqueous ammonia when a mixtures of oxides and hydroxides is precipitated. This is dissolved in hydrochloric acid and some hydrogen sulphide is passed to precipitate out copper sulphide. From the filtrate gallium is separated by precipitating it as potassium gallium ferrocyanide. This on fusion with sodium hydroxide gives sodium gallate, which is soluble in water. The metal is obtained by electrolyzing the solution made slightly acidic with sulphuric acid using platinum electrodes.

Pure Gallium

Gallium of high purity can be obtained by the reduction of pure gallium trichloride and hydrogen in a quartz vessel.

Physical Properties of Gallium

Melting Point	29.8°C
Boiling Point	2403°C
Density	5.9 g/cc
Isotopes	$^{31}Ga_{69}$ $^{31}Ga_{70}$
Electrical Resitivity at 100°C	26 ohms/cm
Standard Entropy	10.2 cal/deg^{-1} mole^{-1}
Heat of Capacity	6.35 cal/deg^{-1} mole^{-1}
Viscosity	1600 Pa/s
Latent Heat of Fusion	19.16 cal/g
Co-efficient of Linear Thermal Expansion (solid)	1.85×10^{-5}
Surface Tension	735 dynes/cm

Chemical Properties

Gallium dissolves in caustic alkalis and is slightly soluble in mercury. Gallium reacts with oxygen to form sesquioxide; it also reacts with other halogens such as chlorine to form chloride. As gallium is a rare element therefore the study of its chemical properties is not very wide.

Compounds of Gallium

Gallium Sesquioxide (Ga_2O_3): This is produced when gallium reacts with oxygen at an elevated temperature.

Gallium + Oxygen ⟶ Gallium Sesquioxide + Gallium Suboxide

$$4Ga + 2O_2 \longrightarrow Ga_2O_3 + Ga_2O$$

With sesquioxide some suboxide is also produced.

Gallium Trichloride ($GaCl_3$) : With chlorine at normal temperature gallium produces gallium trichloride.

Gallium + Chlorine ⟶ Gallium Trichloride

$$2Ga + 3Cl_2 \longrightarrow 2GaCl_3$$

Gallium Phosphide (GaP): It is pale orange transparent crystals produced by the reaction at relatively low temperature and low pressure between gallium suboxide and phosphorous.

Gallium Suboxide + Phosphorous ⟶ Gallium Phosphide + Oxygen

$$2Ga_2O + 2P_2 \longrightarrow 4GaP + O_2$$

This compound is used in semiconductors injection lasers solar cells magneto resistance devices thermistors etc. it is also used in microwave generation.

Gallium Sesquisulphide (Ga_2S_3): This is produced by the action of hydrogen sulphide on gallium sesquioxide.

$$\text{Gallium Sesquioxide} + \text{Hydrogen Sulphide} \longrightarrow \text{Gallium Sesquisulphide} + \text{Water}$$
$$Ga_2O_3 + 3H_2S \longrightarrow Ga_2S_3 + 3H_2O$$

Gallium Sulphate ($Ga_2(SO_4)_3$): This compound of gallium is prepared by the action of sulphuric acid on the gallium metal.

$$\underset{\text{Trichloride}}{\text{Gallium}} + \underset{\text{Acid}}{\text{Sulphuric}} \longrightarrow \underset{\text{Acid}}{\text{Hydrochloric}} + \underset{\text{Sulphate}}{\text{Gallium}}$$
$$3H_2SO_4 + 2GaCl_2 \longrightarrow 6HCl + Ga_2(SO_4)_3$$

It can also be prepared by the action of gallium chloride and sulphuric acid.

$$\underset{\text{Trichloride}}{\text{Gallium}} + \underset{\text{Acid}}{\text{Sulphuric}} \longrightarrow \underset{\text{Acid}}{\text{Hydrochloric}} + \underset{\text{Sulphate}}{\text{Gallium}}$$
$$3H_2SO_4 + 2GaCl_2 \longrightarrow 6HCl + Ga_2(SO_4)_3$$

Detection of Gallium

Gallium is detected by the UV-VIS spectrophotometer.

GERMANIUM

Melting Point : 937°C | Atomic Number : 32

Boiling Point : 2830°C | Atomic Weight : 72.59

History and Occurrence

Germanium was discovered by Wrintler in 1886 in a rare silver mineral Argyroclite, GeS24Ag2S (5% germanium) from Freiberg. It was named as germanium as it was found in Germany. The *Germanite* mineral (Cu, Ge, Fe, ZnCa) (S and As) a copper *iron thiogemanate* discovered in South West Africa in 1916 contains 5 - 6% germanium. and is an important mineral of the element. For long time germanium was a chemist's curiosity due to its rarity, extensive investigations began only after the discovery of germanite.

Germanium has been found in the ash of some smelting cools along with gallium. It is also found in small amounts in the smelting residues of some zinc blende.

Extraction

Germanium is extracted from coal ash as well as from flue dust of waste heat boilers by heating with hydrochloric acid when the volatile tetrachloride (GeC4) distills off gallin ($GaCl_3$) in the solution extracted with ether for its recovery.

For extraction of the metal from germanite, the finely powdered ore is first digested with a mixture of potassium cyanide and charcoal or by heating in a current of hydrogen.

Manufacture of Transistors require ultra pure germanium. This is made by the process of zone refining in which a strip of the metal is heated near one end so that a short zone of molten germanium is produced, the impurities being more soluble in the liquid than in the solid, therefore concentrate in the melt. The molten zone is made to move steadily along the strip from one end to another by a moving heating source so that the impurities eventually collect at one end of the strip, which can be recovered.

Physical Properties

Melting Point	237°C
Boiling Point	2830°C
Density	5.323 gm/cc
Covalent Atomic Radius	1.22Å
Electronegativity (Pauling's Scale)	2.07
Electrical Resistivity at 700°C	1600 ohm/cm
Standard Entropy	10.14 cal/deg^{-1} mole^{-1}
Heat of Capacity	6.24 cal/deg^{-1} mole^{-1}
Hardness on Moh's Scale	6

Chemical Properties and Uses

Germanium is a brittle lustrous metal. It forms octahedral crystals. It is insoluble in hydrochloric acid and dilute sulphuric acid but dissolve in hot concentrated sulphuric acid evolving sulphur dioxide it also dissolve in nitric acid and aqua regia.

According to its position in between silicon and tin, the properties of germanium follows both in many respects. Thus it has a diamond lattice structure like silicon; it forms a series of volatile tetrachloride with oxygen to form the dioxide and with alkalis to form the germanites.

The gemanites and thiogermanites are similar with the corresponding silicon compounds GeO_2 is feebly acidic oxide; like silicon dioxide, germanium dioxide however crystallizes in two forms similar to tin oxide.

Germanium is oxidized to the hydrated dioxide by concentrated nitric acid as in tin. Germanium is an intrinsic semiconductor, used in the manufacture of diodes.

Compounds of Germanium

Germanium like silicon, forms a series of volatile hydrides of the general formula Ge_xH_{2n+2} for example GeH_8, Ge_2H_6 and Ge_3H_8. They resemble the silicon hydrides, but they are not as flammable as silanes, they are, however oxidized to germanium dioxide and water by oxygen.

The germanes are prepared by treating (magnesium germanium) with hydrochloric acid. The evolved gasses are cooled in liquid air when the germanium hydrides condensed to a liquid form which they are separated by fractionation (as silanes). Monogermane, GeH_4 as gas is formed by the action of ammonium bromide in liquid ammonia on magnesium germanide.

Magnesium Germanide	+	Ammonium Bromide	⟶	Manganese Bromide	+	Mono-germane	+	Ammonia
Mg_2Ge	+	$4NH_4Br$	⟶	$2MgBr_2$	+	GeH_4	+	$4HNO_3$

It is conveniently made by reducing GeCl with lithium aluminium hydride in ether.

Germanium Tetrachloride	+	Lithium Aluminium Tetrahydrate	⟶	Mono-germane	+	Lithium Monochloride	+	Aluminium Trichloride
$GeCl_4$	+	$LiAlH_4$	⟶	GeH_4	+	LiCl	+	$AlCl_3$

Germanium Dioxide (GeO_2): It is formed by the strong heating of germanium in a current of oxygen, or by oxidizing the metal with concentrated nitric acid. A white solid, it melts at 1086°C after gradual softening, it solidifies from the melt as a glass (silica glass). It crystallizes in two forms, which are same as quartz and Cassiterite. Sparingly soluble in water, it dissolves in hydrogen chloride forming germanium trichloride, weakly acidic oxide like silica, it also dissolves in alkalis forming germanites, e.g. sodium germanium dioxide, which are similar to silicon tetrafluoride is formed by the action of hydrofluoric acid on germanium dioxide.

Germanium Dioxide	+	Hydrogen Fluoride	⟶	Germanium Tetrafluoride	+	Water
GeO_2	+	4HF	⟶	GeF_4	+	$2H_2O$

It forms H_2GeF_6 with water. Fluorogemanates, e.g. potassium fluorogermanate (K_2GeF_6) ammonium fluorogermanate are similar with other corresponding silicates.

Germanium Tetrachloride ($GeCl_4$): It is a colourless fuming liquid with a boiling point of 83°C, is formed by burning the metal in chlorine or heating GeO_2 with fuming hydrochloric acid. It is slowly hydrolyzed by water precipitating by hydrated dioxide. In concentrated hydrochloric acid it forms chlorogermanic acid (H_2GeCl_4).

Germanium Disulphide (GeS): It is formed as a white precipitate in passing hydrogen sulphide into a solution of germanium tetrachloride.

Gemanium Tetrachloride	+	Hydrogen Sulphide	⟶	Germanium Sulphide	+	Hydrogen Chloride
$GeCl_4$	+	$2H_2S$	⟶	GeS_2	+	4HCl

It is a colourless liquid, insoluble in water but soluble in benzene.

Tetraphenyl Germanium [$Ge(C_6H_5)_4$]: It is a solid with a melting point of 233°C. It is made by the reaction of sodium with a mixture of $GeCl_4$ and C_6H_5Br.

Nitrogen Compounds of Germanium

Germanium tetrachloride reacts with liquid ammonia forming germanium amide.

Germanium Tetracloride	+	Ammonia	⟶	Germanium Amide	+	Ammonium Chloride
$GeCl_4$	+	$6NH_3$	⟶	$Ge(NH)_2$	+	$4NH_4Cl$

Heating to 150°C in an inert atmosphere of nitrogen, it forms germanium diamide

Germanium Amide	$\longrightarrow$	Germanium Diamide
$Ge(NH)_2$	$\longrightarrow$	$GeNH_4$

On further heating it produces germanium nitride (GeN_4)

Germanium Amide	$\longrightarrow$	Germanium Diamide
$Ge(NH)_2$	$\longrightarrow$	$GeNH_4$

It disproportionate at higher temperature giving black germanium chloride. It is very sensitive to moisture and atmosphere.

Germanium Dichloride	+	Water	$\longrightarrow$	Germanium Oxide	+	Hydrogen Chloride
$2GeCl_4$	+	H_2O	$\longrightarrow$	GeO	+	2HCl
						(reaction with moisture)

Germanium Dichloride	+	Oxygen	$\longrightarrow$	Germanium Oxide	+	Germanium Tetrachloride
$2GeCl_4$	+	O_2	$\longrightarrow$	GeO	+	$GeCl_4$
						(reaction with air)

Germanium Dihalides are strong reducing agents.

Germanium Chloroform ($GeHCl_3$): A colourless fuming liquid is formed when germanium dichloride is allowed to react with hydrogen chloride at temperature of 40°C.

Germanium Dichloride	+	Hydrogen Chloride	$\longrightarrow$	Germanium Chloroform
$GeCl_2$	+	HCl	$\longrightarrow$	$GeHCl_3$

Germanium Hydroxide [$Ge(OH)_2$]: It is produced as a yellowish red precipitate by the action of caustic soda on germanium chloroform.

Germanium Chloroform	+	Sodium Hydroxide	$\longrightarrow$	Germanium Hydroxide	+	Sodium Chloride	+	Water
$GeHCl_3$	+	3NaOH	$\longrightarrow$	$G_3(OH)_2$	+	3NaCl	+	H_2O

When heated germanium hydroxide produces germanium oxide, when further dissolved in sodium hydroxide, it produces sodium germanite (Na_2GeO_2).

Germanous Sulphide (GeS): Reddish brown compound is formed by heating the disulphide in hydrogen. It is reduced to germanium on further heating. It is best prepared by reducing a suspension of germanium disulphide in hydrogen chloride with hypophosphorous acid.

Germanium Sulphide	+	Hydrogen Chloride	+	Hypophos-phorous Acid		Water	$\longrightarrow$	Germa-nium Chloride	+	Metaphos-phoric Acid	+	Hydrogen Sulphide
GeS_2	+	2HCl	+	H_3PO_2	+	H_2O	$\longrightarrow$	$GeCl_2$	+	H_3PO_3	+	$2H_2S$

Neutralizing the solution formed and passing hydrogen sulphide.

Detection of Germanium

The germanium is detected by the spectrophotometrical analysis.

ARSENIC

Melting Point : 817°C | Atomic Weight : 74.9816
Boiling Point : 610°C (sublimate) | Atomic Number : 33

History and Occurrence

Arsenic was recognized by Brandt in 1733. In the ancient period *Realgar (AsS)* and *Orpiment* (As_2O_3) was well-known and were used as pigments. Olympiodoros in 15th century B.C. describes about the arsenic oxide (As_2O_3) as white arsenic, which was obtained by roasting arsenic sulphide in air. The element itself was obtained as a sublimate and was used for whitening copper.

Arsenic is available in the free form in the nature and also found as the compounds, known as ores. They are *White Arsenic* (As_2O_3) *Nicolite (NiAs) Chloanthite* $(NiAs_2)$ *Nickel* Glance, *Percolite* $(CaHAsO_4 2H_2O)$ *Erthyrite, or Cobalt Bloom* $[Co(AsO_4)_2 8H_2O]$ and *Mimetite* $[3Pb_3(AsO_4)2PbCl_2]$. Natural sulphur, iron pyrites and other sulphide ores often combine arsenic and sulphuric acid. Arsenical pyrites may contain 1% As_2O_3 , coal smoke especially in yellow fogs, may contain arsenious sulphide or oxides, from pyrites in the coal. Traces of arsenic occur in some mineral water in the soil (1-60ppm) in the adult human being (0.01mg). American Tobacco (6-30ppm) contains unusually large amount of arsenic.

In roasting minerals, e.g. Cobahite, in metallurgical treatment fumes of arsenious oxide may be evolved and condense in flues as powder.

This is obtained in larger amounts by roasting arsenical ores such as mispickel in a current of air. Most of the arsenious oxide used is obtained from flue dust in the U.S.A. Mexico and Sweden could supply the world requirement of As_2O_3 The crude oxide is sublimed in iron pots to form white arsenic (popularly called simple arsenic) the commonest arsenic compound.

Most of the arsenic is used in alloys with lead and copper; as arsenite in weed killer and sheep dips and calcium lead and copper arsenate in spraying solutions, also as oxide, as a poison and for preserving skin. It is also used for glass making, for removing colour, wood preserving and in medicine also.

Extraction

Arseno pyrites on roasting yield the volatile arsenious oxide (As_2O_3) which condenses as a white solid.

$$\underset{\text{Cobaltite}}{4CoAsS} + \underset{\text{Oxygen}}{9O_2} \longrightarrow \underset{\text{Cobalt Oxide}}{4CoO} + \underset{\text{Sulphur Dioxide}}{4SO_2} + \underset{\text{Arsenious Oxide}}{2As_2O_3}$$

Arsenic is obtained by heating arsenious oxide and charcoal together in a crucible.

$$\underset{\text{Arsenious Oxide}}{As_2O_3} + \underset{\text{Carbon}}{3C} \longrightarrow \underset{\text{Arsenic}}{2As} + \underset{\text{Carbon Monoxide}}{3CO}$$

Arsenic collects as a sublimate while carbon monoxide escapes. It can also be produced by heating Arseno pyrites in the absence of air. Arsenic sublimates and separates out.

$$\text{Arseno Pyrites} + \text{Oxygen} \longrightarrow \text{Ferrous Sulphide} + \text{Arsenious Oxide}$$
$$4FeAsS + 3O_2 \longrightarrow 4FeS + 2As_2O_3$$

Physical Properties

Melting Point	817°C
Boiling Point	610°C
Density	5.73 g/cc
Atomic Volume	13.1 cc
Atomic Radius	2.2Å
Ionization energy	231.0 Kcal/mole
Oxidation State	– 3, + 3 + 5
Electronegativity	2.0
Latent Heat of Fusion	5.1 Kcal/mol
Crystal Radius	0.47Å
Heat of Fusion	5.97 cal deg^{-1} mol^{-1}
Standard Entropy	8.4 cal deg^{-1} mol^{-1}
Electrical Resistivity	32 microohm
Viscosity	2.2×10^{10} N/m^2

Chemical Properties

Arsenic as a Metalloid : The difference between metals and non-metals is very critical. There is not a single physical or chemical cause, which will clearly specify the difference. In fact these two classes merge into each other without any distinction. But in few cases one element behaves both as a metal also a non-metal. These elements are thus called *metalloids.* The most important example of metalloid is arsenic, whose metallic and non-metallic properties will appear almost equally balanced as will appear from the fact described below:

Non-metallic Properties of Arsenic

1. Arsenic, like sulphur and exhibits allotropisim. One of the allotropes namely "yellow arsenic" is distinctly non-metallic in character. It does not possess any metallic lusture.
2. It oxidizes As_2O_3 and As_2O_3 are both acidic in character and dissolve in alkalis to form salts called the arsenates and the arsenites respectively.
3. Arsenic forms a hydride with the formula (AH_3) called arsine, which is similar to ammonia and phosphene.
4. Arsenic further resembles non-metals in the behaviour towards acids. It is unattacked by non-oxidizing acids, such as hydrochloric acid and sulphuric acid when treated with concentrated nitric acid, it gets oxidized first to arsenious acid (H_3AsO_3) and than to arsenic acid (H_3AsO_4).

Arsenic as Metal

1. Arsenic is a steel grey brittle solid with a dull metallic lusture and a mild conductor of electrical current. Its density, 5.73 is comparable with that of many metals.
2. The behaviour of arsenic compounds in aqueous solutions is very similar to that of the compounds of many metals. When hydrogen sulphide is passed through an arsenious salt solution a fine yellow precipitates of arsenious sulphide is produced.

More Chemical Properties

In chemical properties, arsenic resembles non-metals.

1. When heated in hydrogen arsenic burns to give arsenic oxide.

$$\text{Arsenic} + \text{Oxygen} \longrightarrow \text{Arsenic Oxide}$$
$$2As_2 + 3O_2 \longrightarrow 2As_2O_3$$

2. Finely powdered arsenic burns spontaneously in chlorine forming arsenic trichloride, ($AsCl_3$). It also combines with other halogens to form trihalides. However, it does not form the pentahalides. Only exception is arsenic pentafluoride (AsF_5).
3. Arsenic combines with sulphur to form the compounds of arsenic disulphide (As_2S_3) arsenic trisulphide (AsS) and arsenic pentasulphide (As_2S_5).
4. Like non-metals, non-oxidizing acids do not affect arsenic. When treated with an acid having oxidizing properties. (e.g. nitric acid and sulphuric acid), it is oxidized to arsenious acid, then to arsenic acid.

$$\text{Arsenic} + \text{Nitric Acid} \longrightarrow \text{Arsenious Acid} + \text{Nitrogen Dioxide} + \text{Water}$$
$$2As + 6HNO_3 \longrightarrow As_2O_3 + 6NO_2 + H_2O$$
$$\text{Arsenious Acid} + \text{Nitric Acid} + \text{Water} \longrightarrow \text{Arsenic Acid} + \text{Nitrogen Dioxide}$$
$$As_2O_3 + 4HNO_3 + H_2O \longrightarrow 2H_3ArO_4 + 4NO_2$$

Compounds of Arsenic

Arsenic resembles phosphorous in most of its properties, but phosphorous is a non-metal and arsenic bears a slight acidic property. The common and the stable form of arsenic are distinctly metallic. Besides, arsenic is incapable of forming arsenium compounds.

> Arsenic is incapable of forming arsenium compounds.

Like phosphorous arsenic generally forms two series of compounds arsenious arsenic compounds, in which the element shows a valency of three and five respectively.

Hydrides

The following halides of arsenic are known, they are all covalent compounds. In all its halides excepting the pentafluoride, arsenic shows a valency of three only. The formulas and the common physical properties of all halides are tabulaled below:

Halides	Formula	Colour	Melting Point	Boiling Point	Sp. Gravity
Trifluoride	AsF_3	Colourless liquid	–8.5°C	60.4°C	–2.67 (at 0°C)
Pentafluoride	AsF_5	Colourless liquid	– 80°C	– 53°C	—
Chloride	$AsCl_3$	Colourless liquid	– 13°C	130.2°C	2.205
Bromide	$AsBr_3$	Colourless crystal	31°C	221°C	3.66(15°C)
Di-iodide	$A5I_2$	Red crystal	130.2°C	378°C	—
Tri-iodide	AsI_3	Red crystal	140.7°C	394-414°C	4.4

Arsenic forms two hydrides arsenic trihydride (arsenic or arseniuretted hydrogen) (AsH_3) which is gaseous and a solid dihydride (As_2H_2).

Arsine (AsH_3) : It is formed by the action of nascent hydrogen in any arsenic compound. When a dilute sulphuric acid containing with little arsenious oxide in solution is treated with zinc, the hydrogen evolved thus has a very unpleasant odour like garlic and burns with a lilac coloured flame. Due to the presence of arsine the reaction is utilized as a sensitive test for arsenic also known as Marsh's Test for arsenic. It is also formed at the cathode by the electrolysis of an arsenious oxide solution. Pure arsine can be produced by passing the gaseous mixture of arsine and hydrogen, through a tube filled with liquid air, arsine is condensed as a colourless liquid with a boiling point of -53°C. It is however, most conveniently prepared by the action of aluminium arsenide and water. The direct union of arsenic and aluminium prepares aluminium arsenide, the chemical mixture is heated in a closed crucible, after the reaction aluminium arsenide is dissolved in water when arsine is produced.

$$\text{Aluminium} + \text{Arsenic} \longrightarrow \text{Aluminium Arsenide}$$

$$Al + As \longrightarrow AlAs$$

$$\text{Aluminium Arsenide} + \text{Water} \longrightarrow \text{Aluminium Hydroxide} + \text{Arsine}$$

$$AlAs + H_2O \longrightarrow Al(OH)_3 + AsH_3$$

Arsine is a very poisonous gas, colourless in nature with a garlic odour. It is decomposed into its elements with a decomposition of a black shinning mirror of arsenic, when the gas is gently heated glass tube. A similar decomposition occurs when moist arsine is exposed to light. It is scarcely affected by water, acid and alkalis; with dilute silver nitrate it gives a black precipitate of metallic silver with the formation of arsenious acid in the solution.

$$\text{Arsine} + \text{Silver Nitrate} + \text{Water} \longrightarrow \text{Arsenious Acid} + \text{Nitric Acid} + \text{Silver}$$

$$AsH_3 + 6AgNO_3 + 3H_2O \longrightarrow H_3AsO_3 + 6HNO_3 + 6Ag$$

A solution of mercuric chloride, when treated with arsine, first turns yellow due to the formation of AsH ($HgCl_2$); on further treatment it gives a brown precipitate of As ($HgCl_3$) which changes finally to the black As_2Hg_3.

$$\text{Arsine} + \text{Mercurous Chloride} \longrightarrow \text{Arsenious Mercurous Chloride} + \text{Hydrogen Chloride}$$

$$2AsH_3 + 4HgCl_2 \longrightarrow 2AsH(HgCl)_2 + 4HCl$$

Arsine	+	Mercurous Chloride	$\longrightarrow$	Arsenious Mercurous Chloride	+	Hydrogen Chloride	+	Chlorine
$2AsH_3$	+	$6Hg_2Cl_2$	$\longrightarrow$	$2AsH(HgCl)_2$	+	$4HCl$	+	Cl_2
Arsine	+	Mercurous Chloride	$\longrightarrow$	Arsenious Mercuride	+	Hydrogen Chloride	+	Chlorine
$2AsH_3$	+	$18Hg_2Cl_2$	$\longrightarrow$	$2AsH(HgCl)_2$	+	$18HCl$	+	$9Cl_2$

The composition of arsine can be detected by decomposition of a known volume of the gas by heating and measuring the volume of hydrogen formed. From the ratio of the volumes of the two gases, with the application of Avogadro's Hypothesis and from the density of arsine the molecular formula AsH_3 can be deducted.

The solid hydride (AS_2H_2) is formed as a brown amorphous powder by the addition of a solution of arsenious chloride in hydrochloric acid to an ethereal solution of stannous chloride.

Arsenious Chloride	+	Ether Stannous Chloride	+	Hydrochloric Acid	$\longrightarrow$	Solid Arsine	+	Stannous Tetrachloride
$2AsCl_3$	+	$4SnCl_2$	+	$2HCl$	$\longrightarrow$	As_2H_2	+	$4SnCl_3$

It is insoluble in cold water, but is decomposed by boiling water with liberation of hydrogen.

Solid Arsine	+	Water	$\longrightarrow$	Arsenious Acid	+	Hydrogen
As_2H_2	+	$6H_2O$	$\longrightarrow$	$2H_3AsO_3$	+	$4H_2$

The hydride (As_2H_2) undergoes slow but spontaneous decomposition at room temperature into arsenic and hydrogen.

Holides	Formula	Colour	Melting Point	Boiling Point	Sp. Gravity
Trifluoride	AsF_3	Colourless liquid	–8.5°C	60.4°C	–2.67 (at 0°C)
Pentafluoride	AsF_5	Colourless liquid	– 80°C	– 53°C	—
Chloride	$AsCl_3$	Colourless liquid	– 13°C	130.2°C	2.205
Bromide	$AsBr_3$	Colourless crystal	31°C	221°C	3.66(15°C)
Di-iodide	$A5I_2$	Red crystal	130.2°C	378°C	—
Tri-iodide	AsI_3	Redcrystal	140.7°C	394-414°C	4.4

Halides

The following halides of arsenic are known, they are all covalent compounds. In all its halides, excepting the pentafluoride, arsenic shows a valency of three only. A dihalide is also known. The formula and the common physical properties of all these halides are tabulated below.

Arsenic burns in fluorine to form a mixture of tri and penta-fluorides. Arsenic trifluoride is obtained as a colourless fuming liquid by heating a mixture of arsenious oxide reacts with hydrogen fluoride.

Excepting the tri-chloride of arsenic all the other halides are soluble in water.

The trifluoride distils over. The pentafluoride can be obtained from the trifluoride by distillation of a mixture of the latter with antimony pentafluoride and bromine.

Arsenic Trifluoride + Antimony Pentafluoride + Bromine ⟶ Antimony Tetrabromofluoride + Arsenic Pentafluoride

$$AsF_3 + 2SbF_5 + Br_2 \longrightarrow 2SbBrF_4 + AsF_5$$

The distillate is collected in a receiver cooled in liquid air.

Arsenic Trichloride ($AsCl_3$): This can be made by the action of hydrochloric acid on arsenic oxide, or by the direct combination of the elements. It is most conveniently prepared by distilling a mixture of arsenious oxide with sulphur monochloride in a current of chlorine.

Arsenious Oxide + Sulphur Pentafluoride + Chlorine ⟶ Arsenious Trichloride + Sulphur Dioxide

$$2As_4O_6 + 3S_2Cl_2 + 9Cl_2 \longrightarrow 8AsCl_3 + 6SO_2$$

It is a poisonous fuming liquid and is hydrolyzed by water.

Arsenious Trichloride + Water ⟶ Arsenious Oxychloride + Hydrogen Chloride

$$AsCl_3 + 2H_2O \longrightarrow As(OH)_2Cl + 4HCl$$

Arsenious Oxychloride + Water ⟶ Arsenious Oxide + Hydrogen Chloride + Water

$$As(OH)_2Cl + 4H_2O \longrightarrow As_4O_6 + 2HCl + 6H_2O$$

Arsenic tribromide and arsenic triiodide are formed by heating arsenic with a solution of bromine or iodine in carbon disulphide. The triiodide can also be prepared as a precipitate by adding a solution of potassium iodide to a solution of arsenious oxide in concentrated hydrochloric acid. The triiodide is not readily hydrolyzed.

A di-iodide (As_2I_4) can also be obtained by heating the elements together in a sealed tube at 250°C. It forms as red crystals soluble in CS_2 but it is decomposed by water into arsenic and arsenic triiodide.

Arsenic Diiodide ⟶ Arsenic Triiodide + Arsenic

$$2As_2l_4 \longrightarrow 4Asl_3 + 2As$$

Oxides and Oxyacids

Arsenious Oxide (As_4O_6): It is also known as white arsenic, it is formed when arsenic bums in air, it is prepared commercially by roasting arsenical ore (See Fig. 17.9). It exists in three different forms on amorphous and two crystalline octahedral and monoclinic.

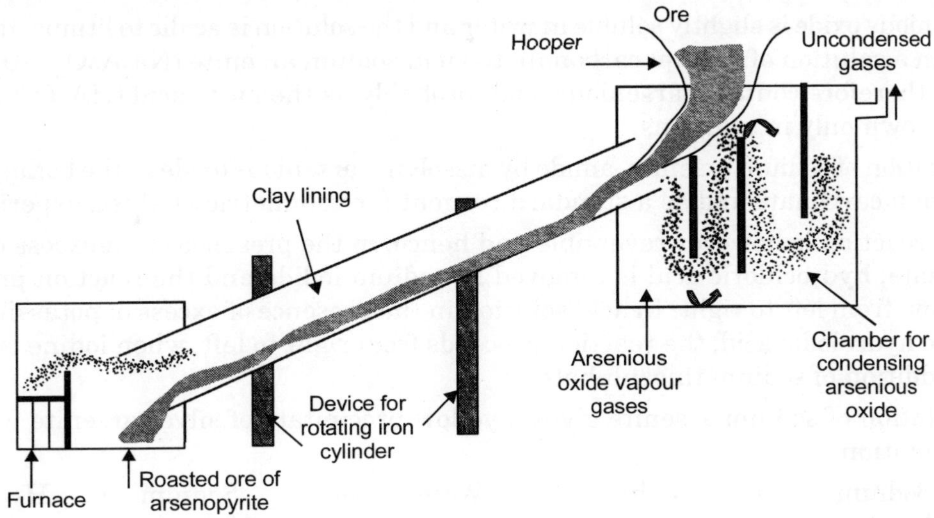

Fig. 17.9. The Industrial preparation of Arsenious Oxide from Arsenopyrite

The transition temperature for the octahedral and monoclinic modification lies at about 280°C. The change is however, very slow.

The vapour density of arsenious oxide lies between 550°C to 1450°C corresponds with the molecular composition, As_4O_6. Determination of the molecular weight of the substance from the freezing point of the solution in nitrobenzene also leads to the same conclusion.

Arsenious oxide is easily oxidized to the pentoxide, arsenic acid or and acetate, by various oxidizing agents such as ozone, hydrogen peroxide chlorine, bromine, iodine nitric acid, aqua regia and alkaline hypochlorite for e.g.

Arsenious Oxide	+	Chlorine	+	Water	⟶	Arsenic Acid	+	Silicon Chloride
As_4O_6	+	$4Cl_2$	+	$10H_2O$	⟶	$4H_3AsO_4$	+	$8HCl$

Copper sulphate in Fhelling's Solution oxidizes arsenious oxide to arsenate and is itself reduced to red cuprous oxide.

Arsenious oxide is reduced to arsenic when heated with charcoal or potassium cyanide and when treated with an acid solution of stannous chloride: e.g.

Arsenious Oxide	+	Hydrogen Chloride	+	Stannous Chloride	⟶	Arsenic	+	Stannous Tetrachloride	+	Water
As_4O_6	+	$12HCl$	+	$6SnCl_2$	⟶	$2As_2$	+	$6SnCl_4$	+	$6H_2O$

Reduction of arsenic also occurs when arsenious oxide is boiled with hydrochloric acid with copper foil.

Arsenious Oxide	+	Hydrogen Chloride	+	Copper	⟶	Arsenic	+	Cuprous Chloride	+	Water
As_4O_6	+	$12HCl$	+	$12Cu$	⟶	$2As_2$	+	$12CuCl$	+	$6H_2O$

Arsenious oxide is slightly soluble in water and the solution is acidic to litmus. It dissolves readily in a solution of sodium carbonate to yield sodium arsenite (Na_3AsO_3). An aqueous solution therefore contains arsenious acid, probably as the meta-acid ($HAsO_2$) arsenious acid is known only in solutions.

A solution of sodium arsenite, amide by dissolving arsenious oxide in the boiling solution of sodium bicarbonate and as a standard reagent for volumetric analysis experiment.

This reaction is however reversible and hence, in the presence of an excess of sodium bicarbonate, hydrochloric acid is removed as sodium iodide and the reaction proceeds to completion from left to right. In acid solution in the presence of excess of potassium iodide i.e. of the hydroiodic acid, the reaction proceeds from right to left, when iodine is removed by the addition of sodium thiosulphate.

A solution of sodium arsenite gives a yellow precipitate of silver arsenite with silver nitrate solution.

$$\underset{\text{Arsenite}}{\text{Sodium}} + \text{Iodine} + \text{Water} \longrightarrow \underset{\text{Arsenite}}{\text{Sodium}} + \underset{\text{Iodide}}{\text{Hydrogen}}$$

$$Na_3AsO_3 + I_2 + H_2O \longrightarrow NA_3AsO_4 + 2HI$$

$$\text{Sodium Arsenite} + \text{Silver Nitrate} \longrightarrow \text{Silver Arsenite} + \text{Sodium Nitrate}$$

$$Na_3AsO_3 + AgNO_3 \longrightarrow AgAsO_3 + 3NaNO_3$$

The precipitate is soluble in acetic acid. With copper sulphate solution sodium arsenate gives a bright green precipitate of cupric arsenite (Scheelie's Green) $CuHAsO_3$. This is used as insecticide and as a pigment. When precipitated copper arsenite is treated with dilute acetic acid, it forms the brilliant green pigment, called Shweinfurt's green or Paris Green, $Cu(C_2H_3O_2)_2\ 3Cu(AsO_2)_2$.

Arsenic Pentoxide (As_2O_3): When arsenious oxide is dissolved in concentrated nitric acid arsenic acid is formed with the evolution of oxides of nitrogen.

The solution on concentration and cooling deposits crystals of arsenic acid ($2H_3AsO_4$ H_2O), this is orthoarsenic acid. On heating, it first looses the water of crystallization and then at a temperature of about 200°C pentoxide, As_2O_5 is formed.

$$\text{Orthoarsenic Acid} \longrightarrow \text{Arsenic Pentaoxide} + \text{Water}$$

$$2H_3AsO_4 \longrightarrow As_2O_5 + 3H_2O$$

Pyroarsenic Acid (As_2O_7): This is formed as an intermediate product. Arsenic pentoxide forms a white delequescent solid. At red heat it melts and decomposes with the formation of arsenious oxide.

$$\text{Arsenic Pentoxide} \longrightarrow \text{Arsenious Oxide} + \text{Oxygen}$$

$$2As_2O_5 \longrightarrow As_4O_6 + 2O_2$$

Arsenic pentoxide dissolves slowly in water to form arsenic acid, in alkalis to form arsenates.

Arsenic acid is an oxidizing agent. It liberates iodine from potassium iodide in acid solution.

$$\text{Arsenic Acid} + \text{Hydrogen Iodide} \longrightarrow \text{Arsenious Acid} + \text{Iodine} + \text{Water}$$
$$H_3AsO_4 + 2HI \longrightarrow H_3AsO_3 + I_2 + H_2O$$

Arsenates are reduced to arsenites by the action of sulphur dioxide or less readily by hydrogen peroxide.

Like phosphoric acid, arsenic acid is also a tribasic acid and forms three series of salts. Primary, Arsenious, secondary, Arsenate and tertiary, Arsenite.

Arsenates are similar to orthophosphates and have indistinguishable formulae. The orthoarsenates exist both in solid and in solutions, but pyro, and met-arsenates are known only in solid state, these compounds are obtained by heating ortho salts of arsenic.

$$\text{Sodium Arsenate} \longrightarrow \text{Sodium Pyroarsenate} + \text{Water}$$
$$2Na_2HAsO_2 \longrightarrow Na_4As_2O_7 + H_2O$$

Common sodium arsenate is used in calico printing.

Ammonium molybdate and concentrated nitric acid give with arsenates, on boiling a yellow precipitate of ammonium arseno molybdate, related to that given by phosphates. With silver nitrate solution arsenates gives a chocolate brown precipitate of silver arsenate and ammonia. A magnesia mixture gives as with phosphates a white crystalline precipitate of magnesium ammonium arsenate [$Mg(NH_4)AsO_4.6H_2O$] with arsenate solutions.

Sulphides

Arsenic forms three well-known sulphides *arsenic disulphide, arsenic trisulphide, arsenic pentasulphide.*

Arsenic Disulphide (As_2S_2): It is found in nature as the mineral Realgar. It is also prepared in the laboratory by the sublimation of sulphur with excess of arsenic, or by distilling iron pyrite with arsenical pyrites.

$$\text{Arsenic Pyrites} + \text{Iron Pyrites} \longrightarrow \text{Ferrous Sulphide} + \text{Arsenic Disulphide}$$
$$2FeAsS + 2F_3S_2 \longrightarrow 4FeS + As_2S_2$$

It is a red brittle crystalline solid, which fuses at 300°C.

Arsenic Trisulphide (As_2S_3): It is found native as the mineral, named as Orpiment, which is used as pigment. It is also prepared in the large scale by the sublimation of arsenious oxide with sulphur. In the laboratory it is prepared by precipitating an aqueous solution of arsenic trichloride with hydrogen sulphide. When precipitated from an acid solution it is obtained as a yellow solid.

$$\text{Arsenic Trichloride} + \text{Hydrogen Sulphide} \longrightarrow \text{Arsenic Trisulphide} + \text{Hydrogen Chloride}$$
$$2AsCl_3 + 3H_2S \longrightarrow As_2S_3 + 6HCl$$

Arsenic trisulphide is soluble in alkali sulphides to form thioarsenates.

$$\text{Arsenic Trisulphide} + \text{Sodium Sulphide} \longrightarrow \text{Sodium Thioarsenate}$$
$$As_2S_3 + 2Na_2S \longrightarrow 2Na_3AsS_3$$

It dissolves alkalis, ammonia or even in ammonium carbonate solution to form a mixture of arsenate and thioarsenate.

Arsenic Trisulphide	+	Sodium Hydroxide	⟶	Sodium Thioarsenate	+	Sodium Arsenate	+	Water
As_2S_3	+	6NaOH	⟶	Na_3AsS_3	+	Na_3AsO_3	+	$3H_2O$

From all these solutions arsenic trisulphide is reprecipitated on acidification. In alkali polysulphides, e.g. ammonium polysulphide $(NH_4)_2S_2$ it forms thioarsenates. Arsenic trisulphide is used as a yellow pigment and as an insecticide; both arsenic trisulphide and disulphide are used in pyrotechnic. Bengal fire is the mixture of realgar sulphur and nitre.

Arsenic Pentasulphide (As_2S_5) : It is obtained as a yellow precipitate by rapidly passing hydrogen sulphide into a warm solution of arsenate or arsenic acid.

Sodium Arsenate	+	Hydrochloric Acid	+	Hydrogen Sulphide	⟶	Arsenic Pentafluoride	+	Sodium Chloride	+	Water
$2Na_3AsO_4$	+	6HCl	+	$5H_2S$	⟶	As_2S_5	+	6NaCl	+	$8H_2O$

Like the trisulphide it is soluble in solutions of alkali sulphides and forms thioarsenates with alkalis hydroxides or ammonia solution it forms a mixture of thioarsenates. It is oxidized by nitric acid. Thioarsenates are also obtained by dissolving arsenious sulphide in alkali or ammonium polysulphides. Arsenic sulphide is reprecipitated from these solutions on acidification.

Detection of Arsenic

All the water-soluble arsenious compounds are virulent* poisons. Hence, the detection of traces of arsenic has acquired a special importance. These are some tests, which are being discussed below:

Marsh Test: This is the most sensitive test of arsenic, in performing this test a current of hydrogen, produced by the action of pure arsenic free zinc and pure dilute sulphuric acid. The hydrogen thus formed reacts with the arsenic present in the sample forming arsine gas, which is further purified by a filter of lead acetate to remove the hydrogen sulphide and then it is dried in the presence of calcium chloride. The arsine passes through a glass tube where it is heated, the gas is decomposed to arsenic and hydrogen on the inner walls of the tube as a shinning mirror, and from this mirror the quantity is then assumed. The arsenic is soluble in sodium hypochlorite solution.

Arsenic	+	Sodium Hypochloride	+	Water	⟶	Arsenic Acid	+	Sodium Chloride
$2As_2$	+	10NaOCl	+	$6H_2O$	⟶	$4H_3AsO_4$	+	10NaCl

The film insoluble in tartaric acid makes a bright yellow deposit when treated with ammonium sulphide, due to the formation of arsenic trisulphide As_2S_3. In all these reactions it differ from antimony mirror or file, produced similarly by an antimony compound.

Fleimn's Test: In this test hydrogen is generated by the action of aluminium tungling's on potassium hydroxide, this alkline mixture has no effect on arsenates and antimony

*Virulent: Deadly but slow.

compounds. The arsine is produced only from the arsenious compounds in this case and is detected by holding a silver nitrate paper on the mouth of the reaction vessel which turns black due to the deposition of the metallic silver, because the produced arsine reduces the silver nitrate to metallic silver.

Reinsh's Test: This test is based on the deposition of metallic arsenic on copper foil boiling hydrochloric acid solution containing any arsenic compound. If the copper foil is afterwards washed dried and then heated in a dry tube a white crystalline sublimate of arsenious oxide is formed on the walls of the tube.

Gutzeit's Rest: This test us used to detect arsenic in drinking water. The amount of arsenic in the 100 g of water. The amount of arsenic present in 100gm of water should not be more than 50mg, or the total percentage is .05%, the reaction apparatus is shown below in (Fig 17.10).

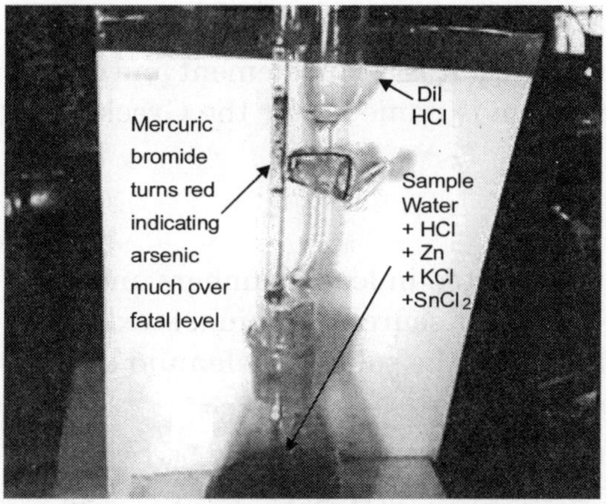

Photo graph showing the detection of arsenic in author's Lab

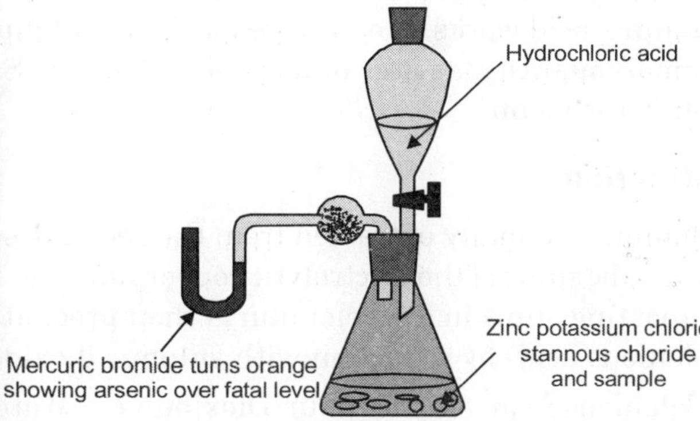

Fig. 17.10. The Lab set up for Arsenic Detection

The reagents required are lead acetate solution, potassium chloride (1.5ml) hydrochloric acid (3ml) stannous chloride (.5ml) and mercurous bromide and zinc (arsenic free).

At first the potassium chloride and zinc is added to the reaction vessel some lead acetate solution is soaked in the cotton plug and kept in the bulb tube, some mercuric bromide is kept in the V-tube. Water containing hydrochloric acid is poured in the vessel through the funnel, hydrogen formed with the reaction of zinc and hydrochloric acid, reacts with arsenic producing arsine. This gas passes through the lead acetate plug where it is made hydrogen sulphide free, this arsine reacts with the mercuric bromide producing a red coloured pigment indicating the presence of arsenic in the water above normal.

Estimation of Arsenic

Arsenic is generally estimated by precipitation of ammonium arsenomolybdate from boiling solution (phosphates give a similar precipitate but in cold). It can also be estimated by the precipitation as a magnesium ammonium arsenate, $NI_4MgAsO_46H_2O$ and then, as in the case of phosphorous, by ignition to magnesium pyroarsenate, $Mg_2As_2O_7$. Arsenates are distinguished from phosphates by the formation of the chocolate, coloured precipitate of silver nitrate with silver nitrate solution, while the silver phosphate is yellow coloured precipitate, silver arsenate is also soluble in nitric acid as well as in ammonia. Further more arsenic after reduction with sulphur dioxide solution gives a yellow precipitate of arsenious sulphide, when treated with hydrogen sulphide.

SELENIUM

Melting Point : 215°C Atomic Number : 34
Boiling Point : 685°C Atomic Weight : 78.96

History and Occurrence

Berzzelius first discovered selenium in 1817 as a red deposit in the lead chambers of sulphuric acid works. It occurs generally in sulphur pyrites. It is a rare element and occurs in small quantity, as ores of metallic selenides. Selenium is named after the Greek word 'Selini' the moon.

Extraction

Selenium is usually extracted from the red deposits, collected in lead chambers and from the anodic slims of the electrolytic copper refining. From these sources selenium is extracted by roasting slims in air, selenium is then precipitated from the solution selenium dioxide (selenious acid) by reduction with sulphur dioxide.

Selenious Acid + Sulphur Dioxide + Water ⟶ Sulphuric Acid + Selenium

$$H_2SeO_3 + 2SO_2 + H_2O \longrightarrow 2H_2SO_4 + Se$$

Physical Properties

Selenium in the periodic table lies between sulphur and tellurium and like sulphur selenium exists the allotropic properties.

(i) ***Amorphous Selenium :*** Red Amorphous selenium is a dark red powder obtained by a chemical reaction in aqueous solutions, e.g. reduction of selenious acid by sulphur chloride on the solution of potassium cyanide and selenium.

(ii) ***Vitreous Selenium :*** An opaque, almost black lustrous solid, is made by quenching molten selenium in cold water. Above 70°C it quickly changes to metallic selenium. Amorphous selenium is slightly soluble in carbon disulphide.

(iii) ***Colloidal Selenium :*** It is made as red crystals by evaporating a solution of red amorphous selenium in carbon disulphide. The crystals are metastable and pass into metallic selenium. Monoclinic selenium is soluble in carbon disulphide and its crystals contain 8-membered rings of Se_8.

(iv) ***Metallic Selenium :*** It is the stable form of selenium, if heated for sometime at 200°C, it becomes silver grey in colour. It is insoluble in carbon disulphide. It forms hexagonal crystals; similar to tellurium and in this form it consists of a zigzag chain of selenium atoms. The grey form melts at 277°C and its specific gravity is 4.80. It has a metallic appearance and is a feeble conductor of electricity.

Liquid selenium boils at 635°C to dark red vapour whose density falls with rise of temperature, at lower temperature Se_8 and Se_6 molecules only and at 2000°C Se_6 mostly. Its molecular weight in yellow phosphorous and in carbon disulphide corresponds to Se_8.

Physical Properties

Melting Point	215°C
Boiling Point	685°C
Density	4.80 g/cc
Heat of Fusion	1.5 Kcal/g
Entropy	10.0 cal deg^{-1} mol^{-1}
Compressibility	12.2×10^{-16}
Magnetic Susceptibility	0.32×10^{-6}
Co-efficient of thermal Expansion	3.79×10^{-5}
Crystalline Structure	Hexagonal
Electrical Resistance	8×10^{-6} ohm cm

Chemical Properties

Selenium though stable in normal temperature but reacts vividly with oxygen when heated forming selenium chloride. Selenium also reacts with hydrogen forming hydrogen selenide.

Selenium + Oxygen ⟶ Selenium Oxide

$$Se + O_2 \longrightarrow SeO_2$$

Selenium + Hydorgen ⟶ Hydrogen Selenide

$$Se + H_2 \longrightarrow H_2Se$$

Selenium combines with alkali and other light metals e.g. aluminium forming selenides and selenades.

Selenium + Sodium Hydroxide ⟶ Sodium Selenide + Sodium Selenade + Water

$$3Se + 6NaOH \dashrightarrow Al_2Se_3 + Na_2SeO_3 + 3H_2O$$

Aluminium + Selenium ⟶ Aluminium Selenide

$$2Al + 3Se \longrightarrow Al_2Se_3$$

Aluminium reacts with water producing hydrogen selenide.

Aluminium Selenide + Water ⟶ Hydrogen Selenide + Aluminium Hydroxide

$$Al_2S_3 + H_2O \longrightarrow 3H_2Se + 2Al(OH)_3$$

Selenium reduces hot silver nitrate solution producing silver silinide and selenic acid.

Selenium	+	Silver Nitrate	+	Water	⟶	Silver Selenide	+	Selenic Acid	+	Nitric Acid
3Se	+	$4AgNO_3$	+	$3H_2O$	⟶	$2Ag_2Se$	+	H_2SeO_3	+	$4HNO_3$

Compounds of Selenium

Selenium Dioxide (SeO_2): This compound is formed when selenium burns in air. It is most conveniently obtained by oxidizing the element in nitric acid and evaporating the solution of selenious acid, thus obtained, to dryness.

Selenium	+	Nitric Acid	⟶	Selenious Acid	+	Water	+	Nitrogen Dioxide
Se	+	$4HNO_3$	⟶	H_2SeO_3	+	H_2O	+	$4NO_2$

Selenious Acid	⟶	Selenium Dioxide	+	Water
H_2SeO_3	⟶	SeO_2	+	H_2O

It is prepared also by the direct combination of selenium and oxygen at a very high temperature, as said earlier.

Selenium	+	Oxygen	⟶	Selenium Dioxide
Se	+	O_2	⟶	SeO_2

Selenium Dioxide is a white crystalline substance it sublimes on heating at a temperature of 345°C. It is also known as selenious acid anhydride as it produces selenious acid when reacting with water.

Water	+	Selenium Dioxide	⟶	Selenious Acid
H_2O	+	SeO_2	⟶	H_2SeO_3

Selenium dioxide is an active organic oxidant.

Selenium Trioxide (SeO_3) : This compound is produced when selenium dioxide reacts with oxygen.

Selenium Dioxide	+	Oxygen	⟶	Selenium Trioxide
$2SeO_2$	+	O_2	⟶	$2SeO_3$

Selenium trioxide is a pale yellow solid which when heated gets decomposed into selenium dioxide and oxygen. Selenium trioxide is regarded as the selenic anhydride because it produces selenic acid when reacts with water.

Selenium Trioxide	+	Water	⟶	Selenic Acid
SeO_3	+	H_2O	⟶	H_2SeO_4

Hydrogen Selenide (H_2Se): It is like hydrogen sulphide, a colourless gas and is also poisonous in nature. It is obtained by the action of dilute hydrochloric acid on ferrous selenide.

Ferrous Selenide + Hydrochloric Acid ⟶ Ferrous Chloride + Hydrogen Selenide

$$FeSe + 2HCl \longrightarrow FeCl_2 + H_2Se$$

The gas possesses an unpleasant odour. It is very unstable and is readily decomposed into its elements on heating.

Hydrogen Selenide ⟶ Hydrogen + Selenium

$$H_2Se \longrightarrow H_2 + Se$$

Hydrogen Selenide is much less stable than hydrogen sulphide as shown in the reaction below:

Hydrogen Selenide + Sulphur ⟶ Hydrogen Sulphide + Selenium

$$H_2Se + S \longrightarrow H_2S + Se$$

The gas in air forming selenium dioxide and water.

Hydrogen Selenide + Oxygen ⟶ Selenium Dioxide + Water

$$2H_2Se + 3O_2 \longrightarrow 2SeO_2 + 2H_2O$$

When passed through the metallic salt solutions the gas precipitates insoluble selenides. The selenides in general resembles sulphides.

Selenium Hexafluoride (SeF_6) : Like sulphur hexafluoride it is a colourless stable gas, quite unaffected by water, since the maximum co-valency of the sulphur and selenium is six, SeF_6 is octahedral like SF_6 this compound is produced when selenium reacts directly in fluorides.

Selenium + Fluorine ⟶ Selenium Hexafluoride

$$Se + 2F_2 \longrightarrow SeF_6$$

Selenium Tetrafluoride (SeF_4) : It is colourless liquid boils at 105.5°C. It is prepared by passing a fluorine-nitrogen mixture over selenium at 0°C, or by heating silver fluoride with selenium chloride at 50°C

Selenium + Fluorine ⟶ Selenium Tetrafluoride

$$Se + 2F_2 \longrightarrow SeF_4$$

Selenium Tetrachloride + Silver Fluoride ⟶ Silver Chloride + Selenium Tetrafluoride

$$SeCl_4 + 4AgF \longrightarrow 4AgCl + SeF_4$$

Selenium Chloride ($SeCl_4$): It is formed by treating selenium with excess of chlorine or by heating selenium dioxide with phosphorous pentachloride.

Selenium Dioxide + Phosphorous Pentachloride ⟶ Phosphorous Oxychloride + Selenium Tetrachloride

$$SeO_2 + PCl_5 \longrightarrow 2POCl_3 + SeCl_4$$

It forms colourless crystals with melting points of 305°C, it sublimes at about 196°C and the vapour is dissociated at 190°C.

Selenium Tetrachloride	$\longrightarrow$	Selenium Monochloride	+	Chlorine
$SeCl_4$	$\longrightarrow$	Se_2Cl_2	+	$3Cl_2$

Selenium Tetrabromide ($SeBr_4$): It is prepared by the reaction of selenium and bromine.

Selenium	+	Bromine	$\longrightarrow$	Selenium Tetrabromide
Se	+	$2Br_2$	$\longrightarrow$	$SeBr_4$

It forms yellow crystals and dissociates at 70°C to 80°C. The tetrahalides are hydrolyzed by water.

Selenium Monochloride	$\longrightarrow$	Selenium Tetrachloride	+	Selenium
$2Se_2Cl_2$	$\longrightarrow$	$SeCl_4$	+	3Se

Selenium Monobromide (Se_2Br_2): It is a red liquid, is less soluble than selenium monochloride, to which is similar in properties. Selenium dichloride and selenium dibromide exists only in vapour state. Selenium difluoride is not yet known.

Selenium Oxyfluoride ($SeOF_2$): It is a colourless fuming liquid and it is produced by heating $SeOCl_2$ with AgF at 140°C

Silver Fluoride	+	Selenium Oxychloride	$\longrightarrow$	Silver Chloride	+	Selenium Oxyfluoride
2AgF	+	$SeOCl_2$	$\longrightarrow$	2AgCl	+	$SeOF_2$
Selenium Oxyfluoride	+	Silicon Dioxide	$\longrightarrow$	Silicon Tetrafluoride	+	Selenium Dioxide
$2SeOF_2$	+	$SeiO_2$	$\longrightarrow$	SiF_4	+	$2SeO_2$

Selenium Oxychloride ($SeOCl_2$): It is colourless liquid made by the action of selenium dioxide dissolved in carbon tetrachloride on selenious chloride or by the practical hydrolysis of selenium chloride.

Selenium Dioxide	+	Selenium Tetrachloride	$\longrightarrow$	Selenium Oxychloride
SeO_2	+	$SeCl_4$	$\longrightarrow$	$2SeOCl_2$

Selenium Tetrachloride	+	Water	$\longrightarrow$	Selenium Oxychloride	+	Hydrogen Chloride
$SeCl_4$	+	H_2O	$\longrightarrow$	$SeOCl_2$	+	2HCl

The liquid has a di-electric constant (46.2) and is also a good solvent.

Selenium Oxybromide ($SeOBr_2$): It is made by distilling selenium oxychloride with sodium bromide.

Selenium Oxychloride	+	Sodium Bromide	$\longrightarrow$	Selenium Oxybromide	+	Sodium Chloride
$2SeOCl_4$	+	2NaBr	$\longrightarrow$	$SeOBr_2$	+	2NaCl

Selenious Acid (H_2SeO_3): It is obtained in colourless hexagonal prisms by evaporation of a solution of selenium dioxide in water or selenium in hot nitric acid. It forms normal and acid selenates e.g. Potassium Selenate and Potassium Hydrogen selenate and forms hetro poly acids with vanadic molybdic and uranic acids. Selenious acid is much weaker than sulphurous acid and is also much feebler than the sulphates. Selenious acid and selenates are easily reduced to red selenium by sulphur, hydrogen sulphide or organic matter in dust.

Selenium dioxide + Sulphur ⟶ Sulphur Dioxide + Selenium

$$SeO_2 + S \longrightarrow SO_2 + Se$$

Selenious Acid + Hydrogen Sulphide ⟶ Selenium + Sulphur + Water

$$H_2SeO_3 + 2H_2S \longrightarrow Se + 2S + 3H_2O$$

Selenious Acid + Sulphur Dioxide + Water ⟶ Selenium + Sulphuric Acid

$$H_2SeO_3 + 2SO_2 + H_2O \longrightarrow Se + 2H_2SO_4$$

Selenic Acid (H_2SeO_4): It is very similar to sulphuric acid. It forms colourless hexagonal crystals, which melts at 57°C to an oily liquid. The acid is hydroscopic and evolves heat with water. It forms the crystal hydrates, H_2SeO_4 with a melting point of 26°C and H_2SeO_4 $4H_2O$ m.p. -51.7°C. it is made by oxidation of selenium dioxide or selenic acid in water - oxidizing agents may be chlorine, bromine, chloric acid or potassium permanganate.

Selenium + Water + Chlorine ⟶ Selenic Acid + Hydrogen Chloride

$$Se + H_2O + 3Cl_2 \longrightarrow H_2SeO_4 + 6HCl$$

Silver Selenide + Water + Bromine ⟶ Selenic Acid + Silver Bromide

$$Ag_2SeO_3 + H_2O + Br_2 \longrightarrow H_2SeO_4 + 2AgBr$$

Evaporation of the solution until 205.5°C gives 95% selenic acid, which on heating breaks up. Drying over concentrated sulphuric acid in a vacuum dissector gives the 97.4% acid which on strong cooling forms colourless hexagonal crystals. Electrolytic oxidation of selenious acid also gives selenic acid.

Concentrated selenic acid chars organic matter like sulphuric acid. Selenic acid is much powerful than sulphuric acid.

The dilute acid oxidizes hydrogen iodide quantitatively. The concentrated acid attacks hydrogen chloride liberating chlorine.

Selenic Acid + Hydrogen Chloride ⟶ Selenious Acid + Water + Chlorine

$$H_2SeO_4 + 2HCl \longrightarrow H_2SeO_3 + H_2O + 3Cl_2$$

Hot and concentrated selenic acid dissolves copper to form cupric selenate and gold to form auric selenate.

Selenic Acid + Gold ⟶ Auric Selenate + Selenium Dioxide + Water

$$6H_2SeO_4 + 2Au \longrightarrow Au_2(SeO_4)_3 + 3SeO_2 + 6H_2O$$

A mixture of selenic acid and hydrochloric acid and hydrochloric acid acts like aqua regia dissolves also platinum, the dilute acid acts like aqua regia dissolves zinc.

Zinc	+	Selenic Acid	⟶	Zinc Selenate	+	Hydrogen
Zn	+	H_2SeO_4	⟶	$ZnSeO_4$	+	H_2

But iron receives protective coating of selenium when treated with the acid and is not dissolved.

The selenates are similar to sulphates as said earlier and also form alums. They are somewhat more soluble (0.08 per g litre at 25°C). Selenic acid is SO_2 - the selenium is filtered dried and weighted; the method is useful for estimation of selenates in presence of tellurates.

Uses of Selenium

Selenium is greatly used in the electronics industry as an element of semi-conductor and making rectifiers, selenium with another element called cesium constitutes photocell. These photocells are now being used for non-conventional energy sources.

The equation on which the cesium-selenium photocell works upon is

Cesium	+ Selenium	+ Photon	⟶	Selenium Cesium	+	Photoelectric Energy
2Cs	+ 2Se	+ ½ Photon	⟶	4SeCe	+	Photoelectric Energy

Detection and Estimation of Selenium

A solution containing selenium compound is first treated with hydrochloric acid 30% of the volume and selenium is precipitated with sulphur dioxide, hydroxylamine hydrochloride is also used with or without sulphur dioxide.

Selenium is estimated gravimetrically by weighing a deposit of selenium after drying the sample at 105°C, with an ordinary balance, this method is applicable to minimum 5g selenium content. The accuracy of the result is 0.5% in pure solution, up to 10mg/ml.

RUBIDIUM

Melting Point : 39.0°C	Atomic Number : 37
Boiling Point : 696°C	Atomic Weight : 85.4678

History and Occurrence

In 1866 Bunsen and Krichhoff discovered rubidium in the mineral lepidolite. The spectrum of this element has new lines in violet, blue, green, yellow and red. Two red lines, lying in the outer most portions of the red solar spectrum, were especially, hence the name *rubidium* meaning dark red was suggested for the name of this new element.

Associated with other alkali elements, rubidium occurs widely distributed in nature but in small proportions. The known mineral deposits rich enough to produce large amounts of rubidium are a very few. Lepiodolite a complex Lithia-Mica, which is the best source of the rubidium contain from a trace to more than 3% of rubidium monoxide.

Camellite and pollucite are other alkali minerals that contain small amounts of rubidium. Other natural sources of traces of element are salt wells, seawater plants and animal organisms. In this U.S. deposits the rubidium mineral are found in California, South Dakota, New Mexico and Marine.

Extraction

Rubidium is not produced commercially at a large scale; a small amount of the metal can be produced by the electrolysis or by the reduction of a rubidium salt. The metal is usually purified by distillation. Small amount rubidium for lab use is prepared by modification of *Hackpilli's reduction method.* Dry rubidium chloride is heated with calcium in an evacuated tube to 500°C and the rubidium formed is distilled off. Rubidium Carbonate is reduced by magnesium in the presence of hydrogen in an iron tube, where metallic rubidium is obtained. In 1940 Russian investigators reported that, rubidium can be obtained by reducing rubidium carbonate or hydroxide to the metal at a temperature of 800°C in vacuum. The electrolysis of fused rubidium chloride with graphite anode and iron cathode was carried out by Bunsen in this initial preparation of rubidium in 1861. A 30% yield of the metal has also been obtained by the electrolysis of fused rubidium hydroxide using a magnesite diaphragm to separate the products.

Physical Properties

Rubidium is a soft, silver white metal, closely related to potassium, it ignites spontaneously in dry oxygen as well as in air burns with a bluish flame.

Physical Properties

Melting Point	39°C
Boiling Point	688°C
Density	1.53 g/cc
Viscosity	6.25 milipoise
Vapour Pressure of Liquid at 280°C	1 mm
Specific Heat	0.082 cal/g
Thermal Conductivity	0.07 cal/(s)(sqcm)(cm)(°C)
Heat of Fusion	6.144 cal/g
Heat of Vapourization	212 cal/g
Heat of Capacity	7.27 cal deg^{-1} mol^{-1}
Standard Entropy	16.6 cal deg^{-1} mol^{-1}
Electrical Resistance	11.0 ohm cm

Chemical Properties

In chemical properties rubidium resembles potassium and cesium. It is more readily oxidized than potassium and upon exposure to air, becomes covered with a grey blue oxide film containing a mixture of the monoxide of rubidium (Rb_2O) and the peroxide (Rb_2O_2) as well as the super oxide (RbO_2). It ignites spontaneously in a vacuum or under an inert liquid such as benzene.

Rubidium reacts with water even in the form of ice at -108°C to form rubidium hydroxide. At room temperature the reaction is violent with the hydrogen igniting. Rubidium dissolves in liquid ammonia and also combines with ozone. It unites with bromine and chlorine directly with a flame.

Compounds of Rebidium

Rubidium Oxide (Rb_2O): When rubidium metal is exposed in air a bluish grey coat is produced on the surface of the metal it self, it consists of the oxide of metal. It also consist of the super oxide of the metal and also the rubidium peroxide (Rb_2O_2).

Rubidium	+	Oxygen	⟶	Rubidium Superoxide
Rb	+	O_2	⟶	RbO_2
Rubidium	+	Oxygen	⟶	Rubidium Peroxide
2Rb	+	O_2	⟶	Rb_2O_2
Rubidium	+	Oxygen	⟶	Rubidium Oxide
2Rb	+	O_2	⟶	$2Rb_2O$

Rubidium Hydroxide (RbOH): This compound is formed when water vigorously reacts with rubidium oxide.

Rubidium Oxide	+	Water	⟶	Rubidium Hydroxide
$2Rb_2O$	+	H_2O	⟶	RbOH

Rubidium Carbonate ($RbCO_3$): It occurs as a colourless crystals. It can be produced from the reaction of rubidium sulphate and barium hydroxide.

Rubidium Sulphate	+	Barium Sulphate	⟶	Rubidium Carbonate	+	Barium Sulphate
Rb_2SO_4	+	$Ba(OH)_2$	⟶	RbOH	+	$BaSO_4$
Rubidium Hydroxide	+	Ammonium Carbonate	⟶	Rubidium Carbonate	+	Ammonium Hydroxide
RbOH	+	NH_4CO_3	⟶	$RbCO_3$	+	NH_4OH

The rubidium compounds are very much hydroscopic.

Rubidium Chloride (RbCl): It crystallizes from water in well defined colourless cubic crystals, melting point 715°C and boiling point 1340°C and sp. gr. 2.76g/cc. It is soluble approximately about 77 parts of the salt in 100 parts of water at 0°C. Rubidium chloride is a sable salt and when subjected to high temperature it melts and volatilizes without dissociation. It is more volatile than sodium chloride, fact to be born in mind that when analytical tests are done it has been noticed that rubidium chloride is conveniently prepared by the gradual addition of a boiling solution of barium hydroxide to a hot solution rubidium alum in the presence of an indicator called bromothymol blue. This indicator change from orange to blue indicating the completion of the reaction it also indicates the precipitation of aluminium. These precipitates are filtered off and washed, the rubidium sulphate thus obtained is treated with barium chloride and rubidium chloride is obtained finally which is crystallized.

Rubidium Sulphate (Rb_2SO_4): It is obtained form the reaction of rubidium alum and barium hydroxide until all the aluminium is deposited. The point is well indicated by spot test for alkalinity with bromothymol indicator. The filtered solution should give negative result for hot the barium and the aluminium present in the alum and concentration and recrystallization from water obtain pure rubidium sulphate. Rubidium sulphate dissolves 82 parts in 100 parts of water at 100°C and about 36 parts in 100 parts of water at 0°C.

Rubidium Aluminium Sulphate, Rubidium Alum [$RbAl(SO_4)12H_2O$]: This compound crystallizes as colourless octahedral crystals. It has a solubility of 43 parts in 100 parts of water. It is prepared by the action of rubidium sulphate and aluminium hydroxide.

Alloys of Rubidium

Rubidium forms alloys with all the alkali metals except lithium. Only sodium rubidium alloys is present in the solid state and all others are found in liquid state. Cesium rubidium alloys are used in electrical and electronics industries. These are used for making parts in the vacuum tubes or valves. Rubidium also amalgamates with mercury, forming Rb_7Hg_8; Rb_3Hg~; $RbHg_7$ etc. The amalgam has been used in the hydrogenation of vynylacetylene, which is not so suitable platinum compound.

INDIUM

Melting Point : 156°C
Boiling Point : 2075°C

Atomic Number : 49
Atomic Weight : 114 .82

History and Occurrence

Indium was discovered by Reich and Richter (1863) in spectroscopic examination of zinc blende. It gives a dark blue flame colourations. It occurs in all commercial tin. The Bolivian mineral cylindrate consists of 0.1 to 1.0%. The metal is precipitated from solution by zinc and is purified electrolytically. It is soft and not attacked by boiling water. And has been used for plating silver.

Extraction

Numerous methods have been suggested or used for the extraction of indium from concentrates and residues. In one method the zinc and the indium are roasted and leached with a solution of sulphuric acid. The solution is treated with zinc and the indium is precipitated, together with copper silver etc. the precipitate is then treated with sulphuric acid and the silver is removed from the resulting solution by hydrogen sulphide. Indium is then removed from solution by electrolysis. Aluminium can also be used for precipitating indium from acid solution in place of zinc.

Another process is based on the fact that indium can be precipitated selectively from a slightly acid solution as a phosphate. The method includes the precipitation of the indium from the leach solution as indium phosphate, the conversion of indium phosphate to indium hydroxide by treating with strong caustic solution, filtering and heating to form the oxide, followed by the reduction of the oxide to metal. The resulting metallic indium can be further refined electrolytically, if desired.

To recover indium from a zincy-lead metal, the molten bath is treated with a suitable chlorine supplying agent or gaseous chlorine to selectively remove the zinc and indium as chlorides, using a low melting point cover slag and temperature sufficiently low to avoid loss of indium by volatilization. The resulting chloride slag is then leached with dilute sulphuric acid and the indium precipitated with zinc dust. The zinc-indium sponge is then melted and the zinc is removed by means of chlorine.

Physical Properties

Melting Point	156°C
Boiling Point	2075°C
Density	7.31 gm/cc
Specific Heat	0.0652 cal/g
Heat of Fusion	6.807 cal/g
Heat of Vapourization	483 cal/g
Volume of Increase on Fusion	2.5%
Thermal Conductivity	0.09 cal(s) (sqcm) (cm) (°C)
Electrical Resistance	29.0 microohm/cm
Standard Electrical Potential	–0.336V
Mohr's Hardness	1.2
Heat of Capacity	6.55 cal.deg^{-1} mol^{-1}
Standard Entropy	12.5 cal.deg^{-1} mol^{-1}
Electronegativity	1.7
Ionic Potential	3.7
Flame Colour	Blue-violet

Chemical Properties

Indium is unaffected by air at ordinary temperature, but at red-hot condition it burns with a blue flame to form the oxide)In_2O_3).

$$\text{Indium} + \text{Oxygen} \longrightarrow \text{Indium Oxide}$$
$$4In + 3O_2 \longrightarrow 2In_2O_3$$

The metal dissolves in mineral acid but remains unaffected by potassium hydroxide or boiling water. When heated in the presence of halogen direct combination takes place.

$$\text{Indium} + \text{Chlorine} \longrightarrow \text{Indium Trichloride}$$
$$2In + 3Cl_2 \longrightarrow 2InCl_3$$

When heated with sulphur, two elements combine with incandescence by an electric arc.

Indium	+	Sulphur	$\longrightarrow$	Indium Sulphide
In	+	S_2	$\longrightarrow$	InS_2

Uses of Indium

(1) The principal use of indium is in the field of surface protection particularly for inhibiting corrosion of cadmium.

(2) Indium is also used for plating (electroplating)

(3) Indium is also used. as a doping material for germanium in semi-conductor manufacture in electronic industry.

Compounds of Indium

Indium Oxide, Indium Sesquioxide, Indium Trioxide (In_2O_3): It is a white or light yellow crystalline solid also available in amorphous form, depending upon temperature.

It is prepared by the action of indium and oxygen at an elevated temperature.

Indium	+	Oxygen	$\longrightarrow$	Indium Oxide
4In	+	$3O_2$	$\longrightarrow$	$2In_2O_3$

The amorphous form is soluble in hot acids but the crystalline salt is insoluble. It can be also preared by heating indium hydroxide;

Indium Hydroxide	+	Oxygen	$\longrightarrow$	Indium Oxide	+	Water
2InOH	+	O_2	$\longrightarrow$	In_2O_3	+	H_2O

Indium Chloride, Indium Trichloride ($InCl_2$): It is a white delequescent powder soluble in water. It is prepared by the direct union of the elements.

Indium	+	Chlorine	$\longrightarrow$	Indium Oxide
4In	+	$3Cl_2$	$\longrightarrow$	$2In_2Cl_3$

Or, it is produced by the action of hydrochloric acid on indium.

Indium	+	Hydrochloric Acid	$\longrightarrow$	Indiuim Chloride	+	Hydrogen
2In	+	6HCl	$\longrightarrow$	$2InCl_3$	+	$3H_2$

It is soluble in water or alcohol, with a melting point 586°C and sublimes at 300°C.

Indium Phosphate ($InPO_4$) : Indium phosphate is prepared by the action of indium chloride and sodium phosphate. It is insoluble in water.

Indium Chloride	+	Sodium Phosphate	$\longrightarrow$	Indium Phosphate	+	Sodium Chloride
$2InCl_3$	+	Na_3PO_4	$\longrightarrow$	$InPO_4$	+	3NaCl

Indium Phosphide (InP): It is a brittle metallic mass with a melting point of 1070°C, slightly soluble in mineral acids. It is prepared by the direct union of phosphorous and indium

Indium	+	Phopshorous	$\longrightarrow$	Indium Phosphide
2 In	+	2P	$\longrightarrow$	2InP

It is used in the experimental solar cells, the lenses of special types used in electron microscopes) and also in semiconductors.

Indium Sulphate [$In_2(SO_4)_3$]: The greyish powdered compound is prepared by the action of indium and sulphuric acid.

Indium	+	Sulphuric Acid	$\longrightarrow$	Indium Sulphate	+	Hydrogen
2In	+	$3H_2SO_4$	$\longrightarrow$	$In_2(SO_4)_3$	+	$3H_2$

It is delequescent compound soluble in water.

Indium Nitrate [$In(NO_3)_2$]: This compound is prepared by the action of indium on nitric acid

Indium	+	Nitric Acid	$\longrightarrow$	Indium Nitrate	+	Hydrogen
2In	+	$6HNO_3$	$\longrightarrow$	$In(NO_3)_2$	+	$3H_2$

Indium Hydroxide [$In(OH)_3$]: This is the compound of indium is prepared by the action of sodium hydroxide and indium trichloride.

Indium Trichloride	+	Sodium Hydroxide	$\longrightarrow$	Indium Hydroxide	+	Sodium Chloride
$InCl_3$	+	3NaOH	$\longrightarrow$	$In(OH)_3$	+	6NaCl

Detection of Indium

Indium is best detected by the spectrophotometrical analysis.

ANTIMONY

Melting Point : 630.5°C	Atomic Number : 51
Boiling Point : 1440°C	Atomic Weight : 121.75

History and Occurrence

Black antimony sulphide was used as paint for eyebrows as early as the 320 B.C in Egypt. The custom was in style in ancient Egypt and Mesopotamia. A Chaldean vase made entirely of antimony and belonging to 300 B.C. has also been found excavation. Alchemists largely used antimony and its compounds. The Arabian alchemists were expert in preparation of antimony from its sulphide, the sulphide mineral stibnite was named as *Khul* or *Kohl* by them. Basil Valentine an alchemy writer of early seventeenth century gave an elaborate description of antimony and a large number of its compounds in his prevention. The Triumphal Chariot of Antimony (1664).

The only important ore of antimony is antimony sulphide called *Stibnite (Sb~3)* found mainly in China, Bolivia Czechoslovakia, Yugoslavia, etc. It is also available in Sweden, Borneo and other places. Like arsenic antimony occurs frequently associated with the sulphide ores of many metals, such as copper; silver; lead etc.

Extraction

There are two methods for the extraction of sulphide ore.

1. Precipitation Process : The so-called precipitation process of smelting the sulphide with scrap iron in graphite crucibles when the following reaction occurs.

Antimony Sulphide	+	Iron	$\longrightarrow$	Antimony	+	Ferrous Sulphide
Sb_2S_3	+	3Fe	$\longrightarrow$	2Sb	+	3FeS

2. Roast Reduction Process : This process is carried out in two steps:

***(a) Roasting*:** The first steps is to heat the ore in air (roasting) so as to convert the sulphide into the oxide. If the roasting is done in a good supply of air, the non-volatile tetraoxide, $SbzO_4$ is obtained. The usual process, however is to carry out the roasting in a limited supply of air, when the volatile trioxide, $SbzO_3$ is the chief product. It is volatilised and collected in condensation chamber. The process, therefore, called volatile roasting or simply volatilisation.

Antimony Sulphide	+	Oxygen	$\longrightarrow$	Antimony Trioxide	+	Sulphur Dioxide
Sb_2S_3	+	O_2	$\longrightarrow$	$2Sb_2O_3$	+	$6SO_2$

The plant for the volatilisation roasting of antimony ore is shown in Fig. 17.11. The ore mixed with coke is charged into the roasting furnace shown in the figure. A slow current of air is sucked into the apparatus by mean of an exhauster. The antimony trioxide sublimes into the vertical condensers and is periodically removed through the doors provided at the bottom. The last trace of the oxide is removed by passing the gasses upon hot iron

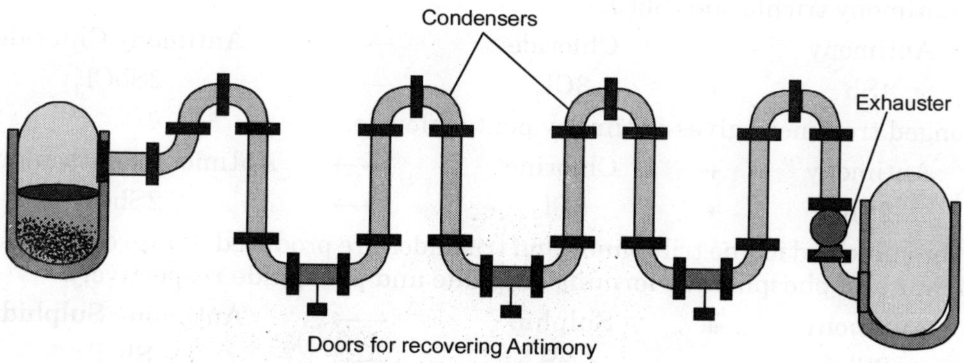

Fig. 17.11. The Extraction of Antimony

The Tower packed with coke, down with water is sprayed. The sulphur dioxide too is dissolved out by water.

Physical Properties

Melting Point	630.5°C
Boiling Point	1440°C
Density	1.97 g/cc (grey antimony)
Atomic Radius	1.36 Å
Crystal Radius	0.62Å
Electronegativity (Pauling's Scale)	1.82
Standard Entropy	10.5 cal deg^{-1} mol^{-1}
Heat of Capacity	6.08 cal deg^{-1} mol^{-1}
Oxidation States	3, 5
Ionic Potential	8.5eV
Electrical Resistivity at 0°C	39.0 ohm/cm
Ionic Radius	2.45Å

Chemical Properties

It is not attacked by air, but if it is hot its gets slowly oxidised; in moist air it gets tarnished. When heated in air or oxygen, it burns with a blue flame and forms antimony trioxide, ($SbZO_3$) and antimony tetraoxide (Sb_2O_4)

$$\text{Antimony} + \text{Oxygen} \longrightarrow \text{Antimony Trioxide}$$
$$4Sb + 3O_2 \longrightarrow 2Sb_2O_3$$
$$\text{Antimony} + \text{Oxygen} \longrightarrow \text{Antimony Tetraoxide}$$
$$4Sb + 4O_2 \longrightarrow 2Sb_2O_4$$

It is not attracted by water but it decomposes steam at red heat. It bums in chlorine forming antimony trichloride ($SbCl_3$).

$$\text{Antimony} + \text{Chloride} \longrightarrow \text{Antimony Chloride}$$
$$4Sb + 3Cl_2 \longrightarrow 2SbCl_3$$

Prolonged treatment gives antimony pentachloride.

$$\text{Antimony} + \text{Chlorine} \longrightarrow \text{Antimony Pentachloride}$$
$$2Sb + 5Cl_2 \longrightarrow 2SbCl_5$$

With bromine and iodine tribromide and triiodides are produced. It also combines readily with sulphur and phosphorous forming sulphide and phosphide respectively.

$$\text{Antimony} + \text{Sulphur} \longrightarrow \text{Antimony Sulphide}$$
$$2Sb + 3S \longrightarrow Sb_2S_3$$
$$\text{Antimony} + \text{Phosphorous} \longrightarrow \text{Antimony Phosphide}$$
$$Sb + P \longrightarrow SbP$$

It is not attracted by dilute acids. The pure metal is attracted by concentrated acid like hydrochloric, which forms the chloride and liberates hydrogen.

$$\text{Antimony} + \text{Hydrochloric Acid} \longrightarrow \text{Antimony Trichloride} + \text{Hydrogen}$$
$$4Sb + 6HCl \longrightarrow 2SbCl_3 + 3H_2$$

It is also attracted by concentrated sulphuric acid, which produces sulphur dioxide.

Antimony	+	Sulphuric Acid	$\longrightarrow$	Antimony Sulphate	+	Sulphur Dioxide	+	Water
2Sb	+	$6H_2SO_4$	$\longrightarrow$	$Sb(SO_4)_3$	+	$3SO_2$	+	$6H_2O$

It does not dissolve in concentrated nitric acid but gets oxidised to antimony pentoxide, $SbZO_5$ and antimonic acid, H_3SbO_4.

Yellow Antimony or Alpha Antimony: - This is obtained by the action of ozonized oxygen on liquid antimony hydride at -90°C

It is unstable and changes readily into the black beta-antimony. It can exist only at low temperature. It is slightly soluble in carbon disulphide.

Black or Beta Antimony: - It is an amorphous black powder and quite stable and changes only gradually to the metallic forms. It is formed directly by rapid cooling of antimony vapour. On heating, it changes rapidly to metallic form.

Explosive Antimony: - It is obtained by the slow electrolysis of a solution of antimony trioxide and hydrochloric acid, using a platinum cathode and antimony anode. An amorphous powder gets deposited to the cathode, which has the appearance of polished graphite rod. The deposit is believed to be a mixture of antimony chloride and alpha-antimony. If the deposit is scratched it explodes violently, at the same time the alpha form gets transferred to more stable beta or metallic form and the temperature rises to 250°C. it has been ascetained that about 19.6 calories of heat per gram of antimony is evolved. Clouds of antimony trichlorides are given off. Hence the term explosive antimony.

Uses of Antimony

The main use of antimony is in the making of different alloys. About half of the worlds production of antimony goes in the manufacture of lead storage batteries because lead consisting of antimony in small amount is harder and resistant to the action of acids than ordinary lead. It is used for colouring pottery cloth and paper. It anhydrides are used as mordants.

The important alloys of antimony are being stated below in the table.

Name	Composition	Uses
Type metal	Lead 82% Antimony 15% Tin	Casting Printing type
Pewter	Tin 90% Copper 2% Antimony 6% Bismuth 2%	Cup Mug etc.
Britannia Metal	Tin 90% Copper 2% Antimony 8%	Cutlery
Babbit Metal	Tin 90% Copper 3% Antimony 7%	Machinery Parts

Compounds of Antimony

Antimonous Oxide, Antimony Trioxide (Sb_2O_3) : It is formed as a white powder by burning the metal in air by roasting stibnite. The purer product is obtained by the hydrolysis of antimony trichloride with hot water.

Antimony Trichloride	+	Water	⟶	Antimonous Oxide	+	Hydrogen Chloride
$2SbCl_3$	+	$3H_2O$	⟶	Sb_2O_3	+	$6HCl$

Antimonous Oxide is almost insoluble in water but soluble in dilute hydrochloric acid. It also dissolves in hot concentrated sulphuric acid to yield antimonous sulphate [$Sb_2(SO)_3$]. Antimony oxide is dissolved by a boiling solution of potassium hydrogen tartrate, forming potassium antimonium tartrate [$K[(SbO)C_4O_6]$] this is largely used in the medicine. In alkali solutions the oxide dissolves to form metaantimonate e.g sodium metaanimonate, when heated tetraoxide $SbzO_4$ at 700°C-900°C. At higher temperatures the tetraoxide decomposes into the trioxide. Antimonous oxide is used in white paints and animals.

Antimony Pentoxide (Sb_2O_5): It is made by oxidizing antimony with nitric acid. It forms a yellow powder, decomposes on heating into trioxide and oxygen. The pentoxide is almost insoluble in water, but reddens litmus. Hydrated antimonypentoxide is formed by the action of hot water on antimony pentachloride. Antimony pentoxide is an acidic oxide and forms salts called antimonates.

Potassium Antimonate, (meta antimonate) ($KSbO_3H_2O$): It is obtained by fusing the pentoxide with potassium hydroxide. It is soluble in boiling water and the solution gives a precipitate of sodium antimonate on the addition of a solution of sodium salt.

Antimony Tetraoxide (Sb_2O_4): It is formed by prolonged ignition of other two oxides at 600°C-700°C, at which the tetraoxide is most stable. It is a white solid not very appreciable soluble in water or acids, but when fused with alkali it gives rise to more or less soluble salts called hypoantimonates, e.g $K_2Sb_2O_5$

Antimony Trisulphide (Sb_2S_3): This compound occurs in nature as sibnite, It is obtained as red precipitate by passing hydrogen sulphide in a solution of antimony trichloride in hydrochloric acid.

Antimony Trichloride	+	Hydrogen Sulphide	⟶	Antimony Trisulphide	+	Hydrogen Chloride
$2SbCl_3$	+	$3H_2S$	⟶	Sb_2S_3	+	$6HCl$

When heated at 200°C in a current of carbon dioxide, the red sulphide changes into the dilute acids, but dissolves in hot concentrated hydrochloric acid. It dissolves in alkali sulphides forming thioantimonates and in hot concentrated solutions of alkalis and alkali carbonates to form a mixture of antimonate and thioantimonates. From these solutions it is reprecipitated by the action by the action of dilute acids. But it is insoluble in ammonium carbonate solution. In alkali polysulphides it dissolves to form thioantimonates. Antimony trisulphide is used as a pigment and in vulcanizing rubber. Mixed with nitre and sulphur it is used as blue fire in pyrotechnic and in manufacture of matches.

Antimony Pentasulphide (Sb_2S_5): It is obtained as brilliant orange powder by the action of hydrogen sulphide on a cold solution of antimony pentoxide in hydrochloric acid. It dissolves in alkali sulphide solutions to form thioantmonates. From these solutions it is reprecipitated by acids

Sodium Thioantimonate ($Na_3SbS_4\,9H_2O$): Which forms pale yellow crystals and is known as Schhppe's Salt.

Antimony Hydride (SbH_3): It is also known, as Stibine formed mixed with hydrogen when a solution of antimony compound is added to zinc and sulphuric acid or by the action hydrochloric acid on an alloy of magnesium and antimony (33%)

Magnesium Stibinite	+	Hydrochloric Acid	$\longrightarrow$	Magnesium Chloride	+	Stibinine
Mg_3Sb_2	+	6HCl	$\longrightarrow$	$3MgCl_2$	+	$2SbH_3$

The gas is washed with water, dried with calcium chloride and phosphorous pentoxide and then passed through a tube cooled with liquid air, stibine condenses to a white solid, pure stibine is produced on warming the solid. It has a peculiar smell and extremely poisonous like arsme. Stibine is sparingly soluble in water and largely in carbon disulphide.

Antimony Trifluoride (SbF_3): It is formed by dissolving antimonous oxide, Sb_2O_3 in hydrofluoric acid and evaporating. It is not readily hydrolyzed by water. It combines with fluorides to form complex compounds e.g. potassium fluoroantimonate, K_2SbF_5

Antimony Pentafluoride (SbF_5): It is obtained as an oily liquid by boiling antimony pentachloride and anhydrous hydrogen fluoride. The product is subsequently purified by fractional distillation.

Antimony Pentachloride	+	Hydrogen Fluoride	$\longrightarrow$	Antimony Pentafluoride	+	Hydrogen Chloride
$SbCl_5$	+	5HF	$\longrightarrow$	SbF_5	+	5HCl

With alkali fluorides it forms complex fluorides $KSbF_6$ and K_2SbF_7.

Antimony Trichloride ($SbCl_3$) : This is prepared by dissolving the trioxide or trisulphide (stibnite) in concentrated hydrochloric acid.

Antimony Trisulphide	+	Hydrochloric Acid	$\longrightarrow$	Antimony Trichloride	+	Hydrogen Sulphide
Sb_2S_3	+	6HCl	$\longrightarrow$	$2SbCl_3$	+	$3H_2S$

With alkali fluorides it forms complex fluorides $KSbF_6$ and K_2SbF_7.

The solution is distilled when water is first removed and then hydrochloric acid finally antimony trichloride distills over and solidifies in the reservoir. It forms white delequescent crystals known as 'butter of antimony' and can be hydrolyzed by water with precipitation of an oxychloride.

Antimony Trichloride	+	Water	$\longrightarrow$	Antimony Oxychloride	+	Hydrogen Chloride
$2SbCl_3$	+	H_2O	$\longrightarrow$	SbOCl	+	2HCl

With excess of heat it is completely hydrolyzed to Sb_2O_3. Antimony oxychloride was formerly called **Powder of Algaroth**.

Antimony Pentachloride ($SbCl_5$): It is obtained as yellow fuming liquid by the action of aqua regia on the antimony trichloride, its vapour suffers partial disproportionation into trichloride and chlorine.

Antimony Pentachloride	+	Potassium Iodide	⟶	Antimony Trichloride	+	Potassium Chloride	+	Iodine
$SbCl_5$	+	2KI	⟶	$SbCl_3$	+	2KCl	+	I_2

Antimony Tribromide ($SbBr_3$): It is formed, as delequescent needles by the action of bromine upon antimony like the trichloride it is hydrolyzed by water to the bromide.

Antimony Triiodide (SbI_3): It is produced by warming powdered antimony with a solution of iodine in carbon disulphide. On filtering and evaporating the solution the iodide separates in the form of red pellets. Two other grayish-yellow modifications are also known. It is decomposed by water with the formation of an oxyiodide.

Detection and Estimation

Compounds of antimony are very poisonous; hence their detection is also very important. In absence of arsenic Marsh Test is carried out. Antimony is estimated by the precipitation of trisulphide and then by drying out at 300°C in a current of carbon dioxide; other heavy metals should be absent.

BISMUTH

Melting Point : 271.3°C	Atomic Number : 83
Boiling Point : 1500°C	Atomic Weight : 208.980

History and Occurrence

There is no evidence that bismuth was known to the ancient chemists. Though Basil Valentine first discovered metallic bismuth, it was recognized as distinct metal only in the middle of the 18th century.

Bismuth occurs in small amounts in the free state in nature in many localities, usually associated with tin and lead. In combination or as ores, *bismuth glance, $BizS_3$* and the *basic antimonate busmuthate, or bismuth spar $(BiO)_2CO_3$*. The most important source in Bolivia, where the metal is found mostly in the native state and associated with tin. Other sources are Canada, China and Australia, where it occurs mainly in lead ores.

Extraction

From ores containing the mixture bismuth, the metal is obtained by liquidation, it means that ores are heated with or without the addition of a flux (sodium carbonate and lime) in inclined iron tubes, when the easily fusible bismuth is run off at the bottom. The oxide and carbonate ores are dissolved in hydrochloric acid and the solution poured into a large volume of water to precipitate the oxychloride, BiOCl, which, is thus freed from the impurities of the other metals. Sulphide ores are roasted at a low temperature in air to volatilize away most of the arsenic and antimony and at the same time to convert bismuth sulphide to oxide. The oxide is then reduced by heating with carbon in the form of sodium carbonate as a flux. The dust forms lead and tin ores and anode slimes form copper and lead refining contain bismuth, a considerable amount of bismuth is recovered from the sources. The metal obtained as described above is refined electrolytically, using a solution of bismuth chloride and hydrochloric acid as electrolyte, in a manner similar to that used in the process of copper.

The metal in a highly pure state is made by recrystallizing bismuth nitrate from strong nitric acid solution heating to convert the nitrate to oxide and then the latter by fusing with potassium cyanide.

Physical Properties

Melting Point	271.3°C
Boiling Point	1500°C
Density	9.80 g/cc
Oxidation States	3, 5
Atomic Radius	1.46Å
Crystal Radius	0.74Å
Electronegativity (Pauling's Scale)	1.67
Standard Entropy	13.6 cal deg^{-1} mol^{-1}
Heat of Capacity	6.1 cal deg^{-1} mol^{-1}
Ionization Potential (1st)	8.0eV
Electrical Resistivity (0°C)	107 microohm/cm

Chemical Properties

1. Bismuth unlike antimony does not exist in allotropic form. It has a distinct metallic lusture (silvery white with a reddish tinge). It is brittle and has relative density equal to 9.8 g/cc. It is a brittle and has low meting point; it conducts electricity fairly like metals.
2. Bismuth like antimony is not attacked by air at ordinary temperature and does not tarnish when heated it bums in air to form bismuth trioxide.

Bismuth	+	Oxygen	⟶	Bismuth Trioxide
4 Bi	+	$3O_2$	⟶	$2Bi_2O_3$

3. It combines with chlorine forming bismuth trichloride.

Bismuth	+	Chlorine	⟶	Bismuth Trichloride
2 Bi	+	$3Cl_2$	⟶	$2BiCl_3$

4. Bismuth like antimony is not attacked by hydrochloric acid. However, it dissolves even indilute nitric acid forming bismuth nitrate.

Bismuth	+	Nitic Acid	$\longrightarrow$	Bismuth Nitrate	+	Nitric Oxide	+	Water
2Bi	+	$8HNO_3$	$\longrightarrow$	$2Bi(NO_3)_3$	+	2NO	+	$4H_2O$

It is not attacked by sulphuric acid but hot and concentrated sulphuric acid reacts and liberates dioxide from it.

Uses

Bismuth is used largely for making alloys of low melting point, some of the alloys melt below 100°C. These alloys are mainly used in making automatic electrical safety devices etc.

Compounds of Bismuth

Bismuth Hydride, Bismuthine (BiH_3): The existence of this compound was first detected by the radioactive isotopes of bismuth (radium C and thorium C) instead of ordinary bismuth. An alloy of the isotope with magnesium when treated with dilute hydrochloric acid give a minute quantity of bismuth hydride mixed with a large amount of hydrogen.

This minute quantity could be detected only because of the radioactivity of the gas this also made it possible to study some of its properties. Later on Paneth prepared the gas in very small amount by using the ordinary bismuth in its alloy with magnesium.

The gas is very unstable and decomposes readily even at ordinary temperatures into bismuth and hydrogen.

Bismuth Hydride	$\longrightarrow$	Bismuth	+	Hydrogen
$2BiH_3$	$\longrightarrow$	2 Bi	+	$2H_2$

Bismuth Trioxide (BiO_3): It is prepared by heating bismuth in air or by heating bismuth carbonate or bismuth nitrate.

Bismuth	+	Oxygen	$\longrightarrow$	Bismuth Trioxide
4Bi	+	$3O_2$	$\longrightarrow$	$2Bi_2O_3$

Bismusth Carbonate	$\longrightarrow$	Bismuth Trioxide	+	Carbon Dioxide
$4Bi(NO_3)_3$	$\longrightarrow$	Bi_2O_3	+	CO_2

Bismusth Nitrate	$\longrightarrow$	Bismuth Trioxide	+	Nitrogen Dioxide	+	Oxygen
$4Bi(NO_3)_3$	$\longrightarrow$	Bi_2O_3	+	$12NO_2$	+	O_2

It is yellow powder, amphoteric and dissolve in alkalis as well as in acids forming bismuthates and bismuth and bismuth salts respectively.

Bismuth Trioxide	+	Hydrochloric Acid	$\longrightarrow$	Bismuth Trichloride	+	Water
Bi_2O_3	+	6HCl	$\longrightarrow$	$2BiCl_3$	+	$3H_2O$

Bismuth Trioxide	+	Sodium Hydroxide	$\longrightarrow$	Sodium Bismuthate	+	Water
Bi_2O_3	+	$6NaOH$	$\longrightarrow$	$2Na_3BiO_3$	+	$3H_2O$

It forms series of hydrates. It is used for glazing porcelain making glazed glass and optical glass.

Bismuth Pentoxide (BiO_5): If the bismuth trioxide is treated with chlorine in dilute potassium hydroxide, as mentioned above it is combined for a long time, a scarlet red precipitate, known as potassium metabismuthate ($KBiO_3$) is obtained. The precipitate after thorough washing is reacted with dilute nitric acid when metabismuthic acid is produced, as scarlet red solid. On drying with sulphuric acid, this changes into bismuth pentoxide.

$$\text{Metabismuthic Acid} \xrightarrow[\text{Sulphuric Acid}]{-\text{Water}} \text{Bismuth Pentaoxide} + \text{Water}$$

$$2HBiO_3 \xrightarrow[H_2SO_4]{-H_2O} Bi_2O_5 + H_2O$$

Bismuth pentoxide is brown powder, which decomposes into bismuth trioxide when heated above 120°C.

Bismuth Pentaoxide	$\longrightarrow$	Bismuth Trioxide	+	Oxygen
Bi_2O_5	$\longrightarrow$	Bi_2O_3	+	O_2

It reacts with hydrochloric acid forming bismuth salts and liberating chlorine and with sulphuric acid liberates oxygen thus behaving like peroxide.

Bismuth Pentaoxide	+	Hydrochloric Acid	$\longrightarrow$	Bismuth Trichloride	+	Water	+	Chlorine
Bi_2O_5	+	10 HCl	$\longrightarrow$	$2BiCl_3$	+	$5H_2O$	+	$2Cl_2$
Bismuth Pentaoxide	+	Sulphuric Acid	$\longrightarrow$	Bismuth Sulphate	+	Water	+	Oxygen
Bi_2O_5	+	$3H_2SO_4$	$\longrightarrow$	$Bi_2(SO_4)_3$	+	$3H_2O$	+	O_2

It forms bismuthate with concentrated alkalis, which shows that it has feebly acidic peroxides.

Bismuth Pentaoxide	+	Sodium Peroxide	$\longrightarrow$	Sodium Bismuthate	+	Oxygen
$2Bi_2O_5$	+	$6Na_2O_2$	$\longrightarrow$	$4Na_3BiO_4$	+	$3O_2$

These compounds are used as oxidizing agents.

Bismuth Trichloride ($BiCl_3$): It is obtained in the anhydrous state by passing chloride over heated bismuth.

Bismuth	+	Chlorine	$\longrightarrow$	Bismuth Trichloride	+	Water
2Bi	+	$3Cl_2$	$\longrightarrow$	$2BiCl_3$	+	$3H_2O$

The solution is evaporated to dryness in a current of carbon dioxide. Bismuth trichloride forms colourless crystals m.p 232°C. It undergoes hydrolysis in water (like antimony chloride) forming bismuth oxychloride (bismuthyl chloride).

Bismuth Trichloride	+	Water	$\longrightarrow$	Bismuth Oxychloride	+	Hydrogen Chloride
$BiCl_3$	+	H_2O	$\longrightarrow$	BiOCl	+	2HCl

Bismuth oxychloride is used as a white pigment under the name 'peast white'. The pentachloride of bismuth is unknown.

Bismuth Nitrate [Bi(NO$_3$)$_3$ 5H$_2$O]: Bismuth nitrate is prepared by dissolving the metal or its oxide or carbonate in dilute nitric acid.

Bismuth	+	Nitric Acid	⟶	Bismuth Nitrate	+	Nitric Oxide	+	Water
2Bi	+	$8HNO_3$	⟶	$2Bi(NO_3)_3$	+	2NO	+	$4H_2O$

Bismuth Trioxide	+	Nitric Acid	⟶	Bismuth Nitrate	+	Water
Bi_2O_3	+	$6HNO_3$	⟶	$2Bi(NO_3)_3$	+	$3H_2O$
Bismuth Carbonate	+	Water	⟶	Bismuth Nitrate	+	Nitric Acid
$Bi(NO_3)_3$	+	H_2O	⟶	$BiONO_3$	+	$2HNO_3$

It is soluble in water when the solution is diluted hydrolysis takes place as in the case of bismuth chloride giving bismuth oxynitrate.

Bismuth Nitrate	+	Water	⟶	Bismuth Oxynitrate	+	Nitric Acid
$Bi(NO_3)_3$	+	H_2O	⟶	$BiONO_3$	+	$2HNO_3$

Bismuth oxynitrate is used in medicines and also for the treatment of diarrhoea and dysentery.

On mixing bismuth nitrate solution with a solution of ammonium carbonate, a white precipitate of basic bismuth carbonate [$2(BiO)_2CO_3H_2O$] results, this is also used for the treatment of diarrhea, etc. This is given to patients requiring X-Ray of Stomach alignments before taking the X-ray photographs.

Bismuth Sulphate [Bi$_2$(SO$_4$)$_3$]: It is obtained by dissolving bismuth trioxide in concentrated sulphuric acid.

Bismuth Trioxide	+	Sulphuric Acid	⟶	Bismuth Sulphate	+	Water
Bi_2O_3	+	$3H_2SO_4$	⟶	$Bi_2(SO_4)_3$	+	$3H_2O$

The solution is evaporated to get the crystals. It is colourless crystalline substance, with hydroscopic character.. It undergoes partial hydrolysis by water giving bismuth oxysulphate.

Bismuth Sulphate	+	Water	⟶	Bismuth Oxysulphate	+	Sulphuric Acid
$Bi_2(SO_4)_3$	+	$2H_2O$	⟶	$(BiO)_2SO_4$	+	$2H_2SO_4$

Bismuth Sulphide (Bi$_2$S$_3$): It occurs in nature as bismuth glance. It is obtained as dark brown crystals when hydrogen sulphide is passed through a solution of bismuth trichloride in hydrochloric acid.

Bismuth Trichloride	+	Hydrogen Sulphide	⟶	Hydrogen Chloride	+	Bismuth Sulphide
$BiCl_3$	+	H_2S	⟶	6HCl	+	Bi_2S_3

Unlike the corresponding sulphide of arsenic and antimony this compound does not dissolve in alkali sulphides.

Test of Bismuth

When hydrogen sulphide gas is passed through a solution of bismuth compound, bismuth sulphide is precipitated, which is not soluble in yellow ammonium sulphide.

TELLURIUM

Melting Point : 450°C Atomic Number : 52

Boiling Point : 997°C Atomic Weight : 128.60

History and Occurrence

Natural tellurium occurs in very small amounts and that it is called aurum paradoxum, because of its lusture. In 1872 Von Reichenstein suspected that the Transylvanian mineral known as white gold ore contained an unknown element (till that time).

Bergman confirmed this suspicion and the element was discovered in 1798 by Klaporth who also explained its importance. It was named tellurium from the Latin word tells, the earth, Berzelius in 1832 pointed out its analogy with sulphur and selenium.

Tellurium is comparatively rare but available native in many parts of the world, e.g Brazil, Bolivia, Colorado and Rumania, also among with selenium in Japanese Sulphur. Important minerals are Tetradymite (Bi_2Te_3) Nagygite or back tellurium $(Au, PbMTe, S, Sb)_3$ (similar mixture of sulphides and tellurides, Chiefly of lead, gold, silver, and antimony, Sylvanaite or graphic tellurium $[(Ag, Au)Te_2]$ and Hessite (Ag_2Te). Gold tellurides are used for extraction of gold in Australia and Colorado. Anode slime of electrolytic copper refining is an important source of tellurium.

Extraction

Tellurium is obtained from the bismuth ore Tetradymite

1. *By fusion with Sodium Carbonate* and then oxidizing the sodium telluride (Na_2Te) formed by passing air through the aqueous solution.

   Tetradynite + Sodium Carbonate ⟶ Bismuth Trioxide + Sodium Telluride + Carbon Dioxide

   $$Bi_2Te_3 + 3Na_2CO_3 \longrightarrow Bi_2O_3 + 3Na_2Te + 3CO_2$$

   Sodium Telluride + Oxygen + Water ⟶ Sodium Hydroxide + Tellurium

   $$2Na_2Te + O_2 + H_2O \longrightarrow 4NaOH + 2Te$$

2. *By Electrolysis of a solution of Tellurium* dioxide in hydrofluoric acid and sulphuric acid, pure tellurium is deposited on a lead cathode. This is the pure form of tellurium.
3. *By Reducing TeO_2 with hydrogen* and distilling under reduced pressure.

Physical Properties

Melting Point	450°C
Boiling Point	997°C
Density	6.24 g/cc at 30°C
Atomic Radius	21cc
Ionic Radius	2.22Å
Standard Entropy	11.88 cal deg^{-1} mol^{-1}
Heat of capacity	6.15 cal deg^{-1} mol^{-1}
Electronegativity	2.0
Electrical Resistance	1.6×10^5 microohm/cm
Oxidation States	–2, +4, +6

Chemical Properties

When heated in air or oxygen tellurium burns with blue flame to form tellurium dioxide. It is unaffected by W ions but dissolves in oxidizing acids like nitric acid. Caustic potash dissolves in oxidizing acids like nitric acid. Caustic potash dissolves it forming a telluride and tellurite.

Tellurium combines with many metals to give tellurides and dark red alkali polytellurides. It dissolves in concentrated sulphuric acid to give a cherry-red solution. On fusion with potassium cyanide tellurium combines with chlorine to give Tellurium tetrachloride and tellurium dichloride, according to the proportion taken. Tellurium reacts with iodoform producing iodide of tellurium (Te_4). Tellurium compounds are poisonous offensive odour.

Uses of Tellurium

Addition of tellurium in small amounts 0.05 - 0.08% to lead gives the metals a greater tensile strength and resistance to acids, particularly sulphuric acid. A tellurium anode with air bubbling through the solution has been recommended for the determination of pH over range 9-12 particularly useful in the presence of the mild oxidizing agents.

Compounds of Tellurium

Hydrogen Telluride (H_2Te): It is formed by the action of dilute hydrochloric acid on aluminium telluride,

$$\text{Aluminium Telluride} + \text{Hydrochloric Acid} \longrightarrow \text{Hydrogen Telluride} + \text{Aluminium Trichloride}$$

$$Al_2Te_3 + 6HCl \longrightarrow H_2Te + AlCl_3$$

The gas can be condensed to a liquid after drying (boiling point is -2.3°C). It has got an offensive smell and is very unstable. It decomposes on exposure to light, particularly when moist air to tellurium. The gas burns in air with a pale blue flame.

Hydrogen Telluride	+	Oxygen	$\longrightarrow$	Tellurium Dioxide	+	Water
$2H_2Te$	+	$3O_2$	$\longrightarrow$	$2TeO_2$	+	H_2O

It is soluble in water to give an acid solution, the solution however decomposes easily. It is stronger acid than hydrogen sulphide and hydrogen selenide. It forms tellurides. The alkali tellurides are soluble in water, but their solutions are very unstable and are readily oxidized with the precipitation of tellurium.

Oxides and Oxy-Acids and Halides of Tellurium

Tellurium, like selenium forms tellurium dioxide (TeO_2) and tellurium trioxide (TeO_3). They give rise to the corresponding oxyacids, tellurous acid (H_2TeO_3) and telluric acid, (H_2TeO_4).

Tellurium Dioxide (TeO_2): It is formed as a volatile colourless solid, when tellurium bums in air, or by evaporating a solution of tellurium in dilute nitric acid to dryness. Tellurium dioxide sublimes without melting and is insoluble in water. It dissolves in 50% nitric acid and also in concentrated alkali to form tellurites. The tellurium dioxide is therefore an amphoteric oxide.

Tellurious Acid (Hz TeO_3): It is obtained as a white powder by the addition of dilute nitric acid to a potassium telluride. It rapidly looses water yielding tellurium dioxide. It is very weak acid.

Tellurium Trioxide (TeO_3): It is obtained as a solid, when telluric acid is heated at 400°C it decomposes into the dioxide and oxygen. It is insoluble in water, but on remaining in contact with the later it slowly passes into solution forming telluric acid.

Telluric Acid (H_2TeO_4): It is best obtained by passing chlorine into water containing finely divided powder of tellurium in suspension. The solution is converted and telluric acid is prepared by the addition of alcohol. The telluric acid forms white crystals of the composition H_6TeO_6. This is often called orthotelluric acid and differs in composition from sulphuric and selenic acid. Telluric acid is soluble in water and also is a very weak acid. When it is heated above 100°C, it looses water to form the polymerized metatelluric acid $(H_2TeO_4)_n$ which changes to the trioxide beyond 220°C. In aqueous solution the metatelluric acid is slowly converted into orthotelluric acid.

The tellurates are formed by passing chlorine into alkline solution of tellurites, e.g.

Potassium Telluride	+	Potassium Hydroxide	+	Chlorine	$\longrightarrow$	Potassium Tellurite	+	Potassium Chloride	+	Water
K_2TeO_3	+	2KOH	+	Cl_2	$\longrightarrow$	K_2TeO_4	+	2KCl	+	H_2O

Tellurium Hexafluoride	+	Water	$\longrightarrow$	Telluric Acid	+	Hydrogen Fluoride
TeF_6	+	$4H_2O$	$\longrightarrow$	H_2TeO_4	+	6HF

But they all contain molecules of water, which cannot be removed without decomposition. They are therefore, the salts of orthotelluric acid. The formula of the potassium salt is therefore should be written as $K_2H_4(TeO_6)$.

Tellurium Hexafluoride (TeF_6): Tellurium fluorine combine with incandescence to form TeF6, which is fairly stable colourless gas. It is however hydrolyzed by water.

Tellurium Tetrachloride ($TeCl_4$): It is formed by the action of excess of chlorine on tellurium. It is extremely hydroscopic.

Detection of Tellurium

Tellurium is detected by passing sulphur dioxide vapour through tellurous acid, when brown tellurium in amorphous form is produced.

CESIUM OR CAESIUM

Melting Point : 28°C	Atomic Number : 55
Boiling Point : 705°C	Atomic Weight : 132.905

History and Occurrence

Bunsen and Krichoff discovered cesium in 1863, during the spectroscopic examination of the residue obtained from the mineral waters of springs examination of the residue obtained from the mineral water of springs at Durkeim Germany. The residue gave two beautiful lines, never observed before, in the blue region of the spectrum and indicated the presence of a new element which Bunsen named as Cesium (Latin cæsicius = blue-grey).

Cesium like rubidium is also a rare element. The chief ores are:

(i) ***Pollucite [Al_2Cs_4 $(SiO_3)_6$]:*** It is chemically called cesium aluminium silicate $Al_2Cs_4(SiO_3)_6$ and contained about 27% of cesium.

(ii) ***Lepiodilite [K(Li, Al_2(Si, Al)4O_{10}(OH, F)$_2$]:*** Which is chemically called potassium mica, contains 0.2 to 0.7% of cesium oxide.

(iii) ***Beryl [$Be_3Al_2(Si_6O_{18})$]:*** Consists of about 3% of cesium.

Extraction

The extraction of cesium is carried out from pollucite, here the mineral is ignited and then with calcium at 900°C. About 65% of the cesium present volatilize. The crude condensate contains small amount of calcium potassium and lithium also rubidium. But all except rubidium can be removed by sublimation in vacuum at 350°C - 400°C. Alternatively the finely powdered mineral is decomposes in a water bath with strong hydrochloric acid. The solution is treated with (A) antimony trichloride, which precipitate the double chloride of antimony and cesium ($3CsC1.2SbCl_3$); or (B) with an excess of ammonium alum crystals [$(NH_4)_2SO_4Al_2(SO_4)_3 24H_2O$]. The liquid is then allowed to crystallize.

The free metal can be obtained *(i)* by electrolysis of the fused or magnesium in an atmosphere of hydrogen or *(ii)* by heating the chloride (CsCl) with calcium.

Physical Properties

Melting Point	28°C
Boiling Point	705°C
Density	1.90 g/cc
Ionic Radius	1.69Å
Metallic Radius	2.35Å
Flame Colour	Violet
Heat of Capacity	7.42 cal deg^{-1} mol^{-1}
Standard Entropy	19.8 cal deg^{-1} mol^{-1}
Electrical Resistance	18.8 microohm/cm
Ionization Potential (1st)	3.89eV

Chemical Properties

Cesium when exposed to air it first melts and then bursts into flame, which is due to its low melting point. It decomposes cold water with great violence forming hydroxide (CsOH) which is the strongest of all the bases,

Cesium Carbonate (Cs_2CO_3): It is a colourless crystalline powder and extremely stable salt and can be heated to high temperature without loss of carbon dioxide. It is extremely hydrscopic and is quite soluble in alcohol cesium carbonate can be made from the hydroxide by the addition of carbon dioxide. It is best prepared in the pure state by converting cesium nitrate to the oxalate using excess of oxalic acid igniting the oxalate to complete conversion to the carbonate addition a limited amount of hot water and then filtering off the carbon and any other insoluble impurities.

Cesium is the element which produces the strongest base.

Selenic acid is more powerful than sulphuric acid

Compounds of Cesium

Four oxides of cesium have been reported they are, Cesium Monoxide (Cs_2O) Cesium Peroxide (Cs_2O_2) Cesium Sesquioxide (Cs_2O_3) and Cesium Superoxide (CsO_2) these are formed with the exposure to the oxygen to the different temperatures.

Cesium Hydroxide (CsOH): It is produced by the addition of hot barium hydroxide solution to cesium chloride. Cesium hydroxide is very much soluble in water and strongest base known, but extremely hydroscopic and rapidly disproportionate, it is best stored in the silver or platinum containers away from the atmosphere.

Cesium Sulphate (CS_2SO_4): It is readily obtained by the addition of barium hydroxide to a solution of cesium alum until all the aluminium is precipitated. The point is very well indicated by spot testing for alternating with bromothymol blue. The filtered solution should give no test for either aluminium or barium and pure cesium sulphate is then obtained by concentration and recrystallization form water cesium alum (cesium, aluminium sulphate [$CsAl(SO_4)_2$ $12H_2O$]. It has a well defined octahedral shape it is mostly soluble in alkaline alums prepared by the fusion of cesium sulphate and aluminium oxide.

Detection of Cesium

It is best detected by spectrophotometrical analysis.

Cesium Chloride (CsCl): It is a crystal of cesium and chlorine or the compound obtained by the reaction of cesium and chlorine or the compound obtained by the reaction of cesium and hydrochloric acid. Cesium chloride is soluble in water to about 270 parts in 100 parts of water at 100°C. It has a melting point of 646°C and a boiling point of 1290°C having a specific. gravity of 3.97.

Cesium Nitrate ($CeNO_3$): It is obtained as well defined hexagonal prisms as a result of the reactions of nitric acid and cesium chloride till the precipitate of chloride is not deposited.

CHAPTER

18

Acid Bases and Salts

Introduction

In this chapter we shall discuss the chemistry of the common acids, their origin and other properties, they are hydrochloric acid, sulphuric acid and nitric acid. We are also going to study the bases alkalis and salts.

Definition of Acid

Now what are acids?

In dictionary word acid means sour in taste (In Latin, *acedous* = sour taste). In chemistry acid is defined as *a compound containing one or more replaceable hydrogen atoms, which may be partially or wholly replaced by a metal or a group of non metal acting as metal, which forms a salt, when the acid acts upon the alkali or base.*

According to Boyle's in his book called *Experimental History of Colours* (1664), he recognized the following general properties of acids :

(*i*) They have sour taste.

(*ii*) They act as solvents

(*iii*) They precipitate sulphur from a solution of liver of sulphur (polysulphides of sulphur).

(*iv*) They turn red many of the blue vegetable colours.

(*v*) They react with alkalis.

On the basis of these Hofmann (1723) and Black (1755) were able to show that carbonic acid is a true acid but a weak one.

(*vi*) Cavendish in 1766 showed that gas is evolved when acid reacts with metals.

(*vii*) Acids can donate number of hydrogen ions (H^+) according to its strength and this is called its basicity, and acid is powerful according to the concentration of the hydrogen base of the acid. This can be well understood by the following example:

$$HCl \longrightarrow H^+ + Cl^- \text{ (Monobasic acid)}$$
$$H_2SO_4 \longrightarrow 2H^+ + SO^-_4 \text{ (Dibasic acid)}$$
$$H_3PO_4 \longrightarrow 3H^+ + PO^-_4 \text{ (Tribasic acid)}$$
$$CH_3COOH \longrightarrow H^+ + CH_3COO^- \text{ (Monobasic acid)}$$
$$HNO_3 \longrightarrow \text{try yourself}$$

Lewis Theory of Acids

A more general theory of acids and bases is known as electronic theory according to this theory *acid is a substance that can accept a pair of electron from a base, in combining with it*. The acids, which obey this order are called *Lewis Acids.*

Examples :

H^+ of an acid can accept a pair of electron.

$$\underset{\text{Acid}}{H^+} \quad + \quad \underset{\text{Base}}{OH^-} \;\ldots\ldots\ldots\; \underset{\text{Water}}{H{:}OH}$$

Proton theory ⇒ Acids are Proton donors

Electronic theory ⇒ Acids are electron pair acceptors

In this chapter we shall discuss however only the usual Arthenius view of hydrogen and hydroxyl ions, as the characteristics of the acids and the bases.

Strong acids, on this basis, are therefore, which are less completely ionized in aqueous solution, liberating hydrogen ions examples hydrochloric acid and nitric acid, etc.

Weak acids are only incompletely ionized in solution giving rise to an equilibrium between the ions and the undissolved molecules for example acetic acid.

$$\text{Acetic Acid} \longrightarrow \text{Acetate Ion} + \text{Hydrogen Ion}$$
$$CH_3COOH \longrightarrow CH_3COO^- \;+\; H^+$$

Only a very small portion of acetic acid molecules suffer dissociation giving H^+ ions in water, a great number of undissociated molecules remain unchanged.

Classifications of Acids

On the basis of molecular constitution acids are classified into two classes:

1. Hydracids and (Oxyacids) or a radical are known as hydracids e.g hydrochloric acid (HCl)hydroiodic acid (HI) hydrobromic acid (HBr) etc.

2. Oxyacids which contain both the oxygen and the hydrogen along with the non-metallic group, for example phosphoric acid (H_3PO_4) nitric acid (HNO_3) sulphuric acid (H_2SO_4) etc. Oxyacids consists of larger portion of oxygen are called 'IC' (pronounce + E) e.g. nitric acid, phosphoric acid. The oxyacids containing lesser portions of oxygen are called 'OUS' acid, e.g nitrous acid, phosphorous acid, sulphurous acid, etc. If the oxyacids are started with per they consist of oxygen more than OUS and IC, e.g., persulphuric acid. If the acids starts with 'HYPO' they consist of oxygen more than OUS but less than IC e.g., hypochlorous acid.

Nitric acid - - - - - - - - - - - - HNO_3
Nitrous Acid - - - - - - - - - - - - HNO_2
Persulphuric Acid - - - - - - - - - - - - $H_2S_2O_8$
Hypochlorous Acid - - - - - - - - - - - - HOCl

Bases

A base is a compound (oxide or hydroxide of a metal) which reacts with acid producing salt and water only and it donates electron while reacting, e.g., Sodium oxide (Na_2O) potassium oxide (K_2O) calcium oxide (CaO) etc.

1. Cupric oxide dissolves in sulphuric acid producing copper sulphate and water.

$$\text{Cupric Oxide} + \text{Sulphuric Acid} \longrightarrow \text{Copper Sulphate} + \text{Water}$$
$$CuO + H_2SO_4 \longrightarrow CuSO_4 + H_2O$$

2. Sodium oxide reacts with hydrochloric acid forming sodium chloride and water.

$$\text{Sodium oxide} + \text{Hydrochloric Acid} \longrightarrow \text{Sodium Chloride} + \text{Water}$$
$$Na_2O + 2HCl \longrightarrow 2NaCl + H_2O$$

These bases when react with water produces alkali, which is chemically, called hydroxides of bases, for example when sodium is introduced to water, it reacts violently with water and produces sodium hydroxide, same thing happens to most of the bases also.

$$\text{Sodium Oxide} + \text{Water} \longrightarrow \text{Sodium Hydroxide}$$
$$Na_2O + H_2O \longrightarrow 2NaOH$$
$$\text{Potassium Oxide} + \text{Water} \longrightarrow \text{Potassium Hydroxide}$$
$$K_2O + H_2O \longrightarrow 2KOH$$
$$\text{Calcium Oxide} + \text{Water} \longrightarrow \text{Calcium Hydroxide}$$
$$CaO + H_2O \longrightarrow Ca(OH)_2$$

The alkali came from the word Al-quili (ashes of the plants), so an alkali can be defined as the compound produced when base reacts with water producing appropriate hydroxide.

But here it must be stated *that all the alkalis are bases but all the bases are not alkalis*, because some bases do not produce hydroxide by reacting with water e.g CuO, FeO, NiO etc. These are called non-alkline bases. Some of the important alkaline bases are sodium hydroxide, potassium hydroxide, calcium hydroxide, etc.

Like acid bases also follow the Lewis theory, they donate the pair of electrons to an acid while reacting with it.

$$NH_3 + H^+ (H_3N : H)^+$$

These bases are called Lewis bases.

Salt

A salt is a compound derived from and acid by replacing its hydrogen atom by a metal. A salt therefore consists of a cation (positively charged ion) and anion (negatively charged ion). Salts are electrolytes. In aqueous solutions salts are general neither acidic nor basic (they are neutral but some salts are feebly alkline or acidic).

Formation of Salts

Salts are produced by the reactions of acids and alkali this reaction has and end point or truly the equilibrium point called the *neutralization point*. So the *neutralization is defined as the process of making an acid neutral by adding alkali or vice-versa, where a salt and water produced.*

e.g., Sodium Hydroxide + Hydrochloric Acid ⟶ Sodium Chloride + Water

$$NaOH + HCl \longrightarrow NaCl + H_2O$$

Indicators

This neutralization is a chemical reaction and cannot be visualized by naked eye; *therefore some chemical rather organic substances are used to visualize this neutralization point, these substances are called indicators*, which have two states of colour one in acid and in the basic state.

Some of the known indicator with their colour changes is give below :

Indicator	Chemical Name	Acid Colour	pH range	Basic Colour
Bromothymol blue	Dybromothymolsulfonphthalein	Yellow	6.0 – 7.6	Blue
Phenol Red	Phenolsulphonphthalein	Yellow	6.4 – 8.2	Red
Neutral Red	2-methyl 3-amino 6-dymethylaminophenazine	Red	6.80 – 8.0	Yellow
Quinoline Blue	o-Cresol sulphonphthalein	Yellow	7.2 – 8.8	Red
Phenolphthalein	3,3-bis(p-hydroxyphnyl)pthalide	Colourless	8.2 – 10	Purple
Indigo Carmine	Sodium Indigodisulphate	Blue	11.6 – 14	Yellow

Classification of Salts

Salts are generally classified into three groups:

(1) Normal Salts (2) Acidic Salts (3) Basic Salts or Amphoteric Salts.

1. ***Normal Salts*** : These salts are formed by the complete removal of the hydrogen atoms e.g., $NaSO_4$, K_2SO_4 or NaCl.
2. ***Acidic Salts*** : These salts are produced by the partial removal of the hydrogen atom e.g., $NaHSO_4$, $NaHCO_3$ (which also goes under further change).
3. ***Basic Salts*** : These salts are the result of reaction of acid with the greater proportion of the base required to form a normal salt are called basic salts, e.g., Pb(OH)Cl.

Properties of Salt

Acid salts are intermediate between acid and normal salts, they contain neither ionizable $(OH)^-$ and $(H)^+$

Examples :

Acid	Acid Salt	Normal Salt
H_2SO_4	$NaHSO_4$	Na_2SO_4
H_2CO_3	$NaHCO_3$	Na_2CO_3
H_3PO_4	NaH_2PO_4 Or Na_2HPO_4	Na_3PO_4

Basic salts are intermediate between bases and normal salts. So they are called Amphoteric salts

Examples :

Base	Basic Salt	Normal Salt
$Fe(OH)_3$	$Fe(OH)(C_2H_3O_2)$	$Fe(C_2H_3O_2)$
	$Fe(OH)_2\ (C_2H_3O_2)$	
	$FeO\ (C_2H_3O_2)$	
$Al(OH)_3$	$Al(OH)SO_4$	$Al_2(SO_4)_3$

Basic salts are sparingly soluble in water, behaving like the corresponding hydroxides.

Hydrolysis

Generally speaking hydrolysis *may be defined as a chemical reaction in which a molecule or ion is converted into two or more molecule or ions by the reaction of water, or more correctly, with H^+, and OH^- ion of water.*

Sodium acetate in aqueous solution reacts alkaline, where as aluminium sulphate under same condition gives acedic reaction. They are therefore said to be undergoing hydrolysis in water. The hydrolytic reaction for these salts are generally represented as :

$$\underset{\text{Sodium Acetate}}{CH_3COONa} + \underset{\text{Water}}{H_2O} \longleftrightarrow \underset{\substack{\text{Acetic Acid}\\ \text{Weak acid}}}{CH_3COOH} + \underset{\substack{\text{Sodium Hydroxide}\\ \text{Strong base}}}{NaOH}$$

$$\underset{\text{Aluminium Sulphate}}{Al_2(SO_4)_3} + \underset{\text{Water}}{6H_2O} \longleftrightarrow \underset{\text{Aluminium Hydroxide}}{2Al(OH)_3} + \underset{\text{Sulphuric Acid}}{3H_2SO_4}$$

The case of sodium acetate is anionic hydrolysis and that of aluminium sulphate is cationic hydrolysis, since both NaOH and H_2SO_4 are completely ionized, where as CH_3COOH and $Al(OH)^+$ remains particularly undissociated bases and those of strong acids with weak bases suffer hydrolysis. For the anions of the former as well as the cations of the later will readily combine with H^+ and OH^- ions of water to form undissociated molecules of weak acids and of weak bases respectively. Excess of hydroxyl or hydrogen ion is thus left in the solution, which therefore becomes alkaline or acedic. In short as all the salts are completely ionized in water by the ions of salt to form undissociated solutions. The hydrolysis of a weak acid and a weak base can be illustrated in the case of ammonium acetate as follows :

$$NH_4^+ + CH_3COO^- + H^+ + OH^- \longleftrightarrow NH_4OH + CH_3COOH$$

If both the acid and the base were equally weak, the solution may show neither an acid nor an alkaline reaction, due to the removal of equal amounts of hydrogen and hydroxyl ions of water. This occurs in the case are aqueous solution of ammonium acetate, as ammonia and acetic acid are aqueously dissociated in solution.

The hydrolytic dissociation of the salts increases in the dissociation of water with temperature.

These are causes of hydrolysis of non-ionized molecules :

$$PCl_5 + 4H_2O \longrightarrow H_3PO_4 + HCl$$

$$CaC_2 + 2H_2O \longrightarrow Ca(OH)_2 + C_2H_2$$

Hydrogen Ion Concentration and pH

We have seen that water is both and acid and a base. It undergoes ionization and form very small but equal amounts of hydrogen and hydroxide ion.

$$H_2O + H_2O \longleftrightarrow H_3O^+ + OH^- \; K_w = H_3O^+ \; OH^- = 1.0 \times 10^{-14} \text{ at } 25°C.$$

The degree of ionization increases as the temperature increases. At 100°C K_w, is about 1×10^{-12} 100 times larger than at 25°C.

Pure or distil water is neutral, i.e. neither acidic nor alkaline because the self ionization of water yields the same number of hydrogen ions and hydroxyl ions at any temperature at 25°C.

$$H_3O^+ = OH^- = K_w = 1.0 \times 10^{-7} \text{ M}$$

When we add acid to the water hydrogen ion concentration (H_3O^+), becomes larger than 1.0×10^{-7} M and the hydroxide ion concentration (OH^-), becomes smaller than 1.0×10^{-7} M *but not zero.* Similarly, when we added base to water (OH^-) becomes greater than 1.0×10^{-7} M and (H_3O^+) decreases *but not zero.* The product of H_3O^+ and OH^- is always a constant, so the value of one or the other concentration in solution is to use a logarithmic scale, such as the negative of the log of the hydrogen ion concentration.

pH is the negative logarithmic of the hydrogen ion concentration

$$pH = -\log(H_3O^+)$$

This equation, log is the notation with a pH meter, with a solution of an indicator or a pH paper. A neutral solution has a pH of 7.0 at 25°C. The pH of an acedic solution is less than 7 and the pH of an alkaline solution is more than 7.

We can use of pOH, the negative of the log of the hydroxide ion concentration to represent (OH^-)pOH = $-\log(OH^-)$ and (OH^-) = 10^{-pOH}.

Normally we do not measure the pOH value they are calculated from the pH values.

At 25°C pH + pOH = 14.00 or pOH = 14.00 −pH

We can show this by starting with the equation (H_3O^+) (OH^-) = K_w. If we take the negative of the log of both sides of this equation we have.

$$-\log(H_3O^+)(OH^-) = \log K_w$$

or, $$(-\log H_3O^+) + (-\log OH^-) = -\log K_w$$

To be sure with the definition of pH and pOH, we will define $-\log K_w$ as pK_w. Then at 25°C temperature we have :

$$pK_w = -\log(1.0 \times 10^{-14}) = (-14.00) = 14.00$$

And $$pH + pOH = pK_w = 14.00$$

The table below shows the pH and the pOH

$(H_3O)^+$	$(OH)^-$	pH	pOH	Remarks
10^{-0}	10^{-14}	1	14	Strongly acidic 1M HCl (hydrochloric acid)
10^{-2}	10^{-13}	2	13	
10^{-3}	10^{-12}	3	12	
10^{-4}	10^{-11}	4	11	Gastric juice, lime juice
10^{-5}	10^{-10}	5	10	
10^{-6}	10^{-9}	6	9	1M Acetic Acid, stomach acid wine
10^{-7}	10^{-8}	7	8	
10^{-8}	10^{-7}	8	7	Coffee
10^{-9}	10^{-6}	9	6	
10^{-10}	10^{-5}	10	5	Neutral, distil water
10^{-11}	10^{-4}	11	4	
10^{-12}	10^{-3}	12	3	
10^{-13}	10^{-2}	13	2	Milk of Magnesia
10^{-14}	10^{-0}	1	14	Household ammonia 1M NaOH (Sodium hydroxide)

This pH is generally detected by a special indicator called *Universal indicator,* which is prepared by the solution of number of indicators, changing the colour at different pH, this is becomes red at pH 1 – 4 nearly orange at pH = 5 – 6, pH = 6.5 – 7.5 green, pH = 8.5 – 9.0 blue and from pH = 9.5 - 14 violet. This will become more clear from the chart shown in the Fig. 18.1. The pH scale was invented by sorensen in (1848) pH means pouver hydrogen (Hydrogen Power).

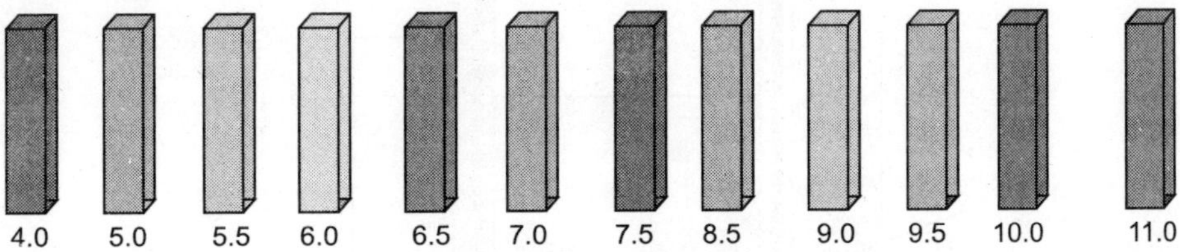

Fig. 18.1. The Coloured pH chart (as Found in the Bottles of the Universal Indiators)

Generally speaking the universal indicators which are commercially available can determine pH from 4 to 11, to detect pH beyond this range we use pH meter. These pH meters are more accurate than indicators and records the pH directly with the help of glass electrode and reference electrode. Now days one push button direct reading digital pH meters are also available. These are more convenient and user friendly, some pH meters also measures the ionization taking place inside the solution under test (SUT).

Actually the pH meter measures the e.m.f of a concentration of cell formed from the reference electrode the test solution and the glass electrode sensitive to hydrogen electrode (see Fig. 18.2).

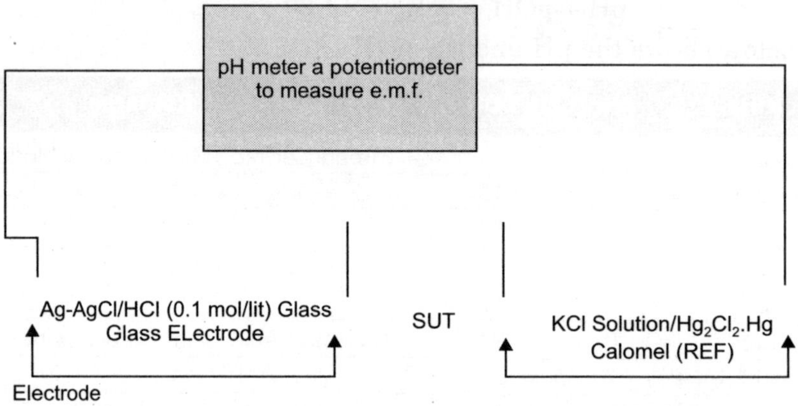

Fig. 18.2. The Diagrammatic Representation of a pH Meter.

Glass Electrode: The glass electrode consists of a thin glass bulb 0.1 mm thick blown on to a hard glass tube of high resistance. In side the bulb a solution of a hydrochloric acid 0.1 mol/lit connected to a platinum wire via a silver-silver chloride electrode, which is reversible to hydrogen ion (see Fig. 18.3).

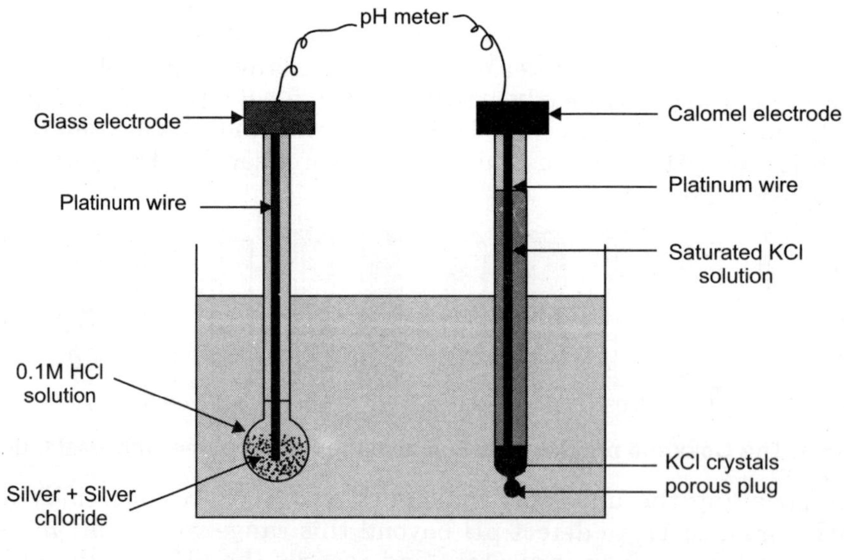

Fig. 18.3. The Electrodes of a pH Meter

A potential developed across the thin glass of the bulb, which depends upon the pH of the solution in which it is immersed; that is when the pH is low the potential is high and if the pH is high the potential is low. Salts, protein and oxidizing agents do not readily affect the potential, so the electrodes can be used in a wide variety of media. The glass electrode in the test solution consists of the half-cell and the measuring circuit is complete by a reference electrode, which is not sensitive to hydrogen ions.

Calomel Electrode: The calomel electrode or reference electrode is used due to its stability and can be easily prepared and the potential with respect to standard hydrogen electrode is currently known.

The e.m.f of the complete cell (E) formed by linking of the two electrodes is

$$E = E_{ref} - E_{glass},$$

where E_{ref} is the potential of the calomel, (which is reference electrode which at normal room temperature is +0.250V) and E_{glass} is he potential of the glass electrode which depends upon the SUT.

Calculation of pH

We shall now calculate the value of pH from different parameters.

Example 1 : Calculation of pH from Hydrogen Ion Concentration.

Stomach acid is a solution of hydrochloric acid and the concentration of hydrogen ion is 1.2×10^{-3}. Calculate the pH of this acid.

Because the hydrochloric acid is highly concentrated acid and therefore the ionization is 100%, it will have the same concentration of 1.2×10^{-3}.

$$\text{Thus we have the pH} = -\log(H_3O)^+$$
$$= -\log (1.2 \times 10^{-3}) = -\{-(0.012)\} = (-1.92).$$

Example 2 : Calculation of pH from Hydrogen Ion Concentration.

Calculate the pH of a 0.125M solution of nitrous acid (a weak acid, 6.5%) ionized.

The present ionization for the nitrous acid is 6.5%. Hence the number of moles of hydrogenium ion in 1 litre of solution is 6.5% of 0.125M, or

$$\frac{6.5}{100 \times 0.125} = .0081 \text{ or } 8.1 \times 10^{-3}$$

$$\therefore \quad pH = -\log(H_3O^+) = -\log(8.1 \times 10^{-3})$$
$$= -(-1.09)$$
$$= 1.09$$

Example 3 : Calculation of pH and pOH.

We have seen that pure water is very difficult to obtain because the water consists of dissolved carbon dioxide. The resulting carbonic acid H_2CO_3 gives the solution a hydrogenium ion concentration of 2.0×10^{-6} that is nearly 20 times greater than that of pure water. Calculate the pH of the solution at 25°C.

The hydrogenium ion concentration is 2.0×10^{-6}

∴ pH is –log (2.0×10^{-6}) = –(–4.69) or 4.69

Students are requested to use log table or scientific calculator for this purpose.

Example 4 : Calculation of Hydrogen Ion Concentration from pH.

Calculate the hydrogen ion concentration of blood, the pH of blood is 7.3

$$\therefore \quad \text{pH is the } -\log(H_3O)^+ = 7.3$$
$$= \log(H_3O)^+$$
$$= -7.3 \text{ or } 10^{-7.3} \text{ or antilog of } -7.3$$
$$= 5.01 \times 10^{-8}\text{M}.$$

Example 5 : Calculation of pOH of a basic solution.

What are pH and the pOH of a 0.0125M dilute solution of KOH?

Potassium hydroxide is a strongly basic compound and ionizes completely in dilute proportions, $(OH)^-$ is 0.0125M

It is known that pOH is the $-\log(OH)^- = -\log.0125 = -(-1.90)$ or 1.90.

Now the pH of the solution can be determined from pOH,

$$\therefore \quad \text{pH} = 14.00 - \text{pOH}$$
$$= 14.00 - 1.90$$
$$= 12.1$$

NUMERICAL EXAMPLES

1. Calculate the pH of a 0.01M solution of acetic acid and which is 1.3% ionized. **(Ans :** pH = 2.88)
2. Calculate the pH and the pOH of a 7.88×10^{-4} solution of HBr a strong acid at 25°C. **(Ans :** pH = 2.10 and pOH = 11.89)
3. Calculate the hydrogen ion concentration of the solution having pH of 3.9. **(Ans :** 4.67×10^{-10})
4. Calculate the pOH and the pH of the solution of 0.1M solution of acetate ion at 25°C. There is a percentage yield of the pH in the reaction of acetate ion in water is 0.0075%. **(Ans :** pH = 5.12 and pOH = 8.88)
5. The pH of a soft drink is 3.82 what is the hydrogen ion concentration. **(Ans :** 1.51×10^{-4})
6. What is the pH of 1×10^{-7} M HCl. **(Ans :** pH = –6)

SOME PHYSICAL AND CHEMICAL REACTIONS OF ACIDS AND BASES

Experiment 1 : In a test tube pour some hydrochloric acid (dilute) and invert it over the thumb and lick the thumb the taste will be sharp sour. Add two drops of litmus (blue) solution, which turns red. Repeat the above experiment with nitric acid and sulphuric acid the result will be same.

The above experiment shows that the acids tastes sour and turns blue litmus red.

Experiment 2 : In three test tubes take three solutions of HCl, HNO_3 and H_2SO_4. Now add one in each tube phenolphthalein, methyl orange and bromocresol green, note the colour change, the tube contains methyl orange turns red. The tube consists of phenolphthalein turns colourless and the tube containing bromocresol green turns yellow. These happen due to the different colour changes of the different indicators.

Experiment 3 : In three test tubes take HCl, NaOH and distil water. Add two drops to each tube universal indicator. Note the colour. The tube containing HCl turns red the tube consists of water turns green and the tube with sodium hydroxide turns blue or violet. The colour change is due to the different pH level of the three solutions.

Experiment 4 : Take 5ml of hydrochloric acid and 5ml of sodium hydroxide solution in two different test tubes and add a little phenolphthalein to the solution of sodium hydroxide solution, now notice the initial colour, which is pink. Now start adding hydrochloric acid slowly. Notice the colour changes from pink to colourless. This is due to neutralization of the alkali with acid. Try to do this experiment with universal indicator and try to bring the green colour.

These experiments are very interesting and students must perform these experiments!

Experiment 5 : In the two test tubes few ml of nitric and hydrochloric acids are kept, no other reagents or apparatus provided to you, identify nitric acid.

The process is very simple, here the skin acts as the reagent which when comes in contact the protein of the skin reacts with nitric acid and forms a yellow pigment. This will not happens in the case of the hydrochloric acid.

Experiment 6 : In three test tubes take solutions of NaOH, KOH and $Ca(OH)_2$. Cover the tubes with thumb one by one and invert them one by one and taste the solutions one by one (as before) the taste is bitter. Now rub the fingers with each other, the fingers are slippery like soap. Now drop red litmus solution to the tubes containing solutions. The solution turns blue. So it is clear that alkalis tastes bitter feels soapy and turns red litmus blue.

Experiment 7 : Take some ammonium hydroxide solution in a flask and ad some phenolphthalein is added to it. Now spray the solution on a white paper, the pink solution turns transparent after sometimes. Because the ammonium hydroxide contains ammonia, which is an alkali, it turns the phenolphthalein solution pink, but as it is a volatile solution it evaporates away living behind the water, which becomes colourless in the presence of phenolphthalein. This is called vanishing colour. The solution of the ammonium hydroxide must be concentrated.

Acidimetry Alkalimetry

This is the analytical measurement consists of the process of determination of the strength of acid and alkali by neutralization of the other and this is done by a very common and easy experiment called burette titration. The name of this experiment is volumetric analysis, the experiment determines the neutralization point by using proper indicator.

Selection of an Indicator

There are some selected indicators for the determination of the neutralization point for a definite strength of acid and alkali:

Name of the Indicator	Type of Acid	Type of Alkali
Phenolphthalein	Strong/Weak	Strong
Methyl Orange	Strong	Weak
Bromothymol Blue	Weak	Weak

Equivalent Weights of Acid Bases and Salts

Equivalent weight and the gram equivalent weight of an acid is defined as the number of parts by weight of it which contains one part by weight of replaceable hydrogen.

Or $$\text{Equivalent weight} = \frac{\text{Molecular Weight}}{\text{Basicity}}$$

Example : 36.5 part of HCl contain 1 part of hydrogen for replace that is its basicity is 1.

∴ Its Equivalent Weight = 36.5/1 or 36.5

Similarly the equivalent weight of sulphuric acid is

$$\frac{\text{Molecular Weight}}{\text{Basicity}} \text{ or } \frac{98}{2} = 49$$

The basicity, equivalent weight and also the gram equivalent weight of some commonly used acids are shown in the table below :

Name of the Acid	Molecular Weight	Basicity	Equivalent Weight	Gram Equivalent
Hydrochloric Acid	36.5	1	36.5	36.5
Sulphuric Acid	98	2	49	49
Nitric Acid	63	1	63	63
Acetic Acid	60	1	60	60
Oxalic Acid	90	1	45	45

Equivalent Weight and Gram Equivalent Weights of Bases

The equivalent weight of a base is the number of parts by weight of it, which neutralizes completely one equivalent part of an acid. Again, the acidity of the base is the number of equivalent of an acid which can be just be neutralized by one molecule of the base. In other words *the acidity of the base is the number of replaceable hydroxyl group concentrated in a single molecule.*

So the equivalent weight of an base is $\frac{\text{Molecular Weight of the Base}}{\text{Acidity}}$

Equivalent of the base expressed in gram is known as *gram equivalent weight or gram mole.*

Molecular weight of NaOH is 40 and its basicity is 1

∴ Its equivalent weight is 40/1 = 40

Similarly the molecular weight of calcium hydroxide is 73, $Ca(OH)_2$ and the acidity is 2.

∴ The equivalent weight is 73/2 = 36.5

The gram equivalent weight and the equivalent weight also the basicity of the common alkalis are stated next.

Alkali	Molecular Weight	Acidity	Equivalent Weight	Gram Molecular Weight
NaOH	40	1	40	40
KOH	56	1	56	56
$Ca(OH)_2$	74	2	37	37
NH_4OH	35	1	35	35

In the case of basic oxide, the equivalent weight of and oxide can be determined as follows :

$$\text{Equivalent Weight of Basic Oxide} = \frac{\text{Molecular Weight of the Oxide}}{\text{Number of the atoms of the molar} \times \text{Valency}}$$

Molecular Weight or Gram Equivalent Weight of a Normal Salt

This can be defined as the *part by weight of a salt, which contains one equivalent of the constituent metal or the radical equivalent of the metal.*

$$\text{Equivalent Weight of the Salt} = \frac{\text{Molecular Weight of the Salt}}{\text{Total Number of Atoms} \times \text{Valency of the Metal}}$$

Example : In 100 parts of sodium chloride NaCl (35.5 + 23) or 58.5 parts by weight (one equivalent) of sodium is present or 100 parts by weight of calcium carbonate contains 40 parts by weight of calcium or two equivalents of calcium.

∴ The equivalent weight of calcium is 100/2 = 50

The equivalent weight of salt expressed in grams will be the gram equivalent weight of the salt.

Salts	Molecular Weight	Number of Metal Atoms × Valency	Equivalent Weight
Na_2CO_3	106	2 × 1	106/2 = 53
$Al_2(SO_4)_3$	342	2 × 3	342/6 = 57
$BaCl_2$	244	1 × 2	244/2 = 122
$CuSO_4$	249.5	1 × 2	249.5/ 2 = 124.75
$AgNO_3$	170	1 × 1	170

Solutions and their Strengths

The strength of the solution or the concentration means that the weight of the solid substance present in the solution. This can be expressed in different ways. A solution that contains a known weight of the solid compound dissolved in a known volume of the solution is called the *standard solution.* Therefore, *the standard solution is a solution of known strength.*

Normal Solution : The solution containing 1 gram equivalent weight of the solid in 100 ml of water. It is indicated by (N).

Example : The gram equivalent weight of the hydrochloric acid is 35.5.

∴One litre of (N) HCl shall contain 35.5 g of hydrochloric acid and the solution is called N solution.

Strength of the Solution in Terms of Normality

The normality of a solution is 1 (or the solution 1N) when it contains one gram equivalent of the solute per litre. If one litre of a solution consists a quantity which is multiple or a fraction of one gm equivalent of the solid substance, the strength of such a solution in terms of normality is expressed in the following chart.

Quantity of solid per litre	Name	Normality Abbreviation	Normal of the Solution
1 g equivalent	Normal	1N or N	1
3 g equivalent	Three normal	3N	3
5 g equivalent	Five normal	5N	5
[illegible] g equivalent	Semi normal	N/2 or 0.5N	135 or .1
[illegible] g equivalent	Centi normal	N/100 or 0.01N	@ or 0.01

∴1 litre of 5N H_2SO_4 solution will contain $49 \times 5 = 245$g of H_2SO_4

1 litre of 2N Na_2CO_3 solution will contain $53 \times 2 = 106$ g of Na_2CO_3

1 litre of N/10 H_2SO_4 solution will contain $49 \times .1 = 4.9$g of H_2SO_4

Molar Solution: A molar solution of a substance is that solution, one litre of which contains one gram molecular weight or one gram molecule of the substance. A molar solution is represented by (M)

> In volumetric analysis generally the normal and the decinormal solutions are used

One gram molecule of H_2SO_4 is 98g

One gram molecule of NaOH is 4g

$$\therefore \text{ Molarity (M)} = \frac{\text{Number of g molecules in solution}}{\text{Number of litres of the solvent}}$$

Relation between Normality and Molarity

Normality and molarity can be changed according to the understated relationship.

$$\text{Molarity} = \text{Normality} \times \frac{\text{Equivalent Weight}}{\text{Molecular Weight}}$$

$$\text{Normality} = \text{Molarity} \times \frac{\text{Molecular Weight}}{\text{Equivalent Weight}}$$

Formal Solution : With a view of expressing the strength of the solutions, it is now been prepared to use formal concentration and is denoted by (F).

A formal solution of *a substance is that the solution of one litre of which contains one gram-formula weight (that is formula expressed in gms) of the substance.*

$$\text{Formality} = \frac{\text{Number of g-formula weight}}{\text{Volume of the solution in one litre}}$$

A formal solution of a substance to be the same as the molecular solution. But one has to clearly understand the difference between the two. If the formula of the compound is the same as it is correct molecule, representing the molecular structure actually present in the solution, then formality and molarity becomes same. Thus 0.1F solution of glucose having the molecular formula of $C_6H_{12}O_6$ is 0.1M solution. But 0.1F solution of NaCl cannot be regarded as its 0.1M solution of NaCl, it is preferably stated as 0.1M respect of Na^+ ions and 0.1M of Cl^- ions.

Molarity and Molal Solutions

Sometimes the concentrations of a solution is expressed in terms of molarity which is the number of moles of the solute dissolved in 1000 g of the solvent. The molal solution of a solution that contains 1 mole of the solid per 1000 g of the solution. The letter M is used for the solution 1 molal.

Percentage Strength

The percentage strength is the number of grams of the solute present in 100ml of the solution. Thus 5% of NaOH will contain about 5.0g of the compound in the 100ml of the solution. In a 10% of HNO_3 there will be 10g of HNO_3 in 100ml of water.

Strength in Terms of g/litre

The strength may also be denoted by the number of g of the solute dissolved in 1 litre of the solution. Thus when one litre of NaOH solution contains 10g of the solute, the strength of the solution is said to be 10g/lit.

Conversion of Strength Units

Percentage Strength to Normality

100 ml of 5% NaOH solution will contain 5 g of NaOH

$\therefore$ 1000 ml of 5% NaOH solution will contain 5 g × 10 = 50 g of NaOH

By definition, a normal solution of NaOH is one, which contains 40 g of it per litre of the solution.

Thus, 40 g of NaOH is present in 1000 ml give 1N NaOH.

$\therefore$ 50 g of NaOH is present in 1000 ml give $\frac{5 \times 10}{40} = 1.25\text{N}$ of NaOH

When the percent strength of a solution is A, its strength in terms of normality is

$$\text{Normality} = \frac{A \times 10}{\text{Gram-equivalent of solute}}$$

From Normality to g/litre

Amount of solute present per litre of 1N sol = g equivalent of solute (as per definition)

∴ Amount of solute present per litre of 2N sol = 2 × g equivalent of solute.

∴ Amount of solute present per litre of $\frac{1}{10}$ N sol = $\frac{1}{10}$ × g equivalent of solute.

∴ Weight in g/litre =Normality × g equivalent weight.

Some Important Principles of Acidimetry and Alkalimetry

(A) We know that a normal solution of a substance contains one g-equivalent of the substance dissolved in one litre of it. Hence, 5N solution and $\frac{1}{10}$ solution will contain respective 5 g equivalent and $\frac{1}{10}$ of 1 g equivalent of the solute per litre.

∴ 1 ml of 5N solution ≡ 5ml of (N) solution ≡ 50 ml of $\left(\frac{1}{10}\right)$ solution and

1 ml of (N) solution ≡ 10ml of $\frac{1}{10}$ (N) solution ≡ 2 ml of N/2 solution ≡ 100ml of N/100 solution.

Thus y ml of x(N) solution ≡ ($y \times x$)ml of (N)solution ≡ x ml of y(N) solution.

∴ Volume of a given solution is inversely proportional to its strength.

(B) We know that in neutralisation reaction acid and alkali react with each other in equivalent contents. One gram-equivalent of any acid will neutralise completely 1 g-equivalent of base. The neutralising power of 1g-equivalent of any acid is the same. This equality is true in the case of any alkali.

∴ 49 g of H_2SO_4 ≡ 40 g of NaOH
40 g of NaOH ≡ 50 g of KOH
50 g of KOH ≡ 53 g of Na_2CO_3
53 g of Na_2CO_3 ≡ 35 g of NH_4OH

40 g of NaOH ≡ 36.5 g of HCl
36.5 g of HCl ≡ 63 g of HNO_3
63 g of HNO_3 ≡ 32.65 g of H_3PO_4

Again 1000 ml (N) acid ≡ 1000 ml (N) alkali
2000 ml (N) acid ≡ 2000 ml (N alkali
10 ml (N) acid ≡ 10 ml (N) alkali
1 ml (N) acid ≡ 1 ml (N) alkali
V ml (N) acid ≡ V ml (N) alkali

∴ Solutions of an acid and a base of the same strength in terms of normality must neutralise each other in equal volume.

∴ $$V_1 \times S_1 = V_2 \times S_2$$

Or *volume of the acid × strength of the acid = volume of the alkali × strength of the alkali*

This is the most important formula of acidimetry and alkalimetry.

Reduction of Strength by Dilution

From the above discussion it is clear that the strength or concentration of a solution of a given weight of the solute is inversely related to the volume of the solution.

$V \propto 1/S$ (where S is the strength of the solution and V = volume).

If the volume of the solution is increased i.e the solution is diluted its strength is proportionately reduced. A solution of higher strength may easily be converted to a solution of any desired strength but of lower strength by dilution.

Let 25 ml of (N) acid is ≡ to 10 ml of alkali with unknown strength. Obviously the alkali solution is stronger than the acid, to reduce the strength of the alkali solution exactly normal (25 – 10) that is 15 ml water is added to every 10 ml of the solution.

Similarly, 10 ml of 2.5(N) solution can be made N/10 by dilution.

10 ml of 2.5(N) solution ≡ (10 × 2.5) ml of (N) solution or 250 ml of N/10 solution.

∴ Addition of (250 – 10) or 240 ml of water is necessary to 10 ml of the solution.

Volume of the water V_2 which must be added to V_1 ml of a concentrated solution of Normality S1 to give a solution of normality S_2 can be derived from the following relation.

$$V_1S_1 = (V_1 + V_2)S_2 = V_1S_1 + V_2S_2$$

Or,

$$V_2S_2 = V_1S_1 - V_2S_2 = V_1(S_1 - S_2)$$

∴

$$V_2 = \frac{V_1(S_1 - S_2)}{S_2}$$

To prepared an approximately N/10 solution of sulphuric acid which is having an equivalent weight 49 grams.

To prepare N/10 we require 49 × [illegible] g of sulphuric acid or 4.9 g of the sulphuric acid.

Now let the purity of the acid be 97%

That is 97 g of pure acid is present in the 100 g of acid.

∴ 4.9 g of pure acid is present in the $\frac{100 \times 4.9}{97} = 5.0517$ g of the acid

$$\text{Now density} = \text{Mass} \div \text{volume}$$

∴

$$\text{volume} = \text{mass} \div \text{density}$$
$$= 5.0517 \div 1.84$$
$$= 2.745 \text{ ml}$$

Similarly the strength of the hydrochloric acid and nitric acid can be calculated. The table below shows some of the N/10 quantities of important acids.

Acid	Gram Equivalent Weight (g)	Density (g/cc)	Per cent strength	Normally	Volume in 1000 m(ml)
H_2SO_4	49	1.84	97	N/10	2.725
HCl	36.5	1.19	97	N/10	3.16
HNO_3	63	1.42	97	N/10	4.6
CH_3COOH	60	1.05	97	N/10	5.89

To prepare the 250 ml of N/10 sodium hydroxide solution.

The g molecular equivalent weight of the sodium hydroxide is 40 g.

To prepare 100ml of solution we require $40 \times 1 = 4$ g

To prepare 1 ml of solution we require $\dfrac{4}{1000 \times 250} = 1$ g

Similarly we can calculate the required amounts of all the other alkalis and they are shown in the table below :

Alkali	Gram Equivalent Weight	Normality	Mass in 1000 ml
NaOH	40	N/10	4 g
KOH	56	N/10	5.6 g
$Ca(OH)_2$	74	N/10	7.4 g

VOLUMETRIC ANALYSIS ACIDIMETRY ALKALIMETRY

PRACTICAL CHEMISTRY

Experiment : Determination of Strength of Acid by Burette titration

Theory : A known strength of alkali solution is allowed to triturate with an acid solution of unknown strength, in the presence of an indicator.

Apparatus : 50 ml burette, 25 ml pipette, 100 ml volumetric flask and 250 ml conical flask and clamp stand.

Reagents : 250 ml N/10 sodium hydroxide solution and 100 ml given acid of hydrochloric solution, bromocresol green indicator(pH 3.8 to 5.4)

Procedure : The acid solution is filled in the burette and 25 ml of N/10 solution of sodium hydroxide solution is taken in the flask, 2 drops of indicator is added to the alkali solution, this indicator changes from blue to yellow in alkali to acid. Then the acid solution is allowed to fall dropwise, until the blue colour changes from blue to just yellow. This experiment is repeated 3 times with the fresh reagents provided.

Observations : The reading of the burette is noted and tabulated below :

Number of Experiments	Volume of Alkali	Burette Reading		Volume of acid solution	Mean of the Volume
		Initial	Final		
1	25 ml	0	1 ml	1 ml	
2	25 ml	0	1 ml	1 ml	1 ml
3	25 ml	0	1 ml	1 ml	

So, 25 ml of alkali solution = 1 ml of acid solution.

Calculation : From the Law of Normality

Volume of Acid × Strength of Acid = Volume of Alkali × Strength of Alkali

Let the Strength of acid be S

$$\therefore 1 \text{ ml} \times S = 25 \text{ ml} \times 1N/10 \text{ or } S = \frac{25\times1(N/10)}{1} = 25N/10 = .25N,$$

$$\text{the acid in 250 ml is } \frac{35.5\times.25}{4} = 2.28125 \text{ g.}$$

Inference : From the above experiment we are able to determine the strength and as well as the exact amount of acid content in the solution.

Precaution : Before use the burette and the stopcock must be checked for leak, also should be carefully operated so that there is no seepage of extra acid. Always keep acid in the burette. Avoid parallax error.

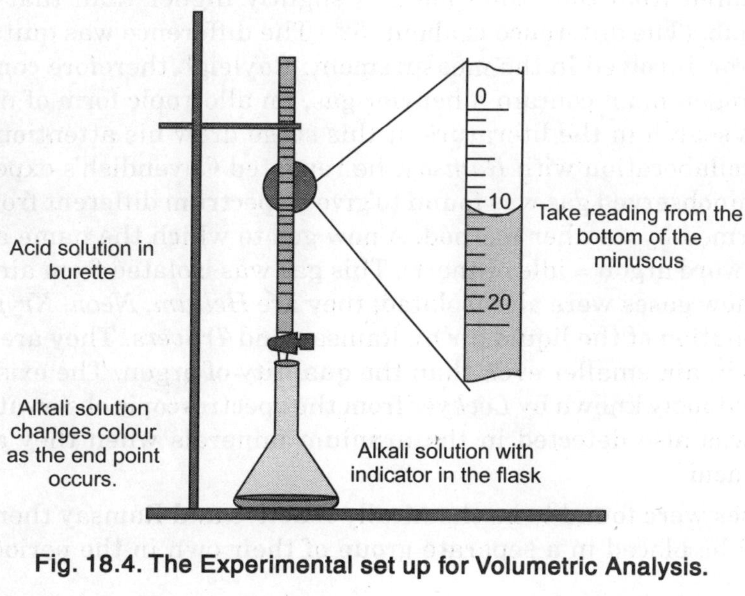

Fig. 18.4. The Experimental set up for Volumetric Analysis.

CHAPTER

19

Inert Gases and Uses

Introduction and History

As early as in 1785 Cavendish recognised the presence of a gas in air more inert than nitrogen. For a long time he passed an electric current through the gas mixture of air and oxygen in a sealed bulb with potassium hydroxide solution, as long as there was no diminution of volume, due to the formation of oxides of nitrogen which were absorbed by the alkalis. He found that after removal of oxides by alkali, there still remained a small amount of gas. This gas was amounting to A of the volume of the original air. This observation of Cavendish, however, remained unnoticed, over a century, till 1894 *Lord Rayleigh* during a systematic investigation on the density of gases found that the density of nitrogen prepared from the atmosphere is slightly higher than that of the nitrogen prepared in the lab. (The difference is about .5%) The difference was quite large than the experimental error, involved in the measurement. Rayleigh therefore concluded that the atmospheric nitrogen must contain a heavier gas, an allotropic form of nitrogen or some other new gas. A search in the literature at this stage drew his attention to Cavendish's experiment. In collaboration with *Ramsay,* he repeated Cavendish's experiment and the small residue of unobserved gas was found to give a spectrum different from nitrogen. The result was confirmed by another method. A new gas to which the name argon was given (from the Greek word argon = idle or inert). This gas was isolated form air. Soon after this isolation, other new gases were also isolated; they are *Helium, Neon, Krypton* and *Xenon.* From the fractionation of the liquid air by Ramsay and *Travers*. They are present in very small quantities in air, smaller even than the quantity of argon. The existence of helium however was previously known by *Lockyer* from the spectroscopic observations of the solar atmosphere, it was also detected in the uranium minerals when they are heated with dilute sulphuric acid.

All these gases were found to be chemically inactive and Ramsay therefore suggested that they should be placed in a separate group of their own in the periodic table. In the view of their chemical inactivity, indicating the absence of all chemical affinity and be placed before group I of the strongly electropositive alkali metal and after the group VII of the strongly electronegative halogen elements. This is justified by the arrangement that an abrupt change from a highly negative to a highly positive character, or vice versa. Because of inertness these gases are often called *noble gases* or *inert gases*. Sometimes rare gases due to their scarcity. Table below shows the volume % and Weight % of the five inert elements:

Gas	Volume %	Weight %
Helium	0.0005	0.00007
Neon	0.0018	0.0012
Argon	0.9325	1.2862
Krypton	0.0001	0.0003
Xenon	0.000009	0.00004
Radon (niton)	5×10^{-18}	4×10^{-17}

HELIUM

Melting Point : 272°C
Boiling Point : 268°C

Atomic Number : 2
Atomic Weight : 4

Occurrence

As helium is a product of radioactive disintegration, it is found in small quantities in radioactive rocks. It is also a minor constituent of gases from some mineral springs, mines volcanoes and fumaroles; some natural gas contains significant quantity. Helium is a constituent of the atmosphere, estimated at 1 part in 185,000 to 200,000 parts of air at the earth's surface.

Helium Bearing Natural Gases

Helium bearing natural gases is the only known raw material from which helium can be extracted in large quantities, at reasonable cost. The helium consists of natural gases from different pools vary from quantities that cannot be detected by ordinary analytical methods or mere spectroscopic traces, to 8% or more by volume. Usually the gases of higher helium contents are found in fields that lie over buried granite ridges, such as the deeply buried Amarillo Mountains of the Texas Panhandle and the Rattlesnake Field in New Mexico. Traces of Helium are also available in non-associated natural gases Cambay and Ankleshwar in Gujarat and Namrup in Assam.

Extraction of Helium from Natural Gas

Production and Purification : Helium is separated from natural gas by cooling the gas to a liquification point of its ordinary constituents, but above the liquefication point. However the gas must be first treated chemically to remove small quantities of carbon dioxide and water vapour that would solidify in the low temperature (see Fig. 19.1).

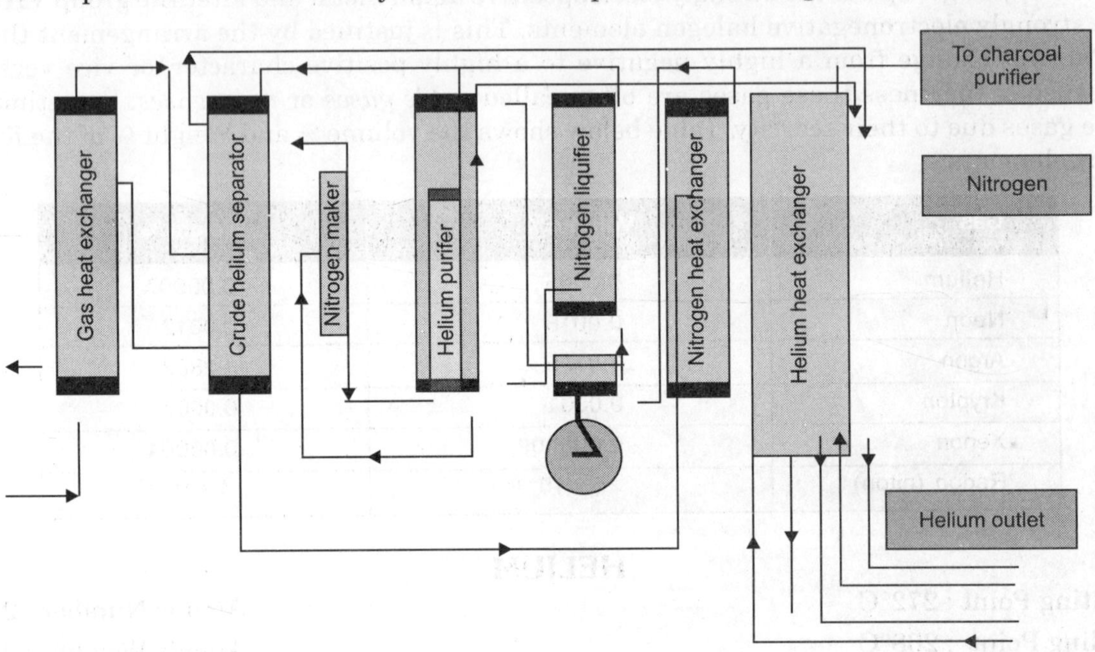

Fig. 19.1. The Extraction of Helium from Natural Gas

The natural gas enters a plant at about 4.136 Kpa pressure directly from the pipeline or it may be compressed if necessary. It is then brought in contact to a bubble tower with a solution of monoethanolamine and dythylene glycol to remove water vapour, carbon dioxide and any hydrogen sulphide that may be present. Next it is passed through activated bauxite to remove any remaining water vapour.

After the preliminary treatment the natural gas, still at a pressure of approximately 4.136 Kpa, enters a heat exchanger where its temperature is reduced to about –140°C by the cold out going residual gas. The partly liquefied natural gas is throttled, from this heat is exchanged, the liquid natural gas is passed into a separator which operates at 1.723 Kpa that is cooled with vapour from liquid nitrogen. Gaseous crude helium is drawn from the top of the separator the liquefied portion of the natural gas (residue gas) is throttled from the bottom of the separator to a pressure of about 482 Kpa and flows back through the heat exchanger. There it chills the in coming natural gas and is vaporised and wormed approximately to the atmospheric temperature. It can then flow (or is compressed if necessary) into a pipeline for transportation to fuel gas markets (except at Navajo point where the residue gas consists of large amounts of nitrogen and is discharged to the atmosphere. The helium extraction process increases the heating value of the natural gas by removing the carbon dioxide and helium and a part of nitrogen and sweetens the gas in process. The gas is compressed in three steps in total; to 4.31 Kpa cooled with water and brought to its temperature of liquefication in an expansion engine cycle of the Claude type.

The crude helium drawn off in the first stage (production cycle) consists of about 60% helium and about 40% of nitrogen. This mixture is compressed in 4 steps of about 391 Kpa cooled with water driven with lump of potassium hydroxide and passed without substantial reduction of pressure, through heat exchanging and gas liquification equipment, which is one of the part of the production cycle. Under the high pressure and the temperature of about –190°C, produced by a bath of liquid nitrogen boiling at approximately atmospheric pressure, most of the nitrogen in the oxide helium condenses to a liquid and the helium is drawn off as a gas with a purity of 98.3%.

The helium is then passed and cooled by rectified nitrogen –156°C to –184°C. The charcoal absorbs virtually all of the remaining impurities and yields helium with purity of better than 99.9%. The entire process of rectification and purification is continuous. The residual time of the natural gas form entering equipment to discharge into pipeline, is less than 1 minute.

Physical Properties of Helium

Melting Point	–272.2°C
Boiling Point	–268.9°C
Density (at 0°C)	0.1785 g/cc
Refractivity (n–1) × 10^{-6} at 0°C and 760 mm	
At 4800Å	35.05
At 5769Å	34.82
At 6438Å	34.86
Viscosity	186 centipoise
Heat of Vapourization	22 cal/mol
Critical Temperature	–267°C
Critical Pressure	2.26 atm
Critical Density	0.069 g/cc
Thermal Conductivity	340 cal/(sec)(sqcm)(°C/cm)
Isotope Number	4, 3
Ionisation Potential	24.5V
Excitation Potential	20.91V

Uses of Helium

- Helium is used mainly in filling lighter aircraft such as air ships weather balloons etc. Although it is twice as heavy as hydrogen its lifting power (which is determined by the difference between the density of balloon) is not a serious disadvantage in view reasons of safety because unlike hydrogen helium is non-flammable.

- Since helium, like nitrogen, is not suitable in blood even at even high pressure a mixture of 80% oxygen and 20% helium is used in modern diving apparatus instead of nitrogen or ordinary air. Due to high-pressure divers suffer from coisson's disease due to the dissolution of nitrogen in the blood and immediately it vapourises when the diver comes back to the surface. A similar oxygen-helium mixture is used for the treatment of the asthma or any other respiratory disease also for carbon monoxide poisoning.
- Helium has also a number of scientific use e.g liquid helium (Bp. -268°C) is used to provide low temperature for studying some phenomenon, which occur near the absolute zero temperature (Cryogenics). It is also used in gas thermometers for very low temperature readings.
- Helium is also used for the creation of an inert atmosphere for protection of newly formed metals to protect them forming producing oxides, in steel and other metal producing plants.
- In resent years researches on semiconductors are being taking place, which require liquid helium, Helium is also used for fusion experiments in nuclear chemistry.
- Helium is also as one of the chief ingredients of helium-cadmium laser.
- In one sentence it can be stated that helium is required in the many scientific applications, including inorganic chemistry.

NEON

Melting Point : 248.7°C | Atomic Number : 10

Boiling Point : 245.9°C | Atomic Weight : 20

Occurrence and Extraction

Neon is separated from the crude matter containing helium and nitrogen and also is freed from hydrogen, which is present in small amounts in the atmosphere and accumulates with the other non-condensable matters. Nitrogen is removed from the crude mixture by liquification under pressure and finally by low temperature adsorption in activated carbon. Some neon is lost because of its solubility in liquid nitrogen. Hydrogen is removed by reaction with copper oxide and the gas is dried to remove the water formed. The purified neon helium mixture is separated into its components relative adsorption of neon on activated carbon at low temperature.

Physical Properties

Melting Point	–248.7°C
Boiling Point	–245.9°C
Density at 0°C	0.90035 g/lit
Refractivity (n–1) × 10^6 at 0°C and 760 mm At 4800Å	67.31

At 5769Å	67.10
At 6438 Å	67.02
Viscosity at 0°C	297 centipoise
Heat of Fusion	80 cal/mol
Heat of Vapourization	400 cal/mol
Critical Temperature	–228°C
Critical Pressure	26.86 atm
Critical Density	0.484 g/lit
Thermal Conductivity at 0°C	110 cal/(sec) (sqcm) (°C/cm)
Solubility in Water at 20°C	0.015 ml gas/gm of water at STP
Isotope Number	20, 22
Ionisation Potential	21.5V

Uses of Neon

Neon has a characteristic property of giving an orange-reed glow in a discharge tube at low pressure (2 mm) and a voltage of 1000V. It is used for advertisement purpose. If mercury vapour is mixed with neon it glows with a bluish green colour.

Variety types of colours can be obtained by changing the gas mixture, which is shown in the table given below :

Types of Mixtures	Types of Tube	Colour of Light
Neon	Colourless	Orange Red
Neon	Light Red	Deep Red
Neon-Argon	Colourless	Ripple Effect
Neon-Argon	Colourless	Light Blue
Mercury Vapour	Green	Light Green
Same as Above	Purple	Pinkish Blue

Neon has a remarkable property of carrying high current when a voltage of 200V is applied. It is therefore used for safety devices as lightning arresters. Neon is also used in the laser as one of its gaseous constituent.

ARGON

Melting Point : 189.3°C
Boiling Point : 185.89°C

Atomic Number : 18
Atomic Weight : 39.95

Extraction of Argon

Argon is generally extracted from the mixture of oxygen and argon, this mixture is fed at A into the middle of a rectifying column shown in Fig. 19.2.

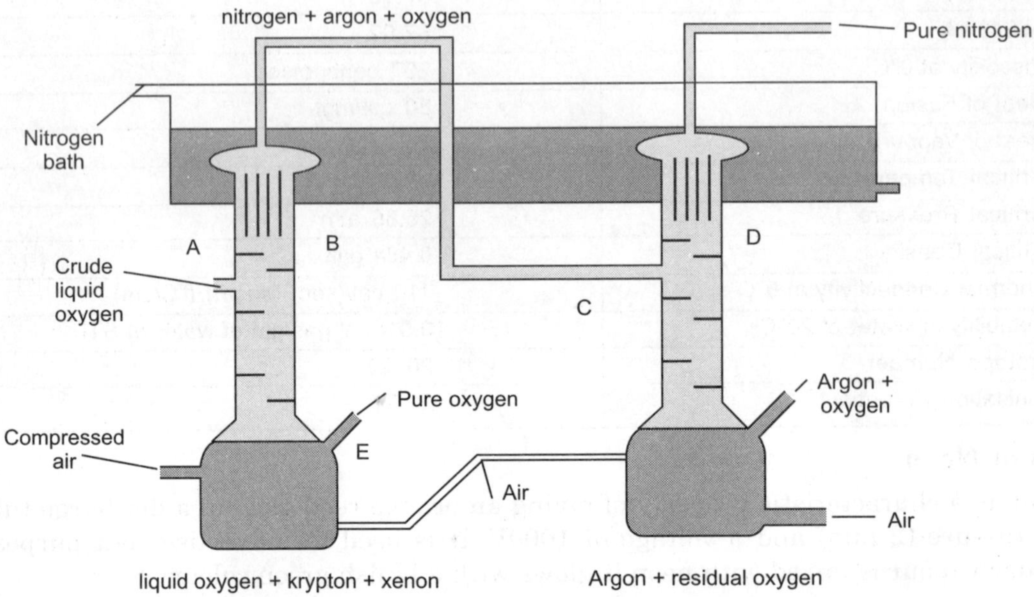

Fig. 19.2. The Extraction of Argon

This gaseous mixture consisting roughly of 47% of nitrogen 47% of argon and 6% of oxygen is lead at C into the middle of another column has also a series of tubes (D) which are kept cool in a bath of liquid nitrogen. Nitrogen gas present in the mixture cannot liquefy under these conditions and therefore, passes out from the column A. While argon mixed with residual oxygen come out from the bottom of the column. This is evaporated by the heat of compressed air passing through a coil and the gas consisting of argon and oxygen escapes from F as shown. The oxygen present is removed by passing it over heated copper and the pure argon is obtained.

Physical Properties

Melting Point	–189.3°C
Boiling Point	–185.89°C
Density	1.7839g/lit
Refractivity, $(n - 1) \times 10^6$ at 0°C and at 760 mm	
At 4800Å	35.04
At 5769Å	34.82
At 6438Å	34.86
Viscosity at 0°C	210 centipoise
Heat of Fusion	281 cal/mol

Heat of Vapourization	1557 cal/mol
Critical Temperature	–122.5°C
Thermal Conductivity at 0°C	0.038 ml gas at STP/g of water
Transitivity (relative to air)	0.79
Isotope Mass Number	40, 36
Ionisation Potential	15.7V

KRYPTON AND XENON

Extraction

From the first tower of the separation of argon krypton and xenon are separated with oxygen. As they are less volatile they accumulate at the bottom of the tower (see Fig. 19.3).

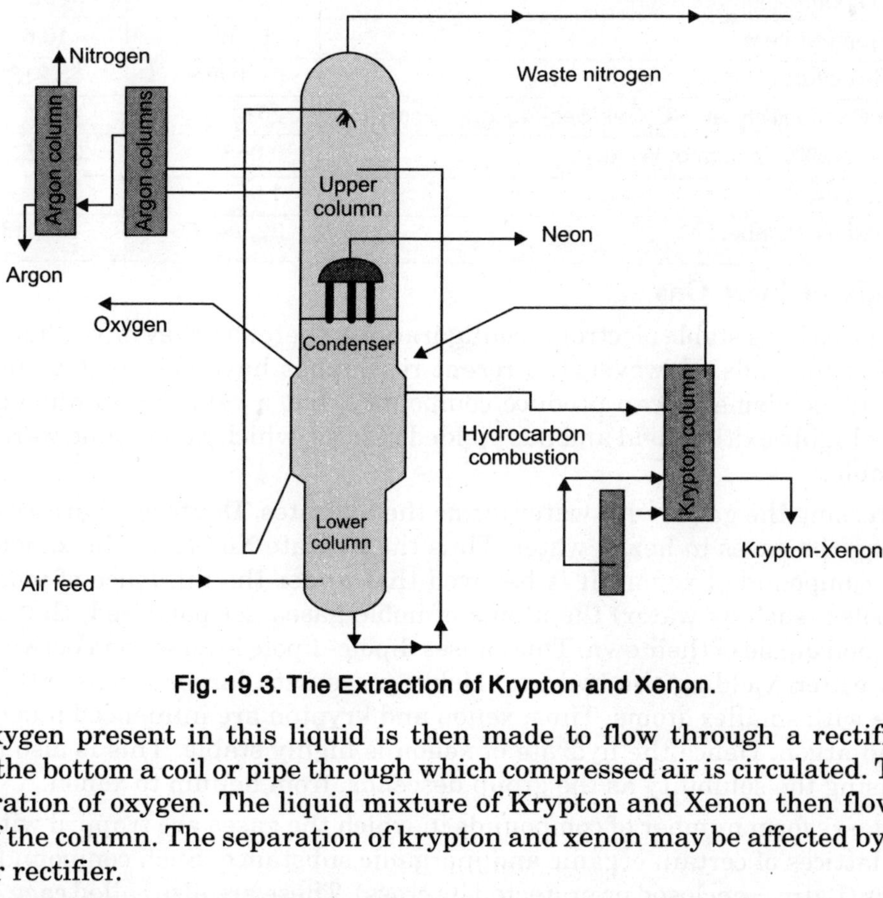

Fig. 19.3. The Extraction of Krypton and Xenon.

The oxygen present in this liquid is then made to flow through a rectifier column resting at the bottom a coil or pipe through which compressed air is circulated. This causes the evaporation of oxygen. The liquid mixture of Krypton and Xenon then flows down at the base of the column. The separation of krypton and xenon may be affected by treatment on another rectifier.

Quite often the mixture of the two gases is used as such and their separation is not considered necessary.

Physical Properties

Property	Krypton	Xenon
Melting Point	–157.1°C	-111.8°C
Boiling Point	–152.9°C	-108°C
Density (gm/lit)	3.708	5.851
Refractivity, (n–1) × 10^6 at 0°C and at 760 mm		
At 4800 Å	431.8	712.8
At 5769 Å	427.6	702.9
At 6438 Å	425.8	698.7
Viscosity at 0°C (micropoise)	230	210
Heat of Fusion (cal/mol)	360	548
Heat of Vapourization (cal/mol)	2310	3202
Critical Temperature	–63.8°C	+16.6
Critical Pressure	54.1 atm	58.218 atm
Thermal Conductivity at 0°C Cal/(Sec)(sqcm)(°C/cm)	20+	10+
Solubility at 20°C (ml/gm of Water)	0.063	0.125
Transmissivity	1.95	5.45
Isotope Mass Number	82, 84, 86	129, 131, 132

Compounds of Inert Gas

The noble gases have stable electronic configurations; therefore they have almost negligible amount of compounds. However, the recent researches have shown that under certain conditions these elements can produce compounds, but a very few in number. This has been a very highly exiting field and has yielded results, which at one time were considered as improbable.

Compressing the gases with water forms the hydrates. Deuterates are also formed by compressing the gases in heavy water. Thus the hydrate $Xe.6H_2O$, the xenon hydrate is the stable compound of xenon. It is believed that under the influence of molecules with strong dipoles (such as water) the atoms of noble gases get polarised, that is acquire a weak included dipole of their own. This causes dipole-dipole intersection between the noble gases and water, yielding the hydrate. Noble gases with larger atoms influenced more than those with smaller atoms. Thus, xenon and krypton are influenced much more than helium and argon. Hence the hydrate of xenon is highly stable. This is also the cause of the increasing the solubility as the group descends, from helium to xenon.

Inert gases form number of compounds in which the gases are trapped within cavities of crystal lattices of certain organic and inorganic substance. Such compounds are called Catharates (Latin = enclosed or protected by cross). These are also called cage compounds.

Xenon combines directly with fluorine. Its compounds with oxygen are obtained with fluorides. Xenon chloride is known but it is not unstable.

Some of the definite stable compounds of xenon are stated below in table:

Oxidation State of Xenon	Compounds of Xenon	
	Formula	Names
II	XeF_2	Xenon Difluoride
IV	XeF_4, $XeOF_2$	Xenon Tetrafluoride, Xenon Oxyfluoride
VI	XeF_6, $XeOF_4$, XeO_3	Xenon Hexafluoride, Xenon Oxytetrafluoride, Xenon Trioxide
VIII	XeF_8, XeO_4	Xenon Octafluoride, Xenon tetraoxide

Fluorides of Xenon

Xenon Difluoride (XeF_2): It is formed (1) by photochemical combination of xenon and fluorine under influence of mercury vapour.

$$\text{Xenon} + \text{Fluorine} \xrightarrow{\text{Mercury vapour}} \text{Xenon Difluoride}$$

$$Xe + F_2 \longrightarrow XeF_2$$

(2) A better method can also be used, by heating a mixture of xenon and fluorine in a nickel tube, which also acts as a catalyst. When the apparatus is cooled the compound is obtained. (The ratio of the Xe:F = 3:1)

$$\text{Xenon} + \text{Fluorine} \xrightarrow{\text{Nickel}} \text{Xenon Difluoride}$$

$$Xe + F_2 \xrightarrow{400°C} XeF_2$$

Excess of xenon is used to avoid formation of Xenon Tetrafluoride.

Alternatively the mixture of the xenon and fluorine is also subjected to an electrical discharge, to get xenon difluoride.

1. Xenon difluoride is a colourless, crystalline solid which melts at 140°C
2. It dissolves in hydrogen fluoride without reaction.
3. It reacts with hydrogen to give hydrogen fluoride and xenon.
4. It reacts with water evolving oxygen.

$$\text{Xenon Difluoride} + \text{Water} \longrightarrow \text{Oxygen} + \text{Xenon} + \text{Hydrogen Fluoride}$$

$$XeF_2 + H_2O \longrightarrow O_2 + 2Xe + 4HF$$

5. It oxidises iodine in the presence of Boron Tetrafluoride to give Iodine Monofluoride

$$\text{Xenon Difluoride} + \text{Iodine} \xrightarrow{BF_3} \text{Iodine Monofluoride} + \text{Xenon}$$

$$XeF_2 + I_2 \longrightarrow 2IF + Xe$$

6. It is also strong oxidizing agent. It can oxidise HCl to Cl_2

Xenon Tetrafluoride (XeF_4): It can be prepared by (1) heating a mixture of xenon and fluorine, in the ratio of 1:5 through a nickel tube, at a temperature of 410°C under a pressure of 5 to 6 atm.

Xenon + Fluorine $\longrightarrow$ Xenon Tetrafluoride

$$Xe + 2F_2 \longrightarrow XeF_4$$

It can also be prepared by passing an electrical discharge through a mixture of xenon and fluorine in the molecular ratio 1:2 at –80°C

Properties

1. It is colourless, crystalline solid which melts at 107°C. It sublimes readily
2. It dissolves in hydrogen fluoride without any reaction.
3. It is reduced by hydrogen at 117°C.
4. It reacts with boron trichloride to form boron trifluoride.

Xenon Tetrafluoride + Boron Trihloride $\longrightarrow$ Boron Trifluoride + Xenon + Chlorine

$$3XeF_2 + 4BCl_3 \longrightarrow 4BF_3 + 3Xe + 6Cl_2$$

5. It undergoes disproportionation in water giving a highly explosive xenon trioxide.

Xenon Tetrafluoride + Water $\longrightarrow$ Xenon + Xenon Trioxide + Hydrogen Fluoride + Fluorine

$$2XeF_4 + 3H_2O \longrightarrow Xe + XeO_3 + 6HF + F_2$$

6. It forms an dual compound with antimony pentafluoride.

Xenon Tetrafluoride + Antimony Antimony $\longrightarrow$ Xenon Tetrafluoride. Pentafluoride

$$XeF_4 + SbF_5 \longrightarrow XeF_4.SbF_5$$

Xenon Oxyfluoride ($XeOF_2$): It is prepared by slow and partial hydrolysis of XeF_4 at a low temperature.

Xenon Tetrafluoride + Water $\longrightarrow$ Xenon Oxyfluoride + Hydrogen Fluoride

$$XeF_4 + H_2O \longrightarrow XeOF_2 + 2HF$$

It is not a very stable compound.

Xenon Hexafluoride (XeF_6): It is prepared by the reaction of xenon with excess of fluorine in the ratio 1:20 in a nickel vessel under a pressure of about 50 atm.

Xenon + Fluorine $\longrightarrow$ Xenon Hexafluoride

$$Xe + 3F_2 \longrightarrow XeF_6$$

It is also obtained by the reaction of chlorine and xenon tetrafluoride under pressure.

Xenon Tetrafluoride + Fluorine $\longrightarrow$ Xenon Hexafluoride

$$XeF_4 + F_2 \longrightarrow XeF_6$$

It is also prepared by the electrical discharge through a mixture of xenon and fluorine at a low temperature at a ratio of 1:3.

Properties

1. It is a solid crystalline substance with a melting point of 46°C.
2. It is most volatile of all the fluorides of xenon. Its vapour has a greenish-yellow colour.
3. It dissolves in hydrogen fluoride giving a solution, which is a good conductor of electricity. This is attributed to the formation of ions as under:

Hydrogen Fluoride	+	Xenon Tetrafluoride	⟶	Xenon Pentafluoride	+	Hydrogen Difluoride
HF	+	XeF_6	⟶	XeF_5^+	+	HF_2^-

Xenon thus differs from its di and tetra fluorides as it dissolves in hydrogen fluoride without reaching.

4. It is highly reactive even with quartz.

Silicon Dioxide	+	Xenon Tetrafluoride	⟶	Xenon Oxytetrafluoride	+	Silicon Tetrafluoride
SiO_2	+	$2XeF_6$	⟶	$2XeOF_4$	+	SiF_4

5. It reacts with hydrogen like all other fluorides of xenon. Thus

Xenon Tetrafluoride	+	Hydrogen	⟶	Xenon	+ Water	+	Hydrogen Fluoride
$XeOF_4$	+	$3H_2$	⟶	Xe	+ H_2O	+	4HF

6. It undergoes slow hydrolysis with atmospheric moisture producing highly explosive xenon trioxide (XeO_3).

Xenon Oxytetrafluoride ($XeOF_4$): This compound of xenon can be prepared by partial hydrolysis of xenon hexafluoride.

Xenon Hexafluoride	+	Water	⟶	Xenon Oxytetrafluoride	+	Hydrogen Fluoride
XeF_4	+	H_2O	⟶	$XeOF_4$	+	2HF

It can also be prepared by heating xenon and fluorine in a molecular ratio of 1:4 in the presence of excess oxygen at about a temperature of 230°C. It can be purified by vacuum distillation.

Properties

1. It is a colourless compound melting point at –40°C.
2. On treatment with hydrogen it is reduced to xenon.

Xenon Oxytetrafluoride	+	Hydrogen	⟶	Xenon	+ Water	+	Hydrogen Fluoride
$XeOF_4$	+	$3H_2$	⟶	Xe	+ H_2O	+	4HF

3. It reacts with water to form another oxyfluoride, xenon dioxydifluoride, in which the xenon remains the same oxidation state 6.

Xenon Oxytetrafluoride	+	Water	$\longrightarrow$	Xenon Oxytetrafluoride	+	Hydrogen Fluoride
$XeOF_4$	+	H_2O	$\longrightarrow$	XeO_2F_4	+	SiF_4

4. It also reacts with silica forming same compound.

Xenon Dioxydifluoride (XeO_2F_2): This compound is prepared by the reaction of water with xenon tetrafluoride as stated earlier. It can also be prepared by the reaction of xenon tetraoxide and xenon oxytetrafluoride at a very low temperature about -78°C

This is a colourless compound and easily decomposes with atmospheric moisture.

Xenon Hexafluoride	+	Water	$\longrightarrow$	Xenon Trioxide	+	Hydrogen Fluoride
XeF_6	+	$3H_2O$	$\longrightarrow$	XeO_3	+	6HF

Properties

Xenon trioxide is a white non-volatile compound and is soluble in water. It is quite stable when present in aqueous solution. However, it explodes violently when dry. It can be detonated just by rubbing or pressing it produces nearly same effect like TNT. This has created lot of interest in the preparation of this compound.

It is also used as a powerful oxidizing agent. The advantage of this compound is that it produces xenon as the by-product during oxidation and does not make the reaction complicated. When dissolved in aqueous sodium hydroxide it undergoes disproportionation. Xenon gas is evolved and sodium perxenate is produced, which can be recovered from the solution.

Xenon Trioxide	+	Sodium Hydroxide	+	Water	$\longrightarrow$	Xenon	+	Oxygen	+	Sodium Perxenate
$2XeO_3$	+	4NaOH	+	$6H_2O$	$\longrightarrow$	Xe	+	$2O_2$	+	$Na_4XeO_68H_2O$

Sodium perxenate	+	Hydrochloric Acid	$\longrightarrow$	Sodium Chloride	+	Xennic Acid
Na_4XeO_6	+	4HCl	$\longrightarrow$	4NaCl	+	H_4XeO_6

Xennic Acid	$\longrightarrow$	Xenon Trioxide	+	Water	+	Oxygen
H_4XeO_6	$\longrightarrow$	$2XeO_3$	+	H_2O	+	O_2

The acid solution of sodium perxenate, therefore, has strong oxidizing properties. It can convert manganous ion into permanganic ion.

Xenon Octafluoride (XF_8) : This compound is prepared by the reaction of fluorine and xenon in the ratio of 1:8 in a nickel tube at a considerable high temperature and pressure.

Xenon	+	Fluorine	$\longrightarrow$	Xenon Octafluoride
Xe	+	$4F_2$	$\longrightarrow$	XeF_8

The structure of this compound has not yet been established.

Xenon Oxyhexafluoride ($XeOF_6$) : This compound of xenon is also known. It is formed with xenon tetrafluoride when xenon and fluorine are together heated in air. The oxygen comes form the atmosphere.

Fluorides of Krypton and Radon

Fluorides of Krypton : Resent experiments have shown that krypton can also combine with fluorine under suitable conditions yielding *krypton difluoride (KrF_2)* and *krypton tetrafluoride (KrF_4)*.

Krypton difluoride is obtained by photochemical combination of krypton and fluorine. *Krypton tetrafluoride* is obtained by passing an electrical discharge through a mixture of krypton and fluorine at the temperature of liquid nitrogen.

These fluorides are much less stable than the corresponding compounds of xenon. These are easily hydrolysed by water giving krypton oxygen and aqueous hydrogen fluoride.

Krypton Difluoride	+	Water	$\longrightarrow$	Krypton	+	Oxygen	+	Hydrogen Fluoride
KrF_2	+	$2H_2O$	$\longrightarrow$	Kr	+	O_2	+	4HF
Krypton Tetrafluoride	+	Water	$\longrightarrow$	Krypton	+	Oxygen	+	Hydrogen Fluoride
KrF_4	+	$2H_2O$	$\longrightarrow$	Kr	+	O_2	+	4HF

Radon Difluoride (RnF_2): This compound has been able to be prepared by the above-mentioned methods but it is highly unstable.

OXY-COMPOUNDS OF KRYPTON

Krypton Trioxide ($KrO_3 . xH_2O$): It is obtained in a very small amount by the hydrolysis of krypton tetrafluoride (KrF_4) using ice at a temperature below –30°C. The value of X may be 1, 2 or 3 when compound of the formula (H_2KrO_4, H_2KrO_5 or H_2KrO_6) is produced. Krypton trioxide resembles xenon trioxide in its properties.

CHAPTER

20

Solute and Solutions

Introduction

In this chapter we shall discuss the nature and the chemistry of solutes and also solutions, colloids and emulsions etc.

Solute and Solutions

1. What is solution - If we add some coarse sand, finely powdered sand and few crystals of potassium perman ganate in water contained in three beakers, and shake them well we observe that :

(*i*) Coarse sand settles down immediately, i.e., it cannot form a uniform mixture and immediately settles at the bottom.

(*ii*) Fine sand remains uniformly distributed in the water for some time, the mixture is therefore called suspension of sand and water but after sometime, the sand will settle at the bottom.

(*iii*) Potassium permanganate forms a uniformly coloured mixture with water, from which crystals of potassium permanganate do not separate. On evaporating such a mixture $KMnO_4$ is left as residue. The uniform mixture of potassium permanganate is called a solution.

Thus we can define, *when a substance or solid forms a uniform mixture (homogeneous with the other), may be solid or liquid, the mixture is called solution.*

We can define solute *as the solid or the liquid substance or portion of the solution which dissolves in the liquid or solid forming solution.*

We can also define solvent as the *solid or liquid substance or portion of the solution, which dissolves solute forming solution.*

For example, in the solution of potassium permanganate and water, potassium permanganate is solute and water is solvent. Therefore it is a solid + liquid solution.

Common Characteristics of Solutions

1. ***True Solutions are Molecular Dispersions:*** This means that the solute is present in minute particles, probably the size of the atoms and molecules in the vacant spaces between the molecules of the solvent.
2. ***Heat of the Solution:*** When a solute is dissolved in the solvent, heat may be absorbed (endothermic process) e.g., (ammonium chloride + water) or (evolved reaction) e.g., (sulphuric acid + Water)
3. ***Change in Volume:*** Volume of the solution may be equal or more or less than the sum of the volumes of solute and solvent that is 25 cc of alcohol + 25 cc of water produce 48cc of solution. Ammonium chloride dissolves in water in water and volume of the solution exceeds the sum of volumes of water and ammonium chloride.
4. ***Changes in Freezing Point and Boiling Point:*** A solute lowers freezing point raises boiling point, that is a solvent freezes at lower temperature and at much higher temperature than the solvent.

Types of Solution

Often a solvent a liquid or solid and solute may be liquid solid or gas, for e.g aerated water is a solution of water and carbon dioxide (gas in liquid) the solution of water and alcohol (liquid in liquid) and the alloy of brass (solid in solid).

Saturated Unsaturated and Supersaturated Solution

If we go on adding small quantities of salt to water and gently starring the salt goes on dissolving in water but after a time a limit is reached when no more salt gets dissolved and if added more settles at the bottom of the solution. That is solvent (water) cannot dissolve solute (salt) at this temperature. Such *a solution when no more solute can be dissolved at a given temperature is called saturated solution. Any solution having less solute than the maximum quantity of solute is called an unsaturated solution* and it can dissolve more solute even at this temperature.

Prepare a saturated solution of salt and water at 40°C and cool slowly. As temperature falls salts (excess of saturation limit for any temperature) gradually settles out. However if we cool the solution very gently in a clean vessel, the excess solute may not settle out on cooling, thus we get a solution containing more solute than necessary to make it a saturated solution at a particular temperature. *It is called supersaturated solution when the solution consisting more solute than its saturation level.*

It we stir the supersaturated solution or add a little solute to it, the excess solute immediately settles at the bottom, because such a solution is an unstable equilibrium.

Solubility

There is usually a limit to the extent to which a solute dissolves in a given quantity of solvent.

In a saturated solution, at a given temperature number of grams of solute present in every 100 g of solvent, is called solutions solubility at that temperature.

$$\text{Solubility} = \frac{\text{Weight of Solute in Grams}}{\text{Weight of Solvent in grams}} \times 100$$

Determination of Solubility in Solutions

At Room Temperature: Prepare a saturated solution of the solute at room temperature i.e by starring small quantities of solute in 40 g to 50 cc of solvent till a few grams of solute remains at the bottom. Filter and collect the filtrate in a clean weighed china dish or glass evaporating dish and find weight of solution taken (by difference). Gently evaporate the solvent on a water or sand bath. Avoid separating. Continue heating till whole of solvent has evaporated. Cool the dish in a desicator and weigh. Heat again for few minutes cool and weigh. Repeat till the weight is constant.

Calculation : From knowledge of the weight (1) empty dish (2) dish + solution and (3) dish + residue, we can calculate:

$$\text{Solubility} = \frac{\text{Weight of Solute (Weight of Solute + Dish) - Weight of Dish}}{\text{Weight of Solvent (= Weight of Solution - Weight of Solute)}}$$

At other Temperature : Prepare a saturated solution at 100°C and allow it to cool. When the temperature reaches to 90°C, 80°C, 70°C and 60°C repeatedly pipette out the solutions in different clear vessels each time, after stirring everytime and collect it in china dish, evaporate and weight every time as before.

Solubility Curve and its Application

A graph drawn between solubility and temperature is called solubility curve (see Fig. 20.1).

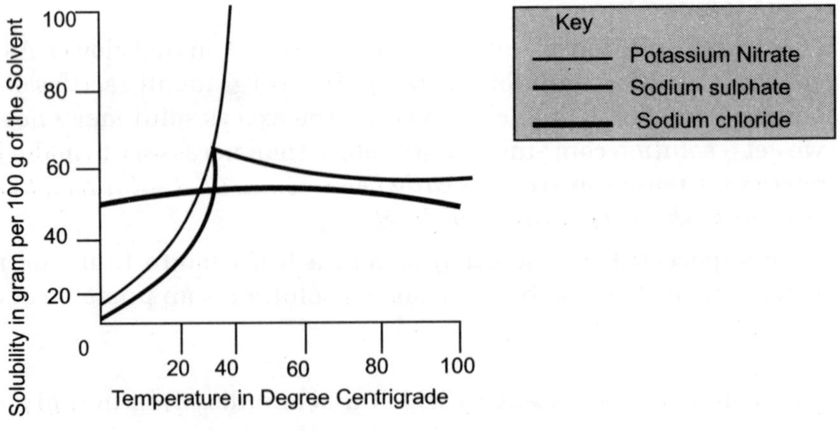

Fig. 20.1. The Solubility Curve

According to the trend of the curve the solubility of the chloride has not being greatly affected by the rise of temperature, while that of potassium nitrate has got a rapid rise.

Utility : (1) From the curves we can read solubility of substance at any temperature.

(2) We can choose that temperature for fractional crystallisation of a substance from a mixture solution that is the temperature at which solubility differs maximum.

(3) Changes in solubility with temperature can be readily noted.

Crystallisation

The process of crystallisation is based on the fact that a solution of a substance saturated at one temperature is not saturated at another temperature. Hence, if a saturated solution at higher temperature is cooled to a lower temperature, the excess solute is deposited. For example, if hot saturated solution of some common substance, e.g. alum, copper sulphate, sugar etc is allowed to cool a quantity of each solute is thrown out of that solution. An examination of the deposited solute shows that they consist of a number of small particles. These particles possess a definite geometrical shape and are bounded by perfectly plane faces. These particles are called crystals. The process by which these crystals are obtained is called crystallisation. The crystals are not only allowed to form a state of solution but also obtained from a state of fusion or vapour. The liquid left after crystals have separated out from it is called mother liquor.

Characteristics of a Crystal

It should possess a definite geometrical shape and size and bounded by perfectly plane faces. The angles between corresponding faces of all the crystals of a given substance are constant and are characteristics of that substance. Solid substances are either crystalline or non-crystalline. The non-crystalline solids are called amorphous, the examples of amorphous solids are coal, pith etc. and the examples of crystalline solid are alum, diamond sugar, common salt etc. Solid crystalline substance break up easily along certain planes, for e.g. mica splits upon into thin sheets, rock salts break up into cubes and rectangles. An amorphous solid on the other hand does not split along any particular plane. When broken by a sharp blow, but fractures with equal ease in any direction.

Fractional Crystallisation

The process of fractional crystallisation is used for the purification and separation of solids by making use of their different solubilities and particularly the fact that the substance increase more rapidly with rise of temperature than those of the other salts. The theory of the process may clearly, be understood from the inspection of the solubility curves the following example will show how the process is carried out.

Let us take 50 g of potassium chlorate in a clean dish it is wormed for about 10 to 15 minutes just above the melting point. The mass is then cooled. It contains undecomposed potassium chlorate, some potassium chloride and potassium perchlorate. About 50 cc of water at 50°C is added to it when all three of the salts are dissolved to give a solution mixture. The solution is then evaporated until a drop crystallises when rubbed on a cold surface, it is then slowly cooled. The crystals of potassium perchlorate are the first to separate because it has the least solubility at 15°C (the solubility of potassium perchlorate, potassium chloride and potassium chlorate at 15°C are, 1.5 g, 36 g and 6.6 g respectively). The first drop of crystal of potassium perchlorate is filtered off and redissolved and again allowed to crystallise by cooling the hot solution, potassium perchlorate can thus be obtained almost free from other two salts. When the crystals of potassium perchlorate are separated the solution is further evaporated and allowed to cool. This time potassium chlorate is separated. These are filtered off and must be recrystallised in order to isolate a purer salt. The process of re-crystallisation has to be repeated several times in order to obtain the crystals in a high degree of purity.

The fractional crystallisation is the process in which only one of the solutes is allowed to crystallise from the solution of two or more different solutes having different solubilities.

Water of Crystallisation

Many substances while crystallising out of aqueous solution yield crystals containing definite amount of combined water. They take up this water from the solution during the process of crystallisation. The water molecules are loosely associated with the crystals are indefinite number, the crystalline molecules of water are removed by some physical means such as heating the substance looses its crystalline shape. *The waters thus associated with the molecule of crystal are called water of crystallisation*, or *water of hydration*. The substance is called hydrates and the same substance having no water molecule is called anhydrous molecule. The process of adding water to the anhydrous salt. The process of adding water to the anhydrous salt is called hydration.

Experiment : Heat a crystal of copper sulphate (blue) in an evaporating dish. Slowly the colour changes from blue to white also the shape of the crystal changes. Now stop heating and add a drop of water, the colour of the crystal changes from white to blue and also the original shape of the crystal is regained.

The chemical analysis of the blue crystal reveals that the compound consists of 5 molecules of water per molecule of the compound, so the copper sulphate with water is called hydrated copper sulphate. Some other examples are $FeSO_4\ 7H_2O$, $Na_2CO_3\ 10H_2O$ etc. It must be born in mind that water is not the essential part of the crystal. There is certain substance, which are crystalline in shape but does not contain any water molecule.

Experiments of the Determination of Water of Crystallisation

The percentage of water of crystallisation can be determined for a large number of substances by strongly heating a weighted amount of the hydrated crystals and measuring the loss of weight. For example, when sodium carbonate is heated strongly on a burner and cooled in a dissicator and then weighted. The process is repeated until the weighed is constant. This gives the weight of the empty crucible (W_1). A quantity of crystal in then placed and weighted again. This gives the weight of the crucible and the hydrated crystal salt. Now the crucible is placed on a tripod and heated keeping the lid slightly open for the ventilation, the heating is continued till the salt is completely dehydrated. The heating and the weighing process are continued till the weight is constant. This indicates that the salt is completely dehydrated. The last weight gives the weight of the dehydrated salt and the weight of the crucible.

Calculation: Let us suppose that

(*i*) Weight of the crucible + lid = W_1 g

(*ii*) Weight of the empty crucible + lid + hydrated salt = W_2 g

(*iii*) Weight of the empty crucible + lid + anhydrous salt = W_3 g

$\therefore$ Weight of the hydrated salt = ($W_3 - W_1$) grams and the weight of the water of crystallisation, = ($W_2 - W_1$) g of the hydrated salt contain ($W_2 - W_3$) g of water of crystallisation.

$$\therefore 100 \text{ g of hydrated salt contains } \frac{(W_2 - W_3)}{(W_2 - W_1)} \times 100 \text{ g of water of crystallisation.}$$

Efflorescence

If a hydrated crystal is left exposed in air, its stability depends on the presence of water vapour in the atmosphere. If the presence of water vapour is less than hydrate, the hydrate will loose the water of hydration to the atmosphere and form a lower hydrate or anhydrous salt. The loss of water from a crystal from or hydrate is usually accomplished by the crumbling of the crystal to the powder. Such spontaneous loss of water of hydration is known as efflorescence.

Efflorescence is the phenomenon in which a hydrate spontaneously releases all or the part of its water of crystallisation to atmosphere, e.g., crystals of washing soda when exposed to atmosphere looses its nine molecules of water of crystallisation to the air and falls to power.

Deliquescence

If the presence of the water vapour in the atmosphere is greater than that of the crystal, the water enters in the crystal, this will continue until both the crystal and the atmospheric water vapour pressure comes to equilibrium. This phenomenon is called deliquescence. In this process the crystal ultimately get converted into liquid.

Deliquescence is present in the following compounds, caustic soda, calcium chloride, magnesium chloride etc.

Henry's Law

The amount of gas dissolving in a liquid depends upon nature of gas, temperature and pressure on the surface of the gas. At higher pressure (aerated water) more gas can be dissolved. It has been observed that weight of a gas dissolved in 1 cc of liquid at any given temperature is proportional to the pressure of the gas. This is called Henry's Law, if 1 cc of water dissolves x g of oxygen at 1 atm pressure. At double this pressure $2x$ g of oxygen can be dissolved in 1cc of water but volume of $2x$ g at 2 atms pressure = volume of x g at atm pressure. Thus we see, that volume of the gas dissolving in given volume of water (liquid) is independent of pressure. Weight will increase but volume dissolved remains same, at all pressures.

Colloids

***Crystalloid and Colloid*:** In the year 1861 *Thomas Graham* found that certain substances such as sugar, urea, etc. diffuse rapidly in solution and also can pass through semi-permeable membranes such as parchment paper or fish bladder. These substances are crystalline in solid state, therefore called *Graham Crystalloids.* On the other hand there are such substances like glue, gelatine, albumen, starch etc. which diffuse in solutions very slowly and cannot also pass through the membranes. These substances are amorphous in the solid and are called *colloids.* (Kola = Greek Glue eidos = like).

Graham made a distinction between crystalloid and colloid as different kinds of substances. But latter investigations confirmed that these distinctions between colloids and crystalloids are nevertheless very infinite. For example: many typical colloids such as albumen can be obtained in the form of crystals and many indisputable crystals like common salt can be obtained as colloids. The same substance can behave like a colloid in some solvent but may behave like crystalloid in different solvent. Common salt behaves like colloid in benzene but forms crystals when dissolved in water. A crystalloid or colloid state of any substance is just like we talk about the solid and the liquid state of the substance. The colloid state is therefore a property of the substance when subjected to proper condition.

Dispersion: When a fine powder of any insoluble substance such as a clay is shaken with water, it is found that the larger particles settles down at the bottom while the finest particles remain in the water as 'suspended' state for a considerable time but eventually they all settle down. *If particles of a solid substance are suspended in a liquid it is called suspension.* Any system in which the finely divided particles of one substance are distributed through another substance is called dispersed system; *the process of subdividing one substance into finely divided particles and scattering them through another substance is called dispersion. The finely divided substance is called* ***dispersed phase*** *and the other substance is called* ***dispersed medium***. For example the suspension of clay particles in water is the dispersion of clay particles in water, the dispersed phase consists of clay particles and the dispersion medium is water. This suspension can be described as the dispersion of the solid particles in liquid.

In Suspension, the particles have diameter larger than 10^{-4} cm. Particles in true solution have comparatively smaller diameter of about 10^{-7} cm and the particles of colloidal solution ranges from 10^{-4}.

Colloidal solution is defined as dispersed system in which the size of the dispersed particles is larger than that in true solution, but less than in course suspension; the size of the particles usually lie between the 10^{-7} to 10^{-4} cm. The dispersed particles do not settle, no matter how long the colloidal solutions may stand and can pass through filter paper, or fish bladder.

Suspension can be described as the dispersion of the solid particles in liquid.

Solid colloidally dispersed in a liquid referred to as suspension of solids. *The liquids, which are dispersed, colloidally liquid is known as Sol*. A sol whose dispersion medium is water is called hydrasol. Bensol, when the dispersion medium is benzene and the alcosol when the dispersion medium is alcohol

Dialysis

The colloidal particles can pass through the pores of filter paper but cannot pass through either membranes like parchment paper or fish bladder. On the other hand the particles of the dissolved substance in true solutions can pass through these membranes *this property is utilised in separating the particles of the dissolved substance by membranes is called dialysis.*

The property is utilised in separating the particles of the dissolved substance by membranes is called dialysis. This is the same theory applied in the kidney dialysis but the process is entirely different.

The apparatus is used for dialysis is called a dyalizer. The mixture of a colloidal substance in the dissolved state such as a solution of potassium iodide and starch is placed in a bag of parchment paper the bag is placed in a reservoir of water. Potassium iodide easily comes out of the bag but the starch being a true colloid cannot come out, which remains in the bag as the pure solution. After a period of time if some chlorine is added in the water it turns yellow and not blue because starch is not present (see Fig. 20.2).

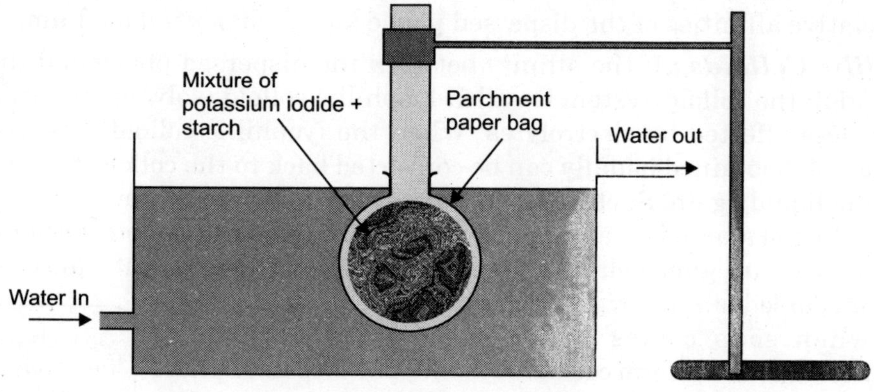

Fig. 20.2. The Dialyzer Apparatus

Electrodialysis

It is a process of dialysis in which an electrical force field is involved. In a mixture of colloid solution and true solution if the substance in a true solution is an electrolyte, then the electrolyte can be separated from the colloidal substance very rapidly. This process is called electrodyalysis (see Fig. 20.3).

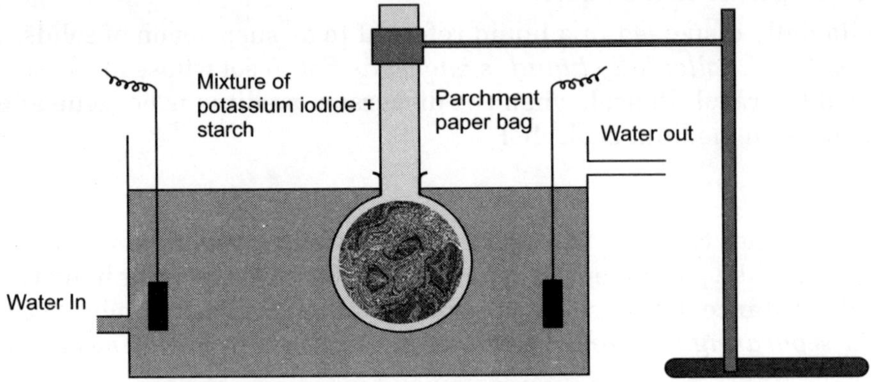

Fig. 20.3. The Process of Electrodialysis.

The colloidal solution is taken in a bag and the bag is placed in water. Then two electrodes are placed in water (in the outer vessel) on the two sides of the bag. When an electric current is passed by means of the electrodes the cations and the anions moves towards the cathode and the anode respectively while the colloidal particles remain in the bag. As a result the dissolved electrolytes in the true solution is separated from the colloidal particles very rapidly. However the use of the electrolyzer for the separation of non-electrolytes like glucose, urea is not possible.

Classification of Colloids

It is convenient to classify colloids into two general groups. This classification is based upon the relative affinities of the dispersed phase for the dispersed medium.

Lyophillic Colloids : If the affinity between the dispersed phase and the dispersed medium is high the colloid system is called lyophillic colloid (solvent linking). Lyophillic colloids are less affected by electrolytes. When the lyophillic colloids are evaporated to dryness, the solid obtained usually can be converted back to the colloid dispersion mearly by adding the liquid again. Such substances are also described as reversible colloids. The colloidal solution of starch in water in an example of a lyophillic colloid. Other examples of lyophillic sols are glue, gum, gelatine, fat and soap. Many lyophillic substances may become colloidally dispersed simply by contact with dispersing liquid, for example when soap comes in contact with water and it spontaneously disperses to form colloidal system. The dispersed phase in lyophillic colloid is also hydrophillic, alcophillic and benzophillic, respectively.

> Colloids are classified into lyophillic and lyophobic types

Lyophobic Colloids : If the affinities of the dispersed phase for the dispersed medium are small, the dispersed phase is said to be lyophobic (solution hating or solution fearing). When the lyophobic colloidal solutions are evaporated to dryness, the solid obtained cannot be brought back into the original colloidal solution once again, therefore these colloids are also known as irreversible colloids. They are precipitated easily be the addition of electrolytes. Colloids of silver chloride in water is the best example of the lyophobic colloid, the other examples are sulphur in water and also metal sulphide in water. The dispersed phase in lyophobic colloids are also called alcohophobic, hydrophobic and also benzophobic.

Preparation of Colloidal Solution

There are two types of methods by which colloids solutions can be prepared. They are *dispersion method* and *condensation method*. In dispersion method, larger particles are made smaller ones usually be grinding in dispersed machines called colloid mills. In condensation method, the particles of molecular dimension condenses to form aggregates of colloidal size. Such methods require chemical particles, giving rise to colloidal particles rise to a colloidal solution.

Some examples of the formation of colloidal solutions are given below :

(i) ***Ferric Hydroxide Solution :*** A solution of ferric chloride is added drop by drop to boiling water in a beaker. Ferric chloride undergoes hydrolysis and froms ferric hydroxide.

$$\text{Ferric Chloride} + \text{Water} \longrightarrow \text{Ferric Hydroxide} + \text{Hydrogen Chloride}$$
$$FeCl_3 + 3H_2O \longrightarrow Fe(OH)_3 + 3HCl$$

Under these conditions hydroxide remains in solutions as colloidal particles forming colloidal solutions of ferric hydroxide. It is then prepared by dialysis.

(ii) ***Gold Solution :*** Several drops of 1% auric chloride are added to distilled water in a test tube, the solution is heated to boiling. Then 2 or 3 ml of a very weak solution of formalein is added to it. Liquid turns to beautiful purple colour, due to the formation of colloidal gold.

$$\text{Auric Chloride} + \text{Formalein} + \text{Water} \longrightarrow \text{Gold} + \text{Hydrogen Chloride} + \text{Formalein}$$

$$2AuCl_3 + 3HC\begin{matrix}\nearrow O \\ \searrow H\end{matrix} + 3H_2O \longrightarrow Au + 6HCl + 3HC\begin{matrix}\nearrow O \\ \searrow H\end{matrix}$$

(iii) ***Colloidal metals, Berdig's method :*** This method is used for the preparation of colloidal solution of metals for example the colloidal solutions of silver is prepared as follows, two silver wires are dipped into distil water taken in a beaker. They are connected to sufficiently powerful source of electric current and they are brought together under the water and an electric arc is struck. Silver is vapourised then complexed by the surrounding cold water into particles of colloid size. The entire liquid then turns brown due to the formation of colloidal solution of silver.

Properties of Colloidal Solution

(a) Optical Properties: *(I) Tyndal Effect*, when a beam of light is allowed to pass through a darkroom, the dust particles floating in air becomes visible as tiny spots of light. When a beam of light is allowed to pass through true solution, the path of the beam is not visible, but when the same beam of light passes through the colloid solution the path of the beam of light becomes visible and the liquid appears turbid. This appearance is due to the *reflection of some of the incident light by the dispersed particles and is known as the Tyndal Effect.*

A strong converging beam of light is sent through a colloidal solution. When the path of the beam is viewed through a microscope (see Fig. 20.4). Whose axis is at right angles to the direction of the beam small flashes of light from the individual particles may be seen against the dark background, through a phase contrast microscope.

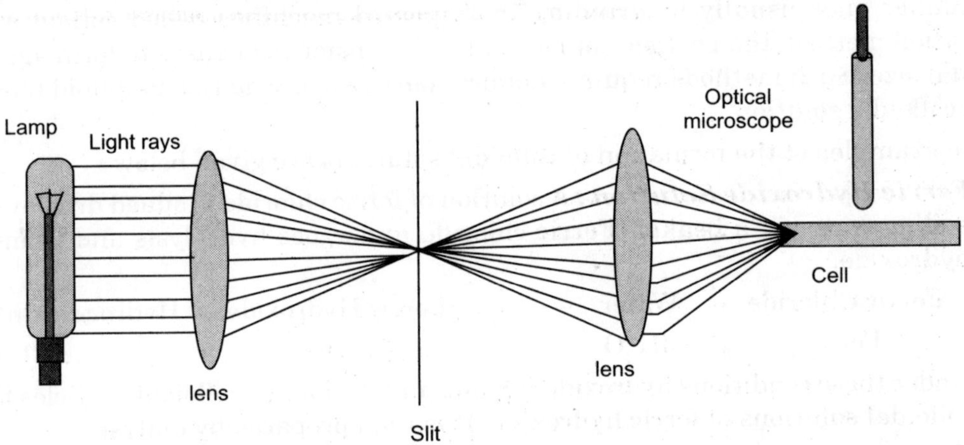

Fig. 20.4. The Experimental set up for the demonstration of Tyndal Effect.

(b) Brownian Movement: When through a phase contrast microscope, the colloidal particles are formed to be in a state of rapid irregular dancing motion. Such irregular movement of colloidal particles is known as Brownian Movement. After the name of the English botanist Robert Brown, who first discovered it in 1927 in pollen grains suspended in water. The smaller the size of the particle the faster is the Brownian Movement (See Fig 20.5).

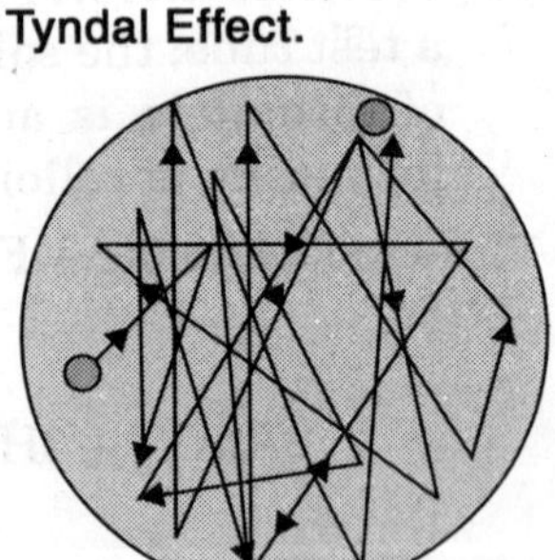

Fig. 20.5. The Brownian Movement.

(c) Electrical Properties: Electrophoresis, nearly all colloidal particles are electrically charged either positive or negative. When two electrodes connected to the poles of a current source of high voltage is dipped in the colloidal solution, the colloidal particles move towards the anode and cathode. This movement of colloidal particles in an electrical field is called electrophoresis (or cataphoresis).

Thus, when an electrical current is passed through an As_2S_3 solution the colloidal particles move towards the anode, proving that As_2S_3 particles bear a negative charge. On the other hand when electric current is passed through ferric hydroxide solution, the particles move towards the cathode indicating that they are positively charged.

All colloidal particles in a given colloidal solution have the same type of charge, solutions of metallic hydroxides generally owing a positive charge while most of the other solutions such as As_2S_3 etc. carry a negative charge. An apparatus, which is used, is half filled with distilled water, about 5 g of urea is added to about 50 ml of colloidal solution of arsenious sulphide to make the solution denser than water. The yellow colloidal solution is then run down slowly by a pipette into the lower part of the 'U' tube. The electrodes are connected to the poles of a current source of a slightly high voltage. After some times the level of arsenious sulphide is found to be full in the cathode side of the 'U' tube and to rise on the anode side. This experiment therefore shows very clearly that arsenious sulphide moves towards the anode and hence the colloid particles of arsenious sulphide solution are negatively charged. This proves that the particles of colloidal ferric hydroxide carry positive charge.

(d) Coagulation of Colloids: If a small amount of electrolyte is added to a colloidal solution the solution first becomes turbid and finally *the dispersed material is separated out of the liquid in the form of a precipitate. This form of precipitate is called coagulation.* The solid, which is separated, is called *coagulam*. For example, when some hydrochloric acid or some salt solution is added, to a yellow solution of arsenious trisulphide, As_2S_3, the arsenious trisulphide separates out of the solution in the form of yellow precipitate.

The coagulation of a colloidal solution is due to neutralisation of the charge on the colloidal particles by the added electrolyte. When an electrolyte is added to a solution it greatly increases the total concentration of ion in it. The colloidal particles absorb ions of opposite polarity. Thus, the initial charge of the colloidal particles is neutralised after which the coagulation of the solution occurs.

Emulsions

Emulsion is defined as a stable colloidal system in which one liquid is dispersed into another liquid. The common example of emulsion is milk is the emulsion of butter fat suspended in water.

If oil such as kerosene is shaken with water an emulsion of oil and water is obtained. Such emulsions are not stable and upon standing separation of the liquids into two layers occur. If certain substances such as soap are added to the oil and water before shaking the emulsions are stabilised. Such substances are called *emulsifiers* or *emulsifying agent*.

> The agents who stabilize the emulsion are called emulsifier.

Uses of colloids : Colloids are of greatest importance in our everyday life and industry. Every individual animal is built up from complex material called protoplasm which is a colloid. Hence biological phenomenon is connected with the processes that take place in the colloidal dispersion of matter. Matter in many colloidal state also plays very important role in agriculture and in dyeing and textile industries such as tanning of leather, in the production of rayon. In the manufacture of paints. Many of our foods are colloids as stated milk is the emulsion of fat and water. In ice-cream casein is present in the colloidal form an is prevented from coagulation by the addition of protective colloids like albumen (white of egg) and gelatine. Many pharmaceuticals are either colloidal suspensions or emulsions. For example colloidal gold is used for intermascular injections also colloidal silver (argyl and protargal) is used in eye treatment. The blue sky is blue due to the scattering of blue light by the dust particles and water particles suspended in air. In soil, there exist various colloidal substances such as the humus colloidal ferric hydroxide, aluminium hydroxide clay etc.

Owing to the presence of these, soluble substance, necessary for the life of plant are retained in the soil instead of being washed away. Thus it would not be wrong to say that colloidal state of matter has been chosen by nature as the vehicle of life.

Differentiation between true solution, colloidal solution and emulsion

True Solution	Colloidal Solution	Emulsion
1. In the true solution the average diameter of the particles are smaller than 10^{-7}	1. In the colloidal solution the average diameter of the dispersed particles lies between the range of 10^{-7} to 10^{-4} cm.	1. Emulsion is the mixture of two liquids.
2. True solutions can pass through the pores of the filter paper and the parchment paper and the membranes of the fish bladder.	2. Colloidal solutions can pass through filter paper but can be detained by parchment paper or fish bladder.	2. Emulsions are not stable, they get separated while standing and they need a substance for stabilisation called emulsifier.
3. True solutions are homogeneous.	3. Colloidal solutions are heterogeneous.	3. Emulsions are heterogeneous and homogeneous.
4. True solutions do not exhibit Brownian Movement and Tyndal Effect.	4. Colloidal solutions exhibit these two phenomenons.	4. Emulsions do not exhibit this phenomenon.
5. True solutions cannot be coagulated.	5. Colloids can be coagulated.	5. Emulsions are unstable gets coagulated.
6. The colour of the solution depends upon the ions or molecules present.	6. The colloidal solution gets its colour from the wavelength of light, which is scattered by the particles.	6. The colour of the emulsions depends upon its ingredients and also on emulsifier.